Microtubule Proteins

Editor

Jesús Avila, Ph.D.

Professor
Centro Biologia Molecular
Consejo Superior de Investigaciones Científicas
Universidad Autónoma de Madrid
Madrid, Spain

CRC Press
Taylor & Francis Group
Boca Raton London New York

CRC Press is an imprint of the
Taylor & Francis Group, an informa business

First published 1990 by CRC Press
Taylor & Francis Group
6000 Broken Sound Parkway NW, Suite 300
Boca Raton, FL 33487-2742

Reissued 2018 by CRC Press

Library of Congress Cataloging-in-Publication Data

Microtubule proteins / editor, Jesús Avila de Grado.
p. cm.
Bibliography: p.
Includes index.
ISBN 0-8493-5527-3
1. Tubulins. I. Avila de Grado, Jesús.
QP552.T82M53 1990
574.87'3--dc19 89-852

A Library of Congress record exists under LC control number: 89000852

Publisher's Note
The publisher has gone to great lengths to ensure the quality of this reprint but points out that some imperfections in the original copies may be apparent.

Disclaimer
The publisher has made every effort to trace copyright holders and welcomes correspondence from those they have been unable to contact.

ISBN 13: 978-1-315-89554-3 (hbk)
ISBN 13: 978-1-351-07464-3 (ebk)

Visit the Taylor & Francis Web site at http://www.taylorandfrancis.com and the
CRC Press Web site at http://www.crcpress.com

PREFACE

Microtubules are involved in several basic and essential cell functions such as chromosome segregation, intracellular organization, axonal transport, motility, and determination of cell shape. All of these functions are based on the capacity of tubulin, the main component of microtubules, to polymerize and depolymerize.

In this book we have summarized our knowledge of tubulin structure and function, with a special emphasis on the analysis of polymerization mechanism. In this way, the factors that promote or decrease tubulin assembly are indicated.

This book is published for postgraduate students and researchers studying microtubules and the poisons which prevent their polymerization. All the contributors would be glad if this book would help to a further development of the study of microtubules.

J. Avila
December 1988

THE EDITOR

Jesús Avila, Ph.D., has been the Director of Centro de Biología Molecular in Madrid, and currently is Research Professor of Spanish Research Council and Professor of the Department of Molecular Biology at the Universidad Autónoma de Madrid.

Dr. Avila received his Ph.D. from Complutense University in Madrid in 1971. After doing postdoctoral work at the National Institutes of Health, Bethesda, Maryland, he was appointed as a staff member of the Centro de Biología Molecular.

Dr. Avila has published more than 100 papers. His current research interest includes the structure-function relationship to microtubule proteins.

CONTRIBUTORS

Jesús Avila, Ph.D.
Professor
Centro Biologia Molecular
Consejo Superior de Investigaciones Científicas
Universidad Autónoma de Madrid
Madrid, Spain

Nicholas J. Cowan, D.Phil.
Professor
Department of Biochemistry
New York University Medical Center
New York, New York

Javier Diaz-Nido, Ph.D.
Centro de Biologia Molecular
Consejo Superior de Investigaciones Científicas
Universidad Autónoma de Madrid
Madrid, Spain

Yves Engelborghs, Ph.D.
Department of Chemistry
University of Leuven
Leuven, Belgium

Ernest Hamel, M.D., Ph.D.
Senior Investigator
Laboratory of Biochemical Pharmacology
Developmental Therapeutics Program
Division of Cancer Treatment
National Cancer Institute
National Institutes of Health
Bethesda, Maryland

María A. Hernandez, Ph.D.
Centro de Biologia Molecular
Consejo Superior de Investigaciones Científicas
Universidad Autónoma de Madrid
Madrid, Spain

Sally A. Lewis, M.A.
Associate Research Scientist
Department of Biochemistry
New York University Medical Center
New York, New York

Luis Serrano, Ph.D.
Centro de Biologia Molecular
Consejo Superior de Investigaciones Científicas
Universidad Autónoma de Madrid
Madrid, Spain

TABLE OF CONTENTS

Chapter 1

DYNAMIC ASPECTS OF MICROTUBULE ASSEMBLY

Yves Engelborghs

TABLE OF CONTENTS

I. INTRODUCTION

This chapter on microtubule dynamics is conceived as a review of the major paradigms that have governed this research field. In this way I hope to transmit some of the thrill of the field, as well as an important consideration: whenever a measurement was made directly instead of indirectly, a surprising observation was made and totally new aspects of the system appeared.

The dynamics of the formation of microtubules can be studied by a variety of techniques. All of them reveal a different aspect of the total process, and a full picture will be obtained when all these pieces can be integrated into a single mechanism. In this chapter the mechanism of microtubule assembly will be divided in sequential conceptual steps. The contribution of the different techniques to the understanding of these steps will be given.

The initial analysis of microtubule assembly *in vitro* is based on the theory of Oosawa,[1] which had been successfully applied to the study of actin polymerization. It remains the basic framework to which regular reference will be made. Therefore, it will be discussed in some detail. Aspects not directly related to microtubule dynamics will only be discussed very briefly and can be found more extensively in excellent reviews.[2-5]

II. COMPONENTS OF THE SYSTEM

Since the structure of the final polymeric product is a source of inspiration for the pathway of assembly, we will describe the microtubule structure here, as well as the basic components of the system.

A. TUBULIN

The major constituent of the microtubule system is the protein tubulin. It is the association product of two different subunits: α and β tubulin. Both have a molecular weight of 50,000 Da and are highly homologous.[6,7] The word tubulin always refers to the α-β heterodimer. This is usually considered as one unit, although the association is only due to noncovalent interactions. Detrich and Williams[8] determined the dissociation constant by equilibrium sedimentation and found a value of $8(\pm 3) \times 10^{-7}$ *M*. This implies that at a tubulin concentration of 2 μ*M* about 24% of the protein is dissociated into the two subunits. This measurement has, however, only been done in one particular buffer condition: in 0.1 *M* PIPES buffer, 1 m*M* $MgSO_4$ at pH 6.9. It is conceivable that the different buffer conditions that influence polymerization also influence dimer formation. This remains to be investigated.

In relation to polymerization it is important to mention that tubulin has two binding sites for GTP, only one of them allows free exchange with nucleotide in solution (the E-site). At this E-site, GTP is hydrolyzed during assembly.[9,10] It has been shown that the E-site is situated on the β subunit with a photoaffinity analogue of GTP,[11] or by direct UV illumination of GTP-tubulin (T-GTP).[12] Tubulin is a rather unstable protein. At room temperature its polymerizability is lost within a couple of hours.

Electrophoretic analysis indicates that tubulin consists of many isoforms.[13] Some of them show tissue specificity and the genetic regulation of their expression is the subject of intensive research (see other chapters in this book). Tubulin is also subjected to post-translational modifications, e.g., detyrosination.[14]

B. MICROTUBULE-ASSOCIATED PROTEINS

During the process of the isolation of microtubules, by repeated cycles of polymerization-depolymerization, certain proteins copurify. These proteins are called the microtubule-associated proteins (MAPs).[15-17] They can be separated from tubulin by ion exchange chromatography,[17] whereby care has to be taken not to remove the necessary Mg^{2+} ions.[18]

Originally these proteins were considered to be impurities of the preparation. However, their role as cofactor for polymerization became soon undisputable.[19] Their association to microtubules *in vivo* was further proven by the use of specific antibodies.[20]

The mixture of MAPs has been divided into two classes on the basis of their molecular weight: the τ-group with a relative molar mass between 58 and 70 kDa and the high molecular weight (HMW) group, subdivided in the HMW_1 (or MAP1) and the HMW_2 (MAP2) group with a molar weight around 350 and 270 kDa, respectively.[21-23] More MAPs have been isolated since then, and their properties and biological distribution are reviewed by Matus.[23] These associated proteins stabilize microtubules and therefore increase the yield of polymerization. They are supposed to interact with other cell organelles. Their affinity for microtubules can be modulated by phosphorylation.[15] Different MAPs can create a different morphology of microtubules in electron microscopic pictures.[24,25] The properties of, for example, MAP2 are extensively described in Reference 26.

C. MICROTUBULES

Microtubules are formed by the reversible association of either pure tubulin, or of the microtubule protein mixture (MTP), i.e., the mixture of tubulin and the MAPs. The earlier structural studies based on electron microscopy and image reconstruction is presented in an excellent review by Amos.[27] A tubular structure is proposed with an outer and inner diameter of about 24 and 14 nm, respectively. Hydrated samples give a slightly larger outer diameter of 30 nm.[28] The length varies considerably according to the polymerization conditions and can reach several micrometers.

In the microtubules two binding domains can be considered: (1) the longitudinal associations which lead to the formation of so-called protofilaments, in which the tubulin molecules are aligned according to their own length axis. (2) In the cylinder these protofilaments are laterally associated while the tubulin subunits are slightly staggered. In this way a closed cylinder is formed with 13 protofilaments. Electron microscopic studies on flagellar outer doublet microtubules, revealed that the whole structure can be described by a series of helices, e.g., a left- handed three-start helix.[29] Two lattices can be defined dependent on the distribution of the subunits in the three-start helix (see Figure 1). In the A lattice, the three-start helix is formed by alternating α and β subunits, while in the B lattice, it is formed by pure α-α or β-β sequences. The flagellar doublets consist of a complete microtubule with A lattice, and a C-shaped sheet (with B lattice) attached to it.[29] *In vitro,* the number of protofilaments is variable, with 14 being predominating. Here the B lattice is usually found, but the spiral of identical subunits is not continuous; instead a seam is present.[30,31] This seam is not necessarily continuous over the whole length of the microtubule.

When MAPs are present, these are organized on the surface of the microtubules in a kind of superlattice.[24,27]

All evidence points to a microtubule structure where all the tubulin molecules are arranged with the same polarity. Therefore, the microtubule itself shows the same overall polarity with one end having only α subunits exposed to the solvent, the other end only β subunits (except for the lateral interfaces of the subunits). Taking the spiral organization of the microtubule into account, a variety of different configurations can be imagined for the detailed structure of the end. Erikson[32] coined the notion of cosy corners: these are the sites on the end where an associating monomer can make multiple contacts at once. Weisenberg[33] presented a theory where the exact form of the ends plays an important role. It is clear that the structure of the ends continuously fluctuates, such that the rate constants measured represent a weighted average of the different configurations. This average can however differ when growth is fast or slow, depending on the time available for the units on the ends to rearrange or not, and in this way the rate constant of growth and dissociation can be influenced.[34]

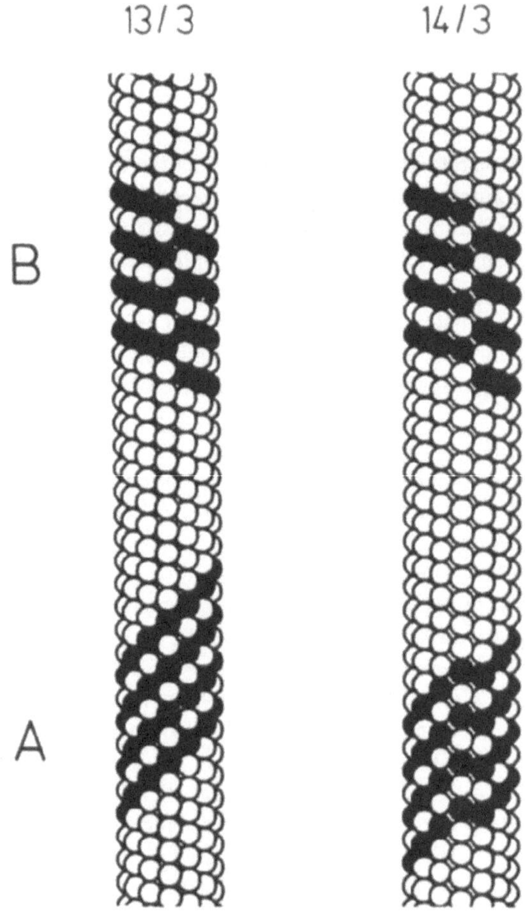

FIGURE 1. Models of possible lattices for microtubules made from 13 and 14 protofilaments. Tubulin ($\alpha\beta$) is shown as a black and empty circle. The A-lattice is shown in the lower part, the B-lattice in the upper part of the models. In the A-lattice the α and the β subunits alternate in the three-start helix. In the B-lattice, α-α or $\beta\beta$ sequences are found, except for the discontinuity. (From Mandelkow, E.-M., Schultheiss, R., Rapp, R., Muller, M., and Mandelkow, E., *J. Cell. Biol.*, 102, 1067, 1986. With permission.)

D. RINGS

In conditions where the formation of microtubules is prevented, e.g., low temperature, presence of calcium ions, or absence of GTP, ring-like oligomers are formed. The size and the shape of these rings varies considerably according to the solvent conditions: pH, ionic strength, and according to the tubulin/MAPs ratio. A detailed electron microscopy and sedimentation study of the rings formed in the presence of MAPs was published by Borisy and co-workers.[35-38] The most prominent form of these rings has a sedimentation coefficient of 30 S. The structure of these double rings, as proposed by the authors, is shown in Figure 2. However, also 18 S rings can be formed at the expense of the 30 S rings, dependent on the pH. Higher aggregates start to appear at pH 6.3 and increase in concentration with decreasing pH. Around pH 6.4 almost uniquely 30 S oligomers are in equilibrium with the dimers and MAPs. Ionic strength is also very important and 0.2 *M* KCl is sufficient to dissociate the 30 S rings completely. Rings have been crystallized by Voter and Erick-

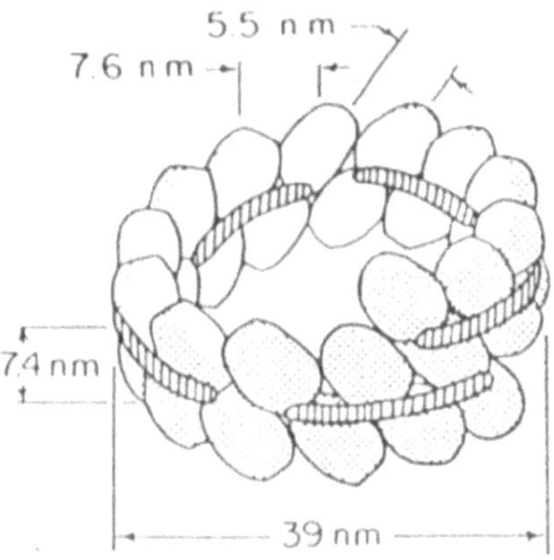

FIGURE 2. A schematic drawing of the ring structure proposed by Scheele and Borisy. Each ovoid represents αβ tubulin; 29 tubulin dimers are arranged in one turn of a helix. The hatched areas are zones where MAPs are supposed to bind. Six tubulin dimers contribute to one MAPs binding domain. (From Scheele, R. B. and Borisy, G. G., *J. Biol. Chem.*, 253, 2834, 1978. With permission.)

son.[39] Rings are in dynamic equilibrium with the constituents, tubulin, and MAPs. Pantaloni et al.[40] measured the half-life for the equilibration of radiolabeled tubulin, with unlabeled tubulin present in the rings. The radioactive tubulin was added in such small concentrations that the dimer ring equilibrium was not disturbed. At 0°C the half-life was 100 min, while this was reduced to 10 min at 22°C. Zeeberg et al.[41] studied nucleotide exchange with rings, and found that the rate-limiting step for exchange of nucleotide is the exchange of tubulin dimers itself. This indicates that nucleotide exchange in the ring state is relatively slow. The rings are very pressure sensitive, and therefore a fast-pressure perturbation allows the study of relaxation behavior of the system.[42] After a fast-pressure drop from 200 to 1 atm the reequilibration can be followed with 90° light scattering. Three relaxation steps are observed in this way for the system at pH 6.4 Two of them are dependent on protein concentration. The identification of these relaxation steps is unfortunately not obvious. The two protein concentration-dependent steps are probably due to the association of tubulin and MAPs with ring intermediates. The concentration-independent step is interpreted as due to ring opening and closure. Indeed, the particle size factor for 90° scattering drops by about 20% upon opening of a single ring, and by 54% upon the opening of a double ring into the single straight rod. Ring formation is slightly exothermic: rings are partially broken down upon warming.

Marcum and Borisy[38] studied the dependence of ring formation on the total microtubule protein concentration. These authors propose a stoichiometry of 5 MAPs to 30 tubulin molecules. However, the system does not follow the description with an all or none equilibrium model:

$$5\ \mathrm{MAPs} + 30\ \mathrm{T} \rightleftharpoons \mathrm{T}_{30} \cdot \mathrm{MAPs}_5 \qquad (1)$$

The concentration dependence of the real ring system is much less pronounced than the theoretical model of full cooperativity predicts.

The process of ring formation with pure tubulin was studied by Frigon and Timasheff[43] for T-GTP and by Howard and Timasheff[45] fot T-GDP. Double rings are formed, the stability of which depends very strongly on the Mg^{2+} concentration, but also on the presence of GDP or GTP. From the concentration dependence of the sedimentation behavior, a model

TABLE 1
Equilibrium Constants for Ring Formation by Pure Tubulin[43-45]

$MgCl_2$ (m*M*)	1[a]	8	n[b]
	For T-GTP		
$Log(K_p)$	91.7	109.35	19.6
$Log(K_2)$	3.37	4.09	0.76
$Log(\Gamma)$	2.41	7.11	5.2
	For T-GDP		
$Log(K_p)$	78.9	119.6	44.8
$Log(K_2)$	3.84	4.16	0.36
$Log(\Gamma)$	−16.2	15.61	34.7

[a] The data at 1 m*M* Mg^{2+} are our own extrapolation on the basis of a linear log(K) vs. log $[Mg^{2+}]$ plot. For GTP only two concentrations are available.

[b] Number of Mg^{2+} ions, obtained from the slopes of these plots, involved in the overall ring formation (from K_p), in dimerization (from K_2) and in ring closure (from Γ).

for ring formation could be deduced: an isodesmic oligomerization is assumed up to the point of ring closure with 26 monomers involved. At high Mg^{2+} ion concentration, e.g., 8 m*M* and higher, rings are much more stable in GDP than in GTP. The difference in stability is largely due to the effect of Mg^{2+} on the step of ring closure. From the concentration dependence of ring closure, it can be deduced that about 35 magnesium ions are taken up. At 1 m*M* Mg^{2+} ring closure is extremely unfavorable with GDP (see Table 1).

A very important conclusion is that oligomer as well as ring formation by pure tubulin is negligible in 1 m*M* Mg^{2+}. However, it becomes very important whenever the magnesium ion concentration is raised to more than 10 m*M*. This is often done to increase the yield of microtubule formation in the absence of MAPs. It should be clear that the mechanism of assembly and disassembly can be different in view of this difference in ring stability (see further for discussion).

III. ASSEMBLY CONDITIONS

In 1968 and 1972 Weisenberg et al.[9,46] determined the essential features for the polymerization conditions:

1. GTP is necessary for polymerization, and it is hydrolyzed. It can be replaced by nonhydrolysable analogues.
2. GDP remains firmly bound in the polymer, while P_i is released.
3. GDP itself is a strong inhibitor.
4. Microtubules are very sensitive to Ca^{2+}. Weisenberg's discovery of this calcium sensitivity is one of the major steps forward in the high-yield preparation of microtubules.

Although it was originally claimed that tubulin did not polymerize on its own, it became soon clear that this largely depends on the solution conditions. Pure tubulin can assemble into microtubules but only at rather high protein concentrations.[47] This can be lowered by the addition of sucrose or glycerol,[48] usually 25% v/v (= 3.4 *M*) of the latter is used. Also Me_2SO (optimally around 8%) stimulates assembly.[49] Different buffers give different results, e.g., PIPES buffers give a higher yield of polymerization as compared to the MES buffer.[50] Glutamate (1 *M*) also stabilizes the polymers very strongly while dissociating the MAPs from the microtubules.[51] The effect of solution components like glycerol or zwitterionic substances on the assembly and the stability of proteins is discussed by Gekko and Timasheff[52] and Arakawa and Timasheff.[53] Another important parameter is the concentration of magnesium ions.[47] These vary from 0.5 to 16 m*M*. High concentrations of Mg^{2+} certainly stimulate assembly, but next to microtubules, other polymers, mostly with S-shaped sheets are formed.[54] This is not necessarily problematic, since the propagation constant of the S-sheets and the microtubules may be very similar. Assembly in phosphate buffer is possible, although interference might be expected from the fact that P_i is liberated during assembly (see further).[55] This wide variety of conditions make comparison of results of different laboratories sometimes difficult.

Ca^{2+} ions dissociate microtubules, and all buffers contain 0.5 to 1 m*M* EGTA to sequester these ions.[46] The ionic strength is very important, and 0.2 *M* is sufficient to prevent any association.

As already mentioned GTP is necessary for assembly. Its concentration varies between 0.1 to a few millimolar. Higher concentrations may be inhibitory.[56] During normal assembly GTP is hydrolyzed and GDP accumulates. Since this is inhibitory, it is better to use a regenerating system, e.g., acetate kinase and acetylphosphate, to keep GTP constant and reduce the GDP concentration to a negligible level.[57] Neither acetylphosphate nor acetate itself have any influence on the polymerization mechanism. Other regenerating systems can be imagined but the absence of interference by the components should always be checked. GTP can be formed from GDP and ATP, since the protein preparation is usually contaminated with NDP-kinase.[58,59] The MAPs can be replaced by other cofactors like polycations. The charge density is very important, and if it is too high a second layer of tubulin can be wrapped around the microtubule to form a so-called duplex microtubule.[60,61]

Tubulin has many binding sites for inhibitors, e.g., colchicine and its analogues, vinca alkaloids, podophyllotoxin, nocodazole, etc. These molecules bind very specifically to tubulin and prevent the formation of correct microtubules.[62-66] One stimulator of assembly with very high affinity is known: taxol.[67] It polymerizes tubulin very efficiently, even without MAPs.

IV. POLYMORPHISM

Depending on the solution conditions, different types of polymers can be formed. Next to the forms already described, tubulin hoops have to be mentioned.[68,69] These large curved sheets are formed in conditions of low protein concentrations. In the presence of Zn^{2+} ions large flat sheets are formed.[70] All these different types of polymers made from the same protein prove that the different solvent conditions result in small variations in surface structure of the protein which are strongly amplified by the high degree of polymerization.

An interesting form of polymer is the structure formed when C-shaped sheets associate laterally to microtubules to form the so-called hooks.[71,72] These structures allow the determination of the structural polarity of the microtubules and can be applied to biological samples.[73]

V. POLYMERIZATION ANALYSIS ACCORDING TO OOSAWA

A. EQUILIBRIUM ANALYSIS

Only the essence of the theory derived by Oosawa[1] is presented here inasmuch as it defines the terminology used. In an open polymerization system, e.g., actin or tubulin, a whole spectrum of polymeric species is formed. The concentration of an i-mer can be defined as a function of the monomer concentration (C_1) as follows:

$$C_i = A \cdot K_p^{-1} \cdot (K_p \cdot C_1)^i \qquad (2)$$

This relation can be understood intuitively: i monomers are involved, which implies i-1 interactions. Except for a small number of initial steps, these are all assumed to be identical, and are therefore all described by the individual association constant K_p. The factor A corrects for the instability of the initial oligomers up to the nucleus (n-mer). The total number concentration ($C_N = \Sigma C_i$) and mass concentration ($C_M = \Sigma i.C_i$) of polymers can be calculated by applying two series expansions on this relation. The following final equations are obtained:

$$C_M = A \cdot C_1/(1 - K_p \cdot C_1)^2 \qquad (3)$$

$$C_N = A \cdot C_1/(1 - K_p \cdot C_1) \qquad (4)$$

A consideration of Relation 3 shows that C_M increases very strongly when C_1 approaches $1/K_p$. This is therefore the upper limit for the monomer concentration and it is called the critical concentration. When the parameter A is extremely small, indicating a very difficult nucleation, a sharp increase in C_M is obtained when the total protein concentration exceeds the critical concentration. The maximum value of A (= 1) is obtained for an isodesmic polymerization, which shows a very gradual increase of C_M with total concentration. Interestingly, the parameter A also describes the destabilization of the ends of the polymer. The number concentration of polymers will be much smaller when nucleation is difficult, as compared to an isodesmic polymerization (see Formula 4). Attempts to determine the parameter A for microtubule nucleation in the presence of MAPs have been made.[74] A value around 0.1 was found at temperatures higher than 30°C, but A decreases to 0.01 around 10°C, indicating that nucleation is not extremely difficult.

It is important to realize that not only an equilibrium mass concentration is defined but also an equilibrium number concentration. The ratio of both concentrations gives the average degree of polymerization: $\langle i \rangle = C_M/C_N$.

Formula 2 also allows the calculation of the length distribution. The ratio $C_{i+1}/C_i = K_pC_1 < 1$ is constant and therefore an exponential length distribution is described.

B. KINETICS

The rate of formation and disappearance of the i-mer can be described as follows:

$$dC_i/dt = k_+ \cdot C_1 \cdot C_{i-1} - k_- \cdot C_i - k_+ \cdot C_1 \cdot C_i + k_- \cdot C_{i+1} \qquad (5)$$

In this relation the rate constant of association (k_+) and of dissociation (k_-) is assumed to be the same for all species, except for the nucleus. Of course it is impossible to follow the formation of each i-mer separately. Generally a parameter proportional to the mass concentration is measured (e.g., turbidity in certain conditions). The time dependence of the mass concentration can easily be obtained by summing up Equation 5 for all the species:

$$\begin{aligned} dC_M/dt &= d\Sigma iC_i/dt = i\Sigma dC_i/dt \\ &= k_+ \cdot C_1 \cdot C_N - k_- \cdot C_N + n \cdot dC_n/dt \\ &= -dC_1/dt \end{aligned} \qquad (6)$$

where Cn is the concentration of the nucleus.

Nucleus formation is usually simplified to a single global process of order n. The rate constants k_n and k_{-n} stand for nucleus formation and disappearance, respectively:

$$dC_n/dt = k_n \cdot C_1^n - k_{-n} \cdot C_n \tag{7}$$

The number concentration C_N is itself the solution of the following rate equation:

$$dC_N/dt = d\Sigma C_i/dt = \Sigma dC_i/dt = dC_n/dt \tag{8}$$

This equation shows that nucleation is the only way for a new polymer molecule (or a new end) to be formed, or to disappear.

Additional mechanisms for the creation of ends, not considered by Oosawa, can be imagined, e.g., the breakage of long polymers. Wegner and Savko[75] assumed such a process of fragmentation to occur in actine, and suggested to describe it by the following additional term in Equation 9:

$$(dC_N/dt)_{fr} = k_{fr} \cdot C_M \tag{9}$$

The dependence on C_M implies that fragmentation can occur at each bond with the same probability. It seems more likely that very long polymers have a higher chance to fragment than shorter ones. When shearing long polymers like actin or microtubules, or subjecting them to sonication, the process of fragmentation is considerably enhanced.

Another process than can influence C_N is annealing of filaments, i.e., the longitudinal association of filaments. This term has been described as

$$(dC_N/dt)_{ann} = k_{ann} \cdot C_N^2 \tag{10}$$

which implies that all i-mers can associate with equal intrinsic probability. This is an oversimplication, as a smaller i-mer will certainly diffuse faster and have a higher chance to associate.[76] It is to be expected that this process is going to be important in all conditions where a very high number concentration is reached, e.g., immediately after shearing or sonication.

A full description of the whole polymerization curve necessitates the integration of the rate equations of all individual species. However, this leads to an infinite set of simultaneous rate equations which cannot be solved. Therefore the simplication was introduced that the growth rate constant (k_+) and the dissociation rate constant (k_-) is the same for all the species with their size greater than n. The number of equations is now limited to n species plus one for C_M. Frieden and Godette[77] obtained a good simulation of the polymerization curves of actin with a numerical integration of such a set of six equations. Wegner and Engel[78] reduced the set to two equations, one for the time dependence of C_N and one for C_M, by assuming a steady state between the initial small oligomers. A similar equation was obtained by Tobacman and Korn,[79] assuming a preequilibrium for the nucleus. Even these two simultaneous equations have to be integrated numerically in order to stimulate the observed polymerization curve. Only by neglecting all the dissociation rate constants, Oosawa could obtain an analytical solution. This equation has therefore only limited applicability.

The theory of one-dimensional polymerization has been further elaborated by De Levie and co-workers,[80,81] and analytical solutions have been obtained in particular situations without neglecting the dissociation rates.

VI. KINETICS OF MICROTUBULE ASSEMBLY

A. NUCLEATION

Although spontaneous nucleation is apparently a process to be avoided in the cell, it occurs *in vitro* and it is interesting in its own. As discussed before, it can be defined as the formation of early oligomers, up to the point where their association rate constants become independent on their size. Nucleation is thus the formation of the first complete motive that is propagated by the addition of a new building block. Is this motive the smallest microtubule that can be formed? This depends on the question of whether the closure of a sheet into the cylinder is a necessary step to obtain the motive. This is not necessarily true. According to Erickson and Pantaloni,[82] it is sufficient to have comparable rates of association in the longitudinal and the lateral direction to obtain fast two-dimensional growth. Therefore, since lateral associations are less stable than longitudinal ones, it is sufficient to make small oligomers, long enough to have an equal probability of asssociation in the lateral and longitudinal direction.

Although the process of growth can be studied separately, whenever nucleation is negligible, e.g., in the approach to steady state, after spontaneous nucleation, or after seeded assembly, this is never possible for nucleation. During spontaneous polymerization, nucleation slows down very rapidly due to its high power dependence on the concentration of the monomer pool, which is rapidly depleted by the growth process. The number concentration of polymers finally formed is the result of integration over time of this competition between growth and nucleation, and it is the most concrete information available about nucleation.

For actin it is relatively easy to imagine how the nucleus looks, because only the association of the first three molecules differs from the subsequent ones. Therefore a stimulation of the full polymerization curve was possible. For tubulin, however, nucleation must be a rather complicated process. As a consequence, most authors determined only the size of the nucleus, from the cooperativity number after spontaneous nucleation. This cooperativity number is very similar to the Hill coefficient for ligand binding. In the case of a polymerizing system, it is the apparent stoichiometry coefficient of the nucleus, and it is derived from the concentration dependence of the lag phase or, of the slope of the linear phase. (This was shown to be the maximal growth rate at the inflexion point of the polymerization curve).[42] The cooperativity number can also simply be derived from the dependence on total protein concentration of the first-order rate constant of approach to steady state, since this reflects the number concentration of ends. For polymerization of tubulin in the presence of MAPs, the cooperativity number was found to be about 2, a surprisingly small number.[83] Pure tubulin in glycerol showed a value as high as 10 to 12.[54] The difference was assumed to be due to the preexistence of oligomers when MAPs are present. The rate-limiting step of nucleus formation is thus the association of two such oligomers. Pure tubulin in Me_2SO also gave a low value.[84] This was interpreted as due to the transient formation of oligomers at the mixing interphases, due to the local high concentrations of Me_2SO. A recent study, however, shows that higher values of n are obtained in Me_2SO, when direct plots or $\log(K_{obs})$ vs. $\log(C_{tot})$ are used.[85]

High-speed sedimentation of microtubule protein resulted in a MAP-free protein fraction that was not able to form microtubules.[86] From this it was concluded that rings are necessary for nucleation. A similar conclusion came from experiments with predissociated rings.[83] When rings were dissociated with the thiol reagent diamide,[87] and when subsequently an excess of mercaptoethanol was added, this protein material was correctly taken up into the microtubules, but was not able to contribute to nucleation. The same results were obtained by increasing and subsequently decreasing the ionic strength.

The X-ray scattering experiments of Mandelkow et al.,[88,89] however, showed that rings are first broken down into smaller fragments prior to the assembly of microtubules. Therefore, it can be concluded that in the presence of MAPs, nucleation starts from small oligomeric fractions, which are formed by the partial dissociation of the rings. Rings are therefore often supposed to be the products of a side path of assembly, whenever microtubules are not formed. Many experiements, however, suggest that rings are the direct products of the dissociation of microtubules.[41,90] In this way they are intermediates in repeated cycles of assembly and disassembly. Especially in the presence of GDP and high Mg^{2+} rings are shown to be rather stable, as mentioned before.[45]

B. INTERMEDIATES

If the nucleus would be the smallest microtubule that can be imagined, only a few species would have to be described.

However, if the nucleus is the association product of two small oligomers, many intermediates can be imagined between this nucleus and complete microtubules. Several attempts were made to visualize these intermediate structures via electron microscopy, carefully trying to preserve the structures in solution. Kirschner and co-workers[91] observed the formation of curly ribbons, with a lateral width smaller than 13 protofilaments. Since the curvature of rings is perpendicular to the curvature of the microtubule surface, a curled spiral could be an intermediate with a gradual change-over of this curvature. An interesting question is whether the sheets grow in the lateral direction and stop at the right width, or whether closure itself is the stop signal for lateral growth. *In vitro*, some flexibility exists about the number of protofilaments involved in a microtubule, suggesting that closure is responsible for arresting growth in the lateral direction.

C. KINETICS OF GROWTH

As indicated before, the kinetics of growth can be studied separately and therefore much more information is available about this process. When microtubule seeds are added to a solution of tubulin, which is made competent to polymerize at the same time, a first-order process of microtubule growth is observed. In terms of the theory of Oosawa, this can be analyzed in a rather straightforward way:

$$-dC_1/dt = k_+ \cdot C_1 \cdot C_N - k_- \cdot C_N \tag{11}$$

This is Equation 6 where the nucleation term has been dropped. This equation is written with the assumption that a polymer has a constant number of growth sites, irrespective of its size. This factor is included in k_+. If C_N is constant, this equation can be integrated to:

$$C_1(t) = (C_0 - C_c) \cdot \exp(-k_+ \cdot C_N \cdot t) \tag{12}$$

The observed first-order rate constant is thus $k_+.C_N$. The average length and the mass concentration allow the calculation of C_N (see earlier). Since it was shown that turbidity is a measure of the mass concentration of microtubules, polymerization kinetics can be followed by this simple technique.[92] First-order growth kinetics were described by Lee et al.[93] These authors demonstrated the sensitivity of the system to ionic strength and the existence of a pH optimum around 6.2 to 6.4 Bryan[94] studied growth of microtubules seeds in the presence of polyanions to sequester the MAPs. His study experimentally demonstrated the essential features of the theory presented above: approach to steady state could be described by a single exponential, the rate constant of which was dependent on the concentration of added seeds, while the extent of polymerization was independent of it. In this way he obtained the value for $k_+ = 4 \times 10^6\ M^{-1}.s^{-1}$ at 30°C (extrapolated to zero polyanion concentration). An extensive study of the different features of this condensation mechanism was done in the absence of polyanions by Johnson and Borisy[95] and a similar rate constant was obtained. After spontaneous nucleation the same growth rate constant was found by Engelborghs et al.[83]

Equation 11 describes a linear relation between the rate of growth ($-dC_1/dt$) and the actual monomer concentration (C_1). Therefore, during the approach to steady state, the derivative of the polymerization curve decreases linearly with the monomer concentration remaining in solution. The equation can also be applied to the initial growth (or dissociation) rate after dilution of microtubules into a monomer solution above (or below) the critical concentration. By extrapolating the dissociation rates to zero monomer concentration, the value of k_- can be calculated. From this extrapolation, Johnson and Borisy obtained a value of $6.8 \pm 0.4\ s^{-1}$ for k_- at 30°C.

This technique can of course be used to identify those factors that influence k_+ and k_- separately. In this way Murphy et al.[96] showed that MAPs only reduce k_-, while leaving k_+ unchanged.

Since rings disappear quite rapidly during spontaneous polymerization, it is often assumed that the process of growth is essentially due to the addition of $\alpha\beta$ dimers to the microtubules. However, Barton and Riazi,[97] Burns and Islam,[98] Gall,[99] and Bayley et al.[100] showed that the approach to steady state can be biphasic, and this is interpreted as an indication for the direct contribution of oligomeric species to the process of growth. In fact the contribution of the fast phase seems to be correlated with the presence of oligomers, and the rate as well as the temperature dependence of the fast step correspond to the process of the dissociation of the rings.[100] MAPs are also shown to disappear very rapidly from solution, and it is suggested that the slow phase is regulated by the redistribution of MAPs on the microtubules already formed.

Heusele et al.,[101] however, claim that the biphasic nature is due to a change in specific turbidity associated with a different polymeric intermediate. Their interpretation is based upon observations using the fluorescent probe DAPI. This molecule shows considerable fluorescence increase upon binding to tubulin. It has a higher affinity for polymers, so that polymerization can be followed. Even polymerization curves that show a biphasic character in turbidity, are described as a single first-order plot using DAPI fluorescence. The authors therefore suggest that DAPI fluorescence is a more accurate measure for the mass concentration of the polymer, irrespective of the exact form of the polymer.

In a recent article, Barton et al.[102] studied the different solution conditions that control the appearance and disappearance of the biphasic nature of the growth curves. High ionic strength (0.2 *M*), low temperature (<30°C), and high Mg^{2+} (>1.2 m*M*) favor the monophasic kinetics.

It seems very likely that rings or fragments contribute to fast early elongation, but an exact description of the process must certainly wait until all the parameters that govern oligomer formation and dissociation are characterized quantitatively.

D. DISASSEMBLY KINETICS

The rate constants for the dissociation of tubulin from the two ends were obtained from the extrapolation of the linear growth rate curve dC_1/dt vs. C_1 to $C_1 = 0$. As with other aspects of microtubule research, the direct measurements gave surprises.

The process of dissociation can be measured, in conditions where assembly is completely inhibited or negligible. A priori, such conditions can be obtained in different ways, e.g., by fast cooling to 4°C, by the fast and irreversible consumption of GTP, using enzymes, by the fast association of unpolymerized tubulin to an inhibitor like colchicine, podophyllotoxin, and possibly Ca^{2+} ions, or simply by extensive dilution below the critical concentration. In all these cases the process of disassembly should be rate limiting. Therefore, the binding of an inhibitor or the turnover of GTP should be very fast. It is also necessary that the agent added or the complex formed should not interfere with the ends itself. Unfortunately this is clearly the case with colchicine and podophyllotoxin, and probably with tubulin-GDP. The inhibitor should not interact directly with the polymer. Unfortunately, most inhibitors do. Preferentially the dissociation should occur at the temperature of interest. Therefore extensive dilution seems to be the safest method, provided it can be done without shearing the microtubules and provided it can give a measurable signal. The former can be controlled by diluting microtubules into tubulin solutions, such that the growth rate reflects the dilution factor. However, this is not possible at extreme dilutions, where the growth rate is of the same magnitude as the rate of spontaneous nucleation. To obtain a measurable signal, even upon extensive dilutions, a cell with a very long pathway was used by Karr and Purich.[103]

In the simple analysis in Oosawa's theory, the dissociation process was described as $k_-.C_N + n.k_{-n}.C_n$. In an irreversible dissociation, despite the fact that all microtubule end up as a nucleus just prior to complete disappearance, the second term can be neglected because it is not the rate-limiting step of dissociation. The first term implies that dissociation occurs only at a constant number of dissociation sites per polymer, independent on its length, with k_- being the sum of the rate constants of individual sites multiplied by their number. It is natural to assume that these sites occur only at the ends.

For the analysis of the full dissociation curve, the length distribution has to be taken into account. As short microtubules disappear earlier than long ones, the number concentration will normally be time dependent. Since the rate of dissociation (the first derivative of the dissociation curve) reflects the number concentration of microtubules present at any moment, the decrease of the dissociation rate after a small time interval dt (second derivative) corresponds to the number of short microtubules that have disappeared during this time interval. Therefore, it is intuitively understandable that the shape of the length distribution is reflected in the shape of the second derivative of the dissociation curve, as pointed out by Johnson and Borisy[95] (see Figure 3). A rigorous mathematical analysis was performed by Purich and co-workers[104,105] and was applied to microtubules. Starting from the initial length distribution the whole dissociation curve could be simulated with a single k_- value. Table 2 shows a set of values obtained by different authors. Dissociation by calcium ions or by cooling also induced endwise depolymerization, but 5 m*M* calcium increased k_- about fivefold.

These data clearly prove that the rate constant of dissociation obtained from direct determinations was much higher than the value calculated from the critical concentration and k_+. Similar deviations were observed by Farrell et al.[106] and cannot be explained on the basis of the simple theory.

The analysis presented so far is based on the assumption that only dimers dissociate from the ends. In fact this point is not yet absolutely clear. Electron microscopy very often shows microtubules with frayed ends.[107] Similar results were presented in the beautiful pictures of Mandelkow, obtained by cryoelectron microscopy.[108] These pictures were made by rapid freezing of microtubules in a state of disassembly induced by slow cooling. In the immediate neighborhood of the microtubule ends, many short and curved oligomers are found. Since lateral associations are endothermic, it is clear that cooling will preferentially destabilize these interactions. The pictures suggest indeed that the longitudinal interactions remain largely intact. The question remains whether isothermal dissociation by dilution also occurs by lateral splitting at the ends and by direct production of oligomers. Fraying of the ends is certainly much less pronounced at high temperature.[108] The mechanism of direct dissociation into rings is also suggested by Howard and Timasheff,[45] on the basis of the stability of pure T-GDP rings. However this is only the case in very high Mg^{2+} concentrations. Zeeberg et al.[41] demonstrated that disassembly in the presence of radioactive monomers leads to the formation of rings that do not contain the label. This is a strong argument in favor of disassembly directly into oligomeric fractions.

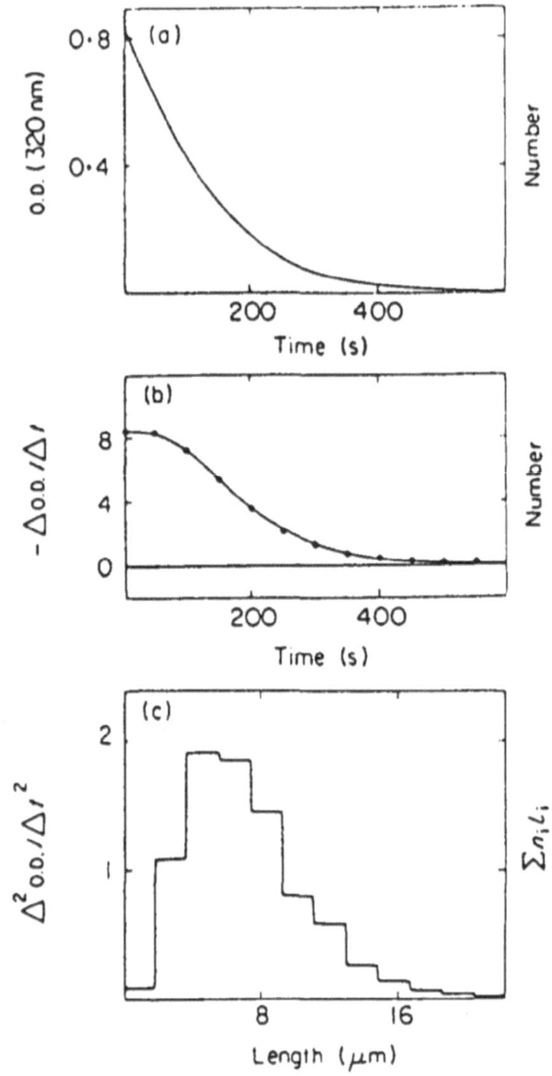

FIGURE 3. Analysis of depolymerization kinetics as influenced by the length distribution of the polymers. (a) Kinetics of dissociation: turbidity vs. time after cooling. Units: O.D. vs. time (s). (b) First derivative, tangent of (a) as a function of time. Units: $10^3 \times \Delta$O.D./s vs. time (s). (c) second derivative of (a) in units $10^3 \times \Delta^2$ O.D./s^2 vs. length. The length scale is related to the time scale by the rate constant $K^- = 0.037$ μm/s. (From Johnson, K. A. and Borisy, G. G., *J. Mol. Biol.*, 117, 1, 1977. With permission.)

E. RELAXATION KINETICS

The steady state can be disturbed by small perturbations of all factors that have an influence, e.g., a small but fast temperature rise or decrease, a fast pressure change, or a small dilution into buffer or tubulin solution. As shown by Engelborghs and Van Houtte,[109] a small temperature jump on microtubules made in the presence of MAPs gives rise to a simple relaxation process. The analysis is based on Equation 11, and assumes that $k_-.C_N$ remains constant. The following observed rate constant or reciprocal relaxation time is derived:

$$\tau^{-1} = k_+ \cdot C_N \qquad (13)$$

TABLE 2
Rate Constants

For Growth and Dissociation of Microtubules (Sum of the Contributions of the Two Ends)[a]

System	Temp (°C)	K_+ ($\times 10^6 M^{-1}s^{-1}$)	K_- (s^{-1})	Ref.
MTP	30	4	—	94
MTP	30	4	7[b]	95
MTP	35	7	7[b]	83
	25	3	5[b]	83
	30	100[b]	113	104
	37	100[b]	154	104
	0	—	166	104
5 m*M* Ca^{2+}	30	—	860	104
Rate Constants for Individual Ends (Tubulin on Axonemes)[c]				
(+) end	30	7.2	17	116
(−) end	30	2.2	7	116
(+) end	37	3.8	(0.4)[b] 340	134
(−) end	37	1.2	(1.1)[b] 212	134

[a] Values for global rate constants, as determined by turbidity.
[b] Values for individual ends as determined by growth of pure tubulin off axonemes.
[c] Indicates that the rate constant was not directly determined but has been calculated from the critical concentration and the other rate constant.

This shows that only information about k_+ is obtained even if a small dissociation is induced.[109] The method can therefore be used to study three effects: factors that influence k_+ directly, factors that influence the number of growth sites on a microtubule end, which is also implicated in k_+, and factors that influence C_N. The method was used to study the effect of tubulin-colchicine on the number of available growth sites on the ends,[64] and on the effect of GDP.[109] It can also be used to follow very easily the reequilibration of the number concentration C_N after fragmentation.

Information about k_- can be obtained from the amplitude or from the initial rate. In fact a continuous spectrum, from relaxation behavior to total disassembly, can be obtained when microtubules are diluted in buffers containing tubulin below the critical concentration. Usually the initial rate of these curves is analyzed, and this should also follow the relation of Equation 11. However, the initial growth rate plots are curved in the negative branch and extrapolate to the k_- values obtained upon infinite dilution.[108,110] The same observations were made for actin.[111] The reason for the curvature will become apparent in the next paragraph.

It is interesting to compare the relaxation behavior of a nucleated system with that of an isodesmic polymerizing protein: glutamate dehydrogenase.[112] Although this protein has rate constants of association and dissociation which are very similar to those of tubulin, the relaxation process occurs in milliseconds instead of minutes. This is due to the much higher number concentration of polymers formed in the absence of nucleation.

F. DISCRIMINATING THE TWO ENDS

The equations presented before described growth in terms of the rate constants for the addition of monomers to and dissociation from one type of growth site on the polymer. However, every linear polymer has two ends, and the rate constants can be different at the

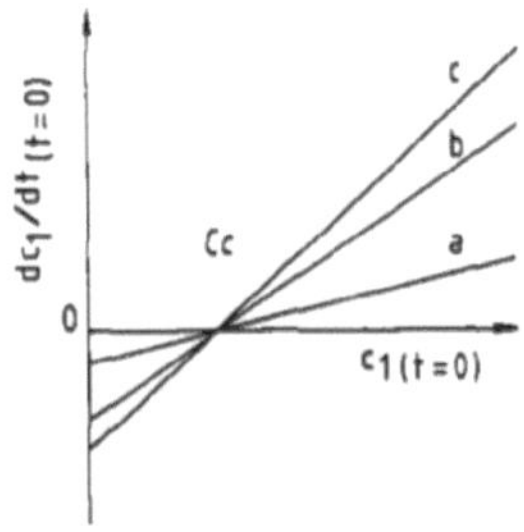

FIGURE 4. Initial growth rate curves of a true equilibrium system. Growth or dissociation is observed after dilution of microtubules in buffer or in monomer solution. The existence of two ends is taken into account. Initial growth rate $dC_l/dt_{(t=0)}$ is plotted vs. the initial monomer concentration $C_{l(t=0)}$. The slope gives $K_+ . C_N$, the intercept $K_- . C_n$. Behavior of a true equilibrium system, where the equilibrium monomer concentration of the two ends is the same, but the kinetics are different. The straight lines cross the abscissa at the point where $C_1 = C_c$. (a) Initial growth rate curve for slow growing end, (b) for the fast growing end, and (c) sum of the two as measured by turbidity.

two ends. The growth rate equation therefore must be written for each end separately, with a single constraint: for a true equilibrium system, the equilibrium constant of growth should be the same at both ends, as pointed out by Wegner.[113]

$$-dC_1/dt = k^a_+ \cdot C_1 - k^a_- = k^b_+ \cdot C_1 - k^b_-$$
$$= 0 \text{ at equilibrium} \tag{14}$$

Here the rate equations are written per end, and the ends are identified with the superscripts a and b.

If the growth rate contants are similar for the two ends, growth will be bidirectional; however, if they differ considerably, growth at one end will dominate and growth will essentially be unidirectional.

It is obvious that turbidity measures the sum of the growth processes at the two ends (see Figure 4). Therefore, the rate constants derived from turbidity measurements are the sum of individual rate constants, and the same type of equations still apply.

The existence of different growth rates at the two ends was demonstrated by Summers and Kirschner[114] by direct observation of microtubules in dark-field light microscopy, the possibilities of which had been shown by Kuriyama and Miki-Noumura.[115]

Individual rate constants for the two ends can only be obtained if the two ends can be identified. For microtubules this was done by growing tubulin on flagellar fragments, by Bergen and Borisy.[116] After growth the flagellar fragment shows two sets of extending microtubules, with one set having an average length much shorter than the other (see Figure 6). The detailed structure of the flagellar fragment allows the unambiguous identification of the ends. By following the growth or the shortening of the extended microtubules, the rate constants for association and dissociation at the two ends can be determined. This is done discontinuously by taking samples for electron microscopy, or continuously by optical (video) microscopy, with dark-field or differential intereference contrast detection. Of course, sufficient measurements have to be done to obtain statistically significant numbers, since individual ends are being measured. The values obtained by Bergen and Borisy using electron microscopy correspond fairly well with previous values: at 30°C k_+ is $7.2 \times 10^6\ M^{-1}.s^{-1}$ and $2.2. \times 10^6\ M^{-1}\ s^{-1}$, respectively, for what is now defined as the (+) and the (−)

end. From tenfold dilution experiments k_- was found to be 17 s^{-1} and 7 s^{-1}, in fair agreement with the extrapolation of the growth curve.

Growth at the two ends can also be observed when two types of tubulin are used which can be recognized by specific antibodies. Rowthwell et al.[117] used seeds made from chicken erythrocyte tubulin and allowed them to grow with chicken brain tubulin. Results consistent with axonemal growth were obtained.

VII. KINETIC EVENTS AT STEADY STATE

Once the steady state is established, turbidity can no longer be used to study the dynamics of the system, since the mass concentration is essentially constant. However, monomers are continuously taken up in the polymer and dissociating from the ends. Since polymerization is accompanied by GTP hydrolysis, GTP is continuously consumed and P_i as well as GDP are produced.

According to the simple theory of Oosawa, all the properties of the steady state can be deduced from the rate constants determined so far. Therefore, no surprises are expected from steady-state studies. In practice, however, the most exciting results were obtained from these studies.

It should be realized that in the simple theory tight coupling between growth and GTP hydrolysis is assumed. Also the steady-state length distribution and number concentration is not necessarily established at the same moment as the steady-state C_M is reached, especially with seeded assembly.

A. INCORPORATION OF PROTEIN OR NUCLEOTIDE

Wegner[113] was one of the first authors to study the dynamic properties of actin at steady state, by quantitating the uptake of a trace amount of labeled actin added at the steady state. Although in these conditions association of monomers is quantitatively compensated by their dissociation, this does not necessarily have to occur at the same single microtubule. For each individual end the chances for an association event are equal to the changes for a dissociation event. Therefore, uptake of labeled protein has the nature of a one-dimensional random walk, and the root mean square average length of a walk at an individual end after a time interval t is

$$(k_+ \cdot C_1 \cdot t)^{1/2} \tag{15}$$

This shows that the uptake of labeled monomers follows a square root dependence on time. Wegner observed a much faster uptake, that could not be accounted for by a simple random walk process, using the value of k_+ obtained from global polymerization kinetics. He concluded that another mechanism must be in operation. In this way he introduced the new concept of treadmilling (initially called head-to-tail polymerization): as a consequence of the (partial) coupling of hydrolysis and polymerization the thermodynamic restriction that both ends should have the same equilibrium constant for dissociation (= critical concentration) can be dropped. If the two ends have a different dissociation constant, a steady-state monomer concentration will be established and the net growth rate at one end will be balanced by the net dissociation rate at the other end (see Figure 5). A direct consequence of this model is the existence of a flux of monomers through the microtubules: the microtubules are treadmilling. Treadmilling was therefore used as a better name for the model (head-to-tail polymerization is used for polar associations like the αβ dimers in a microtubule). This flux equals the net rate of assembly at one end:

$$k_{+1} \cdot C_1 - k_{-1} = k_{+2} \cdot C_1 - k_{-2} \tag{16}$$

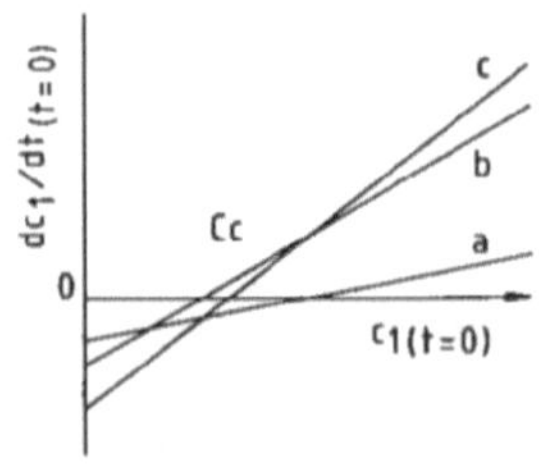

FIGURE 5. Initial growth rate curves for a steady-state system where assembly is coupled to hydrolysis of nucleotides. The two straight lines cross the abscissa in different points. The observed steady-state monomer concentration is obtained when net growth at the b-end is compensated by net dissociation at the a-end, with treadmilling as a consequence. (a) Initial growth rate for the slow end, (b) for the fast end, and (c) sum of the two as measured by turbidity. This line passes through the observed steady-state monomer concentration.

This constant flux implies that the amount of label taken up increases linearly with time (at least for some time). Of course after a certain amount of incorporation, some label will be emitted at the other end. The shorter the polymer is, the faster this will occur, and eventually the label will be distributed statistically. The original analysis of Wegner takes expulsion of label and the influence of the real length distribution of the polymers into account and describes the full course of label uptake. The process of treadmilling was independently shown to occur in microtubules by Margolis and Wilson.[118] These authors did not use labeled protein but radioactive nucleotides. Since GDP is nonexchangeably bound after polymerization, this label can be used instead of a covalent protein label. These authors also devised a procedure which, when interpreted in terms of treadmilling, leads to differential labeling of the two ends: (1) microtubules are first assembled with H^3-GTP and (2) at a steady state C^{14}-GTP is added.[119] The C^{14}-label is essentially only taken up by the growing end. Upon dilution, the loss of the two labels is then assumed to reflect the dissociation kinetics of the two ends separately.

As shown by Equation 14 treadmilling can also be deduced from the separate rate constants of the two ends, as measured in the studies on flagellar fragments, or seeds of different tubulins.[114-117]

Treadmilling was for some time assumed to be a possible mechanism for microtubule-based transport; if the elongating end of a microtubule is also able to push the microtubule, every object linked to the surface of the microtubule will be transported. A mechanism for mitosis on the basis of treadmilling was proposed.[120] Other possible biological roles for treadmilling were suggested by Hill and Kirschner.[121,122] However, recent experiments show that microtubule-based transport is dependent on the action of a different kind of transporter protein: dyneins, kinesins, and similar molecules.[123]

B. HYDROLYSIS

If polymerization and hydrolysis are strictly coupled, the rate of GTP consumption ($V_{hydr.}$) should be

$$k_+ \cdot C_1 \cdot C_N = k_- \cdot C_N = V_{hydr.} \tag{17}$$

In this simple case, dissociation is due only to T-GDP. If coupling is less strict, part of the off rate is also due to T-GTP and hydrolysis is correspondingly smaller:

$$V_{hydr.} = (k_+C_1 - k_-^{(T\text{-}GTP)}) \cdot C_N \tag{18}$$

David-Pfeuty et al.[124] showed that steady-state P_i production was proportional to the number of ends. Carlier and Pantaloni[125] studied P_i production from the beginning of polymerization on, and found that at high protein concentration, uncoupling between fast polymerization, and slow hydrolysis occurs; hydrolysis showed a lag phase with respect to the polymerization curve. (In order to compare the hydrolysis and the polymerization transients, both curves have to be normalized and hydrolysis has to be corrected for steady-state production.)

The existence of lag phase immediately suggests the existence of an inhomogeneous polymer, with a T-GDP core and a T-GTP region or cap at both ends. After the establishment of the steady state for mass concentration, hydrolysis catches up. The question then is, How long is the T-GTP cap at steady state? The size of the cap depends on the following steady-state equation:

$$\begin{aligned} V_{hydr.} &= k_{hydr.}[\text{T-GTP}]_{cap} \\ &= \{k_+ \cdot [\text{T-GTP}] - k_-^{(\text{T-GTP})}\}C_N \end{aligned} \tag{19}$$

The exact formulation of the last term, however, depends on where hydrolysis takes place: over the whole cap or only at the cap/core interface as a consequence of activity, induced by different neighbor interactions. Recent experiments suggest that hydrolysis occurs with a constant maximal rate, indicating hydrolysis at the cap/core interface.[126] Hydrolysis at the interface was also suggested by Caplow and Reid.[127]

Is the existence of such a cap compatible with treadmilling? The answer is yes, provided that fluctuations lead to an occasional exposure of T-GDP.

The coupling of polymerization and hydrolysis has also been studied with different methods by O'Brien et al.[128] These authors never found an uncoupling of hydrolysis and polymerization. They extensively analyzed the technical problems of Pi determinations in the presence of proteins, which can give rise to an underestimation of hydrolyzed GTP. An alternative way to demonstrate directly a possible T-GTP cap was developed by O'Brien et al.[128] and independently by Schilstra et al.[129] In the presence of a fast GTP-regenerating system, the only GDP present in the system is the nucleotide nonexchangeably bound in the polymer. This amount of (radioactive) GDP can easily be determined after precipitation of the protein with perchloric acid, and separation of the nucleotides with high-performance liquid chromatography. In this way both groups find no evidence for a T-GTP cap of an appreciable size.

The kinetics of H^3-GDP incorporation was also studied by Martin et al.[130] and again was found to be much faster than can be accounted for by a simple statistical fluctuation of the polymer length.

The steady-state studies mentioned clearly point to a much faster protein exchange at the ends than expected on the basis of a simple polymerization mechanism. Hydrolysis studies generated the possibility of a T-GTP cap at the ends, which seems to be rather small, and difficult to demonstrate experimentally.

C. DISASSEMBLY

The existence of a T-GTP cap, with an inhomogeneous polymer as a consequence, could explain the much higher dissociation rates of microtubules observed by extensive dilutions. Indeed k_-, obtained by extrapolation of the positive branch of the growth curve, is the dissociation rate constant of the homogeneous T-GTP polymer. However, actual dissociation will remove the T-GTP cap and expose the T-GTP core, which might have a much higher value for k_-. For actin this interpretation is supported by the fact that a complete linear growth rate curve was obtained in the case of ADP-actine.[111] Moreover, the value of k_- of

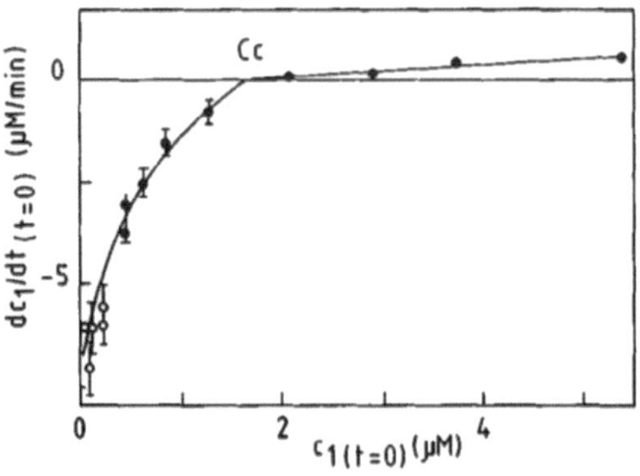

FIGURE 6. Experimental initial growth rate curve. Initial growth rate $dC_1/dt_{(t=0)}$ vs. tubulin concentration. Turbidity is used, such that the sum of the contributions of the two ends is measured. Although the simple theory predicts a straight line, a strong curvature is observed in the negative branch. (●) Points obtained by fourfold dilution into varying tubulin concentrations. (○) Points obtained by more extensive dilution and reduced to the same initial number concentration of microtubules. (From Carlier, M.-F., Hill, T. L., and Chen, Y.-D., *Proc. Natl. Acad. Sci. U.S.A.*, 81, 771, 1984. With permission).

ADP-actine corresponds to the one experimentally determined upon extensive dilution of ATP-actin. For microtubules this control is not possible, since T-GDP is not able to assemble into microtubules.

The exact form of the negative branch of the growth rate curve is also informative. If the T-GTP cap is very small, a very abrupt drop would be expected just below the critical concentration. In fact a rather gradual increase of the dissociation rate is observed (see Figure 6). In order to be able to simulate this gradual curvature Carlier et al.[110] had to assume fast nucleotide exchange at the ends, such that even upon moderate dilutions a fraction of the growth sites on the ends contain GTP and stabilize the end. The same result can be obtained through a different kinetic pathway: the fast binding of a first layer of T-GTP on the T-GDP core with an affinity much higher than the subsequent layers ($K_{as} >> 1/C_c$). Such a possibility was suggested in the case of actin.[111] Such a monolayer is formed very quickly, and an instantaneous equilibration with the T-GTP pool in solution can be assumed. To make an analogy: dissociation as a function of the monomer concentration resembles product formation by an enzyme, partially inhibited by an inhibitor (T-GTP). Caplow and Reid[127] suggested a variant: the binding of the first layer of T-GTP to the core would be nonproductive, but this variant does not fit the positive branch of the growth curve.

D. ROLE OF T-GDP

A question can be asked about the role of T-GDP. Of course T-GDP is incorporated in a microtubule through the pathway of hydrolysis. However, here the role of T-GDP in solution is questioned. A priori three possibilities are open:

1. It could be just an inactive form of tubulin.
2. It could copolymerize with T-GTP.
3. It could be a terminator: bind to a growth site on an end and inhibit further growth (= nonproductive binding).

In any case the following equilibrium with T-GTP will be established:

$$[\text{T-GDP}] = [\text{T-GTP}] \cdot K^{GDP} \cdot [\text{GDP}]/K^{GTP} \cdot [\text{GTP}] \tag{20}$$

where K^{GTP} and K^{GDP} are the binding constant for GTP and GDP respectively, which were shown to be strongly Mg^{2+} dependent.[131] This equilibrium is usually assumed to be very fast and established at any moment, which is certainly a considerable simplification. The three models can be distinguished on the basis of their influence on the critical concentration of T-GTP and on the kinetic properties of the ends. Models (1) and (3) predict that the [T-GTP] will remain the same as the original C_c, while in case (2) the [T-GTP] will decrease.[1] Only model (1) will leave the growth kinetics unchanged, after correction for the coupling with the nucleotide equilibrium.[109] Temperature jump relaxation experiments in the presence of mixtures of GTP and GDP were interpreted as indicative for a termination reaction by T-GDP.[109] Martin and Bayley,[132] however, proposed that the change in the initial rate of the growth relaxation curves could be explained by a temperature dependence of equilibrium (Equation 20), and these authors therefore favor model (1) where the nucleotide composition of the end has no influence. Growth off seed experiments, in the presence of GTP-GDP mixtures, by Carlier et al.,[126] show that the growth rate is simply proportional to the T-GTP concentration. However, the influence of GDP on the dissociation rates is more complicated, and again suggests a nonproductive binding of T-GDP to the ends. The situation of T-GDP is thus not yet clear.

E. DYNAMIC INSTABILITY

Extremely surprising results were obtained by Mitchison et al.,[133-135] when they studied incorporation of labeled protein and located the position of the label in the polymer. This was possible by combining biotin-labeled tubulin with specific antibodies. In this way many new aspects of microtubule assembly and of the properties of cellular nucleation sites *in vitro* were discovered. In order to identify the ends of the microtubules, associated to an organelle, Mitchison and Kirschner also introduced the addition of flagellar fragments as internal standards. Since the two ends of the flagellar fragments show a different length distribution, a comparison with the length distribution of the microtubules at the organelle being studied allows the identification of the exposed end as being of (+) or (−) character.

In this way they studied the nucleation of microtubules from centrosomes.[133] The microtubules showed a narrow length distribution with a single peak corresponding to that of the (+) end of the flagellar fragments. When kinetochores were used as nucleation centers, however, a bimodal length distribution was observed, with peaks corresponding to both the (+) and (−) distributions.[136]

An interesting observation is the fact that the number concentration of microtubules formed on the centrosomes increased in a nonlinear way with the concentration of added tubulin, up to saturation. It is therefore not surprising that the dilution of centrosomes saturated with microtubules leads to a decrease in the number concentration of these microtubules. The same is true for the situation where short microtubules are used as seeds and where a nonequilibrium high number concentration can be obtained. Under these conditions, even Oosawa's theory predicts an evolution to an equilibrium number concentration as well as the length distribution. Therefore, it is to be expected that the number of microtubule ends is continuously decreasing towards a new steady-state number concentration. Surprising is, however, the high speed of this process, and the observation of Mitchison and Kirschner that the average length of the microtubules at the centrosomes was continuously increasing, as if all microtubules present were always growing. The dissociation events, necessary for balancing, must therefore come from the full disappearance of certain microtubules. These new observations did not fit into any of the existing models and the phenomenon was called Dynamic Instability. From now on, the balance of association and dissociation events did not occur between the (+) and (−) ends of the microtubules but between the slow growing of most microtubules and the fast dissociation of a few. McIntosh[137] called this fast dissociation the catastrophy. Kirschner additionally specified that the pos-

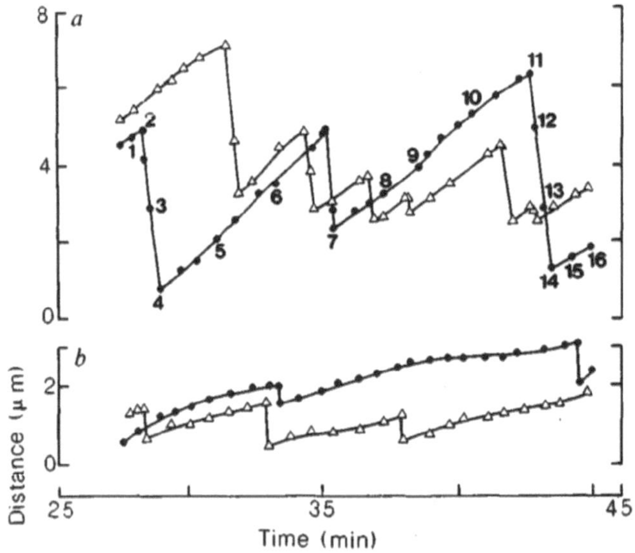

FIGURE 7. Dynamic instability of individual microtubules. *(a)* Length changes at the fast growing end of two microtubules. Notice the short periods of fast dissociation and the long periods of slow growth. The switching from growth to shrinkage and vice versa are rather frequent. *(b)* The behavior of the slow growing end. The same features are present but much less pronounced. (From Horio, T. and Hotani, H., *Nature*, 321, 605, 1986. With permission.)

sibility for switching from the growing population to the dissociating one and vice versa was very small. Stated as such, this model predicts the total disappearance of all microtubules. In fact the decrease of the number concentration was shown to slow down and finally stop. Farrell et al.[138] have also documented this phenomenon. What is the solution of this paradox?

A new surprise and new insight came when Horio and Hotani[139] directly looked at individual microtubules in the video microscope, using dark-field illumination. These authors observed dynamic instability in real time on a single microtubule: a particular end of a microtubule is growing most of the time, and suddenly switches to the dissociation phase and vice versa (see Figure 7). The dissociation rate is fast, but the time spent in it is relatively short. They found thus that the switching from the growing to the shrinking population and vice versa occurred rather frequently, in contrast to the initial picture of dynamic instability. Therefore a new process had to be introduced: the process of rescue, which is the switching of a fast-shrinking microtubule to the growing phase. It is this process which allows the number concentration of microtubules to reach a steady state, despite dynamic instability. Dynamic instability is thus a property of individual microtubules, and not just the property of a population of microtubules. With the video system the two ends can be followed and they behave rather differently; the fast growing end shows frequent switches, the slow growing end rather infrequently. MAPs seems to supress this peculiar behavior.[139]

With dynamic instability of individual microtubules in mind one can now rewrite the global rate equations for the steady state:

$$-dC_1/dt = k_+ \cdot C_1 \cdot \phi \cdot C_N - k_- \cdot (1 - \phi) \cdot C_N \qquad (21)$$

where ϕ is the fraction of microtubules in the growing phase, $(1 - \phi)$ in the shrinking phase. Upon extensive dilution microtubules all end up in the shrinking phase ($\phi = 0$);

upon growing far from the steady state all microtubules end up in the growing phase. In the conditions of Horio and Hotani, ϕ is estimated to be 0.9 in the absence of MAPS and $\phi = 1$ in their presence, since MAPs surpress dynamic instability. The most important question now is clearly which property determines whether an end is growing or shrinking and which process causes the switching from one property to the other. The most likely candidate is of course the presence of a small T-GTP cap, as suggested by Mitchison and Kirschner.[143] An intact cap could be necessary for a growing end, full removal for a fast shrinking end. Since so far only two states were defined the switching should be very fast and cooperative. It is clear that the characterization of the switching process will be an important goal of future microtubule research.

For the dynamics of microtubules, it is the kinetic aspect of the switching process which is very important, i.e., the switching rate or switching frequency. This defines the average length of a single excursion of growth or shrinkage. Together with the length of the microtubules itself, it determines how fast microtubules disappear completely. Since there are only two states,

$$\phi \cdot k_{gs} = (1 - \phi) \cdot k_{sg} \tag{22}$$

where k_{gs} and k_{sg} are the rate constants for switching from growth to shrinkage and rescue, respectively. As a consequence,

$$\phi/(1 - \phi) = k_{sg}/k_{gs} = \tau_g/\tau_s \tag{23}$$

where τ_g and τ_s are the relaxation times (or average duration) of the growing and the shrinking excursions, respectively. From simulation studies it turns out that the rate constants for switching are supposed to be rather steep functions of the free T-GTP concentration.[140,141] Walker et al.[142] observed that especially the rescue frequency showed abrupt transitions with T-GTP concentration. The average length of a growth excursion is defined by the growth rate and relaxation time of the growth state:

$$\langle i \rangle_g = k_+ \cdot C_1 \cdot \tau_g \tag{24}$$

In two recent theoretical studies it was derived that the average length of microtubules grown from nuleation centra, e.g., centrosomes, equals this excursion length.[143,144] This derivation was based on the assumption of a constant τ_g. However, in the early stages of growth, when C_1 is much larger than the final steady-state concentration, τ_g is very large. At the steady-state τ_g is much smaller and it seems possible that the excursion length of growth (as well as the excursion length of shrinking) can be much smaller than the full length of the microtubules.

It is clear that the nonlinearity observed in the negative branch of the growth rate curve, as discussed before, is related to a change of ϕ with the composition of the unpolymerized solution. Clearly, a detailed study of the parameters that determine ϕ is what we are waiting for.

During a small perturbation temperature jump, the value of ϕ should change and cause deviations from a simple first-order relaxation process.[109] Caplow et al.[145] have searched for this type of deviation. However, this is certainly difficult to detect, if the cap is very small and the magnitude of one or a few layers of tubulin. The observed rate constant for the binding of a monolayer is indeed

$$k_{obs^*} = k_+ \cdot C_1 + k_- \tag{25}$$

which is much faster than the observed rate constant for growth:

$$k_{obs} = k_{+} \cdot C_{N} \tag{26}$$

since C_N is usually 1000 times smaller than C_1. Shearing could of course increase this number, but introduces difficulties with the fast-changing number concentration. It is therefore clear that these deviations will be difficult to observe in relaxation experiments.

It should be realized that dynamic instability easily explains the fast uptake of label, when radioactive nucleotide or labeled protein is added at steady state, an important parameter being the length of the excursions. It was the same observation that lead to the suggestion of treadmilling by Wegner.

F. ANNEALING

The fast reduction in number concentration of microtubules obtained after shearing could also be explained by another mechanism that dynamic instability, i.e., the process of annealing. This is especially true since the conditions are very much in favor of annealing: the presence of a high number of small microtubules. The process of annealing was very nicely demonstrated by Rothwell and co-workers,[146] using two types of chicken tubulin: from erythrocytes and from brain. When mixing microtubule seeds made from the two types separately, the number of polymers rapidly reduced and the appearance of block copolymers could be demonstrated using specific antibodies. The size of the blocks was the same as the original seeds. With the same technique, the authors were also able to demonstrate that microtubule seeds covered with MAPs and seeds of pure tubulin anneal without (fast) redistribution of the MAPs.

Caplow et al.[147] demonstrated the existence of annealing in a totally different way. Their experiments consisted in the addition of microtubule seeds to axonemes. The rate of lengthening of the axonemal microtubules was much too high to be accounted for by subunit addition. It could, however, easily be accounted for if direct annealing of the polymers on the axonemes and on the axoneme-associated microtubules was assumed. Especially convincing is the fact that the initial growth rate is proportional to the length of the microtubules added.

It is clear that a rapid reduction in microtubule number concentration can be explained by annealing as well as by dynamic instability. Tempered dynamic instability will lead to reduction in number concentration whenever the average microtubule length is smaller than the theoretical length of the excursion for shrinking. This will be true, for example, after shearing, and a high concentration of polymers is not necessary. The process of annealing can explain the same observations. However, it only contributes in its own appropriate conditions, i.e., when the number concentration is very high. In theory the microtubules do not have to be short for annealing to occur. In practice, however, long microtubules diffuse slower and get rapidly entangled which may make annealing much more difficult. Studies with birefringence also show that microtubules rapidly allign to form anisotropic zones, a process that also will interfere with annealling.[148-150] In conclusion it can be safely stated that the existence of annealing has been firmly established, but it is clear that it only contributes at very high number concentrations of polymer. Even in the case of actin, annealing has been demonstrated.[151]

VIII. OSCILLATIONS

Recently another aspect of dynamic instability has become apparent: the existence of synchronous oscillations in the mass concentration of microtubules. Three groups seem to have discovered these phenomena independently. Carlier et al.[152] observed oscillations with

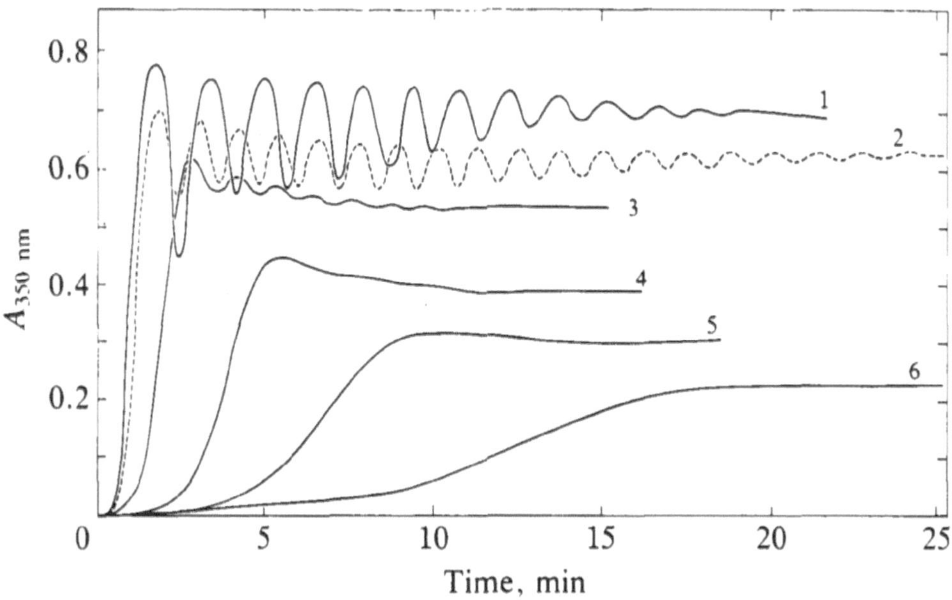

FIGURE 8. Polymerization of pure tubulin as measured by turbidity at 350 nm. At the highest concentrations oscillations are clearly observed. Tubulin concentration in micromolar: (1) 150, (2) 100, (3) 75, (4) 50, (5) 37.5, and (6) 30. Temperature was 37°C, pH 6.8 in 100 m*M* MES, 12 m*M* $MgCl_2$, 1 m*M* EGTA, and 2 m*M* GTP. (From Carlier, M. F., Melki, R., Pantaloni, D., Hill, T. L., and Chen, Y., *Proc. Natl. Acad. Sci. U.S.A.*, 84, 5257, 1987. With permission.)

pure tubulin in normal buffer conditions, except for high magnesium ions (typically around 12 m*M*) and high tubulin concentrations (up to 150 μ*M*). Many control experiments were performed so that it can be concluded safely that these oscillations are not due to a kind of organization in solution as has been observed by other authors[148-150] (see Figure 8).

Pirollet et al.[153] also observed oscillations in the assembly curve at lower protein and slightly more than stoichiometric nucleotide concentration by controlling the speed of the GTP regenerating system. These experiments clearly demonstrate the importance of the GDP and GTP turnover in the oscillations.

Mandelkow et al.[90] have also observed oscillations in the X-ray scattering experiments. These authors demonstrate the formation of small oligomers (size ranging from 3 to 12) in the periods of a low degree of assembly. They propose that these oligomers are the direct product of disassembly, and suggest the importance of their slow breakdown in the cyclic process.

It should be clear that the presence of a slow step alone cannot explain oscillations. Some kind of feedback must be present, and the most likely candidate is the effect of the status of the monomer on the status of the ends and therefore on its own rate of production. This is nicely shown in the numerical simulations of Chen and Hill,[154] or Bayley et al.[141] Upon starting the polymerization with a high concentration of T-GTP, polymers are rapidly formed, having a T-GTP cap, the T-GTP pool is rapidly depleted. Dissociation essentially liberates T-GDP, which is not immediately turned over into T-GTP. Therefore, the T-GTP cap starts to dissociate, leading to an acceleration of dissociation (positive feedback) and the production of a high concentration of T-GDP. After some time enough GDP is exchanged for GTP to recap the dissociating microtubules, which can now rapidly grow again. At very high end concentrations, exchange is slow compared to growth, and oscillations are observed. The switching on the ends should depend in a nonlinear way on the concentration of T-GTP in solution. If polymerization is done at low concentrations, exchange has to be slowed down artificially, by having a low concentration of regenerating system. The simulations presented in Carlier's paper need a rate constant for GDP displacement which is too small to be realistic, compared to the values measured by direct displacement kinetics,[155,156] or by fast enzymatic consumption of GTP.[157] However, if dissociation produces directly oligomers, which exchange their nucleotides more slowly, the necessary slow step is found.

The synchrony in the behavior of the microtubules is gradually lost and this is probably the reason why the oscillations are damped.

IX. *IN VIVO* DYNAMICS

As for the *in vitro* situation, in the cell one can also study overall assembly, overall disassembly, or steady-state kinetic properties of microtubules.

Overall reassembly of microtubules can be studied after first destroying them by cooling, high pressure, or by treatment of the cells with the inhibitor nocodazole of colcemid. These inhibitors bind reversibly, and can be removed very rapidly. In this way, it was shown that microtubules are nucleated at the so-called MTOCs: microtubule organizing centers,[158] i.e., centrosomes in interphase cells, and centrosomes and kinetochores in mitotic cells.[158-161] The growth rate of spindle microtubules was measured and correlated with the rate constants and cellular tubulin pools.[162] Taxol can compete with the cellular nucleation centers and is able to induce microtubules everywhere in the cell, probably by lowering the critical concentration, such that spontaneous nucleation becomes faster than nucleation on MTOCs.[163]

It is also possible to study the kinetics of overall disassembly and use the dissociation rate as a parameter of the dynamic behavior of the cellular microtubules. Total disassembly has to be very fast, e.g., by introduction of nocodazole. In this way Cassimeris et al.[164] could show the existence of about 30% of persistent microtubules next to labile ones.

Small perturbation experiments can be performed by the microinjection of a small amount of labeled tubulin. The incorporation of the label can be followed directly by fluorescence video microscopy or immunological techniques. Assembly competent labeled tubulin was shown to behave in the same way as nonlabeled tubulin *in vivo* as well as *in vitro*.[165-167] The sites of label incorporation after microinjection can be studied. Alternatively, the whole cell can be allowed to come to a new steady state with random distribution of the fluorescently labeled protein within the dynamic pool, and a pattern can be introduced by photobleaching. The disappearance or the movement of the bleached pattern can then be followed. This method is inspired from the technique of fluorescence recovery after photobleaching, orginally developed to study the lateral mobility of fluorescently labeled molecules on surfaces, e.g., biological membranes.[168,169] However, the technique has a much wider applicability in the cell biology, e.g., the diffusion of fluorescently labeled molecules within the cytosol was measured and its size dependence was studied.[170,171] In the microtubule field it was immediately clear that this technique could be used to study treadmilling and other aspects of microtubule assembly. In the framework of treadmilling, it was expected that the dark spot obtained by photobleaching part of the fluorescent microtubule network would persist for some time and move in one direction. However, the spot did not move, and disappeared rapidly with reappearance of fluorescence occurring from all directions.[172-174] It is clear that this type of recovery behavior could not be explained in the classical models of microtubule assembly. It was also shown that the recovery time was about ten times longer in the case of the cytoskeleton in interphase, as compared to the spindle figure. This proves that single microtubules are in a very rapid turnover.

The technique of photobleaching was extended to the resolution of single microtubules by Sammak et al.[175] These authors used antibodies that recognize intact fluorescein, but do not react with the bleached dye. The fate of the bleached zones of individual microtubules was followed in the electron microscope (and consequently by taking samples). In this way

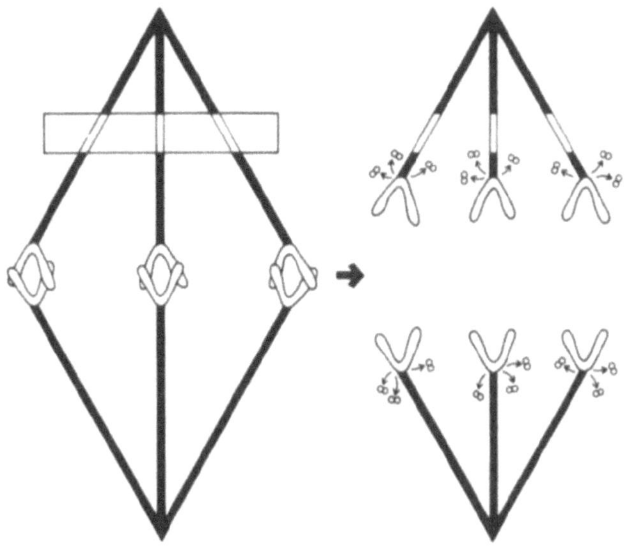

FIGURE 9. Schematic representation of a photobleaching experiment on a spindle figure in anaphase, when chromosomes are transported towards the poles (the centrosomes). Filled fibers represent microtubules with fluorescently labeled tubulin incorporated. The rectangle represents the bleaching spot. The bleached domains on the fibers are represented by open fibers. At the right, chromosomes have moved towards the pole, while the bleached spot has remained stationary with respect to the pole. The kinetochore microtubules shorten by releasing monomers from the (+) end at the kinetochores; separation of the poles is not shown. (From Gorbsky, G. J., Sammak, P. J., and Borisy, G. G., *J. Cell. Biol.*, 104, 9, 1987. With permission.)

it became clear that the behavior of the different microtubules in one spot is totally uncorrelated. This picture can be explained very nicely on the basis of the dynamic instability of individual microtubules, as observed by Horio and Hotani, where each microtubule has the chance to start to dissociate, and after some time to be rescued. It explains why the bleached spot does not move, because fluorescence reappears by random rescue of shrinking microtubules. This was recently confirmed by direct and continuous observation of individual fluorescent microtubules in living cells, by video microscopy.[176] These authors suggest the use of the term tempered dynamic instability for this behavior.

Studying the growth sites, after injection of a small amount of tubulin, either labeled with fluorescein or with biotin, showed that only the ends distal form the centrosomes are open for growing and can be identified as (+) ends. Centrosomes and, to a smaller extent, kinetochores are nucleation centra.[136] Kinetochores can capture microtubules and be moved toward the (+) end in an ATP-dependent manner.[177, 178] Their presence at the (+) end does not interfere with assembly and disassembly of the microtubules bound.

Despite the short lifetime of the bleached zones, photobleaching can be used to distinguish growth or disassembly from sliding of the whole microtubule. In this way it was shown that anaphase movement of chromosomes was accompanied by shortening of the microtubules at the (+) end.[179] These observations were confirmed by *in vitro* experiments on pure centrosome-microtubule-kinetochore complexes, showing that disassembly is sufficient to obtain chromosome movement[180] (see Figure 9). The simplest way to explain this is to assume that kinetochores have an affinity for the side wall of a microtubule with some preference for the end. If they associate to the head of the microtubule, then one would expect inhibition of assembly and disassembly. However, Garel et al.[181] have proposed a mechanism to link movement and disassembly in this case, based on length fluctuations of the polymers.

Interdigitized microtubules in the equatorial zone grow at their (+) end, and show parallel sliding, as demonstrated by the movement of bleached stripes.[182]

Next to dynamic microtubules, always a fraction of stable microtubules is detected.[164] These stable microtubules are also those that contain most detyrosinated microtubules.[183] This is probably more a consequence of their stability, since the enzyme tyr-decarboxylase is only active on the microtubule pool, and not on the pool of unpolymerized tubulin. In fact this latter is a good substrate for a specific tubulin-tyrosine ligase. The role of this modification is not yet clear. It seems unlikely that it signals the age of the microtubules to some unknown sensing mechanism. How these microtubules are made more stable is not yet firmly established. One possibility is the presence of the so-called stop proteins; these are proteins that have a high affinity for the microtubule ends.[184] It is not clear what the functions is of these stable microtubules. It seems logical to think that microtubules which form the tracks for dyein or kinesin based motors have no need for being extraordinary dynamic, apart from the necessity for adaptation to changing cell shapes. This then suggests the possibility that detyrosination might be a signal to the motors. So far no evidence has been presented in that direction.

Speculations about the role of dynamic instability in other aspects of cell functioning, e.g., response to external signals, have been presented by Kirschner and Mitchison.[185] Many aspects of microtubule-based structures and functions in biology are described and illustrated in the book of Dustin.[186]

It is clear that *in vivo* studies on living cells have gained much from the developments in the area of video techniques. Another new video technique has emerged, i.e., nanovoid microscopy, which has been able to resolve single receptor complexes, and will certainly have a profound influence on the future of real time dynamic studies *in vivo*.[187]

X. CONCLUSIONS

The sequence of paradigms that has been described here, and that finely lead to tempered dynamic instability, clearly shows how progress has been made; whenever a new technique was introduced, which allowed a more direct measurement of some properties of the microtubules, and the study of individual microtubules instead of populations, a new picture emerged. The complexity of the system has increased concomitantly, as clearly indicated by the number of rate constants and species involved. These complex mechanisms become difficult to visualize, and computer simulations of proposed mechanisms become indispensable to check their implications in detail.

Dynamic instability of individual microtubules is firmly established. Yet the keystone of its explanation, the T-GTP cap, is more difficult to demonstrate directly. There is, however, no doubt that dynamic instability and oscillations are linked to the hydrolysis of GTP. As suggested by O'Brien et al.,[128] a monolayer of T-GTP can be formed even if polymerization and hydrolysis are tightly coupled, provided hydrolysis takes place at the monomer which is being internalized upon association of the next monomer. Then there will always be one monolayer of T-GTP.

Recently a new entity has been added to the list: the intermediate T-GDP-P_i. Its accumulation is demonstrated for actin and for tubulin.[188,190] In the case of tubulin the affinity of Pi for the empty site on T-GDP is rather low (K_{dis} = 25 m*M*).[188] The inorganic phosphate can be replaced by AlF_4^- and BeF_3^- which bind with high affinity, giving complexes that abolish oscillations and dynamic instability. These species could also form the monolayer of stabilizing monomers at the ends of unstable microtubules. The different behavior of the two ends of microtubules could be due to a different accumulation and properties of these intermediates.

Is tempered dynamic instability the reason why nature spends so much energy: hydrolysis of one GTP for each monomer incorporated in the polymer? The picture of random generation of instable microtubules from centrosomes and selective stabilization of the right ones by capturing the open (+) end, as presented by Kirschner and Mitchison,[185] seems an attractive alibi.

MAPs suppress dynamic instability *in vitro*; nevertheless microtubules *in vivo* display dynamic instability. This suggests the compartementalization of MAPs, an aspect that makes the interpolation from *in vitro* to *in vivo* rather difficult.

Future *in vitro* microtubule research will largely concentrate on these caps and on the rate constants of their formation and disappearance, i.e., rescue and vice versa. For the survival of a microtubule, it can safely be said: to cap or not to cap, that is the question.

ACKNOWLEDGMENTS

I want to thank those authors who have replied to my request of sending their latest (p)reprints. This review gained much from frequent discussions with Dr. P. Bayley, Dr. S. Martin, and Dr. M., Schilstra (MRC, Mill Hill, London), which were made possible thanks to the financial support of an E.E.C. twining grant to our laboratories. Also discussions with Dr. Pantaloni and Dr. Carlier (CNRS, Gif-sur-Yvette) and my own collaborators are acknowledged. I also have to thank the Belgian National Fund for Scientific Research, and the research fund of our university for financial support. The author is Senior Research Associate of the Belgian National Fund for Scientific Research.

REFERENCES

1. **Oosawa, F. and Asakura, S.,** *Thermodynamics of the Polymerization of Protein*, Academic Press, London, 1975.
2. **Kirschner, M. W.,** Microtubule assembly and nucleation, *Int. Rev. Cytol.*, 54, 1, 1978.
3. **Timasheff, S. N. and Grisham, L. M.,** *In vitro* assembly of cytoplasmic microtubules, *Ann. Rev. Biochem.*, 49, 565, 1980.
4. **Purich, D. L. and Kristofferson, D.,** Microtubule assembly: a review of progress, principles and perspectives, *Adv. Prot. Chem.*, 36, 133, 1984.
5. **Wallin, M. and Deinum, J.,** Tubulin, in *Handbook of Neurochemistry*, Vol. 5, Lajtha, A., Ed., Plenum Press, New York, 1983, chap. 4.
6. **Ponstingl, H., Krauhs, E., Little, M., and Kempf, T.,** Complete amino acid sequence of α-tubulin from porcine brain, *Proc. Natl. Acad. Sci. U.S.A.*, 78, 2757, 1981.
7. **Valenzuela, P., Quiroga, M., Zaldivar, J., Rutter, W. J., Kirschner, M. W., and Cleveland, D. W.,** Nucleotide and corresponding amino acid sequences encoded by α and β tubulin mRNAs, *Nature*, 289, 650, 1981.
8. **Detrich, H. W. III and Williams, R. C., Jr.,** Reversible dissociation of the αβ dimer of tubulin from bovine brain, *Biochemistry*, 17, 3900, 1978.
9. **Weisenberg, R., Borisy, G. G., and Taylor, E. W.,** The colchicine-binding protein of mammalian brain and its relation to microtubules, *Biochemistry*, 7, 4466, 1968.
10. **Jacobs, M., Smith, H., and Taylor, E. W.,** Tubulin: nucleotide binding and enzymatic activity, *J. Mol. Biol.*, 89, 455, 1974.
11. **Geahlen, R. L. and Haley, B. E.,** Interactions of a photoaffinity analog of GTP with the proteins of microtubules, *Proc. Natl. Acad. Sci. U.S.A.*, 74, 4375, 1977.
12. **Hesse, J., Maruta, H., and Isenberg, G.,** Monoclonal antibodies localize the exchangeable GTP-binding site in β- and not in α-tubulins, *FEBS Lett.*, 179, 91, 1985.
13. **Field, D. J., Collins, R. A., and Lee, J. C.,** Heterogeneity of vertebrate brain tubulins, *Proc. Natl. Acad. Sci. U.S.A.*, 81, 4041, 1984.

14. **Luduena, R. F.,** Biochemistry of tubulin, in *Mictotubules*, Roberts, K. and Hyams, J. S., Eds., Academic Press, New York, 1979, chap. 2.
15. **Sloboda, R. D., Rudolph, S. A., Rosenbaum, J. L., and Greengard, P.,** Cyclic AMP-dependent endogeneous phosphorylation of a microtubule-associated protein, *Proc. Natl. Acad. Sci. U.S.A.*, 72, 177, 1975.
16. **Borisy, G. G., Marcum, J. M., Olmsted, J. B., Murphy, D. B., and Johnson, K. A.,** Purification of tubulin and associated high molecular weight proteins from porcine brain and characterization of microtubule assembly in vitro, *Ann. N.Y. Acad. Sci.*, 253, 107, 1975.
17. **Weingarten, M. D., Lockwood, A. H., Hwo, S.-Y., and Kirschner, M. W.,** A protein factor essential for microtubule assembly, *Proc. Natl. Acad. Sci. U.S.A.*, 72, 1858, 1975.
18. **Williams, R. C., Jr. and Detrich, H. W., III,** Separation of tubulin from microtubule associated proteins on phosphocellulose. Accompanying alterations in concentrations of buffer components, *Biochemistry*, 18, 2499, 1979.
19. **Sloboda, R. D., Dentler, W. L., and Rosenbaum, J. L.,** Microtubule-associated proteins and the stimulation of tubulin assembly in vitro, *Biochemistry*, 15, 4497, 1976.
20. **Connolly, J. A., Kalninis, V. I., Cleveland, D. W., and Kirschner, M. W.,** Intracellular localization of the high molecular weight microtubule accessory protein by indirect immunofluorescence, *J. Cell Biol.*, 76, 781, 1978.
21. **Cleveland, D. W., Hwo, S.-Y., and Kirschner, M. W.,** Purification of tau, a microtubule associated protein that induces assembly of microtubules from purified tubulin, *J. Mol. Biol.*, 116, 207, 1977.
22. **Fellous, A., Francon, J., Lennon, A.-M., and Nunez, J.,** Microtubule assembly *in vitro*, purification of assembly-promoting factors, *Eur. J. Biochem.*, 78, 167, 1977.
23. **Matus, A.,** Microtubule-associated proteins: their potential role in determining neuronal morphology, *Ann. Rev. Neurosci.*, 11, 29, 1988.
24. **Zingsheim, H.-P., Herzog, W., and Weber, K.,** Differences in surface morphology of microtubules reconstituted from pure brain tubulin using two different microtubule-associated proteins: the high molecular weight MAP2 proteins and tau proteins, *Eur. J. Cell Biol.*, 19, 175, 1979.
25. **Kim, H., Binder, L. I., and Rosenbaum, J. L.,** The periodic association of MAP2 with brain microtubules, *in vitro*, *J. Cell Biol.*, 80, 266, 1979.
26. **Vallee, R. B.,** MAP2 (microtubule-associated protein 2), in *Cell and Muscle Motility*, Shay, J. W., Ed., Plenum Press, New York, 1984, chap. 8.
27. **Amos, L. A.,** Structure of microtubules, in *Microtubules*, Roberts, K. and Hyams, J. S., Eds., Academic Press, New York, 1979, chap. 1.
28. **Mandelkow, E., Thomas, J., and Cohen, C.,** Microtubule structure at low resolution by X-ray diffraction, *Proc. Natl. Acad. Sci. U.S.A.*, 74, 3370, 1977.
29. **Amos, L. A. and Klug, A.,** Arrangement of subunits in flagellar microtubules, *J. Cell Sci.*, 14, 523, 1974.
30. **McEwen, B. and Edelstein, S. J.,** Evidence for a mixed lattice in microtubules reassembled *in vitro*, *J. Mol. Biol.*, 139, 123, 1977.
31. **Mandelkow, E.-M., Schultheiss, R., Rapp, R., Muller, M., and Mandelkow, E.,** On the surface lattice of microtubules: helix starts, protofilament number, seam, and handedness, *J. Cell Biol.*, 102, 1067, 1986.
32. **Erickson, H. P.,** The structure and assembly of microtubules, *Ann. N.Y. Acad. Sci.*, 253, 60, 1975.
33. **Weisenberg, R. C.,** Role of cooperative interactions, microtubule-associated proteins and guanosine triphosphate in microtubule assembly: a model, *J. Mol. Biol.*, 139, 660, 1980.
34. **Hill, T. L.,** Effect of fluctuating surface structure and free energy on the growth of linear tubular aggregates, *Biophys. J.*, 49, 1017, 1986.
35. **Marcum, J. M. and Borisy, G. G.,** Characterization of microtubule protein oligomers by analytical ultracentrifugation, *J. Biol. Chem.*, 253, 2825, 1978.
36. **Vallee, R. B. and Borisy, G. G.,** The non-tubulin component of microtubule protein oligomers, *J. Biol. Chem.*, 253, 2834, 1978.
37. **Scheele, R. B. and Borisy, G. G.,** Electron microscopy of metal-shadowed and negatively stained microtubule protein. Structure of the 30S oligomer, *J. Biol. Chem.*, 253, 2834, 1978.
38. **Marcum, J. M. and Borisy, G. G.,** Sedimentation velocity analysis of the effect of hydrostatic pressure on the 30S microtubule protein oligomer, *J. Biol. Chem.*, 253, 2852, 1978.
39. **Voter, W. A. and Erickson, H. P.,** Tubulin rings: curved filaments with limited flexibility and two modes of association, *J. Supramol. Struc.*, 10, 419, 1979.
40. **Pantaloni, D., Carlier, M. F., Simon, C., and Batelier, G.,** Mechanism of tubulin assembly: role of rings in the nucleation process of associated proteins in the stabilization of microtubules, *Biochemistry*, 20, 4709, 1981.
41. **Zeeberg, B., Cheek, J., and Caplow, M.,** Exchange of tubulin dimer into rings in microtubule assembly-disassembly, *Biochemistry*, 19, 5078, 1980.

42. **Engelborghs, Y., Robinson, J., and Ide, G.,** A pressure relaxation study of tubulin oligomer formation, *Biophys. J.*, 32, 440, 1980.
43. **Frigon, R. P. and Timasheff, S. N.,** Magnesium-induced self-association of calf brain tubulin. I. Stoichiometry, *Biochemistry*, 14, 4559, 1975.
44. **Frigon, R. P. and Timasheff, S. N.,** Magnesium-induced self-association of calf brain tubulin. II. Thermodynamics, *Biochemistry*, 14, 4567, 1975.
45. **Howard, W. D. and Timasheff, S. N.,** GDP state of tubulin: stabilization of double rings, *Biochemistry*, 25, 8292, 1986.
46. **Weisenberg, R.,** Microtubule formation *in vitro* in solutions containing low calcium concentrations, *Science*, 177, 1104, 1972.
47. **Herzog, W. and Weber, K.,** *In vitro* assembly of pure tubulin into microtubules in the absence of microtubule-associated proteins and glycerol, *Proc. Natl. Acad. Sci. U.S.A.*, 74, 1860, 1977.
48. **Shelanski, M. L., Gaskin F., and Cantor, C. R.,** Microtubule assembly in the absence of added nucleotides, *Proc. Natl. Acad. Sci. U.S.A.*, 70, 765, 1973.
49. **Himes, R. H., Burton, P. R., Kersey, R. N., and Pierson, G. B.,** Brain tubulin polymerization in the absence of microtubule associated proteins, *Proc. Natl. Acad. Sci. U.S.A.*, 73, 4397, 1976.
50. **Waxman, P. G., Del Campo, A. A., Lowe, M. C., and Hamel, E.,** Induction of polymerization of purified tubulin by sulfonate buffers, *Eur. J. Biochem.*, 120, 129, 1983.
51. **Hamel, E. and Lin, C. M.,** Glutamate-induced polymerization of tubulin: characteristics of the reaction and application to the large-scale purification of tubulin, *Arch. Biochem. Biophys.*, 209, 29, 1981.
52. **Gekko, K. and Timasheff, S. N.,** Thermodynamic and kinetic examination of protein stabilization by glycerol, *Biochemistry*, 20, 4677, 1981.
53. **Arakawa, T. and Timasheff, S. N.,** Preferential interactions of proteins with solvent components in aqueous amino acid solutions, *Arch. Biochem. Biophys.*, 224, 169, 1983.
54. **Carlier, M. F. and Pantaloni, D.,** Kinetic analysis of cooperativity in tubulin polymerization in the presence of guanosine di- or triphosphate nucleotides, *Biochemistry*, 17, 1908, 1978.
55. **Lee, J. C. and Timasheff, S. N.,** *In vitro* reconstitution of calf brain microtubules: effects of solution variables, *Biochemistry*, 16, 1754, 1977.
56. **Jameson, L. and Caplow, M.,** Effect of guanosine diphosphate on microtubule assembly and stability, *J. Biol. Chem.*, 255, 2284, 1980.
57. **MacNeal, R. K., Webb, B. C., and Purich, D. L.,** Neurotubule assembly at substoichiometric nucleotide levels using a GTP regenerating system, *Biochem. Biophys. Res. Commun.*, 74, 440, 1977.
58. **Jacobs, M. and Huitorel, P.,** Tubulin-associated nucleoside diphosphokinase, *Eur. J. Biochem.*, 99, 613, 1979.
59. **Huitorel, P., Simon, C., and Pantaloni, D.,** Nucleoside diphosphate kinase from brain: purification and effect on microtubule assembly, *in vitro*, *Eur. J. Biochem.*, 144, 233, 1984.
60. **Jacobs, M., Bennett, P. M., and Dickens, M. J.,** Duplex microtubules is a new form of tubulin assembly induced by polycations, *Nature*, 257, 707, 1975.
61. **Kutznetsov, S. A., Gelfand, V. I., Rodionov, V. I., Rosenblat, V. A., and Gulyaeva, J. G.,** Polymerization of purified tubulin by synthetic polycations, *FEBS Lett.*, 95, 343, 1978.
62. **Margolis, R. L. and Wilson, L.,** Addition of colchicine-tubulin complex to microtubule ends: the mechanism of substoichiometric colchicine poisoning, *Proc. Natl. Acad. Sci. U.S.A.*, 74, 3466, 1977.
63. **Sternlicht, H. and Ringel, I.,** Colchicine inhibition of microtubule assembly via copolymer formation, *J. Biol. Chem.*, 254, 10540, 1979.
64. **Lamber, A. and Engelborghs, Y.,** A quantitative analysis of tubulin-colchicine binding to microtubules, *Eur. J. Biochem.*, 109, 619, 1980.
65. **Andreu, J. M., Wagenknecht, T., and Timasheff, S. N.,** Polymerization of the tubulin-colchicine complex: relation to microtubule assembly, *Biochemistry*, 22, 1556, 1983.
66. **Saltarelli, D. and Pantaloni, D.,** Copolymerization of tubulin-colchicine complex and unliganded tubulin in a nonmicrotubular polymer, *Biochemistry*, 22, 4607, 1983.
67. **Schiff, P. B., Fant, J., and Horwitz, S. B.,** Promotion of microtubule assembly *in vitro* by taxol, *Nature*, 277, 665, 1979.
68. **Mandelkow, E.-M., Mandelkow, E., Unwin, N., and Cohen, C.,** Tubulin hoops, *Nature*, 265, 655, 1977.
69. **Mandelkow, E., Schultheiss, R., and Mandelkow, E.-M.,** Assembly and three-dimensional image reconstruction of tubulin hoops, *J. Mol. Biol.*, 177, 507, 1984.
70. **Larsson, H., Wallin, M., and Edstrom, A.,** Induction of a sheet polymer of tubulin by Zn^{2+}, *Exp. Cell Res.*, 100, 104, 1976.
71. **Burton, P. R. and Himes, R. H.,** Electron microscope studies of pH effects on assembly of tubulin free of associated proteins, *J. Cell. Biol.*, 77, 120, 1978.
72. **Mandelkow, E.-M. and Mandelkow, E.,** Junctions between microtubule walls, *J. Mol. Biol.*, 129, 135, 1979.

73. **Heidemann, S. R. and McIntosh, J. R.,** Visualization of the structural polarity of microtubules, *Nature,* 286, 517, 1980.
74. **Engelborghs, Y., Heremans, K. A. H., De Maeyer, L. C. M., and Hoebeke, J.,** Effect of temperature and pressure on polymerization equilibrium of neuronal microtubules, *Nature,* 259, 686, 1976.
75. **Wegner, A. and Savko, P.,** Fragmentation of actin filaments, *Biochemistry,* 21, 1909, 1982.
76. **Hill, T. L.,** Length dependence of rate constants for end-to-end association and dissociation of equilibrium linear aggregates, *Biophys. J.,* 44, 285, 1983.
77. **Frieden, C. and Godette, D. W.,** Polymerization of actin and actin-like systems: evaluation of the time course of polymerization in relation to the mechanism, *Biochemistry,* 22, 5836, 1983.
78. **Wegner, A. and Engel, J.,** Kinetics of the cooperative association of actin to actin filaments, *Biophys. Chem.,* 3, 215, 1975.
79. **Tobacman, L. S. and Korn, E. D.,** The kinetics of actin nucleation and polymerization, *J. Biol. Chem.,* 258, 3207, 1983.
80. **Firestone, M. P., De Levie, R., and Rangarajan, S. K.,** On one-dimensional nucleation and growth of "living" polymers. I. Homogeneous nucleation, *J. Theor. Biol.,* 104, 535, 1983.
81. **Rangarajan, S. K. and De Levie, R.,** On one-dimensional nucleation and growth of "living" polymers. II. Growth at constant monomer concentration, *J. Theor. Biol.,* 104, 553, 1983.
82. **Erickson, H. P. and Pantaloni, D.,** The role of subunit entrophy in cooperative assembly-nucleation of microtubules and other two-dimensional polymers, *Biophys. J.,* 34, 293, 1981.
83. **Engelborghs, Y., De Maeyer, L. C. M., and Overbergh, N.,** A kinetic analysis of microtubules, *in vitro, FEBS Lett.,* 80, 81, 1977.
84. **Robinson, J. and Engelborghs, Y.,** Tubulin polymerization in dimethyl sulfoxide, *J. Biol. Chem.,* 257, 5367, 1982.
85. **Algaier, J. and Himes, R. H.,** The effects of dimethyl sulfoxide on the kinetics of tubulin assembly, *Biochim. Biophys. Acta,* 954, 235, 1988.
86. **Keates, R. A. B. and Hall, R. H.,** Tubulin requires an accessory protein for self assembly into microtubules, *Nature,* 257, 418, 1975.
87. **Mellon, M. G. and Rebhun, L. I.,** Sulfhydryls and the *in vitro* polymerization of tubulin, *J. Cell Biol.,* 70, 226, 1976.
88. **Mandelkow, E.-M., Harmsen, A., Mandelkow, E., and Bordas, J.,** X-ray kinetic studies of microtubule assembly using synchrotron radiation, *Nature,* 287, 595, 1980.
89. **Spann, U., Renner, W., Mandelkow, E.-M., Bordas, J., and Mandelkow, E.,** Tubulin oligomers and microtubule assembly studied by time-resolved X-ray scattering: separation of prenucleation and nucleation events, *Biochemistry,* 26, 1123, 1987.
90. **Mandelkow, E.-M., Lange, G., Jagla, A., Spann, U., and Mandelkow, E.,** Dynamics of the microtubule oscillator: role of nucleotides and tubulin-MAP interactions, *EMBO J.,* 7, 357, 1988.
91. **Kirschner, M. W., Honig, L. S., and Williams, R. C.,** Quantitative electron microscopy of microtubule assembly in vitro, *J. Mol. Biol.,* 99, 263, 1975.
92. **Gaskin, F., Cantor, C. R., and Shelanski, M. L.,** Turbidimetric studies of the in vitro assembly and disassembly of porcine neurotubules, *J. Mol. Biol.,* 89, 737, 1974.
93. **Lee, Y. C., Samson, F. E., Jr., Houston, L. L., and Himes, R. H.,** The *in vitro* polymerization of tubulin from beef brain, *J. Neurobiol.,* 5, 317, 1974.
94. **Bryan, J.,** A quantitative analysis of microtubule elongation, *J. Cell Biol.,* 71, 749, 1976.
95. **Johnson, K. A. and Borisy, G. G.,** Kinetic analysis of microtubule self-assembly in vitro, *J. Mol. Biol.,* 117, 1, 1977.
96. **Murphy, D. B., Johnson, K., and Borisy, G. G.,** Role of tubulin-associated proteins in microtubule nucleation and elongation, *J. Mol. Biol.,* 117, 33, 1977.
97. **Barton, J. S. and Riazi, G. H.,** Evidence for two growth steps in microtubule polymerization, *Biochim. Biophys. Acta,* 630, 392, 1980.
98. **Burns, R. G. and Islam, K.,** Direct incorporation of microtubule oligomers at high GTP concentrations, *FEBS Lett.,* 173, 67, 1984.
99. **Gall, V.,** Double and single exponential kinetics of microtubule assembly, *in vitro, Int. J. Biochem.,* 18, 85, 1985.
100. **Bayley, P. M., Butler, F. M. M., Clark, D. C., Manser, E. J., and Martin, S. R.,** The assembly of microtubule protein *in vitro* — the kinetic role in microtubule elongation of oligomeric framents containing microtubule-associated proteins, *Biochem. J.,* 227, 439, 1985.
101. **Heusele, C., Bonne, D., and Carlier, M. F.,** Is microtubule assembly a biphasic process? — a fluorimetric study using 4′, 6′-diamino-2-phenylindole as a probe, *Eur. J. Biochem.,* 165, 613, 1987.
102. **Barton, J. S., Vandivort, D. L., Heacock, D. H., Coffman, J. A., and Trygg, K. A.,** Microtubule assembly kinetics — changes with solution conditions, *Biochem. J.,* 247, 505, 1987.
103. **Karr, T. L. and Purich, D. L.,** A rapid dilution cuvette for kinetic studies of microtubule disassembly, *Anal. Biochem.,* 104, 311, 1980.

104. **Karr, T. L., Kristofferson, D., and Purich, D. L.,** Mechanism of microtubule depolymerization, *J. Biol. Chem.*, 255, 8560, 1980.
105. **Kristofferson, D., Karr, T. L., and Purich, D. L.,** Dynamics of linear protein polymer disassembly, *J. Biol. Chem.*, 255, 8567, 1980.
106. **Farrell, K. W., Himes, R. H., Jordan, M. A., and Wilson, L.,** On the nonlinear relationship between the initial rates of dilution-induced microtubule disassembly and the initial free subunit concentration, *J. Biol. Chem.*, 258, 14148, 1983.
107. **Erickson, H. P.,** Assembly of microtubules from preformed ring-shaped protofilaments and 6-S tubulin, *J. Supramol. Struc.*, 2, 393, 1974.
108. **Mandelkow, E.-M. and Mandelkow, E.,** Unstained microtubules studied by cryo-electron microscopy — substructure, supertwist and disassembly, *J. Mol. Biol.*, 181, 123, 1985.
109. **Engelborghs, Y. and Van Houtte, A.,** Temperature jump relaxation study of microtubule elongation in the presence of GTP/GDP mixtures, *Biophys. Chem.*, 14, 195, 1981.
110. **Carlier, M.-F., Hill, T. L., and Chen, Y.-D.,** Interference of GTP hydrolysis in the mechanism of microtubule assembly: an experimental study, *Proc. Natl. Acad. Sci. U.S.A.*, 81, 771, 1984.
111. **Korn, E. D., Carlier, M.-F., and Pantaloni, D.,** Actin polymerization and ATP hydrolysis, *Science*, 238, 638, 1987.
112. **Engelborghs, Y.,** Temperature and pressure relaxation studies of protein assembly: tubulin and glutamate dehydrogenase, in *Techniques and Applications of Fast Reactions in Solution*, Gettins, W. J. and Wyn-Jones, E., Eds., D. Reidel, Dordrecht, The Netherlands, 1979, 439.
113. **Wegner, A.,** Head to tail polymerization of actin, *J. Mol. Biol.*, 108, 139, 1976.
114. **Summers, K. and Kirschner, M. W.,** Characteristics of the polar assembly and disassembly of microtubules observed *in vitro* by dark field microscopy, *J. Cell. Biol.*, 83, 205, 1979.
115. **Kuriyama, R. and Miki-Noumura, T.,** Light-microscopic observations of individual microtubules reconstituted from brain tubulin, *J. Cell Sci.*, 19, 607, 1975.
116. **Bergen, L. G. and Borisy, G. G.,** Head-to-tail polymerization of microtubules, *in vitro, J. Cell Biol.*, 84, 141, 1980.
117. **Rothwell, S. W., Grasser, W. A., and Murphy, D. B.,** Direct observation of microtubule treadmilling by electron microscopy, *J. Cell. Biol.*, 101, 1637, 1985.
118. **Margolis, R. L. and Wilson, L.,** Opposite end assembly and disassembly of microtubules at steady state, *in vitro, Cell,* 13, 1, 1978.
119. **Farrell, K. W. and Jordan, M. A.,** A kinetic analysis of assembly-disassembly at opposite microtubule ends, *J. Biol. Chem.*, 257, 3131, 1982.
120. **Margolis, R. L. and Wilson, L.,** Microtubule treadmills — possible molecular machinery, *Nature*, 293, 705, 1981.
121. **Hill, T. L. and Kirschner, M. W.,** Subunit treadmilling of microtubules or actin in the presence of cellular barriers: possible conversion of chemical free energy into mechanical work, *Proc. Natl. Acad. Sci. U.S.A.*, 79, 490, 1982.
122. **Hill, T. L. and Kirschner, M. W.,** Regulation of microtubule and actin assembly-disassembly by associated small and large molecules, *Int. Rev. Cytol.*, 84, 185, 1983.
123. **Vale, R. D.,** Intracellular transport using microtubule-based motors, *Ann. Rev. Cell Biol.*, 3, 347, 1987.
124. **David-Pfeuty, T., Erickson, H. P., and Pantaloni, D.,** Guanosinetriphosphatase activity of tubulin associated with microtubule assembly, *Proc. Natl. Acad. Sci. U.S.A.*, 74, 5372, 1977.
125. **Carlier, M. F. and Pantaloni, D.,** Kinetic analysis of guanosine 5′-triphosphate hydrolysis associated with tubulin polymerization, *Biochemistry*, 20, 1918, 1981.
126. **Carlier, M. F., Didry, D., and Pantaloni, D.,** Microtubule elongation and guanosine 5′-triphosphate hydrolysis. Role of guanine nucleotides in microtubule dynamics, *Biochemistry*, 26, 4428, 1987.
127. **Caplow, M. and Reid, R.,** Directed elongation model for microtubule GTP hydrolysis, *Proc. Natl. Acad. Sci. U.S.A.*, 82, 3267, 1985.
128. **O'Brien, E. T., Voter, W. A., and Erickson, H. P.,** GTP hydrolysis during microtubule assembly, *Biochemistry*, 26, 4148, 1987.
129. **Schilstra, M. J., Martin, S. R., and Bayley, P. M.,** On the relationship between nucleotide hydrolysis and microtubule assembly: studies with GTP-regenerating system, *Biochem. Biophys. Res. Commun.*, 147, 588, 1987.
130. **Martin, S. R., Schilstra, M. J., and Bayley, P. M.,** Dynamic properties of microtubules at steady state of polymerization, *Biochem. Biophys. Res. Commun.*, 149, 461, 1987.
131. **Correia, J. J., Baty, L. T., and Williams, R. C., Jr.,** Mg^{2+} dependence of guanine nucleotide binding to tubulin, *J. Biol. Chem.*, 262, 17278, 1987.
132. **Martin, S. R. and Bayley, P. M.,** Effects of GDP on microtubules at steady state, *Biophys. Chem.*, 27, 67, 1987.
133. **Mitchison, T. and Kirschner, M.,** Microtubule assembly nucleated by isolated centrosomes, *Nature*, 312, 233, 1984.

134. **Mitchison, T. and Kirschner, M.,** Dynamic instability of microtubule growth, *Nature,* 312, 237, 1984.
135. **Kristofferson, D., Mitchison, T., and Kirschner, M.,** Direct observation of steady-state microtubule dynamics, *J. Cell. Biol.,* 102, 1007, 1986.
136. **Mitchison, T. J. and Kirschner, M. W.,** Properties of the kinetochore *in vitro.* I. Microtubule nucleation and tubulin binding, *J. Cell. Biol.,* 101, 755, 1985.
137. **McIntosh, J. R.,** Microtubule catastrophy, *Nature,* 312, 196, 1984.
138. **Farrell, K. W., Jordan, M. A., Miller, H. P., and Wilson, L.,** Phase dynamics at microtubule ends: the coexistence of microtubule length changes and treadmilling, *J. Cell. Biol.,* 104, 1035, 1987.
139. **Horio, T. and Hotani, H.,** Visualization of the dynamic instability of individual microtubules by dark-field microscopy, *Nature,* 321, 605, 1986.
140. **Chen, Y.-D. and Hill, T. L.,** Theoretical treatment of microtubules disappearing in solution, *Proc. Natl. Acad. Sci. U.S.A.,* 82, 4127, 1985.
141. **Bayley, P., Schilstra, M., and Martin, S.,** A minimal model for the assembly and dynamic instability of microtubules, *Biophys. J.,* 53, 29a, 1988.
142. **Walker, R. A., O'Brien, E. T., Pryer, N. K., Soboeiro, M., Voter, W. A., Erickson, H. P., and Salmon, E. D.,** Calculation of transition frequencies for individual microtubules exhibiting dynamic instability behavior, *J. Cell. Biol.,* 105, 29a, 1987.
143. **Mitchison, T. J. and Kirschner, M. W.,** Some thoughts on the partitioning of tubulin between monomer and polymer under conditions of dynamic instability, *Cell Biophys.,* 11, 35, 1987.
144. **Rubin, R. J.,** Mean lifetime of microtubules attached to nucleating sites, *Proc. Natl. Acad. Sci. U.S.A.,* 85, 446, 1988.
145. **Caplow, M., Shanks, J., and Ruhlen, R.,** Analysis of microtubule dynamic instability, *J. Cell. Biol.,* 105, 30a, 1987.
146. **Rothwell, S. W., Grasser, W. A., and Murphy, D. B.,** End-to-end annealing of microtubules, *in vitro, J. Cell. Biol.,* 102, 619, 1986.
147. **Caplow, M., Shanks, J., and Brylawski, B. P.,** Differentiation between dynamic instability and end-to-end annealing models for length changes of steady state microtubules, *J. Biol. Chem.,* 261, 16233, 1986.
148. **Detrich, H. W., III, Jordan, M. A., Wilson, L., and Williams, R. C., Jr.,** Mechanism of microtubule assembly — changes in polymer structure and organization during assembly of sea urchin egg tubulin, *J. Biol. Chem.,* 260, 9479, 1985.
149. **Somers, M. and Engelborghs, Y.,** The spontaneous organization of microtubules in solution, The Cytoskeleton in Cell Differentiation and Development, *ICSU Symp. Ser.,* Vol. 8, Maccioni, R. B. and Arechaga, J., Eds., IRL Press, Oxford, 1987, 91.
150. **Hitt, A. L., Cross, A. B., and Williams, R. C., Jr.,** Anisotropic regions in dilute solutions of microtubules, *J. Cell. Biol.,* 105, Pt2, 29a, 1987.
151. **Murphy, D. B., Gray, R. O., Grasser, W. A., and Pollard, T. D.,** Direct demonstration of actin filament annealing *in vitro, J. Cell. Biol.,* 106, 1947, 1988.
152. **Carlier, M. F., Melki, R., Pantaloni, D., Hill, T. L., and Chen, Y.,** Synchronous oscillations in microtubule polymerization, *Proc. Natl. Acad. Sci. U.S.A.,* 84, 5257, 1987.
153. **Pirollet, F., Job, D., Margolis, R. L., and Garel, J.-R.,** An oscillatory mode for microtubule assembly, *EMBO J.,* 6, 3247, 1988.
154. **Chen, Y.-D. and Hill, T.,** Theoretical studies on oscillations in microtubule polymerization, *Proc. Natl. Acad. Sci. U.S.A.,* 84, 8419, 1987.
155. **Engelborghs, Y. and Eccleston, J.,** Fluorescence stopped flow study of the binding of S^6-GTP to tubulin, *FEBS Lett.,* 141, 78, 1982.
156. **Yarbrough, L. R. and Fishback, J. L.,** Kinetics of interaction of 2-amino-6-mercapto-9-β-ribofuranosylpurine 5′-triphosphate with bovine brain tubulin, *Biochemistry,* 24, 1708, 1985.
157. **Brylawski, B. P. and Caplow, M.,** Rate for nucleotide release from tubulin, *J. Biol. Chem.,* 258, 760, 1983.
158. **Pickett-Heaps, J. D.,** Aspects of spindle evolution, *Ann. N.Y. Acad. Sci.,* 253, 352, 1975.
159. **Osborn, M. and Weber, K.,** Cytoplasmic microtubules in tissue culture cells appear to grow from an organizing structure towards the plasma membrane, *Proc. Natl. Acad. Sci. U.S.A.,* 73, 867, 1976.
160. **Brinkley, B. R., Pepper, D. A., Cox, S. M., Fistal, S., Brenner, S. L., Wible, L. J., and Pardue, R. L.,** Characteristics of centriole- and kinetochore-associated microtubule assembly in mammalian cells, in *Microtubule and Microtubule Inhibitors,* De Brabander, M. and De Mey, J., Eds., Elsevier/North Holland, Amsterdam, 1980, 281.
161. **Witt, P. L., Ris, H., and Borisy, G. G.,** Origin of kinetochore microtubules in chinese hamster ovary cells, *Chromosoma,* 81, 483, 1980.
162. **Goode, D.,** Kinetics of microtubule formation after cold disaggregation of the mitotic apparatus, *J. Mol. Biol.,* 80, 531, 1973.
163. **De Brabander, M., Geuens, G., Nuydens, R., Willebrords, R., and De Mey, J.,** Microtubule assembly in living cells after release from nocodazole block: the effects of metabolic inhibitors, taxol and pH, *Cell Biol. Int. Rep.,* 5, 913, 1981.

164. **Cassimeris, L. U., Wadsworth, P., and Salmon, E. D.,** Dynamics of microtubule depolymerization in monocytes, *J. Cell. Biol.*, 102, 2023, 1986.
165. **Keith, C. H., Feramisco, J. R., and Shelanski, M.,** Direct visualization of fluorescein-labeled microtubules *in vitro* and in microinjected fibroblasts, *J. Cell. Biol.*, 88, 234, 1981.
166. **Wadsworth, P. and Sloboda, R. D.,** Microinjection of fluorescent tubulin into dividing sea urchin cells, *J. Cell. Biol.*, 97, 1249, 1983.
167. **Leslie, R. J., Saxton, W. M., Mitchison, T., Neighbors, B., Salmon, E. D., and McIntosh, J. R.,** Assembly of properties of fluorescein-labeled tubulin *in vitro* before and after fluorescence photobleaching, *J. Cell. Biol.*, 99, 2146, 1984.
168. **Peters, R., Peters, J., Tews, K. H., and Bahr, W.,** A microfluorometric study of the translational diffusion in erythrocyte membranes, *Biochim. Biophys. Acta*, 367, 282, 1974.
169. **Jacobson, K., Derzko, Z., Wu, E.-S., Hou, Y., and Poste, G.,** Measurement of the lateral mobility of cell surface components in single, living cells by fluorescence recovery after photobleaching, *J. Supramol. Struc.*, 5, 565, 1976.
170. **Wojcieszyn, J. W., Schlegel, R. A., Wu, E.-S., and Jacobson, K. A.,** Diffusion of injected macromolecules within the cytoplasm of living cells, *Proc. Natl. Acad. Sci. U.S.A.*, 78, 4407, 1981.
171. **Jacobson, K., Elson, E., Koppel, D., and Webb, W.,** Fluorescence photobleaching in cell biology, *Nature*, 295, 283, 1982.
172. **Salmon, E. D., Saxton, W. M., Leslie, R. J., Karow, M. L., and McIntosh, J. R.,** Diffusion coefficient of fluorescein-labeled tubulin in the cytoplasm of embryonic cells of sea urchin: video image analysis of fluorescence redistribution after photobleaching, *J. Cell. Biol.*, 99, 2157, 1984.
173. **Salmon, E. D., Leslie, R. J., Saxton, W. M., Karow, M. L., and McIntosh, J. R.,** Spindle microtubule dynamics in sea urchin embryos. Analysis using fluorescein-labeled tubulin and measurements of fluorescence redistribution after laser photobleaching, *J. Cell. Biol.*, 99, 2165, 1984.
174. **Saxton, W. M., Stemple, D. L., Leslie, R. J., Salmon, E. D., and McIntosh, J. R.,** Tubulin dynamics in cultured mammalian cells, *J. Cell. Biol.*, 99, 2175, 1984.
175. **Sammak, P. J., Gorbsky, G. J., and Borisy, G. G.,** Microtubule dynamics *in vivo:* a test of mechanisms of turnover, *J. Cell. Biol.*, 104, 395, 1987.
176. **Sammak, P. J. and Borisy, G. G.,** Direct observation of microtubule dynamics in living cells, *Nature*, 332, 724, 1988.
177. **Mitchison, T. J. and Kirschner, M. W.,** Properties of kinetochores *in vitro*. II. Microtubule capture and ATP-dependent translocation, *J. Cell. Biol.*, 101, 766, 1985.
178. **Huitorel, P. and Kirschner, M. W.,** The polarity and stability of microtubule capture by kinetochore, *J. Cell. Biol.*, 106, 151, 1988.
179. **Gorbsky, G. J., Sammak, P. J., and Borisy, G. G.,** Chromosomes move poleward in anaphase along stationary microtubules that coordinately disassemble from their kinetochore ends, *J. Cell. Biol.*, 104, 9, 1987.
180. **Koshland, D. E., Mitchison, T. J., and Kirschner, M. W.,** Polewards chromosome movement driven by microtubule depolymerization, *in vitro*, *Nature*, 331, 499, 1988.
181. **Garel, J.-R., Job, D., and Margolis, R. L.,** Model of anaphase chromosome movement based on polymer-guided diffusion, *Proc. Natl. Acad. Sci. U.S.A.*, 84, 3599, 1987.
182. **Saxton, W. M. and McIntosh, J. R.,** Interzone microtubule behavior in late anaphase and telophase spindles, *J. Cell. Biol.*, 105, 875, 1987.
183. **Schulze, E., Asai, D. J., Chloe Bulkinski, J., and Kirschner, M.,** Posttranslational modification and microtubule stability, *J. Cell. Biol.*, 105, 2167, 1987.
184. **Margolis, R. L., Rauch, C. T., and Job, D.,** Purification and assay of a 145-kDa protein (stop145) with microtubule-stabilizing and motility behavior, *Proc. Natl. Acad. Sci. U.S.A.*, 83, 639, 1986.
185. **Kirschner, M. W. and Mitchison, T.,** Beyond self-assembly: from microtubule to morphogenesis, *Cell*, 45, 329, 1985.
186. **Dustin, P.,** *Microtubules*, 2nd ed., Springer-Verlag, New York, 1984.
187. **Geerts, H., De Brabander, M., Nuydens, R., Geuens, S., Moeremens, M., De Mey, J., and Hollenbeck, P.,** Nanovoid tracking: a new automatic method for the study of mobility in living cells based on colloidal gold and video microscopy, *Biophys. J.*, 52, 775, 1987.
188. **Carlier, M.-F. and Pantaloni, D.,** Direct evidence for ADP-Pi-F-actin as the major intermediate in ATP-actin polymerization. Rate of dissociation of Pi from actin filaments, *Biochemistry*, 25, 7789, 1986.
189. **Humphreys, W. G. and Macdonald, T. L.,** The effects on tubulin polymerization and associated guanosine triphosphate hydrolysis of aluminum ion, fluoride and fluoroaluminate species, *Biochem. Biophys. Res. Commun.*, 151, 1025, 1988.
190. **Carlier, M-F., Didry, D., Melki, R., Chabre, M., and Pantaloni, D.,** Stabilization of microtubule by inorganic phosphate and its structural analogs, the fluoride complexes of aluminum and beryllium, *Biochemistry*, 27, 3555, 1988.

Chapter 2

TUBULIN GENES: STRUCTURE, EXPRESSION, AND REGULATION

Sally A. Lewis and Nicholas J. Cowan

TABLE OF CONTENTS

I. INTRODUCTION

One of the features that makes microtubules interesting objects of study is their remarkable diversity of function. These functions include cell division, where they mediate the segregation of chromosomes during mitosis and meiosis; cell motility, where microtubules are an integral part of both cilia and flagella; intracellular transport, involving the microtubule mediated bidirectional movement of organelles along the axons of nerve cells; and cell shape, as part of the interphase cytoskeleton. In addition, certain highly differentiated cell types contain other specialized forms of microtubules; examples include the marginal band of platelets, a ring of microtubules that circumscribes these cells and that is just within the cell membrane, and the manchette, a pyramidal sheath of microtubules that assembles around the nucleus of the developing spermatid, and that is thought to contribute to the asymmetric shape of the sperm nucleus.

In spite of this remarkable functional diversity, the ultrastructural appearance of microtubules from whatever source is almost invariably the same. A major challenge to molecular biologists interested in the behavior of microtubules is therefore to explain their functional diversity in terms consistent with their ultrastructural similarity. Clearly, a knowledge of what dictates microtubule diversity requires an understanding both of the gene products involved and the way in which their expression is regulated: the expression of different gene products may influence microtubule behavior, structure, and stability but, in addition, the quantitative demand for microtubule proteins must vary considerably from cell type to cell type as well as during progression along different differentiation pathways.

Since the initial construction of cDNA probes encoding α- and β-tubulin (the major soluble proteins of microtubules) from chick brain[1,2] a considerable body of knowledge concerning tubulin polypeptide sequences has emerged. These sequences have been derived from a large spectrum of organisms from the simplest eukaryotes which express only one or a few α- and β-tubulin genes to mammals, which express larger α- and β-tubulin multigene families. This work led to the identification of tubulin isotypes,[3] where an isotype is defined as a tubulin polypeptide that is unique by virtue of one or more amino acid substitutions compared with other tubulins in the same organism, where some or all of these sequence differences are conserved across species. In this chapter, we shall concisely summarize the data that have been obtained on the sequence, genomic organization, and gene structure of α- and β-tubulin genes. We shall discuss the conservation of tubulin isotypes in vertebrates and also the diversity of tubulin sequences in lower organisms in terms of the evolutionary implications of these data for microtubule structure and function.

The expression of the α- and β-tubulin multigene families has been extensively studied both to provide insights into the role of tubulin isotypes in the diversity of microtubule function and as a prerequisite to the study of tubulin gene regulation. The data on the expression and utilization of isotypes in vertebrates at the tissue, cellular, and subcellular level will be discussed. Also discussed will be the expression and genetics of tubulin genes in the life cycles of several lower eukaryotes, which have served as particularly useful models for elucidating some aspects of microtubule biology.

Some progress has been made towards understanding the ways in which tubulin gene expression may be regulated both with respect to maintaining a balance between α- and β-tubulin and maintaining the correct level of tubulin heterodimers within the cell. A summary of this work will be presented, along with an account of the regulation of tubulin gene expression in several lower eukaryotes. Finally, we discuss some future directions for research on the molecular biology of microtubules. This will certainly include a study of the role of tubulin isotypes and a detailed examination of the tubulin molecule to discover the structural determinants of microtubule function.

This chapter is not intended to be an exhaustive review of tubulin genes. We hope to highlight those recent contributions to the field that we consider most important and relevant.

II. STRUCTURE

Since the initial cloning of chicken α- and β-tubulin cDNAs in 1980,[1,2] tubulin genes from a wide variety of organisms have been isolated and characterized. Some of these genes have been cloned to further ongoing genetic and/or regulatory studies (e.g., in *Drosophila Aspergillus, Naegleria, Physarum,* and *Chlamydomonas*). Yeast tubulin genes have been cloned with the hope that the ease of genetic manipulation in that organism will lead to insights into microtubule function. The chicken was chosen as an object of study because of the simplicity of the tubulin multigene family in that organism as compared with other vertebrates. On the other hand we chose to work with mammalian tubulin genes in part because of the intriguing complexity of their α- and β-tubulin multigene families, and in part because of the natural interest that mammalian genes have for humans. This study led to the discovery[4] of processed pseudogenes (reviewed in Reference 5), which are DNA copies of mRNA transcripts integrated into the germ line. Because of the difficulty of identifying transcribed human tubulin genes from amid the sea of pseudogenes, we turned to the mouse. Evolutionary comparisons of tubulin sequences between these two species (and with the chicken) led to important observations on tubulin isotypes, which are tubulin sequences more closely related across than within species. The question of the function of tubulin isotypes has remained with us since their discovery; the progress that has been made toward a solution to this question will be discussed in following sections. In this section, results on tubulin gene structure from the above-mentioned organisms will be summarized. The evolutionary implications of the conservation of tubulin sequences in vertebrates and the diversity of tubulin genes in lower eukaryotes will be discussed.

A. α- AND β-TUBULIN MULTIGENE FAMILIES IN HIGHER EUKARYOTES: CONSERVED ISOTYPES

A large proportion of the genetically encoded diversity among α- or β-tubulins in a given vertebrate species is found to be clustered in the carboxyterminal portion of the polypeptide. These amino acid sequence differences are more or less conserved across species (depending on the isotype in question). Tables 1 and 2 show the amino acid sequence of the carboxyterminal portion of six mouse α-tubulin and six β-tubulin isotypes and compare them to the sequences encoded by genes from other mammalian species, where these data are available, and to sequences from chicken, in those cases where the isotypes are comparable. Tubulin genes that share encoded carboxyterminal motifs also share patterns of expression, and these are indicated in the figures (although they will be discussed in more detail below). This classification differs from the widely cited classification scheme for β-tubulin isotypes presented by Cleveland[25] in three ways. (1) Mβ3 and Mβ4 are classed separately because, although their carboxyterminal sequences are very similar, the amino acid differences between these isotypes in this region and indeed in the entire molecule have been strictly conserved in mammals,[19,22,26] as have their patterns of expression.[19,20,28,29] (2) The patterns of expression that characterize each class are stated more accurately — our work has shown that no β-tubulin in mouse is "constitutive", for example.[31,33] (3) We have grouped the chicken testis isotype into the class containing Mβ3, the β-tubulin that is dominant in mammalian testis, since there seems no reason to distinguish it. The collection of mammalian isotypes displayed in these figures may well be incomplete. However, the task of looking for more isotypes in vertebrates is likely to be unrewarding, since it will only add detail to a picture whose broad outlines are already apparent.

In general, the chicken isotypes are more divergent in sequence and expression from their mammalian counterparts and fit less easily into a classification scheme. This is to be expected from the greater evolutionary distance between mammals and birds. Indeed, if it is true that the collection of β-tubulin genes from chicken is now complete, then it seems

TABLE 1
α-Tubulin Isotypes in Vertebrates

Type	Examples	Expression	Sequence encoded, aa 437 to end	Ref.
1	Mα1 (mouse)	Mainly in brain	VDSVEGEGEEEGEEY	6, 7
	bα1 (human)		....	8, 9
	ILαT1 (rat)		....	10, 11
	Gene I (hamster)		....	12, 13
	Cα1 (chicken)		....	14
2	Mα2 (mouse)	Widely expressed	VDSVEGEGEEEGEEY	6, 7
	Kα1 (human)		....	8
	pT25 (rat)		....	15
	Gene II (CHO)		....	12,13
3	Mα3, Mα7 (mouse)	Testis specific	VDSVEAEAEEGEEY	7
	H2α family (human)		....	7, 15a
	Cα2 (chicken)		T.LF.D.N.A.DS-	16
4	Mα4 (mouse)	Mostly brain and muscle	IDSYEDEDEGEE	7
	Hα44 (human)		....	17
	MTG3α (monkey)		....	17
5	RDαTT1 (mouse)	Expressed in testis postmitotic germ cells at low levels	MGSVEAEGEEEDRDTS CCIMFSSSIGNRHPC	18
6	Mα6 (mouse)	Expressed at low levels in many tissues, possibly mainly in blood	ADSAEGDDEGEEY	7
	Gene III (hamster)		....	12, 13

Note: Amino acid sequences are presented in the one letter code; dots represent identical amino acids and dashes represent absence of amino acids in the sequences, as aligned.

likely that birds and mammals in fact have slightly different tubulin isotype requirements: chickens are missing the brain-specific isotype corresponding to Mβ4, and it is not yet clear if there is a mammalian isotype corresponding to chicken Cβ5.[25]

Drosophila, like vertebrates, has α- and β-tubulin gene families with members that differ in sequence and expression patterns. Recently, it has been shown that some of these sequences are conserved between *D. melanogaster* and *D. hydei*.[35] Thus *Drosophila*, too, has isotypes. A summary of α- and β-tubulin genes in *Drosophila* is provided in Table 3. The *Drosophila* tubulin α4 is the most divergent tubulin yet identified;[36] it shares only about 66% amino acid identity with the other α-tubulins in *Drosophila*.

B. α- AND β-TUBULIN SEQUENCES IN LOWER EUKARYOTES: DIVERSITY

Figures 1 and 2 compare amino acid sequences of α-tubulins and β-tubulins, respectively, from a selection of lower eukaryotes. Residues thought to be involved in GTP binding[44] are underlined. Not surprisingly, these residues are for the most part the same in all the isotypes. The degree of sequence identity between pairs of the sequences presented range from 70 to 89% for β-tubulins and 70 to 86% for the α-tubulins, compared with approximately 40% sequence identity between α- and β-tubulin. Regions of greatest divergence include the 15 or so amino acids at the carboxyterminal end of both α- and β-tubulin, and the area corresponding to amino acids 34 to 50 in Mα1 (Figure 1). All of these regions are known to be on the outside of the microtubule, and in α-tubulin residues in both regions undergo posttranslational modification; the lysine at position 40 can be reversibly acetylated,[45] and the tyrosine is removed from the end of the α-tubulin when polymerized, and added to α-

TABLE 2
β-Tubulin Isotypes in Vertebrates

Type	Examples	Expression	Sequence encoded aa 428 to end	Ref.
1	Mβ1 (mouse)	In platelet and	VRAGLEDSEEDAEEAEVEAEDKDH	19
	Hβ1 (human)	erythroid cell	AK.V..ED..VT....M.P...G.	20
	cβ6 (chick)	lineages	AT.---V..-Y....ASP.-.ET	21
2	Mβ2 (mouse)	Brain abundant	ATADEQGEFEEEEGEDEA	6, 19
	Hβ9 (human)		....	20
	rbt.1 (rat)		....	22
	cβ1/cβ2 (chick)		............GE....	14, 23—25
3	Mβ3 (mouse)	Abundant in testis,	ATAEEEGEFEEEAEEEVA	19
	Hβ2 (human	at lower levels in	....	20
	Cβ3 (chick)	many tissues	................AE	14, 23—25
4	Mβ4 (mouse)	Brain specific	ATAEEGEFEEEAEEEVA	6, 19
	H5β (human)		....	27
	rbt.2 (rat)		....	22, 28, 29
5	Mβ5 (mouse)	Many somatic tis-	ATAEEEEDFGEEAEEEA	6, 19
	HM40 (human)	sues especially	....	3, 30
	rbt.3 (rat)	blood	....	22, 28, 29
6	Mβ6 (mouse)	Neural	ATAEEEGEMYEDDEDESERQGPK	20, 31
	hβ4 (human)		..............E...S....	25
	cβ4 (chicken)		..............E...-....	25, 32

Note: Amino acid sequences are presented in the one letter code; dots represent identical amino acids and dashes represent absence of amino acids in the sequences, as aligned.

tubulin when in αβ heterodimers.[52,53] Inspection of the sequence data shows that not all α-tubulins are substrates for these modifications. It has been shown that acetylated α-tubulin is used to form flagella in *Chlamydomonas*.[54-56] Since all the species listed that have cilia or flagella have a well-conserved sequence surrounding 40lys in their α-tubulin, it seems probable that they, too, use acetylated α-tubulin in these structures. All of the β-tubulin sequences begin with the sequence Met-Arg-Glu-Ile, which has been shown to be involved in the autoregulation of tubulin synthesis[57] (see below).

C. GENOMIC ORGANIZATION AND GENE STRUCTURE

The placement of introns in selected α- and β-tubulin genes and the arrangement of these genes in the genomes of their respective organisms is outlined in Figures 3 and 4. There is considerable conservation of intron placement within mammalian genes, and this extends to chickens, and also to some degree to *Drosophila*. Within the division of fungi, there is also a great deal of similarity in the placement of intervening sequences in β-tubulin genes. However, in some organisms (*Saccharomyces cerevisiae*, and *Trypanosoma brucei*) the β-tubulin genes have no introns at all (like processed pseudogenes in mammals). This is suggestive of an mRNA mediated process, leading to a loss of introns in these species.

Both of the parasitic protozoa, Leishmania and Trypanosomes have unusual arrangements of their tubulin genes involving tandem repeats of α- and β-tubulin genes, either separately or in pairs.[61,62] The α- and β-tubulin genes in all the other organisms so far characterized appear to be unlinked. The amplification of genes in these protozoa may be necessary for providing high levels of α- and β-tubulin. Because of the difference in the arrangement of tubulin genes between Leishmania and Trypanosomes, it seems unlikely that either arrangement is a reflection of the primordial state of ancestral tubulin genes for eukaryotes; rather

TABLE 3
α- and β-Tubulin Istotypes in Drosophila

Gene	Chromosome band[38]	Expression[36-38]	Characteristic amino acid sequences[36,37]	
		α		
D1(α1)	84B3-6	Ubiquitous	QMPSIKTVGGGDDSFN-----------[a]	MDSGDGEGEGA-EEY[b]
D2(α2)	85E6-10	Male specific in adults, also in pupae	H............S.............[a]	I..TTEL..D-....[b]
D3(α3)	84D4-8	Ubiquitous	...........................[a]	...............[b]
D4(α4)	67C4-6	Ovarian nurse cells and early embryos	SLKTKEELTASGS.ASVGHDTSANDAR[a]	L.NAEEG.DEDFD.F[b]
		β		
D1(β1)	56D4-12	Major ubiquitous	Unknown	Unknown
D2(β2)	85D4-7	Testis specific	AG AKY[c]	DEEGEFDEDEEEGGDE-[d]
D3(β3)	60C6-8	Midembryogenesis	SAVTRSSG[c]	.D.FDPEVNG..VE..CI[d]
D4(β4)	97F	Minor ubiquitous	Unknown	Unknown

Note: Amino acid sequences are in the one letter code; dots represent identity with the given sequence, dashes represent gaps introduced for optimal alignment of sequences.

[a] Region corresponding to a.a. 33-49 of D1(α1).
[b] Region corresponding to a.a. 436-450 of D1(α1).
[c] Region corresponding to a.a. 55-59 of D2(β2).
[d] Region corresponding to a.a. 430-445 of D3(β3).

```
                                         *
a  MRECISIHVGQAGVQIGNACWELYCLEHGIQPDG--QMPSDKTIGGGDDSFN--TFFSETGAGKHVPRAVFVDLEPTVI
b     I    I    A V             N         SV     A         SS        IYL
c     V    I    I V                              A                   CI L      V
d     V    I    I V                              A                   CI L      V
e     A C  I    C V    F                      VE A                      L      V
f     V                         G  FPTEN EVHKNNSYLNDGFG       Q  F   SIY     N
g     I  V      T                N   Y NP TASQNS GG S          Q  Y   SIY     N
h     V   N     C         S     K   HLEDGLSKPKGGEEGFS      H  Y  F    IY     N
i     V   N     C         S     KE  HLEDGLSKPKGGEEGFS      H  Y  F    IY     N

a  DEVRTGTYRQLFHPEQLITGKEDAANNYARGHYTIGKEIIDLVLDRIRKLADQCTGLQGFLVFHSFGGGTGSGFT
b                S               V       C   V        S         V         G
c                S        F              A          N         NAV        LG
d                S        F              A          N         NAV        LG
e                S                    V  C          N        Y AV        LG
f  Q     P KD     MV     S       V      S  E   RM  N S                   LG
g  Q    D   D            S       V      V E T K RIG N S                   G
h    N P KD     LS               V R   L D          D       FT   L       LG
i      RFKE      N               V R   V E EE    M   D       FT   L       LG

a  SLLMERLSVDYGKKSKLEFSIYPAPQVSTAVVEPYNSILTTHTTLEHSDCAFMVDNEAIYDICRRNLDIERPTYT
b     L             D C   S                V S  GL   T V   L        L KKS   D  S A
c     L             G TV  S                V S  SL   T V V L          S
d     L             G TV  S                V S  SL   T V V L          S
e  A  L             GYTV  S                V S  SL   T V A L        LT
f  T  L   NME    N Q  V        S          V    A  DN    T       C              E
g  A  L   AME T    Q  V        S          V    A  DLA   T      SC            S E
h     L E  AE        AV        S       TV         A  T          M K      P  SFA
i     L N  YE        AV      L S       TV         A  T            K    G S  SFS

a  NLNRLIGQIVSSITASLRFDGALNVDLTEFQTNLVPYPRIHFPLATYAPVISAEKAYHEQLSVAEITNACFEPAN
b        VA VI   L          IN               M AS          F       L  TV   SS
c         A VI   L          I                M SS    I                 A   S
d         A VI   L          I                M SS    I                 A   S
e           V    L                           V TS                 S  S  V   S
f         A V         A S     N               V  S IV  A  F  SN  Q    Q    Y
g         A V         E S     A               V    IV  A  F  SN  Q    Q    Y
h      N  A V   V       S     N               VS S  L KS  F  SN  S         G
i      G  A VI  V       S     N               VS   IL KKR T  SN  S         G

   QMVKCDPRHGKYMACCLLYRGDVVPKDVNAAIATIKTKRTIQFVDWCPTGFKVGINYQPPTVVPGGDLAKVQRAV
b  M A                     T  V V                       C    A S               L
c  M                          SV                        C
d  M     L                    SV                        C              V
e  M T            M            V            S           C
f           T R   T        I R  Q  VTS  S              I  C    QH   SG    N
g           A R   T        I R  Q  VTT  AR             I  C    QHI  SE     D
h           D     T         TG  QR VEQV N K V L         I  C     AT NSQ   T D
i          TK     N         TR  QR VEQV N K V M         I  C    S I SSE   N D

a  CMLSNTTAIAEAWARLDHKFDLMYAKRAFVHWYVGEGMEEGEFSEAREDMAALEKDYEEVGVDSVEGEGEEEGEEY
b    I       VFS                                  L              AET GDDEA  -  M
c    I  S   G IFS                                 L      F       AE A  A  G
d    I  S   G IFS                                 L      F       AE A  A  G
e    IAN     VF  I       S                        L              AE ADMD   DV
f        S     S         S                        L    R        Q  MDN MY AD
g        S     S         S                        L    R        Q  M VDYM --
h        S     K I R                         T    L    R  I     A  -----YA E  F
i           D  K I Q                         T    L    R  I     A  -----YA--  F
```

FIGURE 1. Comparison of amino acid sequences of α-tubulins from lower eukaryotes. Amino acid sequences are in the one letter code. Only differences with respect to line a, the mammalian template sequence (Mα1), are shown. Dashes represent gaps in the sequence introduced for optimal sequence alignment. The sequences are α-tubulins from (b) *Physarum polycephalum*,[39] (c,d) *Chlamydomonas reinhardi*,[40] (e) *Trypanosoma brucei*,[41] (f,g) *Schizosaccharomyces pombe*,[42] and (h,i) *Saccharomyces cerevisiae*.[43] The residues thought to be involved in GTP binding[44] are underlined in the template sequence, and lysine 40, which can be acetylated,[45] is starred.

```
a  MREIVHIQAGQCGNQIGAKFWEVISDEHGIDPTGTYHGDSDLQLDRISVYYNEATGGKYVPRAILVDLEPGTMDS
b        V V      V       V E     SA   K  T    E  N     VA S              VL
c          G              V                    E  N  F      R     V M
d       CV          S           V      Q       E  N  FD     R    S  I
e        L T         A  QT  G   L AS V N T E   E MN  F   S N      V          A
f        L T         A  QT  G   L GS V N T     E MN  F   S N      V          C
g        L T      V SA  QT  G   L AS I T       E MN  F   G N      V I        A
h       L         V  A  ST A    L SA I   T  A HE LN  F   A        V          A
i      I  S   Y      A   T CG   L FN     HD I KE LN  F   SS  W   S       W I A

a  VRSGPFGQIFRPDNFVFGQSGAGNNWAKGHYTEGAELVDSVLDVVRKEAESCDCLQGFQLTHSLGGGTGSGMGTL
b  I ASSI SM      THA                    E  V        V        IC          L
c    A  Y             T                 I                     VC
d       Y         I                     I      CC             IC
e    A     L                                     R   G        I         A
f    A    EL                              Q V    R   G        I         A
g  L A  N ALY     IY   S    C             N I    R         A  V
h   K  K  NL     IIY       I            AVQ      R   A  A
i    NSAI NL      YI   S   V              A M  I R   G  S     I

a  LISKIREEYPDRIMNTFSVVPSPKVSDTVVEPYNATLSVHQLVENTDETYCIDNEALYDICFRTLKLTTPTYGDL
b          F   M C    M                  I      A  VM
c              M L                              A  CMVL                   F
d      L  Q      M   II              T          S  SM                     F
e          F   M A Y  M                         S   F           M     SN S
f          F   M A    A                        HS   F           M     SN S
g          F   M A    L                        HS   F L  D      IN    SS SD
h   L          M A     A   S             M      S   F       SS IAN   IKS S D
i   F   K  L   M A         T                   HS   F           Q     NQ S

a  NHLVSATMSGVTTCLRFPGQLNADLRKLAVNMVPFPRLHFFMPGFAPLTSRGSQQYRALTVPELTQQVFDAKNMM
b       GV   I A         S        LI        LI Y    A SAMGF           I  SR
c     I  V     C                  LI         V  T                     M
d        VV  I C         S        L          M              G S       MW
e        V     VS        S        L          M         AHHF  VS       M  P
f        V               S    W              V         AYSF  VS       M  P
g        V   I VS        S                   V        S SSF TIS       M  SR
h        V A    SF    E  S                   V     AAI  SSFQ VS       M   N
i   N   SV      S  Y     S        L          V Y    AI   SF           M

a  AACDPRHGRYLTVAAVFRGRMSMKEVDEQMLNVQNKNSSYFVEWIPNNVKTAVCDIPPRGLKMAVTFIGNSTAIQ
b    S         AS T   K  T       HA  T    F        I SS      K M  SA     N C
c  C A         AS L      T                           SS      K    SA
d  Q A         AS L      T                  I      I SS      K           N C
e    S F N     CS I   KV     ED  R                  Q  L S        SS  V
f    S F N     CS I   KV     ED  R I S  Q          IQS L S        SS       S
g  T ANYQN  F  CSTL   KVA    ED  R M   Y            Q  L SM  K     A  V    SV
h  V A            L   KV        IRS  T   A       D  LK   SV  KD   SA       S
i    A   N        F   KV V   EDE HK  S   D          Q    SVA Q  D  A   A   S

a  ELFKRISEQFTAMFRRKAFLHWYTGEGMDEMEFTEAESNMNDLVSEYQQYQDATAEEEEDFGEEAEEEA
b        GL  S                                        E SVDD AMEDDAEA GGAGQNEAVEEF
c   M   V                                               S    GE EG E  A-
d   M R VG    L                                          I   GE D  EQY--
e        G       K                                      GVD   EEY  EAPLEGEE
f       VGD                                             SIS G EEYA E IMEGEE
g     N V N                                   M       E  VSDG GAYDAE G  YEQEE
h   I R LGD  S     M                                  E GID GDEDY IE  KEPLDY
i       VGD  S   K         S     L  S                 E  V DD EVD NGDFG PQNQDEPITENFE
```

FIGURE 2. Comparison of amino acid sequences of β-tubulins from lower eukaryotes. The sequences are presented as in Figure 1. The mammalian tubulin Mβ5 serves as a template for comparison (line a). The sequences are from (b) *Physarum polycephalum*,[46] (c) *Chlamydomonas reinhardi*,[47] (d) *Neurospora crassa*,[48] (e) *Trypanosoma brucei*,[41] (f,g) *Aspergillus nidulans*,[49] (h) *Schizosaccharomyces pombe*,[50] and (i) *Saccharomyces cerevisiae*.[51]

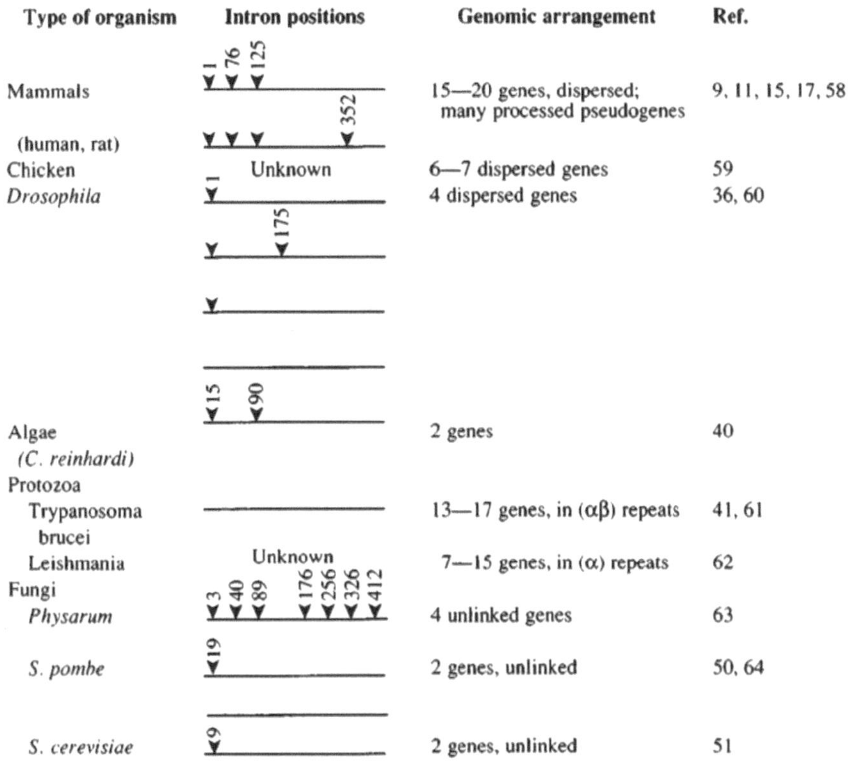

FIGURE 3. Organization of α-tubulin genes. Characteristics of α-tubulin gene families in various organisms are summarized. The arrows indicate positions in which the coding regions of the genes are interrupted by intervening sequences; the numbers above the arrows indicate the last complete codon 5′ to that particular intron.

both seem to be a result of specialized evolutionary pathways in these species. Because of this genomic organization, it also seems likely that the α- and β-tubulin genes of these organisms have undergone concerted evolution (like rRNA genes) and encode similar or identical proteins.

D. EVOLUTIONARY IMPLICATIONS OF TUBULIN GENE STRUCTURE

As we have seen, in higher eukaryotes α- and β-tubulin isotypes are encoded by sizable multigene families, while in lower eukaryotes a smaller number of tubulin polypeptides are expressed. In mouse, for example, six β-tubulin isotypes have been identified.[15,31] The sequence relationships among those isotypes and between them and their counterparts in other species are summarized in Table 4. Four of the isotypes (Mβ2, Mβ3, Mβ4, and Mβ5) are very closely related to each other, yet the differences between them have been preserved since the mammalian radiation, approximately 100 million years ago. Indeed, two of them are closely related to chicken isotypes. The two isotypes most restricted in expression are most divergent in sequence: Mβ6 is primarily neuronal,[31] while Mβ1 is expressed in the erythroid and platelet lineages,[15] where it forms the marginal band of platelets and fetal erythroblasts.[66] Mβ1 (and less so Mβ6) have diverged most over the course of time (see Figure 4, columns 3 and 4); the logical conclusion is that this divergence is a consequence of their narrower range of expression and hence the fewer functional constraints placed upon them. Conversely, the remarkable conservation of the other β-tubulin isotypes would seem to be attributable to functional constraints, i.e., selective pressure, placed upon them in their wider range of roles. What could be the nature of these constraints? The necessity for α-

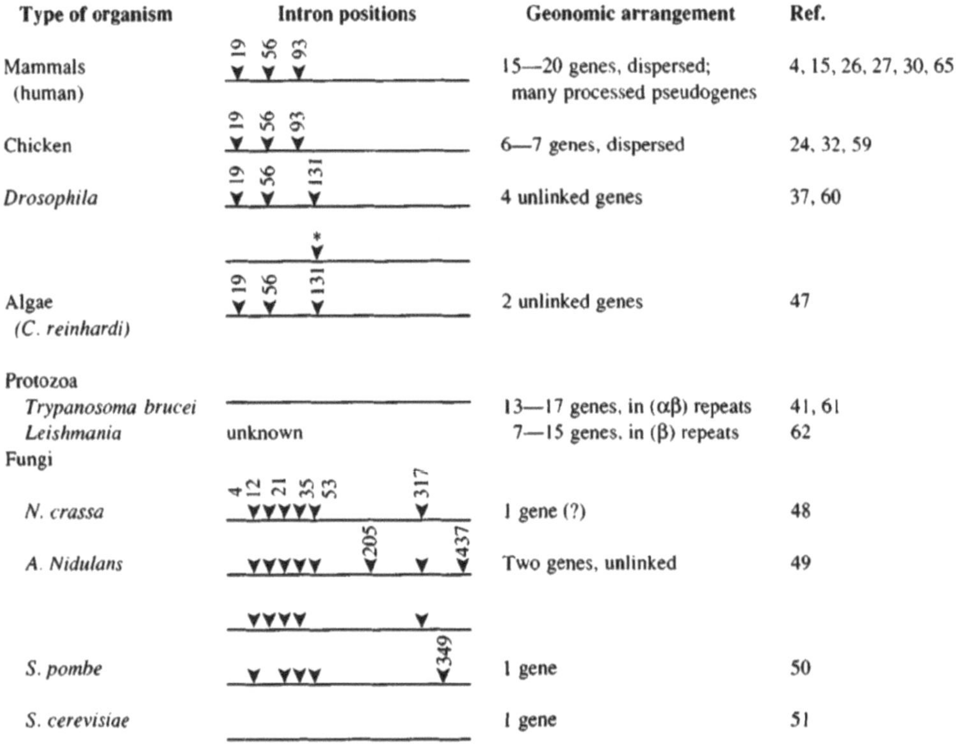

FIGURE 4. Organization of β-tubulin genes. The characteristics of the β-tubulin multigene families in various organisms are summarized. The intervening sequences are marked with arrows, in the appropriate position in which they disrupt the coding sequence. The numbers above the arrows represent the last complete codon 5′ to the intron. * — This intervening sequence occurs after the codon homologous to codon 131 in *Drosophila* β2.

TABLE 4
Amino Acid Sequence Relatedness of Vertebrate β-Tubulin Isotypes

Isotype	Percent aa sequence identity with Mβ5	Percent aa sequence identity with corresponding human isotypes	Percent aa sequence identity with corresponding chicken isotype
Mβ1	78	91	82.5
Mβ2	97	100	99.5
Mβ3	97	100	99.5
Mβ4	96.5	100	—
Mβ5	100	100	—
Mβ6	91	99.5	98.7

and β-tubulin to interact to form microtubules is the most obvious one, yet in mouse the α- and β-tubulins are not expressed in pairs: in fact all coexpressed isotypes copolymerize *in vivo* (see below) and tubulins from many species copolymerize *in vitro*. Thus, this does not place tremendous restrictions on the primary structures of tubulins. Interaction of microtubules with their associated proteins (MAPs) and with modifying enzymes must place further constraints upon α- and β-tubulins. As discussed above the regions of the tubulin molecule that are known to be on the outside of the microtubule are the most variable between isotypes. These are the regions available for tubulin-MAP interaction. Therefore, one might predict that different isotypes would interact differently with various coexpressed (and coevolved) MAPs. This hypothesis has yet to be tested directly.

The α-tubulin isotypes, like the β-tubulins, are more highly conserved when more widely expressed. Therefore, in this case also, isotypic sequence differences would seem to stem from the different functions and interactions of the various isotypes. However, there is one caveat to this argument. The 5′ and 3′ noncoding regions of the vertebrate α- (and to a much lesser extent β) tubulin genes are also conserved: they have a greater degree of nucleotide sequence identity than would be predicted on the basis of neuronal drift operating since the divergence of the species compared.[8,12,13] For example the 5′ noncoding regions of mammalian isotypes represented by Mα1 are 95% identical in mouse, rat, and hamster.*[6-13] This high degree of similarity even extends to silent substitutions in third positions of codons, which are more conserved across species than one would predict based on the rate of mutation in the absence of selection. (It does not, however, extend to the intervening sequences to the same extent.) Either there is some selective pressure acting directly on these nucleotide sequences (i.e., at the level of transcription, mRNA processing, or translation) or the tubulin genes, especially the αs, are mutating especially slowly, compared to the rest of the genome. If the latter is the case, then the nature of the mechanism that could account for the low mutation rate in these dispersed genes, in the absence of selection, is unfathomable. It seems clear from third codon position usage that several of the α-tubulin genes in each species have undergone gene conversion in their coding regions (e.g., Mα1, Mα2, and Mα6 have very similar third codon position usage, as do Mα3 and Mα7).[7] However, gene conversion clearly cannot operate to maintain sequence identity across species. In fact, sequences undergoing concerted evolution (like the 5S RNA genes) tend to be more *dissimilar* across species.[67] Yet, if some process other than selection is stabilizing α-tubulin sequences one could not argue (as we have) that there are different functional constraints on α-tubulins maintaining isotype sequence differences over the course of evolution. However, this argument would not be invalidated for β-tubulins, whose noncoding regions are much less tightly conserved.

The idea that functional constraints create and maintain tubulin isotypes can be applied to the sequences of tubulins from unicellular organisms (Figures 1 and 2); in those lower eukaryotes whose microtubules perform fewer functions (i.e., yeast, *Aspergillus*) the α- and β-tubulin sequences are more dissimilar from the vertebrate sequences. This may be a reflection of the fact that tubulin proteins in these species are under less selective pressure and thus the sequences are more free to drift. There is genetic evidence to support this, in that α-tubulins are functionally interchangeable in yeast,[43,64] and one of the β-tubulin genes is dispensable in *Aspergillus*.[68,69] More tests of this hypothesis are feasible in yeast, due to the ease with which one can introduce mutations into the genome.

III. EXPRESSION

The expression of tubulin isoforms — "spots" on two-dimensional gels — has been a subject of study for many years, with the hope that it would provide a clue to the molecular basis of the bewildering variety of microtubule structures, as well as the role of microtubules in differentiation and development. The cloning of tubulin genes has furthered this research considerably, since it is now possible to identify the "spots" as particular gene products with particular posttranscriptional modifications, and to discover the expression patterns of these genes. In addition, it is possible to express tubulins out of context and to introduce mutant α- and β-tubulins into cells and whole organisms to further explore the relationship between gene structure and microtubule function. In this section, the results of such exper-

* It appeared that the human gene bα1 was an exception, since its 5′ untranslated region showed little homology with its mouse, rat, and hamster counterparts. However, the reported 5′ sequence[9] is transcribed only at very low levels from a weak promoter (pseudopromoter); the 5′ sequence transcribed from the true promoter is 95% identical to the rodent sequences. (Hall and Cowan, unpublished observation.)

TABLE 5
Relative Expression of Tubulin Isotypes in Mouse Tissues

	Brain	Heart	Kidney	Liver	Lung	Muscle	Spleen	Stomach	Testis	Thymus
					$\alpha^{6,7}$					
Mα1	+ + +	:	:	·	+ +	:	:	:	+	:
Mα2	+ + +	+	+	+	+ +	:	+ +	+	+	+ + +
Mα3/7									+ + + +	
Mα4	+	+ +	+	+	:	+ +	+	+	:	+
Mα6	·	:	:	+	:	·	:	+	:	:
					$\beta^{19,20,31}$					
Mβ1				·	·		:			
Mβ2	+ + +	:	+	+	+ +	:	:	+	:	+
Mβ3	:	:	:	:	:	:	:	:	+ + + +	+
Mβ4	+ + +									
Mβ5	+ + +	+	+	+	+ +	+	+ +	+	+	+ + +
Mβ6	+ + +								+	

Note: Symbols represent approximate relative expression levels, + + + + = highest level, : = lowest level, · = trace expression. The mouse α tubulin RDαTT1 is expressed at very low levels in testis, and is not detectable in brain, but other tissues have not been surveyed.[18]

iments will be summarized and discussed in terms of their implications for the functional role of the α- and β-tubulin multigene families.

A. EXPRESSION OF TUBULIN GENES IN VERTEBRATES

1. Overall Patterns of Tubulin Gene Expression

The expression patterns of α- and β-tubulin genes isolated from mouse[7,19,31] (and to a lesser extent chicken,[23] human,[9,26] and rat[28,29]) have been determined by the use of cloned probes to assay levels of mRNA present in the various tissues, at different times during development, and in various cell lines. These probes have been subcloned from 3′ noncoding regions of the genes, because of the high degree of sequence similarity between the coding regions of different α- and β-tubulins. The results of these mRNA blotting experiments on adult tissues for the known tubulin genes of mouse are summarized in Table 5. Inspection of the table shows that each gene has a unique pattern of expression; α- and β-tubulins are not expressed in pairs. Several genes (Mβ1, Mβ6, Mα3/7, RDαTT1) are more restricted in their expression patterns; some mRNAs are found in relatively minor amounts in all tissues in which they are expressed (Mα6, RDαTT1). The abundance of several tubulin mRNA species show dramatic changes over the course of development. For example, the expression of Mα1 decreases manyfold, while that of Mβ4 and Mβ6 greatly increases during the development of mouse brain. Expression of both Mα3/7 and Mβ3 in testis is correlated with the level of spermatogenesis (thus increasing until maturity).

Isotypes that were originally defined by their sequence relatedness are found to be expressed in similar or identical patterns in their respective species. The expression data that exists for rat[28,29] and human[9,26] isotypes corresponding to these mouse isotypes is consistent with that shown in Table 5; thus, the expression patterns as well as the sequences of tubulin isotypes have been conserved over the course of mammalian evolution. The patterns of expression of four chicken β-tubulin isotypes have also been investigated at the mRNA level.[23] cβ3, like its mouse counterpart Mβ3 is expressed mainly in testis, but at a lower level in all tissues examined. cβ2, like its mouse counterpart Mβ2, is expressed at highest levels in brain and lung but might show minor differences in expression in other tissues. The isotype counterparts of cβ1 and cβ4′, if they exist, have not been found in mammals. Thus, in some cases the conservation of characteristic expression patterns of mammalian tubulin isotypes can be extended to encompass chicken isotypes.

2. Subcellular Localization of Tubulin Isotypes: Free Intermingling among Functionally Distinct Microtubules

With the discovery that α- and β-tubulins are both encoded by multigene families, and that each expressed gene frequently encodes a distinct isotype, the prospect arose that tubulin isotypes themselves might contribute to the phenomenon of microtubule diversity. This idea, first formulated as the multitubulin hypothesis,[70] holds in its extreme form that functionally different kinds of microtubule are assembled from distinct tubulin isotypes, thus conferring on different microtubules their unique biochemical and biological properties. The assembly of different kinds of microtubule in this manner would therefore entail the subcellular sorting, or segregation, of isotypes. A modified and less extreme version of the multitubulin hypothesis is that selective enrichment of isotypes, rather than an absolute segregation, could confer different properties on different microtubules.

Genetic experiments in nonvertebrate species (see below) do not lend support to the multitubulin hypothesis. However, the tubulin gene repertoire of those species amenable to convenient genetic manipulation is limited in comparison to that of vertebrate species, and it therefore seemed prudent to test cells derived from these higher organisms for the possibility of subcellular sorting or selective enrichment of tubulin isotypes among their more functionally diverse microtubules. The necessity of testing the multitubulin hypothesis in mammals was particularly apparent because of the strict interspecies conservation of isotype-defining sequences among different mammalian species (Tables 2 and 4), an observation that argues in favor of a functional role of individual isotypes conferred by each unique conserved amino acid sequence.

Examining the distribution (and possible subcellular sorting) of tubulin isotypes in mammalian cells presents a formidable experimental challenge. It requires the development of reagents (i.e., antisera) that can uniquely discriminate among the various tubulin isotypes. This is not entirely straightforward, since tubulins are poor immunogens, and even within the variable carboxyterminal regions there is extensive homology among isotypes. In addition, this region is extremely rich in acidic residues. One way to circumvent these difficulties is to use an immunological tag. This approach was applied in a provocative experiment by Bond et al.,[71] who constructed a chimeric gene consisting of N-terminal chicken β-tubulin sequences coupled to the yeast β-tubulin C-terminal sequence. The extreme divergence and extended nature of the yeast portion of the construct (relative to the corresponding vertebrate C-terminal regions) could be used to follow the fate of the chimeric isotype following transfection into NIH 3T3 cells, in this case with an antibody raised against a synthetic peptide corresponding to the yeast sequences. The result of this experiment was that the chimeric isotype was observed to assemble into both interphase and spindle microtubules in transfected cells; indeed, a permanent cell line expressing the chimeric isotype showed no disruption of normal microtubule function. Although this experiment seems to directly contradict the multitubulin hypothesis, it has been argued[72] that, since the natural C-terminus of the vertebrate β-tubulin isotype was removed (and substituted by "foreign" yeast sequences), the promiscuous behavior of the chimeric molecules was to be expected; interaction with MAPs which might well influence (and perhaps restrict) the behavior of different isotypes could not occur in the absence of the natural C-terminal region.

This issue has been resolved by the development of antisera that discriminate among naturally occurring α- and β-tubulin isotypes. Two approaches have been used to generate such sera. In our laboratory, a method was developed in which cloned fusion proteins

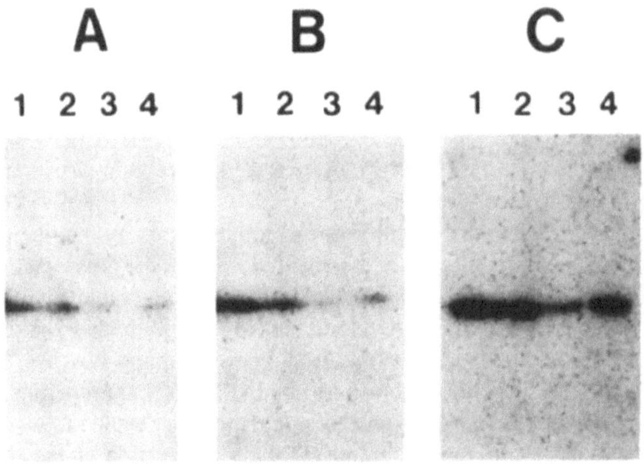

FIGURE 5. Neither spindle nor cytoplasmic microtubles are enriched in any β-tubulin isotype. Western blots of protein extracts containing unpolymerized (lanes 1 and 3) and polymerized (lanes 2 and 4) tubulin were prepared from interphase (lanes 1 and 2) or mitotic (lane 3 and 4) Hela cells. The blots were probed with isotype specific sera to Mβ2(A), Mβ3(B), or Mβ5(C), the coexpressed β-tubulins in Hela cells. (From Lewis, S. A., Gu, W., and Cowan, N. J., *Cell*, 49, 59, 1987. With permission.)

consisting of sequences encoding the amino terminal part of an *Escherichia coli* protein coupled to sequences encoding individual tubulin isotypes were used as either tolerogens or immunogens to generate isotype-specific sera.[66] The method depends on the generation of immune tolerance in animals (rabbits or guinea pigs) to epitopes that are shared among the various tubulin isotypes; this is achieved by coinjection of several purified fusion proteins (encoding isotypes *other* than the isotype of choice) together with cyclophosphamide, an antimitotic drug that generates specific immune tolerance by eliminating clonal expansion of B-cells during the immune response. A specific response to unique epitopes is then elicited by challenging tolerized animals using the purified fusion protein encoding the isotype of choice. Sera specific for Mβ1, Mβ2, Mβ3 + Mβ4, Mβ5, Mβ6,[66] Mα3/7, Mα4, and Mα6[73] have been generated by this procedure.

With these sera in hand, it has been possible to examine the distribution of naturally expressed tubulin isotypes in tissue culture cells in indirect immunofluorescence experiments as well as in semiquantitative experiments in which the tubulin content of interphase and spindle microtubules was assayed by Western blotting (Figure 5). In no case was there evidence for either qualitative or quantitative segregation of isotypes among these different microtubule structures; both seem to contain the same proportion of the isotypes expressed in a given cell type. This free intermingling occurs not only among naturally expressed isotypes; when a foreign isotype is introduced into HeLa cells by transfection, this isotype (e.g., the hematopoietic isotype Mβ1) also coassembles freely into both interphase and spindle microtubules without disruption of function (Figure 6). Essentially identical data have since been reported by Cleveland[74,75] using isotype-specific sera raised with peptide haptens corresponding to C-terminal regions of different β-tubulin isotypes. The free intermingling of available expressed isotypes is not restricted to β-tubulin isotypes; a similar picture emerges upon examination of α-tubulin isotypes, whether naturally occurring[33] or

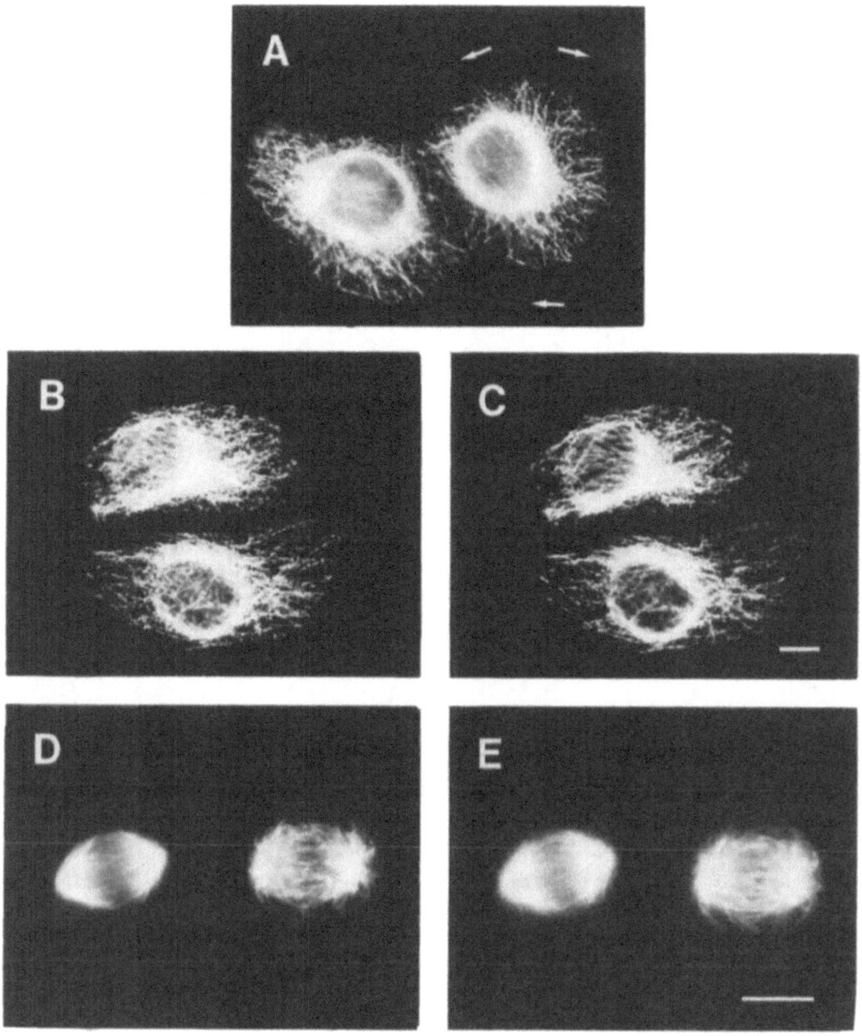

FIGURE 6. The divergent, hematopoietic isotype Mβ1 is efficiently incorporated into spindle and cytoskeletal microtubules when expressed in Hela cells. Cells transfected with an SV40 driven Mβ1 cDNA were analyzed by double label immunofluorescence using an anti-Mβ1 antiserum (A,B,D) and an antiserum that stains all microtubules (C,E). The coincident labeling patterns (B + C, D + E) show that Mβ1 is freely incorporated into all the cell's microtubules. The arrows in panel A point to surrounding untransfected cells, demonstrating the specificity of the antiserum to Mβ1. Bar = 10 μ*M*. (From Lewis, S. A., Gu, W., and Cowan, N. J., *Cell*, 49, 59, 1987. With permission.)

expressed following transfection[73] with sequences encoding foreign α-tubulin isotypes such as the testis-specific isotype Mα3/7.

The above experiments fail to provide any evidence for the subcellular sorting of tubulin isotypes in tissue culture cells; however, this does not rule out the possibility of sorting in the context of normal tissues. To address this issue, we have used our α- and β-tubulin isotype-specific sera to study microtubules in their natural context. For example, the expression of Mβ1 in hematopoietic cells is shown in Figure 7. These data demonstrate that this very divergent isotype is incorporated into spindles, as well as marginal band microtubules.

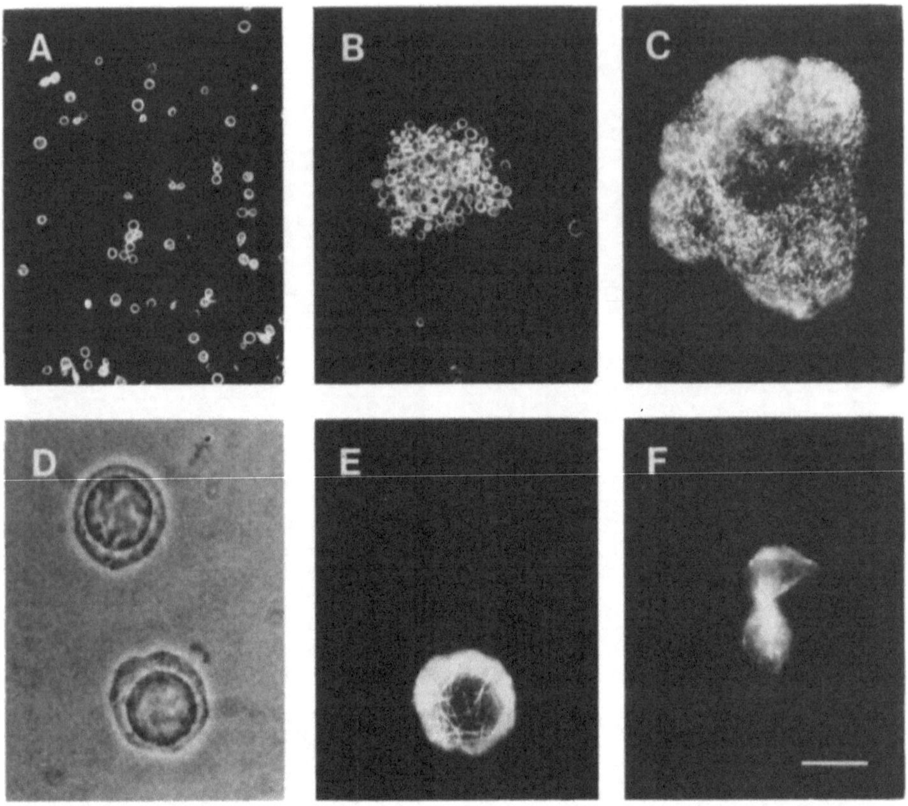

FIGURE 7. The expression and utilization of Mβ1 in vivo. Immunofluorescence of mouse platelets (A + B) shows the location of Mβ1 in the marginal band of these cells. In the precursor cells, megakaryocytes, Mβ1 is compartmentalized, in soluble form, in an extensive network of membranous sacs that will become platelets (C). In fetal nucleated erythroblasts (D), Mβ1 is incorporated into both marginal band microtubules (E) and the spindle (F). Bar, = 10 μ*M*. (From Lewis, S. A., Gu, W., and Cowan, N. J., *Cell*, 49, 59, 1987. With permission.)

The testis presents an opportunity to examine the developing spermatid, a cell type that simultaneously contains two very different kinds of specialized microtubule, namely, the flagellum and the manchette. As in the case of cells in culture, we found no evidence for the subcellular sorting of different tubulin isotypes between these structures,[33] and it seems likely, therefore, that all microtubules are assembled indiscriminately from whatever isotypes are expressed in a given cell type.

This conclusion is clearly in contradiction to all forms of the multitubulin hypothesis. However, we are left with a curious and as yet unresolved paradox: there is free intermingling of tubulin isotypes within a given cell type on the one hand, yet, on the other, there has been rigid interspecies conservation of the amino acid sequences that distinguish one isotype from another, at least since the mammalian radiation. The requirement for isotype-specific interaction with different MAPs, discussed elsewhere in this chapter, could resolve this paradox.

3. Cell Type-Specific Expression

The mRNA blotting experiments outlined above give only a gross picture of the expression of α- and β-tubulin multigene families, since most, if not all, tissues are collections of many cell types. For example, is the low level of Mα6 mRNA observed in most tissues due to contaminating blood cells, or to innervation? Northern blot experiments on mRNA

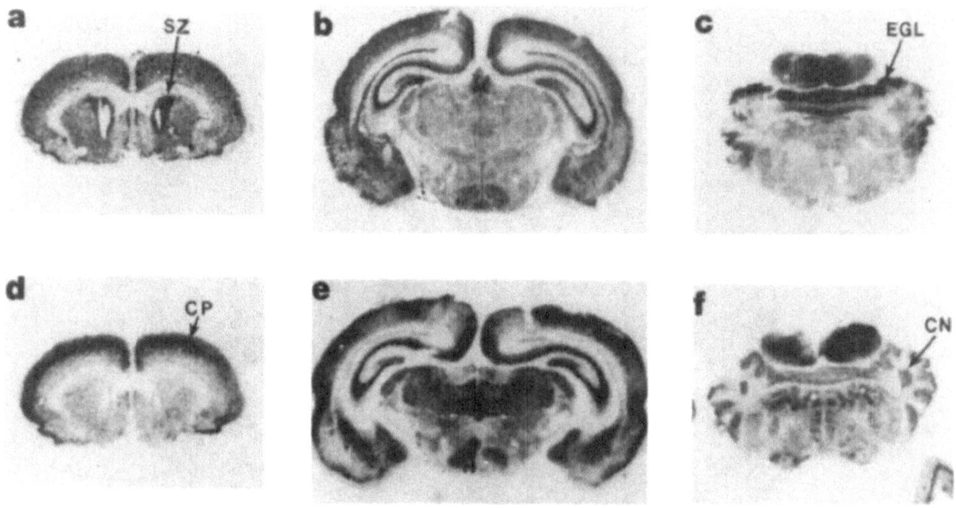

FIGURE 8. Distribution of mRNAs encoding two closely related α-tubulin isotypes in postnatal brain, assayed by in situ hybridization. The mRNA for the isotype T26 (= Mα2) (panels a, b, c) is widely distributed, and is especially prominent in the subventricular zone (SZ) and external germinal layer of the cerebellum (EGL). The mRNA encoding Tα1(=Mα1) (panels d, e, f) is abundant in the cortical plate (CP) and deep cerebellar neclei (CN). (From Miller, F. D., Naus, C. C. G., Durand, M., Bloom, F. E., and Milner, R. H., *J. Cell Biol.*, 105, 3065, 1987. With permission.)

from cell lines and dissociated tissues do not provide a satisfactory approach to this problem, because changes in gene expression often occur in culture. Two methods are useful for the study of the expression of tubulin isotypes *in vivo* at the cellular level: *in situ* hybridization of mRNA with gene specific probes and immunofluorescence experiments with isotype-specific antibodies.

In situ hybridization has been used to study the expression patterns of the rat isotypes corresponding to Mα1 and Mα2, the two major α-tubulin isotypes expressed in brain.[76] Since these isotypes differ by only one amino acid, it would be difficult, if not impossible, to raise distinguishing antisera to study their relative expression at the protein level. The distribution of mRNAs encoding these isotypes in early postnatal rat brain is shown in Figure 8. The mRNA encoding the isotype corresponding to Mα2 is expressed at fairly uniform levels in all cells, while that encoding the Mα1 isotype is expressed in a more restricted pattern, enriched in such regions as the cortical plate, where neurons are actively undergoing extension. In fact, *in situ* hybridization of whole embryo shows the latter isotype is found only in the central and peripheral nervous system, while the former is expressed in all cell types. It seems likely that Mα1 is the major α-tubulin involved in axonal outgrowth.

We have used the isotype-specific sera described above to study the distribution of tubulin isotypes in two tissues, brain and testis, where tubulin is abundantly expressed and where microtubules play a large role in cellular differentiation (neurite outgrowth, spermatogenesis).[31,33] Figure 9 shows indirect immunofluorescence experiments on frozen sections from adult mouse testis. Mα3/7 and Mβ3, the major tubulins expressed in testis (Table 5), are expressed at high levels in germ cells and contribute to all of the microtubule structures (manchette, spindles, flagella, and interphase cytoskeletons) found in these cells. Mβ2 is expressed only in the fibrous capsid of the seminiferous tubules, while Mβ5 is found only (and Mβ6 mainly), in Sertoli cells (Figure 9).

In brain also, each β-tubulin isotype has a unique pattern of expression.[31] The distribution of tubulins in the early cerebellum is summarized in Table 6. Four of the isotypes (Mβ2,

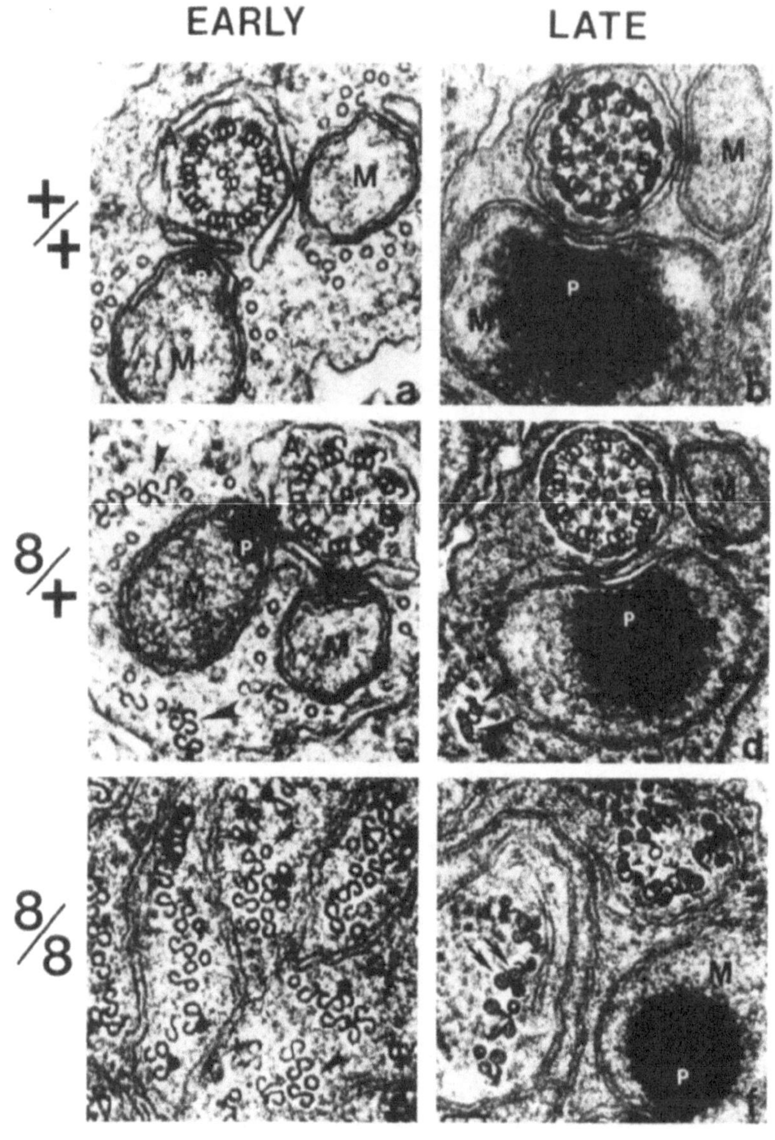

FIGURE 9. Different patterns of expression of β-tubulin isotypes in mouse testis. Double-label immunofluorescence experiments were performed on sections of mouse testis, using isotype specific antisera (a,b,c) and an antitubulin antibody that stains all microtubules (d,e,f). The isotype Mβ2 is found only in the fibrous capsid (a), Mβ3 is seen in the interphase cytoskeleton, manchette, and flagella of germ cells (b), and Mβ5 is found in the parallel arrays of microtubules that make up the Sertoli cell cytoskeleton (C). (From Lewis, S. A. and Cowan, N. J., *J. Cell Biol.*, 106, 2023, 1988.)

Mβ3 + Mβ4 [not distinguished by our antibody], and Mβ5) are present in both neuronal and nonneuronal cells, whereas Mβ6 is found only in neurons. Furthermore, Mβ2 and Mβ6 are expressed in parallel with MAP3 and MAP1a, respectively. The coexpression of these isotypes with MAPs, as well as the unique patterns of expression of all the isotypes, suggest that isotypes might function, in the absence of subcellular segregation, to interact with the cell type-specific MAPs, many of which have presumably yet to be discovered. Coexpressed isotypes and MAPs may coevolve, and this may be the source of the selective pressure which maintains tubulin sequence differences. Obviously, another (but not necessary ex-

TABLE 6
The Localization of β-Tubulin Isotypes in Adult Cerebellum

Location	β-tubulin isotype				
	Mβ1	Mβ2	Mβ3/4	Mβ5	Mβ6
Purkinje cell bodies and dendrites	−	−	−	−	++
Granule cell bodies	−	−	+	+	−
Parallel fibers	−	−	+	+	+
Interneurons, granular layer	−	−	?	?	+
Interneurons, molecular layer	−	−	?	?	+
White matter axons	−	++	−	−	++
Neurons-deep cerebellar nuclei	−	++	−	−	++
Bergman glia	−	++	+	−	−
White matter astrocytes	−	+	++	−	−
Protoplasmic astrocytes	−	−	?	++	−
Oligodendrocytes	−	−	+	−	−

Note: ++, high level of staining; +, intermediate level of staining; −, no detectable staining; ?, no obvious staining but staining of surrounding structures made interpretation difficult.

From Burgoyne, R. D., Cambray Deakin, M. A., Lewis, S. A., Sarkar, S., and Cowan, N. J., *EMBO J.*, 7, 2311, 1988.

clusive) purpose for the tubulin multigene family is to provide an organism with the different amounts of α- and β-tubulin needed in different cell types, and at different times.

B. TUBULIN GENE EXPRESSION IN NONVERTEBRATES SPECIES

Many lower eukaryotes undergo changes during their life cycle that are accompanied by formation of different microtubule-based structures. In some of the cases that have been examined, these changes are reflected in changes in tubulin gene expression. For example, in the slime mold *Physarum polycephalum*, one β-tubulin gene is expressed during the ameobal stage, one in the plasmodial stage, and a third is found in both cell types.[63,77] In the bread mold *Aspergillus nidulans,* one β-tubulin gene, tubC,[68,69] is primarily expressed during the formation of conidia, the asexual spores of this species; the process of conidiation involves making a spore-bearing structure, and requires microtubules for multiple mitoses and for nuclear movement. However, this gene is not essential for conidiation; its role can be taken over by the other β-tubulin gene (which is the major or only β-tubulin expressed during vegetative growth) in this organism. This is shown by normal spore formation in *Aspergillus* mutants bearing nonsense mutations in the tubC gene.[69]

In other cases, changes that occur during the life cycle are not accompanied by changes in tubulin gene expression. For example, both the budding yeast, *Saccharomyces cerevisiae* and the fission yeast *Schizosaccharomyces pombe* have two α-tubulin genes (one major, one minor) and one β-tubulin gene. For both species, either α-tubulin, if expressed in large enough amounts, can serve as the only α-tubulin in the cell during all phases of its life cycle.[64,78] While the minor α-tubulin gene product in yeast is not essential for growth, its absence does lead to poorer spore germination as well as a longer generation time and more chromosome loss in diploid cells,[78] although it is not clear if this is because of the slightly lower levels of α-tubulin in these cells, or because of any property conferred on normal microtubules by this different α-tubulin protein. Likewise, the alga *Chlamydomonas reinhardi* does not use different tubulin genes for distinct purposes. The two α- and β-tubulin genes encode very similar[40] (α) or identical[47] (β) proteins, and in this organism all four genes are coordinately regulated.[79,80] Thus, the growth of a flagellum is not accompanied by changes in the α- and β-tubulin genes expressed. (However, the α-tubulin *is* changed posttranscriptionally, by acetylation, when it forms the flagellar axoneme).[54-56]

Drosophila is closer to vertebrates in that it has larger α- and β-tubulin multigene families which encode conserved isotypes. The expression patterns of these isotypes are summarized in Table 3. Sterile male flies bearing mutations in the β2 gene (the testis specific β-tubulin gene in *Drosophila*) have been used to elucidate the role of this tubulin in the morphogenesis of the many microtubule-based structures in developing sperm.[38,81-83] Dominant mutants in this gene lead to production of aberrant β2-tubulin which disrupts meiosis, nuclear shaping, and axoneme formation, suggesting that this one isotype takes part in all these functions. Recessive mutations in the β2 gene fall into two categories — ones in which the β2 gene product is unstable, so that in homozygotes the germ cells do not have sufficient tubulin to form the meiotic spindle, flagellar axoneme or to shape the sperm nucleous. Other recessive mutants produce stable but aberrant β2-tubulin which causes defects in these structures only at high dosage (i.e., in the homozygote). A particularly interesting example of such a mutant is B2t,[8] where the mutant β-tubulin leads to the formation of S-shaped microtubules in all these microtubule based structures.[84] (Figure 10). While the study of mutations in the β2 gene of *Drosophila* provided the first evidence for the multiplicity of functions of a single tubulin isotype, analysis of mutations in nontubulin genes affecting microtubule function promises to be even more interesting and to provide insights into mechanisms controlling the many fundamental processes in which microtubules play a central role.

IV. REGULATION

A. AUTOREGULATION OF TUBULIN GENES

Tubulin gene regulation is very complex, and occurs on three main levels in eukaryotic cells. α- and β-tubulins are each encoded by multigene families in most organisms, and only a subset of these genes are represented in the mRNA population of a given cell type. Therefore, mechanisms exist for controlling the differential expression of tubulin genes among the different cell types and during the development of multicellular organisms. Furthermore, since α- and β-tubulin form a heterodimer, balanced levels of these two polypeptides must be maintained in each cell. Finally, since the concentration of tubulin heterodimer helps determine the dynamic equilibrium between a nucleated array of slowly growing and catastrophically shrinking microtubules[85] (in dividing cells at least), the overall level of the tubulin heterodimer must be controlled. This control has been shown to be autoregulatory, with the concentration of unpolymerized tubulin affecting the stability of β-tubulin mRNA.

The autoregulation of tubulin synthesis was discovered by Penman and co-workers in 1979.[86] Working with microtubule-destabilizing drugs they found that treatment of cells with either nocadazole or colchicine, both of which lead to the depolymerization of microtubules, causes a rapid decrease in tubulin synthesis (as assayed by ^{35}S-methione incorporation). This specific decrease in α- and β-tubulin synthesis is due to a rapid decrease in translatable tubulin mRNA (assayed by transcription *in vitro* of mRNA from treated and untreated cells). Treatment of cells with vinblastine, which also leads to a loss of microtubules, but through aggregation of the tubulin subunits into paracrystals results in a moderate (approximately twofold) increase in tubulin synthesis. Thus, exposure to drugs which increase the amount of unpolymerized tubulin in a cell leads to a decrease in new tubulin synthesis, whereas use of a drug which decreases the amount of soluble tubulin (vinblastine) leads to slightly increased tubulin synthesis. With the cloning of tubulin mRNAs in 1980, it became possible to investigate the molecular mechanisms underlying this apparent autoregulation of tubulin synthesis based on the amount of free tubulin. This task has primarily been undertaken by Cleveland and co-workers.

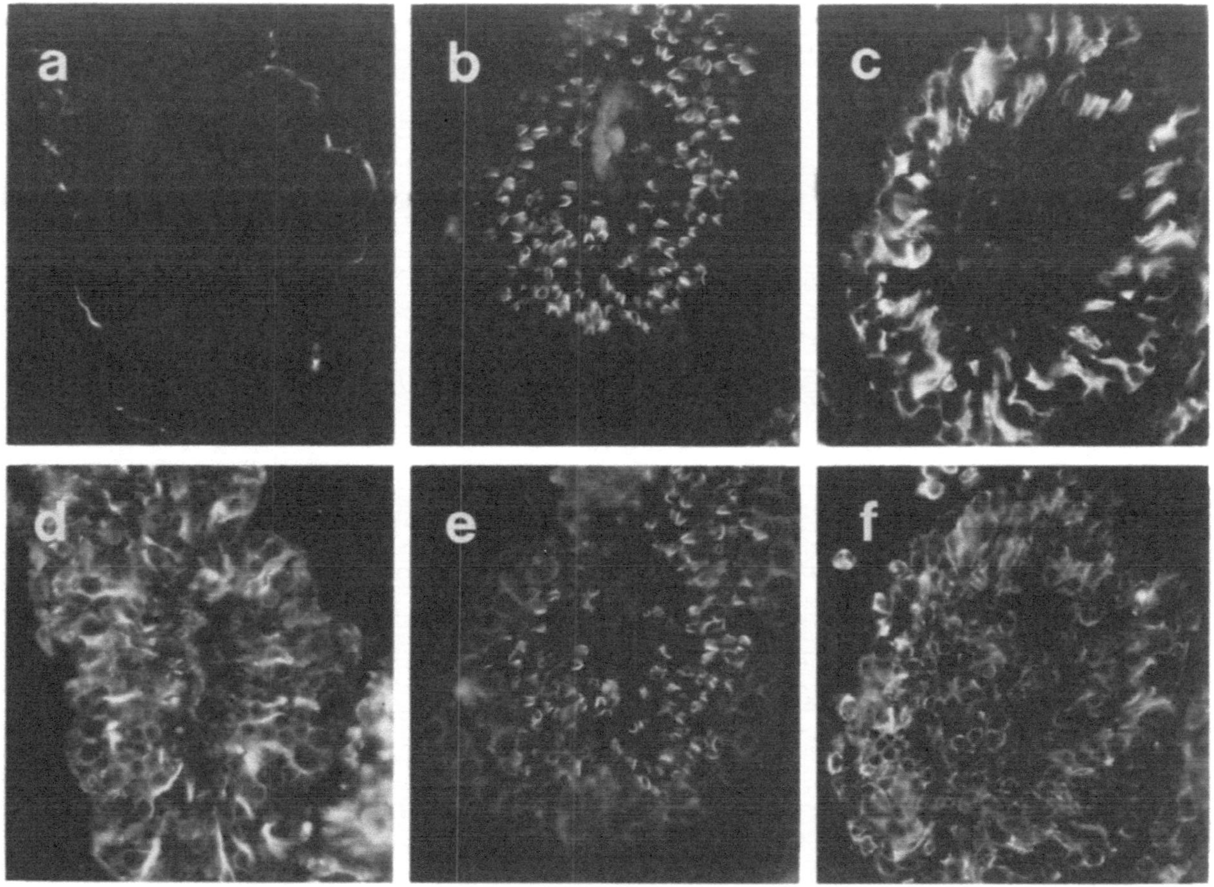

FIGURE 10. Axonemes in spermatids of wild type and mutant *Drosophila*. These electron micrographs show the aberrant microtubules formed in the spermatids of flies that are heterozygous (8/+) and homozygous (8/8) for the mutation $B2t^8$ in the gene encoding β2-tubulin as compared with the wildtype (+/+). Arrows point to these aberrant S-shaped microtubules. (From Fuller, M. T., Caulton, J. H., Hutchens, J. A., Kaufman, T. C., and Raff, E. C., *J. Cell. Biol.*, 104, 385, 1987. With permission.)

To demonstrate that the five- to tenfold repression of tubulin synthesis seen with colchicine and nocadazole is indeed due to an increase in the intracellular pool of the tubulin heterodimer, and not a secondary effect of these very toxic drugs, tubulin was microinjected into culture cells, thereby decreasing the amount of free tubulin; this was likewise shown to decrease tubulin synthesis.[87] In addition, RNA blot analysis confirmed the earlier results that the decrease in tubulin synthesis is a direct result of decreased tubulin mRNA levels.[88] The repression of tubulin synthesis by increasing the free tubulin content is observed even in enucleated cells,[89,90] thus proving that the mechanism responsible for decreasing the levels of tubulin mRNA is cytoplasmic, i.e., specific destabilization or degradation of tubulin mRNA. Gene transfer experiments in which chimeric genes comprised of various β-tubulin gene segments inserted into a thymidine kinase gene show that a gene containing only the first 16 codons of β-tubulin can be autoregulated by the tubulin monomer pool.[91] Sequences required for autoregulation have now been shown to be restricted to the first four codons, Met-Arg-Glu-Ile.[57] These four amino acids are identical in β-tubulins from all species examined (Figure 2). By site-specific mutagenesis of the Arg codon to produce different mRNA sequences encoding the same four initial amino acids, it has been shown that tubulin mRNA destabilization is mediated through the nascent translated tubulin protein, and not via factors binding the mRNA sequence directly. Several experiments[92] with protein synthesis inhibitors support this interesting finding: puromycin and pactamycin, both of which remove mRNA from polysomes, prevent the specific degredation of tubulin mRNA caused by increases in tubulin monomer concentration, while cycloheximide, which stalls mRNA on ribosomes, actually increases the destabilization of tubulin mRNA. Further evidence that only polysome-bound tubulin mRNAs are degraded is provided by gene transfer experiments with a truncated tubulin gene whose transcripts are too short to be found on polyribosomes; this gene is not autoregulated. Thus, it appears that some factor — very possibly the tubulin heterodimer itself — binds to the first four amino acids of β-tubulin as the corresponding mRNA is being translated, and thereby destabilizes the mRNA. This would explain the inverse relationship between tubulin mRNA stability and the concentration of unpolymerized tubulin. The exact mechanism of mRNA destabilization is no doubt under study. Since the concentration of tubulin subunits in part controls the dynamic equilibrium between slowly growing and rapidly shrinking microtubules,[85] this autoregulatory mechanism may be needed for maintaining appropriate levels of heterodimer in a dividing cell. However, the true significance of autoregulation *in vivo* is still unclear. It does not seem to operate effectively to *increase* steady state tubulin concentrations, as demonstrated by the existence of a mutant cell line with 30% decreased levels of the tubulin heterodimer, apparently unbuffered by this type of autoregulation.[93] In addition, this autoregulatory mechanism can affect tubulin concentrations only slowly, since the protein is relatively stable, even in cultured cells.

One indication that autoregulation may be important under more normal physiological conditions comes from experiments on fibroblasts and hepatocytes treated with low levels of the microtubule-depolymerizing drug colcemid, which stably lowers the tubulin polymer mass and increases the pool of soluble tubulin.[94] Autoregulation of tubulin synthesis is observed with these more subtle changes in the concentration of soluble tubulin, although tubulin accumulation is also regulated by a greatly enhanced degradation of subunits in hepatocytes. In this system, the tubulin polymer mass is correlated with the endocytotic degradation of glycoproteins in hepatocytes, and therefore autoregulatory control of the level of tubulin protein could be of physiological importance in the regulation of this process. Some other evidence for the role of autoregulation *in vivo* comes from studies on gene regulation in lower organisms (see below).

Another type of autoregulation is that which keeps α- and β-tubulin polypeptides in balance within a cell. This type of regulation is also nontranscriptional, and occurs by a mechanism of protein degradation. This has been demonstrated by studying cell lines containing zero, one, or several copies of a mutant β-tubulin gene that encodes an electrophoretic variant polypeptide.[95] All three cell lines contained the same amount of total β-tubulin, in spite of an increased amount of mRNA encoding the aberrant β-tubulin in the cell line in which the corresponding gene is amplified. Pulse chase experiments on protein synthesis in these three lines demonstrated that tubulin was being synthesized and degraded faster in the cell line with the amplified β-tubulin gene, and therefore that protein degradation is the mechanism responsible for keeping α- and β-tubulin protein levels in balance. Since both the mutant and normal excess β-tubulin subunits were degraded, the cell line with the amplified mutant gene contained a higher proportion of aberrant β-tubulin, and therefore had a more pronounced mutant phenotype. Thus, the introduction of amplified tubulin genes into cells will be a useful technique for altering the tubulin isotype content of cells or introducing specific tubulin mutants, at the same time lowering the endogenous tubulin content, and studying the effect (if any) on the phenotype of the cell.

The nature of the enzymes involved in balancing α- and β-tubulin concentrations via degradation of excess subunits has not been studied; it is possible that undimerized tubulin is unstable and nonspecifically degraded. This mechanism is probably at work in many species; for example, a mutant assembly-incompetent testis-specific β-tubulin isotype fails to accumulate in the germ cells of affected *Drosophila*, as does the normal, coexpressed α-tubulin, implying that the undimerized α-tubulin subunits are degraded.[81,82] Because of the large number of α- and β-tubulin genes transcribed in various cell types and to varying extents during development, it seems a great simplification that the mechanisms controlling the balance of α- and β-tubulin in each cell and the level of free tubulin heterodimer should be posttranscriptional and operate indiscriminately on all expressed tubulin isotypes.

B. REGULATION OF TUBULIN GENES IN LOWER EUKARYOTES

Some lower eukaryotic organisms have served as particularly simple and attractive model systems for studying tubulin gene function and regulation. The slime mold *Physarum* is useful for studies involving cell cycle regulation of tubulin genes, since the plasmodium form undergoes a naturally synchronous mitosis. The alga *Chlamydomonas* has been used for studying the regulation of tubulin expression during flagellar regeneration, and recently the ciliated embryos of the sea urchin *L. Pictus* have been used in similar studies on the regrowth of cilia. Finally, in the life cycle of the parasitic protozoan *Lieshmania*, the organism undergoes a developmental transformation to a flagellated form, which has been studied with regard to the regulated expression of α- and β-tubulin.

Overall, these studies and others have documented a large variety of ways in which tubulin genes are regulated during processes (such as mitosis and the growth of cilia and flagella) in which microtubules play a vital role. They raise important questions about what signals and controls these processes of which will only be answered by more wide-ranging investigations. The work on tubulin gene regulation in lower eukaryotes has also provided some evidence of an *in vivo* role for the autoregulation of tubulin gene expression via mRNA stability. A more detailed account of tubulin gene regulation in five organisms — *Physarum, Chlamydomonas, L. Pictus, Leishmania, and Tetrahymena* — is provided below.

1. Physarum

The multinucleate plasmodium of *Physarum* uses microtubules only in its mitotic spindles; cytoplasmic microtubules are undetectable. This and the fact that nuclei of the plasmodium undergo mitosis synchronously makes it an excellent system for the study of the relationship between tubulin synthesis and mitosis. Using *in vitro* transcription of mRNA and *in vivo* labeling, it was shown that synthesis of two α- and two β-tubulin isotypes increases coordinately until metaphase, and then falls.[96] This synthesis is regulated by the level of mRNA; α- and β-tubulin mRNA levels rise during G2, peak at metaphase, and fall exponentially. The greater than 40-fold change in tubulin mRNA levels over the cell cycle of *Physarum* is much greater than that found in *Chlamydomonas*[97] (8-fold) or Hela cells[98] (2-fold); this is probably attributable to the fact that in the *Physarum* plasmodium, microtubules are found only during the mitotic phase. Since tubulin mRNA accumulation begins prior to spindle assembly, it seems likely that autoregulation of tubulin mRNA stability by the concentration of free tubulin cannot be responsible for the accumulation of tubulin mRNA levels that is concomitant with spindle disassembly. Finally, since the tubulin protein is relatively stable in *Physarum*, the rapid rise in tubulin synthesis leads only to a two- to fourfold increase in the tubulin concentration of the cell.[96] As it seems probable that this increase in tubulin concentration does not in itself trigger spindle assembly and mitosis, the reasons for the observed sharp peak in tubulin synthesis are unclear.

2. Chlamydomonas

When the alga *C. reinhardi* is deflagellated by mechanical shearing, there is a rapid burst of synthesis of α- and β-tubulin and other flagellar proteins.[79,99] Transcription rates of α- and β-tubulin genes (measured by *in vitro* nuclear runoff assays and *in vivo* labeling) reach their maximum levels 15 min after deflagellation and then decline to basal levels within 1 h.[100] Protein synthesis is not required for this induction, but it is required for the sharp decline in the levels of tubulin mRNAs (and mRNAs encoding other flagellar proteins); in cells treated with cycloheximide tubulin mRNA levels remain elevated for 3 h after deflagellation. Tubulin transcipts are stabilized at early times during flagellar regeneration and destabilized at later times. Since the pool of soluble tubulin is depleted at early times, and replenished after the burst of tubulin synthesis, it is possible that the changes in tubulin mRNA stability seen in this system occur via the autoregulatory mechanism discovered by Penman in cultured cells,[86] and described above. Further evidence for this is the result that when *Chlamydomonas* that have grown flagella in the presence of cycloheximide (thus stabilizing the tubulin mRNA) are induced to resorb these flagella, thereby increasing the intracellular concentration of soluble tubulin, tubulin mRNAs are selectively destabilized.[100]

3. Sea Urchin

Cilia removed from the blastulae of the sea urchin by hypertonic shock can be regenerated up to four times in the presence of protein synthesis inhibitors. This indicates that the embryos have a large store of the structural proteins needed to make cilia, including α- and β-tubulins. In spite of these large stores of tubulin, the embryo of the sea urchin *L. pictus* undergoes a rapid transient increase in α- and β-tubulin gene expression in response to removal of its cilia.[101] This burst of tubulin gene transcription is similar in its rapidity and lack of dependence on protein synthesis to that consequent upon deflagellation in *Chlamydomonas*, described above. Also, like *Chlamydomonas*, the stability of tubulin mRNA sequences in sea urchin embryos seems to depend in an inverse manner on the concentration of soluble tubulin. In sea urchin, the level of free tubulin heterodimer declines through development as the large pools of stored maternal proteins are depleted. The accumulation of tubulin transcripts in response to deciliation varies depending on the stage of the embryo, due to the increasing stability of the α- and β-tubulin transcripts. This demonstrates the existence of an inverse relationship between tubulin mRNA stability and the amount of unpolymerized tubulin. The findings from experiments involving removal of cilia have relevance for gene expression during normal development; during embryogenesis in both *S. purpuratus* and *L. pictus*, there is a transient increase in the transcription of tubulin genes at the time of ciliation, and a slow rise in the level of tubulin mRNA as the pools of soluble protein are depleted.[101] Since the sea urchin embryo has adequate amounts of tubulin needed to produce cilia, the reasons for the regulated increase in tubulin gene expression are unclear.

4. Leishmania

The stage-specific regulation of tubulin gene expression has been studied in two species of the protozoan parasite *Leishmania: L. enrietti* and *L. mexicana*. Surprisingly, different mechanisms seem responsible in the two species for the increased amount of tubulin found in the flagellated promastigotes, as compared with the nonmotile amastigotes. By northern blot analysis and *in vitro* translation of mRNA from the two stages of *L. enrietti*, it appears that increased tubulin synthesis in promastigotes results (at least partially) from increased levels of α- and β-tubulin mRNA.[102] However, mRNA made from the two stages of *L. mexicana* translated *in vitro* produced approximately equal amount of tubulin protein, in spite of the vast differences in the biosynthesis of tubulin in these stages as assayed by *in vivo* labeling.[103,104] Thus, in *L. mexicana*, there is posttranscriptional control of tubulin gene expression, possibly via sequestration of tubulin mRNA in amastigotes.

5. Tetrahymena

The ciliated protozoan Tetrahymena contains two nuclei, the transcriptionally active macronucleus and the inactive micronucleus, which after meiosis and division differentiates into a new macronucleus. It appears that transcription of the single α-tubulin gene in this organism might be activated by somatic rearrangement of DNA sequences — specifically by the deletion of upstream and downstream flanking sequences — during macronuclear differentiation.[105] This would be an interesting addition to the large repertoire of mechanisms by which lower organisms regulate tubulin gene expression.

V. CONCLUSIONS

The α- and β-tubulins of all higher eukaryotes thus far examined are encoded by multigene families. The reason for the existence of these multigene families is as yet unknown, though it seems clear that the various isotypes encoded by the different family members can fully assemble *in vivo* into functionally diverse kinds of microtubule. One contributing factor to the existence of multiple copies of expressed α- and β-tubulin genes might be the varied demand for tubulin subunits in different cell lineages and at different stages of development. Such varying needs could conveniently be met by a series of gene duplication events followed by divergence in elements that regulate levels of transcription. While such a scheme seems entirely plausible, it does not help to explain the observed divergence of tubulin coding sequences, since changes in levels of transcription are presumably dictated by promoter and/or upstream regulatory elements and not by coding sequences themselves. The existence of a range of α- and β-tubulin isotypes, and their interspecies conservation over hundreds of millions of years, must therefore be explained in terms of a functional requirement based on isotype differences, but unrelated to their capacity to coassemble into microtubules. This functional requirement could involve specific interactions with MAPs, so that the striking variation in cell type-specific patterns of tubulin isotype expression could reflect the concomitant cell type-specific expression of different MAPs. In that event, encoded isotypes may have evolved to specifically or preferentially interact with MAPs in those cell types in which they are expressed. This hypothesis can be tested in several ways. Biochemically, the interaction of MAPs and tubulin isotypes can be studied directly *in vitro* using cloned sequences or peptides. Genetically, tubulin isotypes can be substituted for each other *in vivo*, and their ability to perform the same functions assayed. This is most easily done against a null background; for example, the existence of a mutant strain of *Drosophila* missing the testis specific β-tubulin isotype β2 makes possible experiments to introduce sequences encoding other isotypes into this strain, under the control of the β2 promotor.[38] Such isotype substitution experiments have been performed in lower organisms (yeast and *Aspergillus*) but would be even more informative in *Drosophila*, which possesses true tubulin isotypes and a wider range of functionally distinct microtubules. Both molecular biological and genetic methods will be important for the isolation and functional analysis of undiscovered MAPs — via hybridization with already cloned MAPs, or by the isolation of nontubulin mutants or revertants affecting microtubule function. The study of MAPs and their interaction with tubulin isotypes should help to clarify the role of tubulin genes in microtubule biology.

Another avenue for future research on tubulin genes involves experiments with genetically engineered tubulin mutants, directed toward the elucidation of functional domains of α- and β-tubulins. Initial experiments on yeast α- and β-tubulins have shown that an α-tubulin with 3, 9, and 17 amino acid insertions in its amino-terminal variable domain can still serve as the only α-tubulin in yeast, without effect on mating, meosis, or mitosis.[106] It seems that sequence variability in this region of the molecule exists because it can be tolerated. Studies with engineered mutant β-tubulin and chicken/yeast hydrid β-tubulins in cultured cells suggest that the carboxyterminal portion of the molecule is not critical for its assembly or stability.[107] More experiments along these lines will refine our knowledge of the structural determinants of tubulin function.

REFERENCES

1. **Cleveland, D. W., Kirschner, M. W., and Cowan, N. J.,** Isolation of separate mRNAS for α- and β-tubulin and characterization of the corresponding *in vitro* translation products, *Cell,* 15, 1021, 1978.
2. **Cleveland, D. M., Lopata, M. A., MacDonald, R. J., Cowan, N. J., Rutter, W. J., and Kirschner, M. W.,** Number of evolutionary conservation of α- and β-tubulin and cytoplasmic β- and γ-actin genes using specific cloned cDNA probes, *Cell,* 20, 95, 1980.
3. **Hall, J. L., Dudley, L., Dobner, P. R., Lewis, S. A., and Cowan, N. J.,** Identification of two human β-tubulin isotypes, *Mol. Cell. Biol.,* 3, 854, 1983.
4. **Wilde, C. D., Crowther, C. E., Cripe, T. P., Gwo-Shu Lee, M., and Cowan, N. J.,** Evidence that a human β-tubulin pseudogene is derived from its corresponding mRNA, *Nature,* 292, 83, 1982.
5. **Wilde, C. D.,** Pseudogenes, *Crit. Rev. Biochem.,* 9, 323, 1986.
6. **Lewis, S. A., Lee, M. G.-S., and Cowan, N. J.,** Five mouse tubulin isotypes and their regulated expression during development, *J. Cell. Biol.,* 101, 852, 1985.
7. **Villasante, A., Wang, D., Dobner, P., Dolph, P., Lewis, S. A., and Cowan, N. J.,** Six mouse α-tubulin mRNAs encode five distinct tubulin isotypes: testis-specific expression of two sister genes, *Mol. Cell. Biol.,* 6, 2409, 1986.
8. **Cowan, N. J., Dobner, P. R., Fuchs, E. V., and Cleveland, D. W.,** Expression of human α-tubulin genes: interspecies conservation of 3′ untranslated regions, *Mol. Cell. Biol.,* 3, 1738, 1983.
9. **Hall, J. L. and Cowan, N. J.,** Structural features and restricted expression of human α-tubulin gene, *Nucleic Acids Res.,* 13, 207, 1985.
10. **Lemischka, I. R., Farmer, S., Racaniello, V. R., and Sharp, P. A.,** Nucleotide sequence and evolution of a mammalian α-tubulin messenger RNA, *J. Mol. Biol.,* 15, 101, 1981.
11. **Lemischka, I. R. and Sharp, P. A.,** The sequences of an expressed rat α-tubulin gene and a pseudogene with an inserted repetitive element, *Nature,* 300, 330, 1982.
12. **Elliot, E. M., Okayama, H., Savangi, F., Henderson, G., and Ling, V.,** Differential expression of three α-tubulin genes in Chinese hamster ovary cells, *Mol. Cell. Biol.,* 5, 236, 1985.
13. **Eliot, E. M., Henderson, G., Sarangi, F., and Ling, V.,** Complete sequence of three α-tubulin cDNAs in Chinese hamster ovary cells: each encodes a distinct isoprotein, *Mol. Cell. Biol.,* 6, 906, 1986.
14. **Valenzuela, P., Quiroga, M., Zalduar, J., Rutter, W. J., Kirschner, M. W., and Cleveland, D. W.,** Nucleotide and corresponding amino acid structure encoded by α- and β-tubulin mRNA, *Nature,* 289, 650, 1984.
15. **Ginzberg, I., Behar, L., Givol, D., and Littauer, U. Z.,** The nucleotide sequence of rat α-tubulin: 3′ end characteristics and evolutionary conservation, *Nucleic Acids Res.,* 9, 2691, 1981.

15a. **Wilde, C. D., Chow, L. T., Wefald, F. C., and Cowan, N. J.,** Structure of two human α-tubulin genes, *Proc. Natl. Acad. Sci. U.S.A.,* 79, 96, 1982.

16. **Pratt, L. F., Okamura, S., and Cleveland, D. W.,** A divergent testis-specific α-tubulin isotype that does not contain a coded C-terminal tyrosine, *Mol. Cell. Biol.,* 7, 552, 1987.

17. **Dobner, P. R., Kislanskis, E., Wentworth, B. M., and Villa-Komaroff, L.,** Alternative 5′ exons either provide or deny an initiator methionine codon to the same α-tubulin coding region, *Nucleic Acids Res.*, 15, 199, 1986.
18. **Hecht, N. B., Distel, R. J., Yelick, P. C., Tanhuser, S. M., Driscoll, D. E., Goldberg, E., and Jung, K. K.,** Localization of a highly divergent mammalina testicular α-tubulin that is not detectable in brain, *Mol. Cell. Biol.*, 8, 996, 1988.
19. **Wang, D., Villasante, A., Lewis, S. A., and Cowan, N. J.,** The mammalian β-tubulin repertoire: hematopoietic expression of a novel β-tubulin isotype, *J. Cell. Biol.*, 103, 1903, 1986.
20. **Cowan, N. J., Lewis, S. A., Sarkar, S., and Gu, W.,** Functional versatility of mammalian β-tubulin isotypes, in *The Cytoskeleton in Cell Differentiation and Development*, Maccioni, R. and Arechaya, J., Eds., ICSU Press, Miami, 1987, 157.
21. **Murphy, D. B., Wallis, K. T., Machlin, P. S., Ratrie, H., and Cleveland, D. W.,** The sequence and expression of the divergent β-tubulin in chicken erythrocytes, *J. Biol. Chem.*, 262, 14304, 1987.
22. **Farmer, S. R., Bond, J. F., Robinson, G. S., Mbandgollo, D., Fenton, M. J., and Berkowitz, E. H.,** Differential expression of the rat β-tubulin multigene family, in *Molecular Biology of the Cytoskeleton*, Borisy, G. G., Cleveland, D. W., Murphy, D. B., Eds., Cold Spring Harbor Laboratory, Cold Spring Harbor, NY, 1984, 333.
23. **Havercroft, J. C. and Cleveland, D. W.,** Programmed expression of β-tubulin genes during development and differentiation of the chicken, *J. Cell. Biol.*, 99, 1927, 1984.
24. **Sullivan, K. F., Havercroft, J. C., and Cleveland, D. W.,** Primary structure and expression of a vertebrate β-tubulin gene family in *Molecular Biology of the Cytoskeleton*, Borisy, G. G., Cleveland, D. W., Murphy, D. B., Eds., Cold Spring Harbor Laboratory, Cold Spring Harbor, NY, 1984, 321.
25. **Sullivan, K. F. and Cleveland, D. W.,** Identification of conserved isotype-defining variable region sequences for four vertebrate β-tubulin peptide classes, *Proc. Natl. Acad. Sci. U.S.A.*, 83, 4327, 1986.
26. **Lewis, S. A., Gilmartin, M. L., Hall, J. L., and Cowan, N. J.,** Three expressed sequences within the human β-tubulin multigene family each define a distinct isotype, *J. Mol. Biol.*, 182, 11, 1985.
27. **Lee, M. G.-S., Loomis, C., and Cowan, N. J.,** Sequence of an expressed human β-tubulin gene containing ten *Alu* family members, *Nucleic Acids Res.*, 12, 5828, 1984.
28. **Bond, J. F. and Farmer, S. R.,** Regulation of tubulin and actin mRNA production in rat brain: expression of new β-tubulin mRNA with development, *Mol. Cell. Biol.*, 3, 1333, 1983.
29. **Bond, J. F., Robinson, G. S., and Farmer, S. R.,** Differential expression of two neural cell-specific β-tubulin mRNAs during rat brain development, *Mol. Cell. Biol.*, 4, 1313, 1984.
30. **Lee, M. G.-S., Lewis, S. A., Wilde, C. D., and Cowan, N. J.,** Evolutionary history of a multigene family: an expressed human β-tubulin gene and three processed pseudogenes, *Cell*, 33, 477, 1983.
31. **Burgoyne, R. D., Cambray Deakin, M. A., Lewis, S. A., Sarkar, S., and Cowan, N. J.,** Differential distribution of β-tubulin isotypes in cerebellum, *EMBO J.*, 7, 2311, 1988.
32. **Sullivan, K. F. and Cleveland, D. W.,** Sequence of highly divergent β-tubulin gene reveals regional heterogeneity in the β-tubulin polypeptide, *J. Cell. Biol.*, 99, 1754, 1984.
33. **Lewis, S. A. and Cowan, N. J.,** Complex regulation and functional versatility of mammalian α- and β-tubulin isotypes during the differentiation of testis and muscle cells, *J. Cell. Biol.*, 106, 2023, 1988.
34. **Lopata, M. A., Havercroft, J. C., Chow, L. T., and Cleveland, D. W.,** Four unique genes required for β-tubulin expression in vertebrates, *Cell*, 32, 713, 1983.
35. **Michiel, F., Falkenberg, D., Mueller, A. M., Itinz, U., Otto, U., Bellman, R., Glaetzer, K. H., Brand, R., Bialojan, S., and Renkawitz Pohl, R.,** Testis specific β2-tubulins are identical in *Drosophila melanogaster* and *Drosophila hydei* but differ from the ubiquitous β1 tubulin, *Chromosoma*, 95, 387, 1987.
36. **Theurkauf, W. E., Baum, H., Bo, J., and Wensink, P. C.,** Tissue-specific and constituitive α-tubulin genes of *Drosophila melanogaster* code for structurally distinct proteins, *Proc. Natl. Acad. Sci. U.S.A.*, 83, 8477, 1986.
37. **Rudolph, J. E., Kimble, M., Hoyle, H. D., Subler, M. A., and Raff, E. C.,** Three *Drosophila* β-tubulin sequences: a developmentally regulated isoform (β3), the testis specific isoform (β2) and an assembly-defective mutation of the testis-specific isoform ($B2t^8$) reveal both an ancient divergence in metazoan isotypes and structural constraints for β-tubulin function, *Mol. Cell. Biol.*, 7, 2231, 1987.
38. **Raff, E. C.,** Genetics of microtubule systems, *J. Cell. Biol.*, 99, 1, 1984.
39. **Monteiro, M. J. and Cox, R. A.,** Primary structure of an α-tubulin gene of *Physarum polycephalum*, *J. Mol. Biol.*, 193, 427, 1987.
40. **Silflow, C. D., Chisolm, R. L., Conner, T. W., and Ranom, L. P. W.,** The two α-tubulin genes of *Chlamydomonas reinhardi* code for slightly different proteins, *Mol. Cell. Biol.*, 5, 2389, 1985.
41. **Kimmel, B. E., Samson, S., Wu, J., Hischberg, R., and Yarbrough, L. R.,** Tubulin genes of the African trypanosome *Trypanosoma brucei rhodesiense:* nucleotide sequence of a 3.7 kb fragment containing genes for α- and β-tubulins, *Gene*, 35, 237, 1985.
42. **Toda, T., Adachi, Y., Hiraoka, Y., and Yanagida, M.,** Identification of the pleiotropic cell division cycle gene *NDA2* as one of two different α-tubulin genes *Schizosaccharomyces pombe*, *Cell*, 37, 233, 1984.

43. **Schatz, P. J., Pillus, L., Grisafi, P., Solomon, F., and Botstein, D.,** Two functional α-tubulin genes of the yeast *Saccharomyces cerevisiae* encode divergent proteins, *Mol. Cell. Biol.,* 6, 3711, 1986.
44. **Mandelkow, E. M., Hermann, M., and Ruhl, U.,** Tubulin domains probed by limited proteolysis and subunit specific antibodies, *J. Mol. Biol.,* 185, 311, 1985.
45. **LeDizet, M. and Piperno, G.,** Lysine 40: An acetylated site of *Chlamydomonas* α-tubulin, *J. Cell. Biol.,* 105, 278a, 1987.
46. **Burland, T. G., Paul, E. C. A., Oetliker, M., and Dove, W. F.,** A gene encoding the major β-tubulin of the mitotic spindle in *Physarum polycephalum* plasmodia, *Mol. Cell. Biol.,* 8, 1275, 1988.
47. **Youngbloom, J., Schloss, J. A., and Silflow, C. D.,** The two β-tubulin genes of *Chlamydomonas reinhardtii* code for identical proteins, *Mol. Cell. Biol.,* 4, 2686, 1984.
48. **Orbach, M. J., Porro, E. B., and Yanofsky, C.,** Cloning and characterization of the gene for β-tubulin from a benomyl-resistant mutant of *Neurospora crassa* and its use as a dominant selectable marker, *Mol. Cell. Biol.,* 6, 2452, 1986.
49. **May, G. S., Tsang, M. L.-S., Smith, H., Fidel, S., and Morris, N. R.,** Aspergillus beta-tubulin genes are unusually divergent, *Gene,* 55, 231, 1987.
50. **Hiraoka, Y., Toda, T., and Yanagida, M.,** The NDA3 gene of fission yeast encodes β-tubulin: a cold-sensitive nda3 mutation reversibly blocks spindle formation and chromosome movement in mitosis, *Cell,* 39, 349, 1984.
51. **Neff, N. F., Thomas, J. H., Grisafi, P., and Botstein, D.,** Isolation of the β-tubulin gene from yeast and demonstration of its essential function *in vivo, Cell,* 33, 211, 1983.
52. **Gundersen, G. G., Khawaja, S., and Bulinksi, J. C.,** Postpolymerization detyrosination of α-tubulin; a mechanism for subcellular differentiation of microtubules, *J. Cell. Biol.,* 105, 251, 1987.
53. **Argarana, C. E., Barra, H. S., and Caputto, R.,** Release of [^{14}C]-tyrosine from tubulinyl-[^{14}C]-tyrosine by brain extract. Separation of a carboxypeptidase from tubulin tyrosine ligase, *Mol. Cell. Biochem.,* 19, 17, 1978.
54. **L'Hernault, S. W. and Rosenbaum, J. L.,** *Chlamydomonas* α-tubulin is post-translationally modified in the flagella during flagellar assembly, *J. Cell. Biol.,* 97, 258, 1983.
55. **L'Hernault, S. W. and Rosenbaum, J. L.,** *Chlamydomonas* α-tubulin is post-translationally modified by acetylation on ϵ-the amino group of a lysine, *Biochemistry,* 24, 473, 1985.
56. **L'Hernault, S. W. and Rosenbaum, J. L.,** Reversal of the post-translational translational modification on *Chlamydomonas* flagellar alpha tubulin occurs during flagellar resorption, *J. Cell. Biol.,* 100, 457, 1985.
57. **Yen, T. J., Gay, D. A., Pachter, J. S., and Cleveland, D. W.,** Autoregulated changes in stability of polyribosome-bound β-tubulin mRNAs are specified by the first 13 translated nucleotides, *Mol. Cell. Biol.,* 8, 1224, 1988.
58. **Cowan, N. J.,** Tubulin genes and the diversity of tubulin function, in *Oxford Surveys on Eukaryotic Genes,* Vol. 1, McLean, N., Ed., Oxford University Press, New York, 1984, 36.
59. **Cleveland, D. W., Hughes, S. H., Stubblefield, E., Kirschner, M. W., and Varmus, H. E.,** Multiple α and β tubulin genes represent unlinked and dispersed gene families, *J. Biol. Chem.,* 256, 3130, 1981.
60. **Sanchez, F., Natzle, J. E., Cleveland, D. W., Kirschner, M. W., and McCarthy, B. J.,** A dispersed multigene family encoding tubulin in *Drosophila melanogaster, Cell,* 22, 845, 1980.
61. **Thomashow, L. S., Milhausen, M., Rutter, W. J., and Agabian, N.,** Tubulin genes are tandemly linked and clustered in the genome of *Trypanosoma brucei, Cell,* 32, 35, 1983.
62. **Landfear, S. M., McMahon-Pratt, D., and Wirth, D.,** Tandem arrangement of tubulin genes in the protozoan parasite, *Leishmania enrietti, Mol. Cell. Biol.,* 3, 1070, 1983.
63. **Schedl, T., Owens, J., Dove, W. F., and Burland, T. G.,** Genetics of the tubulin gene families of Physarum, *Genetics,* 108, 143, 1984.
64. **Adachi, Y., Toda, T., Niwa, O., and Yanagida, M.,** Differential expression of essential and non-essential α-tubulin genes in *Schizosaccaromyces pombe, Mol. Cell. Biol.,* 6, 2168, 1986.
65. **Wilde, C. D., Crowther, C. E., and Cowan, N. J.,** Diverse mechanisms in the generation of human β-tubulin pseudogenes, *Science,* 217, 549, 1982.
66. **Lewis, S. A., Gu, W., and Cowan, N. J.,** Free intermingling of mammalian β-tubulin isotypes among functionally distinct microtubules, *Cell,* 49, 59, 1987.
67. **Hood, L., Campbell, J. H., and Elgin, S. C. R.,** The organization expression, and evolution of antibody genes and other multigene families, *Ann. Rev. Genetics,* 1, 305, 1975.
68. **May, G. S., Gambino, J., Weatherbee, J. A., and Morris, N. R.,** Identification and functional analysis of β-tubulin genes by site-specific integration in *Aspergillus nidulans, J. Cell. Biol.,* 101, 712, 1985.
69. **Weatherbee, J. A., May, G. S., Gambino, J., and Morris, N. R.,** Involvement of a particular species of β-tubulin (B3) in conidial development in *Aspergillus nidulans, J. Cell. Biol.,* 101, 712, 1985.
70. **Fulton, C. and Simpson, P. A.,** Selective synthesis and utilization of flagellar tubulin. The multitubulin hypothesis, in *Cell Motility,* Goldman, R., Pollard, T., and Rosenbaum, J., Eds., Cold Spring Harbor Laboratory, Cold Spring Harbor, NY, 1976, 987.

71. **Bond, J. F., Fridovich-Keil, J. L., Pillus, L., Mulligan, R. C., and Solomon, F.,** A chicken-yeast chimeric β-tubulin protein is incorporated into mouse microtubules *in vivo*, *Cell*, 44, 461, 1986.
72. **Cleveland, D. W.,** The multitubulin hypothesis revisited: what have we learned?, *J. Cell. Biol.*, 104, 381, 1987.
73. **Gu, W., Lewis, S. A., and Cowan, N. J.,** Generation of antisera that disriminiate among mammalian α-tubulins: introduction of specialized isotypes into cultured cells results in their coassembly without disruption of function, *J. Cell. Biol.*, 106, 2011, 1988.
74. **Lopata, M. A. and Cleveland, D. W.,** *In vivo* microtubules are copolymers of available β-tubulin isotypes: localization of each of six vertebrate β-tubulin isotypes using polyclonal antibodies elicited by synthetic peptide antigens, *J. Cell. Biol.*, 105, 1707, 1987.
75. **Joshi, H. C., Yen, T. J., and Cleveland, D. W.,** *In vivo* coassembly of a divergent β-tubulin subunit (cβ6) into microtubules of different function, *J. Cell. Biol.*, 105, 2179, 1987.
76. **Miller, F. D., Naus, C. C. G., Durand, M., Bloom, F. E., and Milner, R. H.,** Isotypes of α-tubulin are differentially regulated during neuronal maturation, *J. Cell. Biol.*, 105, 3065, 1987.
77. **Burland, T. G., Schedl, T., Gull, K., and Dove, W. F.,** Genetic analysis of resistance to benzsimidazoles in Physarum: differential expression of beta-tubulin genes, *Genetics*, 108, 123, 1984.
78. **Schatz, P. J., Solomon, F., and Botstein, D.,** Genetically essential and non-essential α-tubulin genes specify functionally interchangeable proteins, *Mol. Cell. Biol.*, 6, 3722, 1986.
79. **Silflow, C. D. and Rosenblum, J. L.,** Multiple α and β tubulin genes in *Chlamydomonas* and regulation of tubulin mRNA levels after deflagellation, *Cell*, 24, 81, 1981.
80. **Brunke, K. J., Young, E. E., Buchbinder, B. U., and Weeks, D. P.,** Coordinate regulation of the four tubulin genes of *Chlamydomonas reinhardi*, *Nucleic Acids Res.*, 10, 1295, 1982.
81. **Kemphues, K. J., Raff, R. A., Kaufman, T. C., and Raff, E. C.,** Mutation in a structural gene for β-tubulin specific to testis in *Drosophila melanogaster*, *Proc. Natl. Acad. Sci. U.S.A.*, 76, 3993, 1979.
82. **Kemphues, K. J., Raff, E. C., Raff, R. A., and Kaufman, T. C.,** Mutation in a testis-specific β-tubulin in *Drosophila:* analysis of its effect on meiosis and map location of the gene, *Cell*, 21, 455, 1979.
83. **Kemphues, K. J., Kaufman, T. C., Raff, R. A., and Raff, E. C.,** The testis-specific β-tubulin subunit in *Drosophila melanogaster* has multiple functions in spermatogenesis, *Cell*, 31, 655, 1982.
84. **Fuller, M. T., Caulton, J. H., Hutchens, J. A., Kaufman, T. C., and Raff, E. C.,** Genetic analysis of microtubule structure: a β-tubulin mutation causes the formation of aberrant microtubules *in vivo* and *in vitro*, *J. Cell. Biol.*, 104, 385, 1987.
85. **Kirschner, M. W. and Mitchison, T.,** Beyond self-assembly; from microtubules to morphogenesis, *Cell*, 45, 329, 1986.
86. **Ben Ze'ev, A., Farmer, S. R., and Penman, S.,** Mechanisms of regulating tubulin synthesis in cultured mammalian cells, *Cell*, 17, 319, 1979.
87. **Cleveland, D. W., Pittenger, M. F., and Feramisco, J. R.,** Elevation of tubulin levels by microinjection suppress new tubulin synthesis, *Nature*, 305, 738, 1983.
88. **Cleveland, D. W., Lopata, M. A., Sherline, P., and Kirschner, M. W.,** Unpolymerized tubulin modulates the level of tubulin mRNAs, *Cell*, 25, 537, 1981.
89. **Caron, J., Jones, A. L., Rall, L. B., and Kirschner, M. W.,** Autoregulation of tubulin synthesis in enucleated cells, *Nature*, 317, 648, 1985.
90. **Pittenger, M. F. and Cleveland, D. W.,** Retention of autoregulatory control of tubulin synthesis in cytoplasts: demonstration of cytoplasmic mechanism that regulates the level of tubulin expression, *J. Cell. Biol.*, 101, 1941, 1985.
91. **Gay, D. A., Yen, T. J., Lau, J. T. Y., and Cleveland, D. W.,** Sequences that confer β-tubulin autoregulation through modulated mRNA stability reside within exon 1 of β-tubulin mRNA, *Cell*, 50, 671, 1987.
92. **Pachter, J. S., Yen, T. J., and Cleveland, D. W.,** Autoregulation of tubulin expression is achieved through specific degradation of polysomal tubulin mRNAs, *Cell*, 51, 283, 1987.
93. **Boggs, B. and Cabral, F.,** Mutations affecting assembly and stability of tubulin: evidence for a nonessential β-tubulin in CHO cells, *Mol. Cell. Biol.*, 7, 2700, 1987.
94. **Caron, J. M., Jones, A. L., and Kirschner, M. W.,** Autoregulation of tubulin synthesis in hepatocytes and fibroblasts, *J. Cell. Biol.*, 101, 1763, 1985.
95. **Whitfield, C., Abrtaham, I., Ascherman, D., and Gottesman, M. M.,** Transfer and amplification of a mutant β-tubulin gene results in colcemid dependence: use of the transformant to demonstrate regulation of β-tubulins subunit level by protein degradation, *Mol. Cell. Biol.*, 6, 1422, 1986.
96. **Schedl, T., Burland, T. G., Gull, K., and Dove, W. F.,** Cell cycle regulation of tubulin RNA level, tubulin protein synthesis and assembly of microtubules in Physarum, *J. Cell. Biol.*, 99, 155, 1984.
97. **Ares, M. and Howell, S.,** Cell cycle stage-specific accumulation of mRNAs encoding tubulins and other polypeptides in *Chlamydomonas*, *Proc. Natl. Acad. Sci. U.S.A.*, 79, 5577, 1982.
98. **Bravo, R. and Celis, J. E.,** A search for differential polypeptide synthesis throughout the cell cycle, *J. Cell. Biol.*, 84, 795, 1980.

99. **Lefebvre, P. A., Silflow, C. D., Wieben, E. D., and Rosenblum, J. L.,** Increased levels of mRNAs for tubulin and other flagellar proteins after amputation or shortening of *Chlamydomonas* flagella, *Cell,* 20, 469, 1980.
100. **Baker, E. J., Keller, R. L., Schloss, J. A., and Rosenbaum, J. L.,** Protein synthesis is required for rapid degradation of tubulin mRNA and other deflagellation-induced RNAs in *Chlamydomonas reinhardi, Mol. Cell. Biol.,* 6, 54, 1986.
101. **Gong, Z. and Brandhorst, B.,** Stimulation of tubulin gene transcription by deciliation of sea urchin embryos, *Mol. Cell. Biol.,* 7, 4238, 1987.
102. **Landfear, S. M. and Wirth, D. F.,** Control of tubulin gene expression in the parasitic protozoan *Leishmania enriettii, Nature,* 309, 716, 1984.
103. **Fong, D., Wallach, M., Keithly, J., Malera, P., and Chang, K. P.,** Differential expression of mRNAs for α and β tubulin during differentiation of the parasitic protozoan *Leishmania mexicana, Proc. Natl. Acad. Sci. U.S.A.,* 81, 5782, 1984.
104. **Wallach, M., Fong, D., Chang, K. P.,** Post-transcriptional control of tubulin biosynthesis during Leishmanial differentiation, *Nature,* 299, 650, 1982.
105. **Callahan, R. C., Shalke, G., and Gorovsky, M. A.,** Developmental rearrangements associated with a single type of expressed α-tubulin gene in Tetrahymena, *Cell,* 36, 441, 1984.
106. **Schatz, P. J., Georges, G. E., Solomon, F., and Botstein, D.,** Insertion of up to 17 amino acids into a region of α-tubulin do not disrupt function, *Mol. Cell. Biol.,* 7, 3799, 1987.
107. **Fridovich-Keil, J. L., Bond, J. F., and Solomon, F.,** Domains of β-tubulin essential for conserved functions *in vivo, Mol. Cell. Biol.,* 7, 3792, 1987.

Chapter 3

STRUCTURE AND FUNCTION OF TUBULIN REGIONS

Luis Serrano and Jesús Avila

TABLE OF CONTENTS

I. TUBULIN STRUCTURE

Since tubulin was first identified as the principal component of microtubules, the analysis of the relationship between its structure and function and the regulation of its polymerization capacity is among the main objectives in microtubule research.

Several approaches have been used independently or in combination to analyze these questions. One approach takes advantage of the structural homology between certain proteins with similar functions and the existence of several empirical methods to predict the secondary structure of a protein from its primary sequence. When the sequences of α and β tubulin subunits from pig brain[1-3] or chicken[4] were obtained, a significant degree of homology between those peptides and those of other proteins was found.[1,5,6]*

A. SIMILARITIES WITH OTHER PROTEINS

Similarities were found between actin and five regions of α-tubulin comprising aa 57-68, 95-125, 147-155, 239-254, and 299-312, which correspond, respectively, to actin aa 21-32, 289-321, 144-157, 173-187, and 278-291.[1,2,7] Since an actin-like interaction of tubulin with myosin[8] has been described, this similarity could be of functional significance. There is also a slight similarity between aa 192-244 of α-tubulin and a fragment of the globular head of myosin (aa 1-47).[1,2] The final carboxy terminal residues of α- and β-tubulin have high identity with the first 16 amino acids of troponin T[1,2] and also with calcium-binding proteins.[9,10] In α-tubulin, there are two regions (aa 301-388) and (aa 247-260) similar to two regions (aa 56-147 and 271-284) of cAMP-dependent kinase.[1,2] Finally, there are also several tubulin regions with similarities to GTP- or ATP-binding proteins.[1]

B. LOW-RESOLUTION STRUCTURE

Microtubule structure has been studied by electron microscopy and X-ray diffraction which allows to make some models for the shape and arrangement of tubulin subunits.[11-14] In these models it appears that although only ~40% of the residues are conserved in both subunits the general shape is similar. Tubulin subunits appear to be elongated with a maximum diameter of 4 nm, and what is more important they appeared to be composed of three main domains.[13]

II. CHEMICAL MODIFICATIONS

The chemical modifications of certain residues of tubulin have been used to determine central amino acids in the binding of drugs or GTP to tubulin or in its polymerization capacity (for a review, see Reference 15). Although the role of certain residues in the tubulin properties mentioned above is clear, in few cases the exact determination of the residue involved in the binding of a molecule to tubulin has been described. Cross-linking experiments show that cysteines 239 and 354 may be involved both in the microtubule assembly and in the interaction of tubulin with colchicine,[16,17] while cysteines 12 and 201 or 211[18] could be also involved in tubulin assembly, vinblastine and maytansine binding but not in colchicine binding.[19,20] The most interesting feature of these results is that they give some information about the three-dimensional structure of tubulin, indicating regions that being far away in the tubulin sequence could be near to each other in the folded molecule.

Chemical modification of lysine residues of tubulin by methylation indicates that Lys-394 of the α-subunit is a critical residue in tubulin polymerization[21,22] and its modification completely abolished tubulin polymerization. This Lys residue is specially reactive probably due possibly to the close vicinity of His 393 and may be involved in the formation of a cluster of basic residues probably on an α-helix critical for microtubule assembly.[22]

* A third tubulin subunit, α-tubulin, has recently been described. (Oakley and Oakley, *Nature*, 338, 662, 1989.)

Finally *in vivo* expression of a cloned α-subunit gene in bacteria[23] together with *in vitro* results of chemical reactivity[21] indicated that the amino group of the amino-terminal α subunit is probably involved in a salt-bridge interaction in the dimer, important for tubulin assembly.[21]

III. LIMITED PROTEOLYSIS

Another approach which has been used to study the structure and functional domains of tubulin is limited proteolysis of the protein. Limited proteolysis of purified proteins with known functional characteristics has proved to be a useful method for the analysis of functional domains of many proteins. To obtain coherent results, it is very important to fix specific conditions for reproducible patterns of protein cleavage. At difference from exhaustive proteolysis, in limited proteolysis, the conformation of the substrate may be more important than the specificity of the protease. In controlled proteolysis, experimental conditions such as pH, ionic strength, temperature, or the presence of molecules which alter the conformation of the substrate (i.e., GTP, MAPs, Ca^{2+}, colchicine in the case of tubulin) determine cleavage sites. These conditions must be optimized for the conservation of the native conformation of the substrate in order to retain its functional activity.

A. MAPPING OF PROTEOLYTIC FRAGMENTS

In limited proteolysis, correct mapping of the proteolytic fragments on the protein, is essential for the analysis of protein domains and functions. The first groups to apply limited proteolysis to brain tubulin did not determine the positions of the proteolytic fragments.[24,25] Recently, a number of methods have been used to locate the different proteolytic fragments on the α- and β-subunits. The first approach takes advantage of the specific chemical cleavage of peptides on polyacrylamide gels[26] by using different chemicals, such as, for example, the cleavage of Asp-Pro bonds by formic acid. In α tubulin, there is only one Asp-Pro bond, at position 306-307, while in β tubulin there are two, at positions 31-32 and 306-307 (Ponstingl's numeration (1)). Using this method, the approximate cleavage sites of subtilisin on tubulin subunits,[27,28] *S. aureus* V8 protease on β-subunit, proteinase K on α- and β-subunits and chymotrypsin on the β-subunit, have been mapped.[29,30]

A second procedure takes advantage of the existence of two enzymes, one which specifically adds a tyrosine residue to the C-terminal of α-tubulin,[31,32] and the other which specifically removes it.[33] This process allows labeling of the carboxy terminal of α-tubulin with (^{3}H)- or (^{14}C)-tyrosine. The fragments produced by trypsin or subtilisin[27,34,35] digestion of α-tubulin were first mapped using this method. The third method involves the use of the dye known as "stains-all". Those proteins or protein regions which bind calcium, are phosphorylated, or acidic, stain blue.[36] When "stains all" is added other proteins or protein regions are stained pink, or yellow if they contain attached lipids.[36] Only the last 40 aa of both tubulin subunits stain blue with this dye, while the rest of the protein is pink-stained.[9,37] This characteristic has allowed the location of all proteolytic fragments contained in the carboxy terminal. It has been used to locate the proteolytic fragments produced by other proteases such as pronase or papain.[37]

The specific assignment of the cleaved residues was achieved by sequencing certain residues from the amino[6] or carboxy terminal[38] amino acids of the proteolytic fragments. These data indicate that chymotrypsin cleaves brain β-tubulin at residues Tyr281-Arg282[6] and probably at residues Phe_{446}-Glu_{447} (Serrano, unpublished results). Subtilisin cleaves both α- and β-subunits between residues Glu_{417}-Phe_{418}.[38] Trypsin cleaves α-tubulin between residues Arg_{339}-Thr_{340},[6] producing two fragments of 35 and 8 kDa. However, when tubulin is digested with trypsin for shorter periods or under other experimental conditions,[34] a 20-kDa fragment containing the C-terminal of the α-subunit also appears,[29] which is equivalent

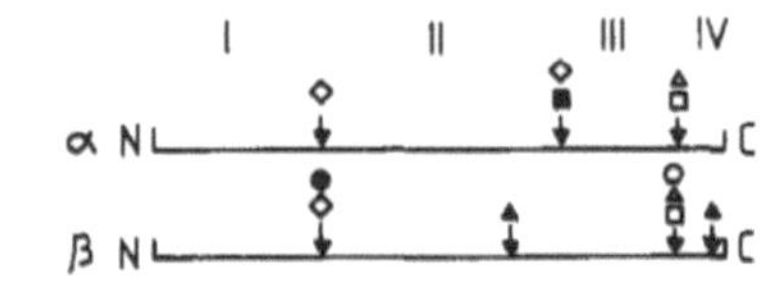

FIGURE 1. Structural map of α- and β-tubulin subunits obtained by limited proteolysis. The points for the cleavage of tubulin by trypsin (■), chymotrypsin (▲), subtilisin (□), pronase (◇), *S. aureus* V8 protease (○), and proteinase K (△) are indicated. Regions referred to as I, II, III, and IV indicate putative functional domains of the molecule.

to that obtained by clostripain digestion of α-tubulin.[29] Since clostripain only cleaves at Arg residues, the formic cleavage of the trypsin tubulin α fragment of 35 kDa produces a new fragment of 33 kDa,[30] (the Asp-Pro bond on α-tubulin is located at residues 306-307), it can be concluded that trypsin also cleaves α tubulin at Arg_{320}-Gly_{321}. When tubulin is incubated with colchicine prior to trypsin digestion, another point of cleavage[34] at Arg 390 appears.[7] Finally, Maccioni and Seeds have suggested that using charcoal extracted tubulin deprived of its bound nucleotide, chymotrypsin also cleaves β- and α-subunits probably at residues Phe_{133}-Glu_{134}.[39]

In Figure 1, a map of the main cleavage sites of the proteases on nonpolymerized brain tubulin is shown. This map is only applicable, however, to brain tubulin, since tubulin from other sources, such as chicken erythrocyte, may have distinct cleavage map (Serrano et al., unpublished results).

IV. INTERACTION BETWEEN TUBULIN SUBUNITS

The results obtained from the limited proteolysis of tubulin with the proteases described above are in perfect agreement with the structural predictions derived from tubulin homologies with other nucleotide binding proteins. Since proteases cleave native proteins on regions of high mobility[40] (regions which are usually located between protein domains), these results would indicate the existence of four domains on tubulin subunits: amino acids 1-133, 134-320 or 339, 340-417, and 418 to the C-terminal. In β-tubulin the existence of a cleavage site at amino acids Tyr_{281}-Arg_{282} and the resistance of its C-terminal region (aa 320-418) to proteolysis suggests that at least part of this region is implicated in the interaction between α- and β-subunits. This is consistent with crosslinking assays of trypsin or chymotrypsin digested tubulin.[6,30,41] These experiments used different crosslinkers with different lengths,[30] suggesting that the interaction between α- and β-subunit occurs between aa 1-320 of α-subunit and 284-417 of the β-subunit[6,30] and probably involves the regions around amino acids 320-340 of the β-subunit.[7]

In taxol-polymerized microtubules, the digestion of tubulin α-subunits by trypsin or chymotrypsin is prevented,[25,42] while substilisin digestion of tubulin at Phe_{418}-Glu_{419} in both subunits, is not prevented by microtubule polymerization.[43] Moreover, the last 40 amino acids of each subunit are exposed and can be detached from the polymer by subtilisin digestion.

Monoclonal or polyclonal antibodies recognize discrete epitopes on both subunits and are able to recognize polymerized tubulin, together with limited proteolysis, to determine the exposed tubulin regions on the assembled microtubule. The epitopes recognized by these antibodies on α- and β-tubulin subunits have been mapped using limited proteolysis with different enzymes[44,45] or after exhaustive proteolysis.[46] There are some discrepancies between data obtained from these two approaches. After limited proteolysis, monoclonal antibody MD1A (Amersham) recognizes an epitope between aa 339-418 of the α-subunit, while monoclonal antibody DM1B (Amersham) recognizes an epitope in the same region in the β-subunit.[45] Monoclonal antibody YolI/34 recognizes an epitope between aa 1-417 of the α subunit, and does not recognize any other proteolytic fragment (aa 1-320, 320-417, 1-135, 136-417).[45] Polyclonal antibody CAABCO recognizes an epitope located between aa 129-150 of both subunits.[45] When the reaction of some of these antibodies was studied on fragments obtained by exhaustive tryptic digestion, or CNBr fragments,[46] it was found that antibody DMIA recognizes a region around aa 426-451 of the α subunit, antibody MD1B recognizes a region around aa 416-430 of the β-subunit and antibody Yol1/34 recognizes an epitope comprising aa 414-422 of α-subunit.[46] These discrepancies could be explained in the case of DM1A and DM1B antibodies by the very low reactivity found against the smaller peptides described,[46] compared to that of intact α- and β-subunits or subtilisin digested tubulin subunits (aa 1-417 or tryptic or chymotryptic fragments (aa 339-453 and 282-455 of α- and β-subunits, respectively.[45] In this case, it is possible that the epitope recognized by these antibodies has been partially destroyed by exhaustive digestion, and that the remaining portion (aa 426-451 and 416-430) may be recognized by the antibody with lower affinity. In the case of the Yol 1/34 antibody, there are no real discrepancies, as this antibody recognizes a peptide including aa 403-422 of the α-subunit[46] and subtilisin digested tubulin (aa 1-417) with high affinity.[45] The lack of recognition of the tryptic fragment of α-tubulin (aa 340-453) could be due to a conformational rearrangement which masks the epitope after renaturation in nitrocellulose sheets. More recently the epitope recognized by another monoclonal antibody in α-tubulin has been determined and it corresponds to a sequence located at the amino terminal region.[47]

Limited proteolysis has been used, not only to determine structural features of the tubulin molecule, but also to analyze functional domains of tubulin involved in the interaction with molecules such as colchicine, MAPs, GTP, and calcium or in the regulation of tubulin polymerization, or in microtubule morphology.

V. COLCHICINE BINDING REGION

The first interaction studied was that of colchicine with tubulin. Cochicine binds to one site in tubulin with high affinity, Kd = 5 μM.[48] The binding of colchicine to this high-affinity site renders tubulin incapable of being incorporated into the polymers.[48,49] Using cross-linkable colchicine derivatives, the binding site has been alternatively assigned either to the α[50] or to α- and β-subunits.[51,52] However the colchicine derivative used to assign the colchicine binding site to tubulin was bromocolchicine, a very potent alkylating agent.[52] Thus, the results of studies using this compound must be analyzed with caution. Limited proteolysis of tubulin with trypsin, has shown that the high affinity colchicine site is located between aa 339-390 of the α-subunit.[7,34,53] Recently, by studying the inhibition of cross-linking between Cys_{239} and Cys_{354} of β-tubulin by colchicine, Little and Ludueña[54] have demonstrated that part of the colchicine binding site may also be located on β-tubulin. However, the studies of Deinum and Lincoln using spin label titration and paramagnetic resonance[55] suggest that the colchicine binding site is not near the GTP, Mg^{2+} and Zn^{2+} binding sites nor the reactive cysteines of α- and β-subunits and results suggest that colchicine-induced conformational change of tubulin prevents cysteine cross-linking.

VI. MICROTUBULE-ASSOCIATED PROTEIN BINDING SITES

The tubulin binding site for different microtubule-associated proteins was also analyzed by limited proteolysis of tubulin with subtilisin.[28,35,56] The results obtained indicate that the carboxy-terminal domain of both α- and β-subunits (aa 418-C-terminal) are implicated at the interaction with tau, MAP_2 and possibly MAP_1. Littauer et al.[57] have confirmed this result, at least for the β-subunit, and have described another region near the amino terminal of the α-subunit (aa 1-75) which may be involved in the interaction with tau factor. They have also shown that aa 444-450 (Postingl's numeration (1)) of the β-subunit are critical for MAP_2 and tau interactions.[57]

VII. GTP BINDING SITE

Since the discovery that tubulin is a GTPase protein, a big number of studies on the tubulin GTP binding site have appeared. The majority of this work used modified GTP molecules capable of covalent binding to the tubulin dimer. In this review we will deal only with those reports which indicate which subunit binds the exchangeable GTP or which have localized the tubulin residue implicated in the interaction with GTP. The tubulin dimer contains one nonexchangeable nonhydrolizable molecule of GTP and can bind another molecule of either GTP or GDP (see the chapters of Engelborghs and Hamel and References 58 to 63). It has been suggested that the exchangeable GTP is bound to the β-subunit, while the nonexchangeable GTP is bound to the α-subunit.[64,65] By nuclear magnetic resonance analysis it has been shown that the exchangeable GTP is complexed to a Mg^{2+} molecule[66] as is the case for other GTP binding proteins.[67] The results of Ludueña et al.[19] indicate that the binding site of GTP is affected by the crosslinking of cysteines 12 and 201 or 211 of the β-subunit, suggesting that the GTP binding site could comprise the cleft that contains these two cysteines. Cross-linking studies using modified GTP molecules or UV irradiation indicate that the ribose moiety of GTP could bind to Lys 105 or 124 of α- and β-subunits,[68] the guanine base could interact with some residues around Tyr 281 of the β-subunit[69] and with some residues present in the first 90 residues of β-subunit.[70]

Increasing knowledge about the structure and properties of nucleotide-binding proteins, has allowed a search of the α- and β-tubulin sequences for homologies with these proteins. The first homologies found comprised residues 58-80 and 135-157 of both subunits. The first was homologous to a region of other GTP-binding proteins implicated in the interaction with the guanine base,[71] while the second is homologous to the so-called "phosphate binding loop" which seems to be implicated in the interaction with the α- and β-phosphates of nucleotides in other proteins.[1,71] It has recently been found that the sequences of α- and β-subunits containing residues 177-183 and 240-246 were homologous to regions of other nucleotide binding proteins implicated, respectively, in the interaction with the ribose and the guanine-base.[6] More recently Sternlicht et al.[72] published a report indicating that tubulin contains the three characteristic sequence elements conserved in all known GTP-binding proteins,[73] and proposed a three-dimensional model for the tubulin GTP-binding domains.[6] However this model may be the matter of discussion, as the method used to generate these three conserved elements included the reversal of tubulin amino acid sequences, i.e., DAKN → NKAD or KGHYTEG → GETYHGK. On the other hand the authors who found the conserved fragment sequence for GTP binding proteins, conclude that such a fragment was not found in tubulin.[73] Finally, there are two other regions of tubulin which are homologous to other nucleotide binding proteins. The first one comprises residues 196-217[74] and it is homologous to a region in ATPases and adenylate kinases[75] which at least in adenylate kinase, may be implicated in the magnesium.[75] The other tubulin region comprises residues 101-120 of both tubulin subunits, and is homologous to the "phosphate loop" of other GTP-binding proteins[76] and thus equivalent to the tubulin region comprising residues 135-157. A summary of these homologies is shown in Table 1. The presence of these six regions in the tubulin sequence homologous to other nucleotide binding proteins, together with experimental data, allowed certain groups to propose hypothetical three dimensional models for the GTP-binding domain of tubulin.[72,74,75,77] However, although there is certainly sequence homology between tubulin and other nucleotide binding proteins which may give us some information about its three-dimensional structure, a rigorous analysis of the different

TABLE 1
Similarity Between Nucleotide Binding Proteins and Tubulin

A

Tb-α	58	A	G	K	N	—	V	P	R	A	V	F	V	D	L	E	P	T	V	I	D	E	V	R	T
Tb-β	56	G	N	K	Y	—	V	P	R	A	I	L	V	D	L	E	P	G	T	M	D	S	V	R	S
EF-G	129	A	N	K	Y	K	V	P	R	I	A	F	V	N	K	M	D	R	M	G	A	N	F	L	K
P 21	104	K	D	S	D	D	V	P	M	V	L	V	G	N	K	C	D	L	A	A	R	T	V	E	S
									β																

B

Tb-α	101	N	N	Y	A	R	—	G	H	Y	T	I	G	K	E	I	I	D	L	V	L	D
Tb-β	99	N	N	W	A	K	—	G	H	Y	T	E	G	A	E	L	V	D	S	V	L	D
EF-Tu	12	V	N	V	G	T	I	G	H	V	D	H	G	K	T	T	L	T	A	A	I	T
EF-G	10	R	N	I	G	I	S	A	H	I	D	A	G	K	T	T	T	T	E	R	I	L
			β														α					

C

Tb-α	137	V	F	H	S	F	G	G	G	T	G	S	G	F	T	S	L	L	M	E	R	L
Tb-β	135	L	T	H	S	L	G	G	G	T	G	S	G	M	G	T	L	L	I	S	K	I
P 21		Y	K	L	V	V	V	G	A	G	G	V	G	K	S	A	L	T	I	Q	L	I
LDH	193	S	T	C	A	V	F	G	L	G	G	V	G	L	S	V	I	M	G	K	A	A
Ad. Kin	9	K	I	I	F	V	V	G	G	P	G	S	G	K	G	T	Q	C	E	K	I	V
Bov. ATPasal	152	K	I	G	L	F	—	G	G	A	G	V	G	K	T	V	F	I	M	E	L	I
				β													α					

D

Tb-α	176	Q	V	S	T	A	V	V	E	P	Y
Tb-β	174	K	V	S	D	T	V	V	E	P	Y
HB	25	G	I	D	N	V	I	L	E	R	Q
GR	43	G	A	R	A	A	V	V	E	S	H
GPDH	38	A	Q	V	V	A	V	N	D	P	F
						β					

E

Tb-α	196–217	E	H	S	D	C	A	F	M	V	D	N	E	A	I	Y	D	I	C	R	R	N	L
Tb-β	194–215	E	N	T	D	E	T	Y	C	I	D	N	E	A	L	Y	D	I	C	F	R	T	L
ATPase-β	234	E	G	R	D	V	L	L	F	V	D	N	I	Y	R	Y	T	L	A	G	T	E	V
ATPase-α	272	R	G	E	D	A	L	I	I	Y	D	D	L	S	K	Q	A	V	A	Y	R	Q	I
Ad. Kin.	109	I	G	Q	P	T	L	L	L	Y	V	D	A	G	P	E	T	M	T	K	R	L	L
									β										α				

F

Tb-α	239	T	A	S	L	R	F	D	G	A	L
Tb-β	237	T	T	C	L	R	F	D	G	Q	L
GDH	263	D	F	S	F	E	V	I	G	R	L
GR	150	P	H	I	L	I	A	T	G	G	M
HB	153	D	Y	I	A	G	C	D	G	F	H
					β						

Note: Tb: tubulin, EF-G: elongation factor G, P 21: ras protein, EF-Tu: elongation factor Tu, LDH: lactato dehydrogenase, GR: glutathione reductase, GPDH: glyceraldehyde 3-phosphate dehydrogenase, and GDH: glutamato dehydrogenase.

TABLE 2
Comparison between the Fingerprint Sequences of the Supersecondary Structures that Contains the Phosphate-Binding Loop of Several Nucleotide-BindingProteins and Tubulin

	β-Sheet	Loop	α-Helix
GTP	p h — h h G	— — — —	G K T/S T/S h — — h — h
	(T) (S) (A)		(T)
Kinases	— — — — h G	— G — F/Y	G — h h — — — h — —
Dehydrog.	p h — h — G	— G — —	G — — — h — — h — —
ATPases	— — — h h G	— — G —	G K T h — h — — h h
	(T)		(G) (T) (T)
Tub 101	— p h — — G	— — — — —	G K/A — h h — — h h —
137	h — — h S G	— — G —	G h T T/S h h — — p h
			(G)

Note: p: polar and h: hydrophobic

families of nucleotide-binding proteins results in the appearance of certain "fingerprint" sequences typical of each family (Serrano and Valencia, unpublished) (Table 2) which are not found in tubulin. Moreover in all the nucleotide-binding proteins whose three-dimensional structure is known, there is no sequence located before the "Rossman" fold comprising the "phosphate loop"[78] able to interact directly with the bound nucleotide. However, the experimental data[70] and homology data from tubulin research[71] suggest that the region comprising residues 58-80 could be implicated in the nucleotide binding.

In summary, tubulin does not seem to belong to any known family of nucleotide-binding proteins, and consequently, the similarities with these families must be the result of a convergent evolutionary process. Although it may be suggested that the GTP-binding domain of tubulin is located between the first 300 amino acid residues, in light of the recent experimental data, it is not yet possible to make a three-dimensional model. Finally, recent works indicating that tubulin could bind ATP[79] suggest that there are two possible β-sheet-loop-α-helix supersecondary structures composing a "phosphate loop". In any case the three-dimensional structure of tubulin will have to be known in order to be able to describe the tubulin-nucleotide interaction.*

VIII. CALCIUM BINDING SITE

The last factor whose tubulin binding site has been studied by limited proteolysis is calcium a cation that binds to tubulin affecting its assembly.[9,10,80-82] The results obtained indicate that both α- and β-subunits contain a high-affinity calcium binding site located between aa 418 and the carboxy terminal. Comparison of the amino acid sequence of this tubulin region with other calcium-binding proteins of the calmodulin family reflect a certain degree of similarity.[9] In erythrocytes, the binding site in the β-subunit appears to be missing (L. Serrano et al., submitted for publication).

* Recently, the binding site of the guanine moiety of GTP on β tubulin has been located on the residues 63 to 77. (Linse and Mandelkow, *J. Biol. Chem.*, 263, 15205, 1989.)

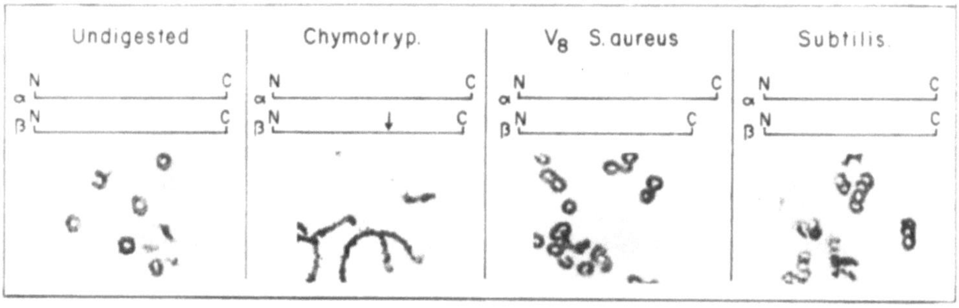

FIGURE 2. Changes in the morphology of tubulin polymers after removal of defined regions by limited proteolysis. Shown are the polymers obtained from undigested tubulin and tubulin cleaved with chymotrypsin, *S. aureus* V8 protease, and subtilisin.

IX. LIMITED PROTEOLYSIS AND TUBULIN POLYMERIZATION (FIGURE 2)

Since the most important feature of tubulin is its capacity to polymerize into microtubules, it is of interest to analyze the effect of limited proteolysis on its assembly capacity. To polymerize, tubulin requires a minimal concentration which is lower in the presence of MAPs.[83-86] The effect of limited trypsin or chymotrypsin proteolysis on the capacity of tubulin to polymerize has been argued. While some groups[6,25] have found only a small negative effect, others[29,39] find that digestion with either of these proteases rapidly diminishes tubulin polymerization capacity. The use of differing polymerization conditions or different temperature (15 to 30°C)[25] for proteolysis, may contribute to this variation. Cleavage of tubulin at aa 320 of α- or 281 of the β-subunit, may increase the thermal motion of tubulin, so that denaturalization at higher digestion temperatures may occur more quickly than at lower temperatures. This hypothesis is also supported by the fact that cleaved tubulin can be cycled only once, and it is not subsequently able to polymerize.[7] The fact that proteolytic fragments of the α-subunit[6] can be incorporated into the polymer in the first cycle may contradict the observed gel filtration separation of the α-trypsin fragment (aa 339-455) from the rest of the molecule.[34] However, those results may be explained as this separation is achieved when tubulin (1) is incubated with colchicine prior to trypsin digestion, (2) the digestion is performed at 30°C, and (3) digested tubulin is diluted to a concentration of less than 0.6 mg/ml in the chromatographic procedure. Under these conditions, only a small percentage of the trypsin α-C-terminal fragment is separated from the rest of the molecule.[7]

X. ROLE OF THE CARBOXY TERMINAL ON TUBULIN POLYMERIZATION

The stimulation of the polymerization capacity of tubulin upon subtilisin digestion which results in the removal of aa 418 to the C-terminal of both subunits has been shown.[9,43,56,87,88] Since the C-terminal domain of tubulin is implicated in the interaction with MAPs and calcium,[9,35,56] and the removal of this domain by subtilisin decreases tubulin's critical concentration for polymerization to a concentration similar to that of tubulin in the presence of MAPs, it can be suggested that the C-terminal domain of tubulin is a regulatory domain which promotes tubulin polymerization, when it is (1) neutralized by MAPs or polycations or (2) it is removed by proteolysis. However, polymerization diminishes when tubulin binds calcium through this domain. Bhattacharyya et al.[88] have shown in slightly different proteolytic conditions, that subtilisin cleaves first the β-subunit (βs) producing an αβs dimer which forms aggregates which are insensitive to cold or to the presence of phodophylotoxin, and are polymerization independent of GTP or GDP. A further digestion of the α subunit present in the aggregates renders an αsβs dimer capable of being subjected to additional polymerization-depolymerization cycles. These polymers are sensitive to cold and phodophylotoxin and are assembled only in the presence of GTP.

There are two other proteases, proteinase K and *S. aureus* V8 protease, that preferentially (proteinase K) or exclusively (*S. aureus* V8) cleave β-tubulin at a position similar to that of subtilisin.[29] The critical concentration of the tubulin digested by these two proteases is higher than that found for subtilisin-digested tubulin, but much lower than that of undigested purified tubulin.[29]

However, the data which indicate that *S. aureus* V8 protease-digested tubulin (αβs) does not form aggregates like those found by Bhataccharyya et al.[88] for partially subtilisin-digested tubulin (αβs), suggests that the exact point of cleavage on the β-subunit by *S. aureus* V8 protease and subtilisin may differ.

A. SITE-DIRECTED MUTAGENESIS AND TUBULIN POLYMERIZATION

Recently, the possibility of site-directed mutagenesis of cloned genes and posterior transfection of eukaryotic cells with these modified genes has allowed the analysis of the role of certain residues or regions in protein structure and function. This approach has been used by the groups of Salomon and Botstein for both α-[89] and β-tubulin subunits.[90] A series of deletion and insertion experiments showed that the amino-terminal region of yeast α-tubulin, comprising residues 33 to 50 could tolerate insertions of 3 to 17 aa and is still able to carry all the normal functions of this α-tubulin in yeast. These authors have also found that insertions of 2 to 4 residues on positions, of a chimeric chicken-yeast gene[91] result either in no expression of the protein, due probably to incorrect folding, or in no incorporation into microtubules. However, the insertion of four residues at position 438 or the deletion of residues 430 to 444 of the same gene, resulted in the incorporation of the mutated proteins on microtubules.

The construction of four different chimeric genes in which the yeast β-tubulin sequences extended further into the amino-terminal region of the chicken β-2 tubulin gene,[91] also gives some insight on tubulin structure and function. The yeast β-tubulin gene region comprising residues 344 to the C-terminal differs in 31 residues from its chicken counterpart, including nonconservative changes in one of the most conserved regions in β-tubulin genes (residues 375 to 430). When this yeast sequence is included in the chimeric gene together with the chicken sequences containing the remaining β-subunit molecule, the product is stable and polymerizes as well as the nonchimeric products of β-tubulin genes. However, when the yeast sequence is extended until residue 282 (47 aa differences), the stability of the chimeric proteins decreases at least fivefold, although it polymerizes normally. When the yeast sequence is extended to residue 201 (64 amino acid differences), the instability of the chimeric protein is so high that it is almost undetectable and the small amount remaining does not polymerize. Finally, when the chimeric gene contains the yeast sequences comprising residues 201 to 282 and 344 to the C-terminus, the stability of the protein is restored to the level of the chimeric gene containing residues 282 to the C-terminus of the yeast sequence, but its polymerization capacity is not. What conclusions can be gathered from these results? First, that there are two tubulin regions, comprising residues 33 to 50 of the α-subunit and residues 430 to the C-terminus of the β-subunit, which allow a high level of variability, including insertions and deletions. This does not mean that they cannot have a functional *in vivo* role but rather that they seem to be unimportant in the correct folding and the polymerization capacity of tubulin. Second, for the rest of the tubulin molecule, there is a high conformational pressure so that even small insertions completely disrupt β-tubulin polymerization capacity. These results are in good agreement with those obtained by limited proteolysis, except in one case. This apparent exception is the β-tubulin mutant P2,[91] in which four residues were inserted close to the subtilisin cleavage position. This mutant protein showed decreased protein stability and was unable to polymerize. This exception is only apparent, however, since it has been demonstrated that, although the carboxy terminal domain of both α- and β-tubulin subunits (residues 408 to the C-terminus) is not implicated in tubulin-tubulin interactions, it is able to regulate tubulin polymerization activity as an inhibitor unless it is neutralized by MAPs or cleaved out by proteolysis.[27] It is thus not difficult to imagine that the insertion of four residues into a highly conserved region (residues 375 to 430) would produce a conformational change capable of rendering the molecule unable to polymerize.

B. REGIONS INVOLVED IN MICROTUBULE SHAPE

Another important feature of tubulin is the morphology of the assembled polymers. In the presence of MAPs and subtilisin *in vitro*, forms cylindrical tubes with 13 to 14 protofilaments. In the absence of MAPs and in the presence of 10% DMSO, hooked (with few hooks) and normal microtubules can be observed.[92] Digestion of tubulin with proteases which cleave at or near *Arg* 320 of the α-subunit or at Tyr 281 of the β subunit does not affect the morphology of the polymer.[29] However, cleavage at the last 40 to 50 aa of both subunits result in changes of microtubule morphology.[29]

Deletion of approximately 10 residues from the C-terminal of the β-subunit produces a different arrangement of tubulin subunits into assembled polymers, resulting in the formation of sheets with no curvature and presenting a great number of protofilaments (>20); these polymers resembled those tubulin sheets induced by zinc.[29] Deletion of approximately 40 aa from the carboxy terminal of the β-subunit results in a moderate increase in the number of hooks, which can grow in opposite directions on the closed microtubule. Deletion of approximately 40 aa from the α- and β-subunit results in an important increase in lateral interactions between polymers, with the appearance of a large proportion of hooked polymers, with many interrelated hooks (Figure 2).

When the removal of the carboxy terminal of α- and β-subunits was equivalent (subtilisin digestion), all the hooks of a single polymer grew in the same direction. However, a preferential removal of the carboxy-terminal of the β-subunit (proteinase K digestion), resulted in hooks growing in opposite directions, yielding disorganized structures (Figure 2).

These results indicate that the carboxy terminal region of tubulin is a regulatory domain of tubulin polymerization and microtubule shape which controls lateral interactions between polymers, curvature, arrangement of the subunits on the polymer and directionality of the interactions between polymers.[93]

Recently, the sequence analysis of a Drosophila β-tubulin mutant, which polymerizes *in vivo* producing S-shaped microtubules similar to the ones obtained by digestion with *S. aureus* V8 protease of tubulin β-subunit, has shown that responsible for this abnormal shape in Drosophila microtubules was a single change of Glu 288 for Lys.[94] These data, as well as those obtained after tubulin digestion with proteases which cleave the C-terminal, and the fact that, upon subtilisin digestion of tubulin, there is a conformational change which affects the colchicine-binding site,[95] indicate that the carboxy terminal of tubulin probably controls microtubule shape in an indirect way.

C. STRUCTURE OF TUBULIN-CARBOXY TERMINAL DOMAIN

As is clear from the experimental data, the final fifty amino acid residues of both the α- and β-subunits seem to be extremely important for tubulin function. Consequently, it will be of interest to know more about the possible structure of this domain. The experimental data indicate that (1) it binds calcium and MAPs, (2) by circular dichroism and in certain solvents, it has a high α-helix content, it is subject to at least two posttranslational modi-

TABLE 3
Heptapeptide Sequence Typical of an α–α Helix-Helix Core-Core Interaction Present in the Carboxy-Terminal of α and β Tubulin

α							β						
a	b	c	d	e	f	g	a	b	c	d	e	f	g
F	S	E	A	R	E	D	F	T	E	A	E	S	N
M	A	A	L	E	K	D	M	N	D	L	V	S	E
Y	E	E	V	G	V	D	Y	Q	Q	Y	Q	D	A
h	p	p	h	p	p	p	h	p	p	h	p	p	p

Note: h: hydrophobic and p: polar.

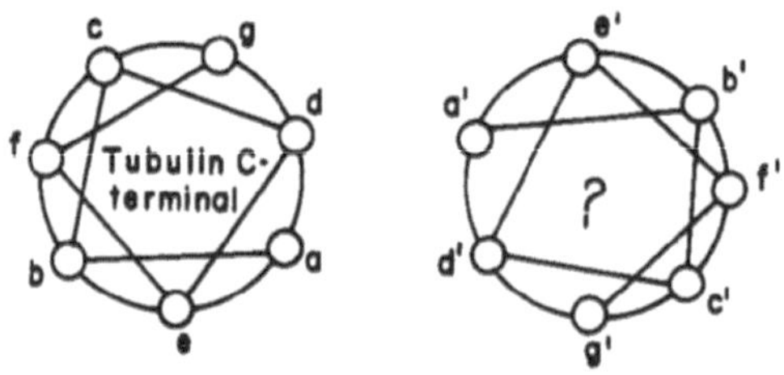

FIGURE 3. Diagrammatic representation of the interaction between a heptapeptide of the tubulin carboxy terminus and a corresponding heptapeptide in the tubulin-binding domain of a microtubule-associated protein.

fications and is exposed to the solvent. By comparison of the known sequences of both α- and β-tubulin subunits, it appears to be composed of two regions, one comprising residues 410 to 438 which is one of the most conserved regions of tubulin (conserved from yeast to man), and another from residue 439 to the C-terminus, which is one of the most variable, though always acidic, regions of the molecule. Secondary structure prediction by a combination of methods indicates that the first region is composed mainly of α-helix with a short sequence stretch between residues 422 to 427 which also has a high probability of being a "loop" or random-coil[9] and the second region which must be mainly in the so-called "random coil" conformation. If one looks for sequence fingerprints characteristic of supersecondary structures, one finds that the first region has the typical repeated heptapeptide sequence, characteristic of and α-helix which binds to other α-helices in an α-α coil-coil conformation (Table 3), similar to that found in intermediate filaments,[96] myosin,[97] or fibrinogen.[98] As it seems that the carboxy-terminal domain of tubulin does not interact closely with the rest of the molecule (it is removed by limited proteolysis), it may thus interact with other molecules, probably MAPs (Figure 3).

Considering this data and the fact that calcium binds to the C-terminal domain of tubulin, it can be suggested that this region may be composed of an α-helix rod of 21 residues, followed by a random coil sequence; in the presence of calcium this rod would probably be disrupted in the middle to allow the formation of a loop to support calcium binding. In the absence of calcium and in the presence of positive-changed molecules, the rod remains interact. This model may explain the fact that MAPs may abolish calcium sensitivity of tubulin.

XI. POSTTRANSLATIONAL MODIFICATIONS

As mentioned in the chapter of Lewis and Cowan, tubulin microheterogeneity has been found at the gene level for both α- and β-subunits. This initial genetic microheterogeneity may be amplified by posttranslational modifications, some common to other proteins such as phosphorylation and others specific to tubulin such as tyrosination-detyrosination.

A. TYROSINATION-DETYROSINATION

This was the first posttranslational modification to be described for tubulin and it was described for its α-subunit. The first report of this modification was published by Barra et al.,[99] followed by the work of Flavin and co-workers.[100] Tubulin α-subunit is synthesized with a tyrosine at its C-terminal in almost all the genes[101] except one ubiquous in monkey[102] and mouse[101] and a chicken testis specific[103] sequenced in vertebrates while in invertebrates. Although the majority of the α-genes codify for a tyrosine at the C-terminal there are more exceptions and some α-genes could be found that do not code for this tyrosine,[104] or they code for a phenylalanine[105,106] or even for a methionine[107] instead of the tyrosine. This C-terminal tyrosine can be removed by an specific carboxypeptidase[108,109] or added back by a specific tyrosine ligase.[32,100,110,111] Both enzymes have been purified[109,112] and their activity on soluble and polymerized tubulin assayed. Tubulin carboxypeptidase copurifies with microtubules[32] and prefers polymerized tubulin to monomeric soluble tubulin.[113,114] Tyrosine ligase interacts preferentially with monomeric tubulin and more specifically with a region of β-tubulin subunit hindered on microtubules.[115] This explains how it catalyzes[99,110] the incorporation of the tyrosine on detyrosinated monomeric α-tubulin rather than on polymerized tubulin.[32,100,116] The ligase activity has been found in various vertebrates,[118,119] invertebrates,[120] and protozoans.[121] In adult vertebrate brain tissue the α-tubulin subunit contains 0.3 mol of tyrosine per mol of tubulin[117] but this ratio varies at different stages of brain development[122-124] specifically in the period of axonal development[125] as well as in early general development.[124] The existence of a monoclonal (YL1/2)[44] and of a polyclonal[126] antibody against tyrosinated and detyrosinated α-tubulin[126] has allowed the first insight into the function of this posttranslational modification. The coexistence of two cytoplasmic and mitotic[126] populations of microtubules, containing detyrosinated (Glu-Mt) or tyrosinated (Tyr-Mt) microtubule, has been found in the cells of organisms ranging from human to protozoa.[126-130] The proportion of each population can change during the cell cycle,[127,131] cell differentiation,[130,132,133] cell aging,[120] or as a response to external stimuli.[134,135] However, detailed analysis by electron microscopy has revealed that generally a single population of either tyrosinated or detyrosinated tubulin cannot be found in a unique microtubule *in vivo*.[126] The analysis of these two populations using different methods (drugs, cold depolymerization, microinjection, etc.) has revealed that Glu-microtubules are generated from Tyr microtubules and are consequently older than Tyr-microtubules,[137] and that Glu-microtubules are generally more stable and less dynamic (with longer half-lives) than Tyr-microtubules.[137-139] This *in vivo* results confirms the *in vitro* hypothesis that the tyrosine ligase activity acts over monomeric tubulin while the carboxypeptidase acts on polymerized tubulin. However the *in vivo* properties of both populations differ while it is well known that the *in vitro* behavior of both tubulins is similar,[116,140] except for subtle differences in the association of the well-known brain microtubule associated proteins.[140] The *in vivo* data indicate that detyrosination occurs equally in every cell microtubule of a cell, thus ruling out the idea that detyrosination is itself a mechanism to stabilize microtubules.[139] Clearly nothing is yet known about the function of this posttranslational modification of tubulin, but a mechanism which is retained from protozoa to man, clearly suggests that it might play a fundamental role in the *in vivo* function of tubulin.

B. ACETYLATION

The acetylation of the α-subunit of tubulin was first found in *Chlamydomonas*[141] in the processes of flagella regeneration. During flagella regeneration, the α-subunit of tubulin is acetylated in the flagellar matrix,[142-144] a process that is reversed during flagella regeneration.[145] As in the case of tubulin tyrosination, the existence of a monoclonal antibody which specifically recognizes the acetylated form of the α-subunit,[146] has facilitated the study of this posttranslational modification. The α-tubulin subunit is acetylated not only in the flagella of protozoa[141-150] but in cilia and flagella of many different organisms ranging from protozoa to man.[146,151] Acetylated α-tubulin has been found generally in very stable microtubules.[142-144] Only in the mitotic spindle there are not found acetylated microtubules.[152-154] Acetylated microtubules comprise between 10 to 40% of the total cytoplasmic microtubule population, with some exceptions in which few or none are found.[155] In general the cytoplasmic acetylated microtubules are curlier and wavier, are concentrated near the cell center, and are more stable and less dynamic than nonacetylated microtubules.[129] However as in the case of detyrosination, acetylation neither *in vitro*[155] or *in vivo*[129] seems to be the cause of microtubule stabilization, but rather a consequence of this process.[129] Consequently, although stable microtubules are acetylated, untyrosinated, or both, the distribution of both acetylated and untyrosinated microtubules is not always identical.[129]

In vitro analysis of α-tubulin acetylation shows that the specific enzyme which acetylates tubulin, has greater affinity for the polymeric form of α-tubulin than for the monomeric form,[155] although it is able to acetylate the monomeric form to a significant extent. That acetylated tubulin is found *in vivo* in the polymeric form,[129] and it implies that this enzyme must be preferentially associated to microtubules, while the deacetylase must be soluble.[155]

Purified chymotryptic fragments from exhaustive digestions of acetylated tubulin are recognized by the antibody against acetylated α-tubulin. It has been shown that the acetylated residue is a lysine, on α-subunit located at position 40 of the *Chlamydomonas* α-tubulin sequence.[156] This position is conserved in all α-tubulins sequenced, except in those from *Saccharomyces pombe, S. cerevisiae,*[157,158] a testis-specific isoform for chicken[103] and an egg and embryo-specific isoform from Drosophila.[106] Moreover, this lysine is located on one of the regions of α-subunit, which are more different of β-subunit (aã 34-49). Littauer et al.[57] have found that, besides the carboxy-terminal domain of tubulin, the region comprised of aã 1-75 of α-subunit appears to be involved in the interaction with tau factor. Thus, these results suggest that the acetylation of α-subunit might probably affect the interaction with tau or other associated.

C. PHOSPHORYLATION

Tubulin phosphorylation has been reported both *in vivo* and *in vitro*. Early *in vivo* studies showed that tubulin obtained from mammalian brain or embryo muscle cells was phosphorylated in its β-subunit, exclusively at a serine residue[159-162] probably located at the carboxy terminal.[162] However, both α- and β-tubulin subunits have been found phosphorylated in coated vesicles from brain.[163]

In neuroblastoma N115 cells, one isotubulin was phosphorylated exclusively at serine, when the neuroblastoma cells were induced to differentiate forming neurites. The phosphorylated subunit has been found mainly in the polymerized fraction.[164] These authors have also reported the presence of phosphorylated tubulin in other cell lines, of neural and nonneural origin.[164] Phosphorylation of tubulin *in vitro* has awakened a controversy since early experiments indicated that, although tubulin could be phosphorylated *in vitro* with an endogenous kinase, the specific activity was very low (0.1 to 0.3 ml phosphate per mole dimer), tryptic peptides were different from the *in vivo* labeled peptides, and phosphorylated tubulin was found mainly with aggregated tubulin.[165] These results suggest that phosphorylation of tubulin induced by the microtubule associated kinase mainly takes place on denatured protein.[166]

It has recently been shown that undenaturing tubulin is a good substrate for several kinases.[167-174] A calcium-calmodulin-dependent protein kinase is able to phosphorylate tubulin at both α- and β-subunits.[167,172,173] The β-subunit is phosphorylated at both serine and threonine residues, while the α-subunit is phosphorylated only at serine residues.[172] Analysis of tubulin phosphorylation using limited proteolysis has shown that it produces a conformational change on the tubulin molecule which results in a drastic inhibition of tubulin polymerization capacity,[168] as well as in an inhibition of the interaction of tubulin with MAPS.[167] In optimal phosphorylation conditions, 2.6 moles phosphate per milliliter of tubulin can be incorporated at the carboxy-terminal domain of α- and β-tubulin subunits (aā 418 to the end).[167]

Experiments with liposomes obtained from neutral lipids have demonstrated that phosphotubulin, phosphorylated with the above enzyme, enhance its incorporation into liposomes, and that this incorporation could be reversed by phosphatase treatment.[174] It is interesting to note that both proteins, one of the calcium-calmodulin kinase subunits and tubulin, are abundant in brain membranes[175] and that tubulin phosphorylated in the α- and β-subunits has been found associated to coated vesicles in brain.[163] Tubulin is also a substrate for the insulin receptor-associated kinase, which phosphorylates tubulin and other associated proteins exclusively on tyrosine residues.[168,170,171] Analysis of the polymerization capacity of tubulin following its phosphorylation with insulin receptor kinase shows the existence of two tubulin populations with different behaviors. One population, in which phosphatase is present mainly at the carboxy terminal tyrosine located in the α-subunit, is unable to polymerize; while the remaining phosphorylated tubulin (with no tyrosine in its carboxy terminal) showed and enhanced polymerization capacity.[168] This latter population incorporates phosphate at different residues of α- and β-tubulin subunits.[168] These *in vitro* results may be correlated with some *in vivo* observations showing changes in the microtubule lattice upon the addition of insulin to cultured cells.[176]

These data suggest that, at the local level within cells, phosphorylation of tubulin plays a very important role in reversibly destabilizing microtubules. A regulation system of phosphorylation-dephosphorylation might be important in the modulation of the different processes in which microtubules are involved.

However, none of these kinases seem to be responsible for the phosphorylation of a single β-isotubulin in neuroblastoma cells[164,177] or in brain.[159] It has been recently observed that casein-kinase II is able to phosphorylate β-tubulin in the same way as is seen *in vivo*, indicating that this kinase could be responsible for the *in vivo*, phosphorylation.[169] By exhaustive *S. aureus* V8 digestion of the *in vivo* phosphorylated β-tubulin it has been found that the phosphorylated residue is present in the minor neuronal specific β-isotubulin (Serrano et al., to be published and Reference 178). *In vitro* phosphorylation of tubulin by casein-kinase II,[169] as well as the *in vivo* phosphorylation data,[165,177] indicate that β-tubulin is preferentially phosphorylated in the polymeric form. As is the case with the other two posttranslational modifications, tubulin phosphorylation seems to be subsequent to microtubule stabilization, indicating that it is not the phosphorylation itself which stabilizes the microtubule. Nothing is known about the consequences of such a posttranslational modification.

In conclusion, current data on posttranslational tubulin modification suggest that detyrosination, acetylation, and phosphorylation seem to occur on the polymerized form of tubulin, and are generally associated with microtubule stabilization, either in cytoplasmic microtubules or upon differentiation of specific stable organelles. These modifications seem, however, to be responsible for stabilization, rather than a consequence of it. Although much work has been done on these modifications, little is known about their *in vivo* role.

REFERENCES

1. **Ponstingl, H., Krauhs, E., Little, M., Kempft, T., Hofer-Warbinek, R., and Ade, W.,** Aminoacid sequence of α and β tubulin from pig brain: heterogeneity and regional similarity to muscle proteins, *Cold Spring Harbor Symp. Quant. Biol.*, 46, 191, 1982.
2. **Ponstingl, H., Krauhs, E., Little, M., and Kempft, T.,** Complete amino acid sequence of α-tubulin from porcine brain, *Proc. Natl. Acad. Sci. U.S.A.*, 78, 2757, 1981.
3. **Krauhs, E. M., Little, J., Kempf, R., Hofer-Warbineck, W., Ade, W., and Ponstingl, H.,** Complete aminoacid sequence of β-tubulin from porcine brain, *Proc. Natl. Acad. Sci. U.S.A.*, 78, 4156, 1981.
4. **Valenzuela, P., Quiroga, N., Zaldivar, J., Rutter, W. J., Kirschner, M. W., and Cleveland, D. W.,** Nucleotide and corresponding aminoacid sequences encoded by α and β tubulin mRNAs, *Nature*, 289, 650, 1981.
5. **Sternberg, M. and Taylor, W. R.,** Modelling the ATP binding site of oncogene products, the epidermic growth factor receptor and related proteins, *FEBS Lett.*, 175, 387, 1984.
6. **Mandelkow, E. M., Hermann, M., and Rühl, U.,** Tubulin domains probed by limited proteolysis and subunit-specific antibodies, *J. Mol. Biol.*, 185, 311, 1985.
7. **Serrano, L.,** Dominios Estructurales y Funcionales del Dímero de Tubulina: Implicaciones en la Regulación de la Polimerización y Morfogénesis del Microtúbulo, doctoral thesis, University Autónoma de Madrid, 1985.
8. **Puszkin, E., Puszkin, S., and Aledorf, L. M.,** Colchicine binding protein from human platelets and its effect on muscle myosin and platelet myosin-like thrombosthrenin M, *J. Biol. Chem.*, 246, 271, 1971.
9. **Serrano, L., Valencia, A., Caballero, R., and Avila, J.,** Localization of the high affinity calcium binding site on tubulin molecule, *J. Biol. Chem.*, 261, 7076, 1986.
10. **Fong, K. C., Babitch, J. A., and Anthony, F. A.,** Calcium binding to tubulin, *Biochem. Biophys. Acta*, 77, 1210, 1988.
11. **Amos, L. A.,** Structure of microtubules, in *Microtubules*, Roberts, K. and Hyams, J. S., Eds., Academic Press, London, 1979.
12. **Crepau, R. H., McEwen, B., and Edelstein, S. J.,** Differences in α and β polypeptide chains of tubulin resolved by electron microscopy with image reconstruction, *Proc. Natl. Acad. Sci. U.S.A.*, 75, 5006, 1978.
13. **Beese, L., Stubbs, G., and Cohen, C.,** Microtubule structure at 18 Å resolution, *J. Mol. Biol.*, 194, 257, 1987.
14. **Beese, L., Strubbs, G., Thomas, J., and Cohen, C.,** Structure of microtubules with reduced hydration, *J. Mol. Biol.*, 196, 575, 1987.
15. **Maccioni, R. B., Serrano, L., and Avila, J.,** Structural and functional domains of tubulin, *Bioessays*, 2, 2165, 1985.
16. **Little, M. and Ludueña, R. F.,** Structural differences between brain β1 and β2 tubulins: implications for microtubule assembly and colchicine binding, *EMBO J.*, 4, 51, 1985.
17. **Palanivelu, P. and Ludueña, R. F.,** Interactions of the τ-tubulin-vinblastine complex with colchicine, podophyllotoxin and *N,N'*-ethylenebis(iodoacetamide), *J. Biol. Chem.*, 257, 6311, 1982.
18. **Little, M. and Ludueña, R. F.,** Location of two cysteines in brain β-tubulin that be crosslinked after removal of exchangeable GTP, *Biochem. Biophys. Acta*, 912, 28, 1987.
19. **Ludueña, R. F., Roach, M. C., Trcka, P. P., Mallevais, M. L., and MacRae, T.,** The effect of 2-(4-methyl-l-piperazinylmethyl)acrylophenone dihydrochloride on the alkylation of tubulin, *Arch. Biochem. Biophys.*, 255, 453, 1987.
20. **Roach, M. C. and Ludueña, R. F.,** Different effects of tubulin ligands on the intrachain cross-linking of β1-tubulin, *J. Biol. Chem.*, 259, 12063, 1984.
21. **Sherman, G., Rosenberry, T. L., and Sternlicht, H.,** Identification of lysine residues essential for microtubule assembly, *J. Biol. Chem.*, 258, 2148, 1983.
22. **Szasz, J., Yaffe, M. B., Elzinga, M., Blank, S. G., and Sternlicht, H.,** Microtubule assembly is dependent on a cluster of basic residues in α-tubulin, *Biochemistry*, 25, 4572, 1986.
23. **Yaffe, M. B., Levinson, B. S., Szasz, J., and Stren, H.,** Expression of a human α-tubulin: properties of the isolated subunit, *Biochemistry*, 27, 1869, 1988.
24. **Love, W., Millay, D., and Huston, J. S.,** Properties of disulphide linked tubulin purified on hydroxyapatite and its comparison with intact and dissociated microtubules, *Arch. Biochem. Biophys.*, 207, 300, 1981.
25. **Brown, H. R. and Erickson, H. P.,** Assembly of proteolitically cleved tubulin, *Arch. Biochem. Biophys.*, 220, 46, 1983.
26. **Sonderegger, P., Janssi, R., Gehring, H., Brunschweiler, K., and Christien, D.,** Peptide mapping of proteins bands from polyacrylamide gel electrophoresis by chemical cleavage in gel pieces, *Anal. Biochem.*, 122, 298, 1982.
27. **Serrano, L., de la Torre, J., Maccioni, R. B., and Avila, J.,** Involvement of the carboxy-terminal domain of tubulin in the regulation of its assembly, *Proc. Natl. Acad. Sci. U.S.A.*, 83, 7162, 1984.

28. **Serrano, L., Díez, J. C., Avila, J., and Maccioni, R. B.,** Controlled proteolysis of tubulin by subtilisin: structural changes and functional consequences, *J. Submicros. Cytol.*, 16, 55, 1984.
29. **Serrano, L., Wandosell, F., de la Torre, J., and Avila, J.,** Proteolytic modification of tubulin, *Methods in Enzymol.*, 134, 179, 1986.
30. **Serrano, L. and Avila, J.,** The interaction between subunits in the tubulin dimer, *Biochem. J.*, 230, 551, 1985.
31. **Flavin, M., Kobayashi, T., and Martensen, T. M.,** Tubulin-tyrosine ligase from brain, *Methods Cell Biol.*, 24, 257, 1982.
32. **Thompson, W. C.,** The cyclic tyrosination/detyrosination of α-tubulin, *Methods Cell Biol.*, 24, 235, 1982.
33. **Argañana, C. E., Barra, H. S., and Caputo, R.,** Release of ^{14}C-tyrosin from tubulin-^{14}C tyrosine by brain extract. Separation of a carboxy-peptidase from tubulin-tyrosine ligase, *Mol. Cell. Biochem.*, 19, 17, 1978.
34. **Serrano, L., Avila, J., and Maccioni, R. B.,** Limited proteolysis of tubulin and the localization of the binding sites for colchicine, *J. Biol. Chem.*, 259, 6607, 1984.
35. **Serrano, L., Avila, J., and Maccioni, R. B.,** Controlled proteolysis of tubulin by subtilisin: localization of the site for MAP_2 interaction, *Biochemistry*, 23, 4675, 1984.
36. **Campbell, K. P., MacLennan, D. H., and Jorgensen, A. D.,** Staining of the Ca^{+2}-binding proteins calsequestrin, calmodulin, troponin C and S-100 with the cationic carbocyanine dye "stains all", *J. Biol. Chem.*, 258, 11267, 1983.
37. **Serrano, L., Wandosell, F., Díez, J. C., and Avila, J.,** The combined use of limited proteolysis and differential peptide staining for protein characterization, *J. Biochem. Biophys. Methods*, 12, 281, 1986.
38. **Maccioni, R. B., Serrano, L., Avila, J., and Cann, J. R.,** Characterization and structural aspects of the enhanced assembly of tubulin after removal of its carboxy-terminal domain, *Eur. J. Biochem.*, 156, 375, 1986.
39. **Maccioni, R. B. and Seeds, N. W.,** Limited proteolysis of tubulin: nucleotide stabilizes an active conformation, *Biochemistry*, 22, 1567, 1983.
40. **Fontana, A., Fassina, G., Vita, C., Dalzoppo, D., Zamai, M., and Zambonin, M.,** Correlation between sites of limited proteolysis and segmental mobility. Termolysin, *Biochemistry*, 25, 1847, 1986.
41. **Kirschner, K. and Mandelkow, E. W.,** Tubulin domains responsible for assembly of dimers and protofilaments, *EMBO J.*, 4, 2397, 1985.
42. **Wandosell, F., Villanueva, N., Serrano, L., and Avila, J.,** Analysis of drug-tubulin interaction by trypsin cleavage: comparison for colchicine, podophyllotoxin, griseofulvin, vinblastine and taxol, *Comp. Biochem. Physiol.*, 85, 635, 1986.
43. **Sackett, L. D. and Wolff, J.,** Proteolysis of tubulin and the structure of the tubulin dimer, *J. Biol. Chem.*, 261, 9070, 1986.
44. **Wheland, J., Willingham, M. C., and Sandoval, I. V.,** A rat monoclonal antibody reacting specifically with the tyrosinated form of tubulin. Biochemical characterization. Effects on microtubule polymerization *in vitro* and microtubule polymerization and organization *in vivo*, *J. Cell Biol.*, 97, 1476, 1983.
45. **Serrano, L., Wandosell, F., and Avila, J.,** Location of the regions recognized by five commercial antibodies on the tubulin molecule, *Analyt. Biochem.*, 159, 253, 1986.
46. **Breitling, F. and Little, M.,** Carboxy-terminal regions on the surface of tubulin and microtubule epitope locations of Yol 1/34, DM1A and DM1B, *J. Mol. Biol.*, 189, 367, 1986.
47. **Grimm, M., Breitling, F., and Little, M.,** Location of the epitope for the α-tubulin monoclonal antibody TU-01, *Biochem. Biophys. Acta*, 914, 83, 1987.
48. **Bryan, J.,** Vinblastine and microtubules. II. Characterization of two protein subunits from the isolated crystals, *J. Mol. Biol.*, 66, 157, 1972.
49. **Olmsted, J. B. and Borisy, G. G.,** Microtubules, *Ann. Rev. Biochem.*, 42, 507, 1973.
50. **Schmitt, H. and Atlas, D.,** Specific affinity labelling of tubulin with bromocolchicine, *J. Mol. Biol.*, 102, 743, 1976.
51. **Barnes, L. D., Roberson, G. M., Divaliotis, M. J., and Williams, R. F.,** Labelling of renal tubulin with a photoaffinity analog of colchicine, *Arch. Biochem. Biophys.*, 196, 511, 1979.
52. **Ludueña, R. F.,** in *Microtubules*, Roberts, K. and Hyams, J., Eds., Academic Press, New York, 1979, 66.
53. **Avila, J., Serrano, L., and Maccioni, R. B.,** Regulatory aspects of the colchicine interaction with tubulin, *Mol. Cell. Biochem.*, 73, 29, 1987.
54. **Little, M. and Ludueña, R. F.,** Structural differences between brain β1 and β2 tubulins: implications for microtubule assembly and colchicine binding, *EMBO J.*, 4, 51, 1985.
55. **Deinum, J. and Lincoln, P.,** Binding to tubulin of an alkocolchicine spin probe: interaction with the essential SH groups and other active sites, *Biochem. Biophys. Acta*, 870, 226, 1986.
56. **Serrano, L., Montejo, E., Hernández, M. A., and Avila, J.,** Localization of the tubulin binding site for tau protein, *Eur. J. Biochem.*, 153, 595, 1985.

57. **Littauer, U. Z., Givein, D., Thieranf, M., Ginsburg, I., and Ponstingl, H.,** Common and distinct tubulin binding sites for microtubule associated proteins, *Proc. Natl. Acad. Sci. U.S.A.*, 83, 7162, 1986.
58. **Jacobs, M.,** Tubulin and nucleotides, in *Microtubules*, Roberts, K. and Hyams, J. S., Eds., Academic Press, London, 1979, 255.
59. **Carlier, M. F.,** Guanosine 5′ triphosphate hydrolysis and tubulin polymerization, *Mol. Cell. Biochem.*, 47, 97, 1982.
60. **Geahlen, R. L. and Haley, B. E.,** Use of GTP photoaffinity probe to resolve aspects of mechanisms of tubulin polymerization, *J. Biol. Chem.*, 254, 11982, 1979.
61. **Pantaloni, D. and Carlier, M. F.,** Involvement of guanosine triphosphate hydrolysis in the mechanism of tubulin polymerization. Regulation of microtubule dynamics at steady state by a GTP cap, *Ann. N.Y. Acad. Sci.*, 466, 496, 1986.
62. **Sandoval, I. V. and Weber, K.,** Guanosine 5% (α, β-methylene) triphosphate enhances specifically microtubule nucleation and stops the treadmill of tubulin protomers, *J. Biol. Chem.*, 255, 6966, 1980.
63. **Sandoval, I. V., Jameson, J. L., Niedel, J., McDonald, E., and Cuatrecasas, P.,** Role of nucleotides in tubulin polymerization. Effect of guanosine methylene diphosphate, *Proc. Natl. Acad. Sci. U.S.A.*, 75, 3178, 1978.
64. **Geahlen, R. L. and Haley, B. E.,** Use of a GTP photoaffinity probe to resolve aspects of the mechanism of tubulin polymerization, *J. Biol. Chem.*, 254, 11982, 1979.
65. **Nath, J. P., Eagle, G. R., and Himes, R. H.,** Direct photoaffinity labeling of tubulin with guanosine 5′-triphosphate, *Biochemistry*, 24, 1555, 1985.
66. **Monasterio, O.,** ^{19}F nuclear magnetic resonance measurement of the distance between the E-site GTP and the high affinity Mg^{+2} in tubulin, *Biochemistry*, 26, 6099, 1987.
67. **Miller, D. L. and Weissbach, H.,** Studies on the purification and properties of factor Tu from *E. coli*, *Arch. Biochem. Biophys.*, 141, 26, 1970.
68. **Maccioni, R. B. and Seeds, N.,** Affinity labeling of tubulin's exchangeable guanosine 5′-triphosphate binding site, *Biochemistry*, 22, 1572, 1983.
69. **Nath, J. P., Eagle, G. R., and Himes, R. H.,** Studies on the exchangeable nucleotide-binding site of tubulin, *Ann. N. Y. Acad. Sci.*, 466, 482, 1986.
70. **Linse, K., and Mandelkow, E. M.,** Identification of the GTP binding region of β-tubulin and homologies with nucleotide-binding proteins, *J. Cell Biol.*, 103, 545a, 1986.
71. **Leberman, R. and Egner, U.,** Homologies in the primary structure of GTP-binding proteins: the nucleotide-binding site of EF-Tu and p21, *EMBO J.*, 3(2), 339, 1984.
72. **Sternlicht, H., Yaffe, M. B., and Farr, G. W.,** A model of the nucleotide binding site of tubulin, *FEBS Lett.*, 214, 226, 1987.
73. **Dever, T. E., Glynias, M. J., and Merrick, W. C.,** GTP-binding domain: three consensus sequence elements with distinct spacing, *Proc. Natl. Acad. Sci. U.S.A.*, 84, 1814, 1987.
74. **Serrano, L.,** Tentative model for the GTP-binding domain of tubulin, in *The Cytoskeleton in Cell Differenciation and Development*, Maccioni, R. B. and Arechaga, J., Eds., IRL Press, Oxford, 1987, 83.
75. **Walker, E. J., Saraste, M., Runswick, M. J., Nicholas, J. R., and Gay, N. Y. I.,** Homology of ATP-synthase with ATP-binding proteins, *EMBO J.*, 1, 945, 1982.
76. **La Cour, T. F. M., Nyborg, J., Thrump, S., and Clark, D. F. C.,** GDP binding to elongation factor Tu, *EMBO J.*, 4, 2385, 1985.
77. **Mandelkow, E. M. and Mandelkow, E.,** Domains, antigenicity and substructure of tubulin, in *Microtubules and Microtubule Inhibitors*, De Brabander, M. and De Mey, J., Eds., Elsevier, Amsterdam, 1985, 31.
78. **Wierenga, R. K., Terpstra, P., and Hol, W. G. J.,** Prediction of the occurrence of the ADP-binding βαβ-fold in proteins using an amino acid sequence fingerprint, *J. Mol. Biol.*, 187, 101, 1986.
79. **Zabrecky, J. R. and Cole, D. R.,** Effect of ATP on the kinetics of microtubule assembly, *J. Biol. Chem.*, 257, 4633, 1982.
80. **Solomon, F.,** Binding sites for calcium on tubulin, *Biochemistry*, 16, 358, 1977.
81. **Berkowitz, S. A. and Wolff, J.,** Intrinsic calcium sensitivity of tubulin polymerization. The contribution of temperature, tubulin concentration and associated proteins, *J. Biol. Chem.*, 256, 11216, 1981.
82. **Kirschner, M. W., Williams, R. C., Weingarten, M., and Gerjart, J. C.,** Microtubules from mammalian brain: some properties of the depolymerization products and a proposed mechanism of assembly and disassembly, *Proc. Natl. Acad. Sci. U.S.A.*, 71, 1159, 1974.
83. **Murphy, D. B. and Borisy, G. G.,** Association of high molecular weight proteins with microtubules and their role in microtubule assembly *in vitro*, *Proc. Natl. Acad. Sci. U.S.A.*, 72, 2696, 1975.
84. **Weingarten, M. D., Lockwood, A. H., Hwo, S. Y., and Kirschner, M. W.,** A protein factor essential for microtubule assembly, *Proc. Natl. Acad. Sci. U.S.A.*, 72, 1858, 1975.
85. **Sloboda, R. D., Dentler, W. L., Bloodgood, R. A., Telzer, B. R., Cranette, S., and Rosenbaum, J. L.,** Microtubule associated proteins and the assembly of microtubules *in vitro*, in *Cell Motility*, Goldman, R., Pollard, T., and Rosenbaum, J., Eds., Cold Spring Harbor Laboratory, Cold Spring Harbor, NY, 1976, 1171.

86. **Keates, R. A. B. and Hall, R. H.**, Tubulin requires accessory protein for self-assembly into microtubules, *Nature*, 257, 418, 1975.
87. **Bhattacharyya, B., Sackett, D. L., and Wolff, J.**, Tubulin hybrid dimers and tubulin, *J. Biol. Chem.*, 260, 1028, 1985.
88. **Sackett, D. L., Bhattacharyya, B., and Wolff, J.**, Tubulin subunit carboxy-termini determine polymerization efficiency, *J. Biol. Chem.*, 260, 43, 1985.
89. **Schatz, P. J., Georges, E., Solomon, F., and Botsten, D.**, Insertion of up to 17 aminoacids into a region of α-tubulin do not disrupt function, *Mol. Cell. Biol.*, 7, 3799, 1987.
90. **Fridovich-Keil, F., Bond, J. F., and Solomon, F.**, Domains of β-tubulin essential for conserved functions *in vivo*, *Mol. Cell. Biol.*, 7, 3792, 1987.
91. **Bond, J. F., Fridovich-Keil, J. L., Pilh, L., Mulligein, R. C., and Solomon, F.**, A chicken-yeast chimeric β-tubulin protein is incorporated into mouse microtubules *in vivo*, *Cell*, 44, 461, 1986.
92. **Himes, H. R., Burton, R. P., and Gaito, J. M.**, Dimethylsulfoxide induced self-assembly of tubulin lacking associated proteins, *J. Biol. Chem.*, 252, 6222, 1977.
93. **Serrano, L., Wandosell, F., de la Torre, J., and Avila, J.**, Effect of proteolytic cleavage on tubulin polymer formation, *Biochem. J.*, 252, 683, 1988.
94. **Rudolph, J. E., Kimble, M., Hoyle, H. D., Subler, M. A., and Raff, E. C.**, Three Drosophila β-tubulin sequences: a developmentally regulated isoform β_3, the testis-specific β_2 and an assembly-defective mutation of the testis specific isoform, reveal both an ancient divergence in metazoan isotypes and structural constraints of β-tubulin function, *Mol. Cell. Biol.*, 7, 2231, 1987.
95. **Avila, J., Serrano, L., and Maccioni, R. B.**, Regulatory aspects of the colchicine interactions with tubulin, *Mol. Cell. Biochem.*, 73, 29, 1987.
96. **Steinert, P. M., Steven, A. C., and Roop, D. R.**, The molecular biology of intermediate filaments, *Cell*, 42, 411, 1985.
97. **McLachlan, A. D. and Karn, J.**, Periodic charge distributions in the myosin rod aminoacid sequence match cross-bridge spacing in muscle, *Nature*, 299, 226, 1982.
98. **Doolittle, R. F., Goldbaum, D. M., and Doolittle, L. R.**, Designation of sequences involved in the "coiled-coil" interdomainnal connections in fibrinogen: construction of an atomic scale model, *J. Mol. Biol.*, 120, 311, 1978.
99. **Barra, H. S., Rodriguez, J. A., Arce, C. A., and Caputto, R.**, A soluble preparation from rat brain that incorporates into its own proteins (^{14}C)-arginine by a ribonuclease-sensitive system and (^{14}C)-tyrosine-insensitive system, *J. Neurochem.*, 20, 97, 1973.
100. **Raybin, D. and Flavin, M.**, An enzyme tyrosylating alpha tubulin and its role in microtubule assembly, *Biochem. Biophys. Res. Commun.*, 60, 1384, 1975.
101. **Villasante, A., Wang, D., Dobner, P., Dolph, P., Lewis, S. A., and Cowan, N. J.**, Six mouse α-tubulin mRNAs encode five distinct isotypes: testis specific expression of two sister genes, *Mol. Cell. Biol.*, 6, 2409, 1986.
102. **Dobner, P. R., Kislanskis, E., Wentworth, B. M., and Villakomaroff, L.**, Alternative 5′ exons either provide or deny an initiator methionine codon to the same α-tubulin coding region, *Nucleic Acids Res.*, 15, 199, 1987.
103. **Pratt, L. F., Okamma, S., and Cleveland, D. W.**, A divergent testis specific α-tubulin isotype that does not contain a coded C-terminal tyrosine, *Mol. Cell. Biol.*, 7, 552, 1987.
104. **Helfbentein, E.**, Nucleotide sequence of a macronuclear DNA molecule coding for α-tubulin from the ciliate *Stylonichya lemnae*. Special codon usage: TAA is not a translation termination codon, *Nucleic Acids Res.*, 13(2), 415, 1985.
105. **Schatz, P. J., Solomon, F., and Botsten, D.**, Genetically essential and non essential α-tubulin genes specify functionally interchangeable proteis, *Mol. Cell. Biol.*, 6, 3722, 1986.
106. **Theurkauf, W. E., Braun, H., Bo, J., and Wensink, P. C.**, Tissue specific and constitutive α-tubulin genes of *Drosophila melanogaster* code for different proteins, *Proc. Natl. Acad. Sci. U.S.A.*, 83, 8477, 1986.
107. **Monteiro, M. J. and Cox, R. A.**, Primary structure of an α-tubulin gene of *P. polycephalum*, *J. Mol. Biol.*, 193, 427, 1987.
108. **Aragaraña, C. E., Barra, H. S., and Caputto, R.**, Tubulin tyrosine carboxypeptidase from chicken brain. Properties and partial purification, *J. Neurochem.*, 34, 114, 1980.
109. **Martensen, T.**, Preparation of brain tyrosine tubulin carboxypeptidase, *Methods in Cell Biology*, *Vol.* 24, Prescott, D., Ed., Academic Press, New York, 1982, 265.
110. **Raybin, D. and Flavin, M.**, Enzyme which specifically adds tyrosine to the α chain of tubulin, *Biochemistry*, 16, 2189, 1977.
111. **Flavin, M. and Murofoshi, H.**, Tyrosine incorporation of tubulin, *Methods Enzymol.*, 106, 225, 1984.
112. **Schröeder, H. C., Wehland, J., and Weber, K.**, Purification of brain tubulin-tyrosine ligase by biochemical and immunological methods, *J. Cell Biol.*, 100, 276, 1985.

113. **Thompson, W. C., Deanin, G. G., and Gordon, W. W.**, Intact microtubules are required for rapid turnover of carboxy-terminal tyrosine of alpha tubulin in cell cultures, *Proc. Natl. Acad. Sci. U.S.A.*, 76, 1318, 1979.
114. **Kuman, N. and Flavin, M.**, Preferential action of a brain detyrosinating carboxypeptidase on polymerized tubulin, *J. Biol. Chem.*, 256, 7678, 1981.
115. **Wheland, J. and Weber, K.**, Tubulin-tyrosine ligase has a binding site on beta-tubulin a two domain structure of the enzyme, *J. Cell Biol.*, 104, 1059, 1987.
116. **Arce, C. A., Hallack, M. E., Rodriguez, J. A., Barra, H. S., and Caputto, R.**, Capability of tubulin and microtubules to incorporate and to release tyrosine and phenylalanine and the effect of these aminoacids on tubulin assembly, *J. Neurochem.*, 31, 205, 1978.
117. **Arce, C. A., Rodriguez, J. A., Barra, H., and Caputto, R.**, Incorporation of L-tyrosine, L-phenylalanine and L-3,4-dihydroxyphenilalanine as single units into rat brain tubulin, *Eur. J. Biochem.*, 59, 145, 1975.
118. **Rodriguez, J. A. and Borisy, G. G.**, Modification of the C-terminus of brain tubulin during development, *Biochem. Biophys. Res. Commun.*, 83, 579, 1979.
119. **Preston, S. F., Deanin, G. G., Hanson, R. K., and Gordon, M. W.**, The phylogenetic distribution of tubulin: tyrosine ligase, *J. Mol. Evol.*, 13, 233, 1979.
120. **Gabius, H. J., Graupner, G., and Cramer, F.**, Activity patterns of aminoacyl-tRNA synthetases, tRNA methylases, arginyltransferase and tubulin: tyrosine ligase during development and ageing of *Caenorhabditis elegans*, *Eur. J. Biochem.*, 131, 231, 1983.
121. **Stieger, J., Wyler, T., and Seebeck, J.**, Partial purification and characterization of microtubular protein from *T. brucei*, *J. Biol. Chem.*, 259, 4596, 1984.
122. **Rodriguez, J. A. and Borisy, G. G.**, Modification of C-terminus of brain tubulin during development, *Biochem. Biophys. Res. Commun.*, 89, 893, 1978.
123. **Barra, H. S., Arce, C. A., and Caputto, R.**, Total tubulin and its aminoacetylated and nonaminoacetylated forms during the development of rat brain, *Eur. J. Biochem.*, 109, 439, 1980.
124. **Preston, S. F., Deanin, G. G., Hanson, R. K., and Gordon, M. W.**, Tubulin tyrosine ligase in oocytes and embryons of *Xenopus laevis*, *Dev. Biol.*, 81, 36, 1981.
125. **Cumming, R., Burgoyne, R. D., and Lytton, M. A.**, Immuncytochemical demonstration of alpha-tubulin modification during axonal maturation in the cerebellar cortex, *J. Cell Biol.*, 98, 347, 1984.
126. **Gundersen, G. G., Kalmoski, M. H., and Bulinski, J. C.**, Distinct populations of microtubules. Tyrosinated and nontyrosinated alpha-tubulin are distributed differently *in vivo*, *Cell*, 38, 779, 1984.
127. **Gundersen, G. G. and Bulinski, J. C.**, Distribution of tyrosinated and nontyrosinated alpha-tubulin during mitosis, *J. Cell Biol.*, 102, 1118, 1986.
128. **Sherwin, T., Schneider, A., Sasse, R., Seebeck, T., and Gull, K.**, Distinct localization and cell cycle dependence of COOH terminally tyrosinolated alpha-tubulin in the microtubules of *Trypanosoma brucei brucei*, *J. Cell Biol.*, 104, 439, 1987.
129. **Schulze, E., Asai, D. J., Bulinski, J. C., and Kirschner, M.**, Postranslational modification and microtubule stability, *J. Cell Biol.*, 105, 2167, 1987.
130. **Deanin, G. G., Preston, S. F., Hanson, R. K., and Gordon, M. W.**, On the mechanism of turnover of the carboxy-terminal tyrosine of the alpha chain of tubulin, *Eur. J. Biochem.*, 109, 207, 1980.
131. **Forrest, G. L. and Klevecz, R. R.**, Tyrosil-tubulin ligase and colchicine binding activity in synchronized chinese hamster cells, *J. Cell Biol.*, 78, 441, 1978.
132. **Nath, J. and Flavin, M.**, Tubulin tyrosination *in vivo* and changes corresponding differentiation of cultured neuroblastome-glioma hybrid cells, *J. Biol. Chem.*, 254, 11505, 1979.
133. **Gundersen, G. G. and Bulinski, J. C.**, Microtubule changes in differentiated cells contain elevated levels of a postranslationally modified form of tubulin, *Eur. J. Cell Biol.*, 42, 288, 1986.
134. **Deanin, G. G., Preston, S. F., and Gordon, M. W.**, Carboxy terminal tyrosine metabolism of alpha-tubulin and changes in cell shape: chinese hamster ovary cells, *Biochem. Biophys. Res. Commun.*, 100, 1642, 1981.
135. **Nath, J., Flavin, M., and Schiffmann, E.**, Stimulation of tubulin tyrosinolation in rabbit leukocytes evoked by the chemoattractant formyl-methionyl-leucyl-phenylalanine, *J. Cell Biol.*, 91, 232, 1981.
136. **Geuens, G., Gundensen, G., Nuydens, R., Cornelissem, F., Bulinski, J. C., and De Brabander, M.**, Ultrastructural colocalization of tyrosinated and detyrosinated alpha-tubulin in interphase and mitotic cells, *J. Cell Biol.*, 103, 1883, 1986.
137. **Gundersen, G. G., Khawaja, S., and Bulinski, J. C.**, Postpolymerization detyrosination of alpha-tubulin. A mechanism for subcellular differentiation of microtubules, *J. Cell Biol.*, 105, 251, 1987.
138. **Khawaga, S., Gundersen, G. G., and Bulinski, C.**, Enhanced stability of microtubules enriched in detyrosinated tubulin is not a direct function of detyrosination level, *J. Cell Biol.*, 106, 141, 1988.
139. **Gundersen, G. G., Khagawa, S., and Bulinski, J. C.**, Microtubules enriched in detyrosinated tubulin *in vivo* are less dynamic than those enriched in tyrosinated tubulin, in *The Cytoskeleton in Cell Differenciation and Development*, Maccioni, R. B. and Arechaga, J., Eds., ICSN Press, Paris, 1987.

140. **Kumar, N. and Flavin, M.,** Modulation of some parameters of assembly of microtubules *in vitro* by tyrosination of tubulin, *Eur. J. Biochem.*, 128, 215, 1982.
141. **L'Hernault, S. W. and Rosenbaum, J. L.,** Chlamidomonas tubulin is posttranslationally modified by acetylation on the epsilon-amino group of a lysine, *Biochemistry*, 24, 473, 1985.
142. **Greer, K., Maruta, H., L'Hernaut, S. W., and Rosenbaum, J. L.,** Alpha-tubulin acetyled activity in isolated cell flagella, *J. Cell Biol.*, 101, 2081, 1985.
143. **L'Hernaut, S. W. and Rosenbaum, J. L.,** Chlamidomonas alpha-tubulin is post-translationally modified in the flagella during flagella assembly, *J. Cell Biol.*, 97, 258, 1983.
144. **McKeithan, J. W., Lefevbre, P. A., Silflow, C. D., and Rosenbaum, J. L.,** Multiple forms of tubulin in Polytomella and Chlamydomonas: evidence for a precursor of flagellar alpha-tubulin, *J. Cell Biol.*, 96, 1056, 1983.
145. **L'Hernaut, S. W. and Rosenbaum, J. L.,** Reversal of the post-translational modification on Chlamydomonas flagellar alpha-tubulin occurs during flagellar resorption, *J. Cell Biol.*, 100, 457, 1985.
146. **Piperno, G. and Fuller, M. T.,** Monoclonal antibodies specific for an acetylated form of alpha-tubulin recognize the antigen in cilia and flagella from a variety of organism, *J. Cell Biol.*, 101, 2085, 1985.
147. **Russell, D. G., Miller, D., and Gull, K.,** Tubulin heterogeneity in the trypanosome *C. fasciculata*, *Mol. Cell Biol.*, 4, 779, 1984.
148. **Russell, D. G. and Gull, K.,** Flagellar regeneration of the trypanosome *C. fasciculata* involves post-translational modification of cytoplasmic alpha-tubulin, *Mol. Cell Biol.*, 4, 1182, 1984.
149. **Schneider, A., Shermin, T., Sasse, R., Russell, D. G., Gull, K., and Seebeck, T.,** Subpellicular and flagellar microtubules of *Trypanosoma brucei* contain the same alpha-tubulin isoform, *J. Cell Biol.*, 104, 431, 1987.
150. **Diggins, M. A. and Dove, W. F.,** Distribution of acetylated alpha-tubulin in *Physarum polycephalum*, *J. Cell Biol.*, 104, 303, 1987.
151. **Sasse, R., Glyn, M. C. P., Birkett, C. R., and Gull, K.,** Acetylated alpha-tubulin in *Physarum*. Immunological characterization of the isotype and its usage in particular microtubular organelles, *J. Cell Biol.*, 104, 41, 1987.
152. **Piperno, G., Le Dizet, M., and Chang, X. J.,** Microtubules containing acetylated alpha-tubulin in mammalian cells in culture, *J. Cell Biol.*, 104, 289, 1987.
153. **Le Dizet, M. and Piperno, G.,** Cytoplasmic microtubules containing acetylated alpha-tubulin *C. seinhardi:* spatial arrangement and properties, *J. Cell Biol.*, 103, 13, 1986.
154. **Thomson, W. C., Asai, D. J., and Carney, D. H.,** Heterogeneity among microtubules of the cytoplasmic microtubule complex detected by a monoclonal antibody to alpha-tubulin, *J. Cell Biol.*, 98, 1017, 1984.
155. **Maruta, H., Green, K., and Rosenbaum, J. L.,** The acetylation of alpha-tubulin and its relationship to the assembly and disassembly of microtubules, *J. Cell Biol.*, 103, 571, 1986.
156. **Le Dizet, M. and Piperno, G.,** Identification of an acetylation site of Chlamydomonas alpha-tubulin, *Proc. Natl. Acad. Sci. U.S.A.*, 84, 5720, 1987.
157. **Toda, T., Adachi, Y., Hiraoka, Y., and Yanguida, M.,** Identification of the pleotropic cell division gen NDA2 as one of two different alpha-tubulin genes *S. pombe*, *Cell*, 37, 233, 1984.
158. **Schatz, P. J., Pillus, L., Gusfl, P., Solomon, F., and Botstein, D.,** Two functional alpha-tubulin genes of the yeast *S. cerevesiae* encode divergent proteins, *Mol. Cell. Biol.*, 6, 3711, 1986.
159. **Eipper, B. A.,** Rat brain microtubule protein: purification and determination of covalently bound phosphate and carbohydrate, *Proc. Natl. Acad. Sci. U.S.A.*, 69, 2283, 1972.
160. **Caughlin, B. A., White, H. D., and Purich, D. C.,** Autophosphorylation of brain microtubule protein, *Biochem. Biophys. Res. Commun.*, 92, 89, 1980.
161. **Piras, R. and Piras, M. M.,** Changes in microtubule polymerization during cell cycle of HeLa cells, *Proc. Natl. Acad. Sci. U.S.A.*, 72, 1161, 1975.
162. **Eipper, U. A.,** Properties of rat brain tubulin, *J. Biol. Chem.*, 249, 1407, 1974.
163. **Pfeffer, R. S., Drubin, G. D., and Kelly, D. B.,** Identification of three coated components as alpha and beta-tubulin linked to a phosporylated 50,000 dalton polypeptide, *J. Cell Biol.*, 97, 40, 1983.
164. **Gard, L. A. and Kirschner, M. W.,** A polymer-dependent increase in phosphrylation of beta-tubulin accompanies differentiation of mouse neuroblastoma cell line, *J. Cell Biol.*, 100, 764, 1985.
165. **Ludueña, R. F.,** Biochemistry of tubulin, in *Microtubules*, Roberts, K. and Hymas, J. S., Eds., 1979, 65.
166. **Rappaport, L., Letenier, J. F., Vivion, A., and Nunez, J.,** Phosphorylation of microtubule-associated proteins, *Eur. J. Biochem.*, 62, 539, 1976.
167. **Wandosell, F., Serrano, L., Hernández, M. A., and Avila, J.,** Phosphorylation of tubulin by a calmodulin-dependent protein kinase, *J. Biol. Chem.*, 261, 10332, 1986.
168. **Wandosell, F., Serrano, L., and Avila, J.,** Phosphorylation of tubulin carboxy terminal tyrosine prevents its incorporation into microtubules, *J. Biol. Chem.*, 262, 8268, 1987.
169. **Serrano, L., Díaz-Nido, J., Wandosell, F., and Avila, J.,** Tubulin phosphorylation by casein kinase II is similar to that found *in vivo*, *J. Cell Biol.*, 105, 1731, 1987.

170. **Kadowaki, T., Fujita-Yamaguchi, Y., Nishida, E., Takaku, F., Akiyama, T., Kathura, S., Anamura, Y., and Kasuga, M.,** Phosphoryation of tubulin and microtubule-associated proteins by the purified insulin receptor kinase, *J. Biol. Chem.*, 260, 4016, 1985.
171. **Yoshikawa, K., Usul, H., Imazu, M., Tsukamoto, H., and Takeda, M.,** Comparison of tyrosine protein kinase in membrane fraction from mouse liver and Ehrlich ascites tumor, *J. Biol. Chem.*, 260, 15091, 1985.
172. **Goldenring, J. R., González, B., McGuire, J. S., Jr., and De Lorenzo, R. J.,** Purification and characterization of a calmodulin-dependent kinase from rat brain cytosol able to phosphorylate tubulin and microtubule-associated proteins, *J. Biol. Chem.*, 258, 12632, 1983.
173. **Yamamoto, H., Fukunaga, K., Goto, S., Tanaka, E., and Myamoto, E.,** Ca^{+2} calmodulin-dependent regulation of microtubule-associated protein 2, tau-factor and tubulin, and comparison with the cyclic AMP-dependent phosphorylation, *J. Neurochem.*, 44, 759, 1985.
174. **Kelly, P. T., McGuiness, T. L., and Greengard, D.,** Evidence that the major postsynaptic density protein is a component of a Ca^{+2}/calmodulin-dependent protein kinase, *Proc. Natl. Acad. Sci. U.S.A.*, 81, 945, 1984.
175. **Hargreaves, A., Wandosell, F., and Avila, J.,** Phosphorylation of tubulin enhances its interaction with membranes, *Nature*, 323, 827, 1986.
176. **Bocku, B. J. and Stiles, C. D.,** Regulation of cytoskeletal architecture by platelet derived growth factor, insulin and epidermal growth factor, *Exp. Cell Res.*, 153, 186, 1984.
177. **Edde, B., Jeantet, C., and Gros, F.,** One beta-tubulin subunit accumulates during outgrowth in mouse neuroblastoma cells, *Biochem. Biophys. Res. Commun.*, 3, 1035, 1981.
178. **Ludueña, R. F., Zimmermann, H. P., and Little, M.,** Identification of the phosphorylated beta-tubulin isotype in differentiated neuroblastoma cells, *FEBS Lett.*, 230, 142, 1988.

Chapter 4

INTERACTIONS OF TUBULIN WITH SMALL LIGANDS

Ernest Hamel

TABLE OF CONTENTS

I. INTRODUCTION

Well over half the mass of microtubules is composed of the protein tubulin, and interference with the functions of tubulin in cells can lead to profound disturbances of cellular physiology. Since the mitotic spindle is largely composed of microtubules, one of the most dramatic consequences of microtubule disruption is metaphase arrest. Condensed chromosomes are unable to progress through mitosis because the spindle fails to form. Besides the extensive protein-protein interactions required for microtubule assembly, tubulin interacts as well with numerous small molecules; these latter interactions are important both for understanding the mechanism of microtubule assembly and for devising pharmacologically relevant means for interfering with this process. This latter goal is of particular interest because antimitotic agents already have important, although relatively limited, roles in the treatment of neoplastic and inflammatory diseases. Although tubulin clearly interacts with molecules which are normally thought to represent solvent components (e.g., glycerol,[1-3] glutamate,[4,5] sulfonate buffers[6,7]), this article is restricted to three classes of agent — nucleotides, divalent cations, and antimitotic drugs — which are of most importance in the structure, function, and control of tubulin physiology and offer obvious potential for manipulating the protein to gain a greater understanding of its intracellular roles.

II. NUCLEOTIDES

A. GUANOSINE DI- AND TRIPHOSPHATE

Virtually from the time of its specific identification as the subunit protein of microtubules, tubulin has been known to contain guanine nucleotide.[1,8] The protein consists of two nonidentical protein subunits of nearly identical molecular weight (about 50,000).[9,10] Termed α- and β-tubulin,[11,12] these subunits are similar in isoelectric points (5.3 and 5.4)[13,14] and amino acid compositions and are highly homologous in sequence;[9,10] but they have only been separated preparatively after denaturation and alkylation.[15-17] In addition, there are multiple forms of both subunits (termed microheterogeneity) attributable both to primary sequence variants[9,10] and to posttranslational modifications.[18-20] There is no covalent disulfide bond between the two subunits, although there are probably two intrapolypeptide cystine bridges.[21] Based on evidence from protein cross-linking experiments, it is believed that tubulin exists as an α-β heterodimer rather than as a mixture of α-α and β-β homodimers.[22]

Each heterodimer tightly binds two molar equivalents of guanine nucleotide.[1,8,23-26] Half this nucleotide cannot be removed from the protein without its denaturation or even exchanged with free nucleotide. This nonexchangeable nucleotide is believed to always be bound to

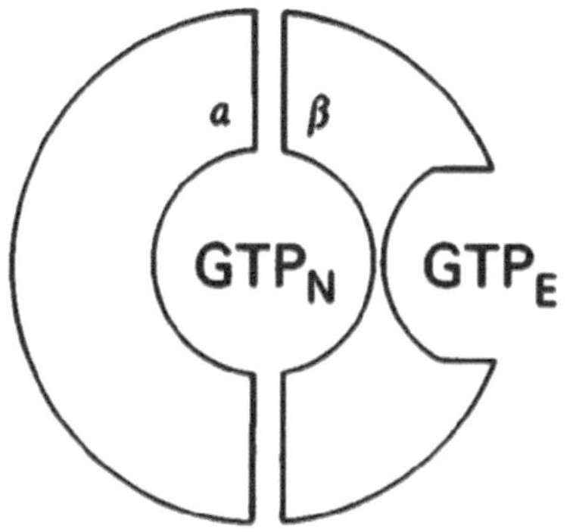

FIGURE 1. A schematic model of the α-β tubulin dimer. This model is meant to stress the continuing elusiveness of the nonexchangeable nucleotide site to systematic study and the probable location of the exchangeable site primarily or entirely on the β-subunit.

the protein as GTP.[1,25,26] Even in cells grown in culture, the turnover time of the nonexchangeable GTP is very slow and comparable to that of tubulin itself.[27] The other half of the bound nucleotide can be rapidly exchanged with free GDP or GTP.[1,23,26,28] The exchangeable nucleotide can be removed at least partially from tubulin by treatment with charcoal,[29,30] alkaline phosphatase,[31] or gel filtration in the presence of guanosine 5′-(β-γ-methylenetriphosphate) (pCH_2ppG).[32] Generally, tubulin polymerization is associated with hydrolysis of exchangeably bound GTP to GDP.[4,29,33-40] Consequently, microtubules contain one molar equivalent each of GDP (in the exchangeable site) and GTP (in the nonexchangeable site); in the absence of exogenous nucleotide, following depolymerization tubulin similarly contains one molar equivalent each of GDP and GTP.[4,41] Although a comparable observation was made initially with a GTP analogue[42] (see Section 3), GTP has been covalently linked to β-tubulin following prolonged exposure of the protein-nucleotide complex to UV light.[43,44] The binding site for the nucleotide was initially localized to the amino terminal half of β-tubulin;[45] but the specific β-tubulin sequence forming the crosslink to the nucleotide has now been more precisely localized — the major portion of the covalently bound radiolabel derived from GTP was found in a peptide corresponding to amino acid residues 155 to 162.[46] Figure 1 presents a simple scheme summarizing these features of the tubulin molecule. There is thus far no evidence localizing the nonexchangeable nucleotide to either subunit, although amino acid sequence considerations may favor its binding to the α-subunit.[47,48]

The nonexchangeable nucleotide has thus far been totally inaccessible to biochemical manipulations, and therefore a discussion of interactions of tubulin with GTP, GDP, and other nucleotides must be restricted to the exchangeable site at the present time. In the discussion which follows, the term "tubulin-GTP" refers to tubulin with GTP bound in the exchangeable site and "tubulin-GDP" to tubulin with GDP in the exchangeable site. If required, "tubulin-O" will be used to indicate tubulin which has been treated in some fashion to reduce or eliminate nucleotide originally bound in the exchangeable site. If tubulin is used, it indicates protein containing unspecified nucleotide in the exchangeable site. In my opinion there has been no convincing evidence presented for the existence of assembly-competent tubulin which does not contain GTP in the nonexchangeable site, and the presence of GTP in this site is assumed throughout the discussion which follows. The binding of nucleotide in the exchangeable site can be considered a more primitive function of the protein than the ability to assemble into microtubules, as charcoal-treated tubulin which fails to polymerize will rebind exogenous GTP.[49] Moreover, when tubulin sulfhydryls[49,50] or histidines[51] are progressively alkylated, loss of polymerizability occurs before nucleotide binding or content is significantly affected. Finally, the binding of nucleotides to tubulin is profoundly affected by divalent cations;[52-54] antimitotic drugs have major effects on tubulin-nucleotide interactions.[55,56] This work will be discussed primarily in Sections III and IV.

The affinities of GDP and GTP for the exchangeable site have been measured and expressed in terms of dissociation constants in many laboratories.[3,26,38,54,57-59] The values obtained for GDP have varied from 0.02 to 1.9 μM and those for GTP from 0.01 to 1.0 μM. The reasons for this variation are not completely clear but may be largely due to considerable variation in experimental conditions and techniques. For example, in our own study of the effects of reaction pH on nucleotide affinity for tubulin, we found that the dissociation constant for GDP at pH 6.0 was about 1.0 μM, but at pH 7.0 was 0.02 μM (an apparent 50-fold range as the pH rises or falls), and the dissociation constants for GTP displayed a similar range.[38] Mg^{2+} concentration has an equally impressive effect, but only on the binding of GTP, as opposed to GDP, to tubulin (see Section III.A).[53,54]

One issue of importance in interpreting such studies is the question of how "exchangeable" is nucleotide bound in the exchangeable site. A number of workers have described experiments in which exogenous nucleotide seemed unable to completely displace exchangeable site nucleotide.[28,34,43,60] This, however, is almost certainly the result of reaction conditions chosen for study.[53,54] In our own laboratory we have prepared tubulin-[8-^{14}C]GDP[41,61] and used it in a wide variety of reaction conditions to study the fate of GDP initially bound in the exchangeable site ("endogenous" GDP). In every case the endogenous GDP in the exchangeable site appeared to be fully exchangeable with exogenous GDP and GTP.[38,53,62,63] Figures 2 to 4 present extensive studies in which this question was addressed, using preparations of tubulin-GDP and tubulin-[8-^{14}C]GDP each containing 1.7 mol of guanine nucleotide per mole of tubulin (half GTP in the nonexchangeable site, half GDP in the exchangeable site). Figure 5 documents the tight binding of nucleotide to tubulin in the stock tubulin-[8-^{14}C]GDP solution (Figure 5A), its release from the protein by urea (Figure 5B), and the complete recovery of the radiolabel in the GDP peak and an equimolar, nonradiolabeled GTP peak (Figure 5C).

In the study presented in Figure 2 the effect of exogenous GTP concentration was examined on the relative concentrations of tubulin-GTP and tubulin-GDP, using paired reaction mixtures containing either tubulin-GDP + [8-^{14}C]GTP or tubulin-[8-^{14}C]GDP + GTP. The study presented in Figure 3 examined the effect of exogenous GDP at fixed initial tubulin-GDP and GTP concentrations on final tubulin-GTP and tubulin-GDP concentrations, with the latter divided into residual endogenous tubulin-GDP as well as tubulin-GDP derived from the exogenous GDP. This experiment required three sets of reaction mixtures: tubulin-GDP + GTP + [8-^{14}C]GDP; tubulin-GDP + [8-^{14}C]GTP + GDP; and tubulin-[8-^{14}C]GDP + GTP + GDP. The study presented in Figure 4 examined the effect of initial tubulin-GDP concentration at a fixed GTP concentration on the relative final amounts of tubulin-GTP and tubulin-GDP, again using paired reaction mixtures containing either tubulin-GDP + [8-^{14}C]GTP or tubulin-[8-^{14}C]GDP + GTP. Total exchangeable nucleotide was essentially constant, and panel B in each figure presents a recalculation of the data both in terms of micromolar tubulin-GTP (solid symbols) and in terms of percent tubulin-GTP (open symbols, ignoring tubulin lacking exchangeable nucleotide). Within experimental error, these findings demonstrate complete exchangeability of radiolabeled (i.e., exchangeable site) nucleotide and yielded an equilibrium constant, under this specific reaction condition, for the exchange of GTP for GDP of 2.7 (dimensionless constant).[63]

One report describing limited exchangeability of nucleotide merits special mention. Caplow and Zeeberg[28] observed with microtubule protein that only tubulin dimer could rapidly exchange its nucleotide. Ring oligomers consisting of tubulin and microtubule-associated proteins were unable to do so. Our own studies have been either with purified, electrophoretically homogeneous tubulin or in reconstituted systems containing this tubulin

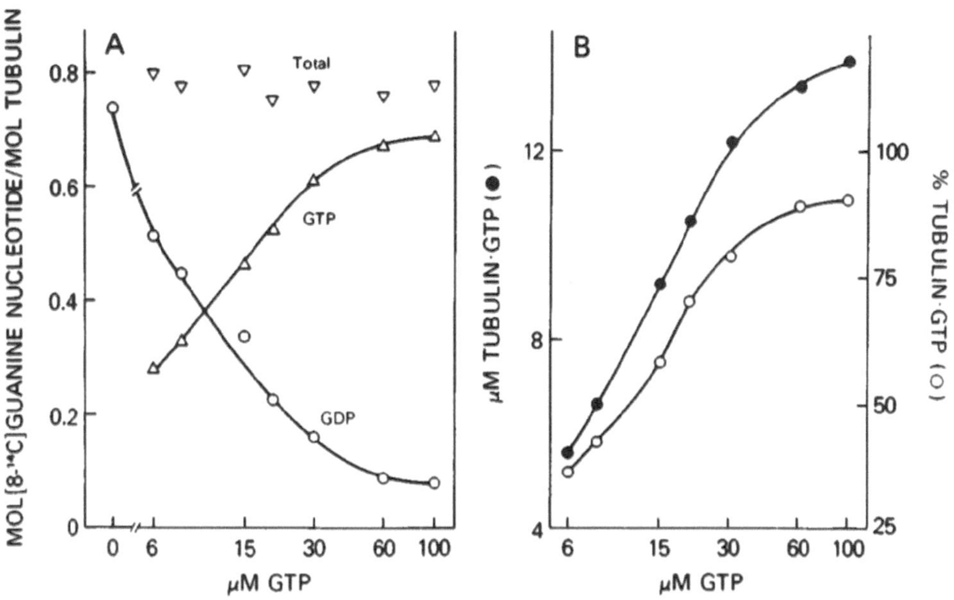

FIGURE 2. The tubulin-GTP-tubulin-GDP equilibrium as a function of GTP concentration. For determination of the tubulin-GTP concentration, 20 μM tubulin-GDP (nonradiolabeled) was mixed with the indicated concentration of [8-^{14}C]GTP. For determination of the tubulin-GDP concentration, 20 μM tubulin-[8-^{14}C]GDP was mixed with the indicated concentration of nonradiolabeled GTP. Both sets of reactions were stopped with maytansine. See Reference 63 for further experimental details.

+ soluble microtubule-associated proteins. Although the latter reaction mixtures contain abundant ring oligomers,[49] we have not been able to demonstrate any inhibitory effect of a soluble microtubule-associated protein preparation on nucleotide exchange (Table 1).[63] In our preparation of microtubule-associated proteins, however, we have observed that a large, but active, subset of these proteins is insoluble.[64] It is perhaps among these relatively insoluble proteins that the factor(s) inhibiting nucleotide exchange will be found.

It has been observed that the combination of high GDP and high Mg^{2+} concentrations favors the formation of ring oligomers of tubulin.[65] We have noted, while performing gel filtration chromatography, that dissociation of tubulin-GDP is markedly affected by the protein concentration, being progressively reduced as the protein concentration rises.[38,41] This is inconsistent with a simple dissociation of nucleotide from protein and suggests cooperative effects in this reaction, most readily rationalized in terms of still poorly defined tubulin oligomers. Spann et al.[66] have presented convincing physical evidence for the participation of nonring oligomeric species as important intermediates in microtubule assembly. Moreover, multiple oligomeric tubulin species have been observed on polyacrylamide gels under nondenaturing conditions,[21,67,68] but it has been argued that these represent artifactual aggregates rather than functionally significant oligomers.[68]

The binding of nucleotides at the exchangeable site is extremely rapid,[28] and reaction rates have not been directly measured. At least four groups, however, have reported quantitative measurements of nucleotide dissociation from the exchangeable site.[69-72] The values of the dissociation rate constant for GDP ranged from 0.68 to 8.4 min^{-1} (corresponding to half-times of dissociation of 1.0 min to less than 5 s), and those for GTP from 0.32 to 9.0 min^{-1} (corresponding to half-times of dissociation of 2.2 min to less than 5 s). Glycerol increased the dissociation half-time of GDP severalfold.[72] Some of these differences may be a consequence of oligomer formation under the experimental conditions chosen for study.

Although a detailed examination of the role of GTP in microtubule assembly is beyond

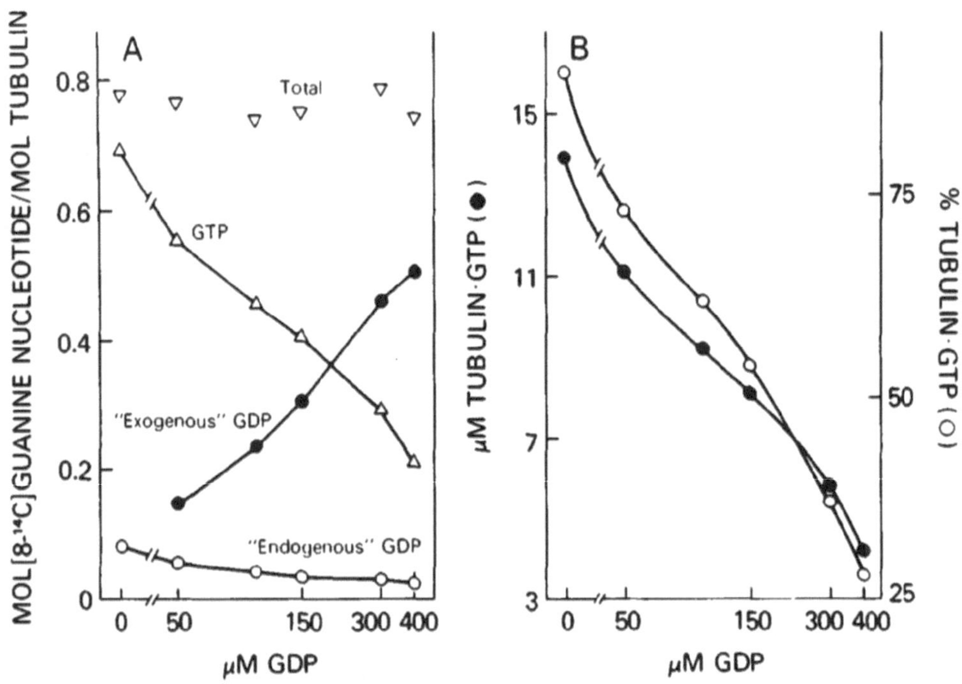

FIGURE 3. The tubulin-GTP-tubulin-GDP equilibrium as a function of GDP concentration. For determination of the tubulin-GTP concentration, 20 μ*M* tubulin-GDP (nonradiolabeled) was mixed with 100 μ*M* [8-^{14}C]GTP and the indicated concentration of nonradiolabeled GDP. For determination of the concentration of tubulin-GDP derived from exogenous GDP, 20 μ*M* tubulin-GDP (nonradiolabeled) was mixed with 100 μ*M* nonradiolabeled GTP and the indicated concentration of [8-^{14}C]GDP. For determination of the concentration of tubulin-GDP derived from endogenous GDP, 20 μ*M* tubulin-[8-^{14}C]GDP was mixed with 100 μ*M* nonradiolabeled GTP and the indicated concentration of nonradiolabeled GDP. The three sets of reactions were stopped with maytansine. See Reference 63 for further experimental details.

the scope of this chapter (such issues as the precise time of GTP hydrolysis relative to polymerization,[7,38,39,41,73,74] possible roles of GTP hydrolysis in subunit flux or treadmilling,[75-78] the postulated role of a cap of tubulin-GTP subunits at microtubule ends as a stabilizing element,[79-82] and rationalization of continuing GTP hydrolysis at the polymerization steady state[4,36,38,41,73,74,83]), interactions of GDP with tubulin cannot be fully appreciated without reference to its inhibitory effects on polymerization and its more variable effects on polymer.

Workers in many laboratories have observed extensive inhibition of microtubule assembly and of polymerization independent of microtubule-associated proteins by GDP.[4,34,35,37,38,53,61,84-90] Increasing concentrations of GDP progressively delay the onset of polymerization (i.e., inhibit nucleation or initiation of microtubule assembly), progressively reduce the rate of microtubule elongation, and progressively suppress the total yield of polymer. Different laboratories have, however, found major differences in the potency of GDP as an inhibitor of polymerization. Most likely these differences simply reflect differences in reaction conditions, for we have found that the potency of GDP as an inhibitor (in terms of the GDP to GTP ratio required) increases as the reaction pH rises,[38] as the Mg^{2+} concentration falls,[53] and as the reaction temperature is reduced.[49] An example of the Mg^{2+} effects is presented in Figure 6.

The effect of GDP on polymer has been examined in several laboratories. One approach has been to add differing mixtures of GDP and GTP to microtubules at steady state, to dilute microtubules into solutions containing GDP or tubulin-GDP, or to add microtubule seeds

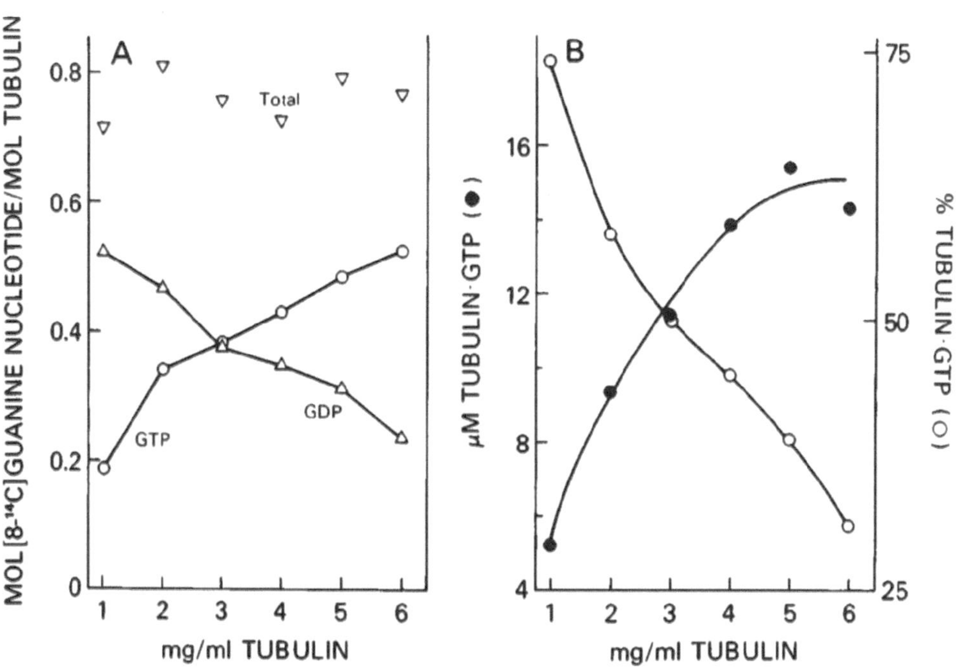

FIGURE 4. The tubulin-GTP-tubulin-GDP equilibrium as a function of tubulin concentration. For determination of the tubulin-GTP concentration, the indicated amount of tubulin-GDP (nonradiolabeled) was mixed with 15 μM [8-^{14}C]GTP. For determination of the tubulin-GDP concentration, the indicated amount of tubulin-[8-^{14}C]GDP was mixed with 15 μM nonradiolabeled GTP. Both sets of reactions were stopped with maytansine. See Reference 63 for further experimental details.

to solutions of tubulin-GDP (although most of this work has actually been performed with microtubule protein rather than purified tubulin).[79,84,85,91,92] A second approach has been to add a large excess of GDP to microtubules at steady state,[34,86] while a third has been to enzymatically degrade residual GTP at the polymerization steady state.[77,93-95] The general conclusions from these studies has been that while tubulin-GTP is absolutely required for the nucleation of microtubules, elongation can occur with tubulin-GDP. Since most workers observed complete or partial stability of microtubules with GDP added at the polymerization plateau (presumably at a steady-state phase of microtubule assembly), they concluded that tubulin-GDP could add to preformed microtubules. In keeping with these observations, direct incorporation of small, but variable, amounts of tubulin-[8-^{14}C]GDP into microtubules under routine reaction conditions has been recently directly demonstrated in our laboratory (also see Reference 37),[62] as well as when pCH_2ppG is used to initiate microtubule assembly.[96] We have found that this direct incorporation of tubulin-GDP into microtubules occurs in a nearly constant proportion throughout an assembly cycle (Figure 7), with the exact proportion determined by precise reaction conditions. The reaction is favored whenever the concentration of tubulin-GDP is increased at the expense of tubulin-GTP.[63]

Addition of high concentrations of GDP has been reported both to have no effect on microtubules[34] and to cause their complete depolymerization.[86] Similarly, enzymatic degradation of residual GTP has been described as having only minor effects on microtubule stability[77,93] and as causing total disassembly.[94] Again, differences may result from differing reaction conditions, in this case possibly a microtubule-associated protein, for Lee et al.[95] described both partial and complete GDP-induced microtubule disassembly in preparations of microtubule protein prepared by very different procedures.

Finally, at various times different laboratories have postulated a third nucleotide-binding

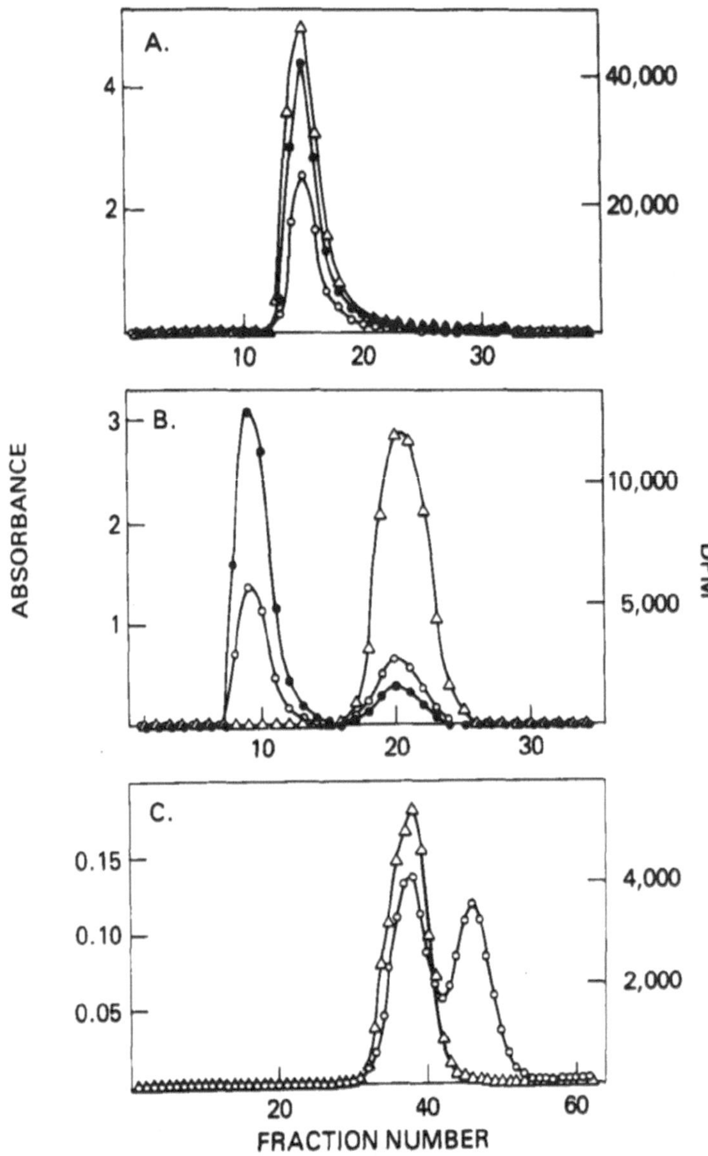

FIGURE 5. Characterization of tubulin-bound guanine nucleotide after three cycles of glutamate-induced polymerization supported by [8-^{14}C]GTP. (A) An aliquot of the tubulin solution chromatographed on a gel filtration column in 1 M glutamate. (B) An aliquot of the tubulin solution chromatographed on a gel filtration column in 8 *M* urea. (C) The nucleotide peak released in 8 *M* urea chromatographed on a DEAE-cellulose column to separate GDP (first peak) and GTP (second peak). In all three panels open circles represent A_{252} (guanine), closed circles represent A_{280} (protein), and triangles represent DPM. All the recovered radiolabeled nucleotide was in the form of [8-^{14}C]GDP. See Reference 41 for further experimental details.

site on tubulin, based on a variety of experimental observations.[86,89,97,98] At the present, in my opinion, there is no compelling evidence to accept any of these hypotheses.

B. GUANINE NUCLEOTIDE ANALOGUES

Again, although a detailed consideration of the mechanism of microtubule assembly and nucleotide involvement in this process is beyond the scope of this chapter, it is impossible

TABLE 1
Microtubule-Associated Proteins do not Alter Nucleotide Exchange on Tubulin

Nucleotide	With MAPs (mol/mol of tubulin)	Without MAPS (mol/mol of tubulin)
GDP	0.22	0.25
GTP	0.52	0.56
GTP (%)	70	69

Note: Reaction mixtures contained 20 μM tubulin-GDP, 15 μM GTP, and, if present, 0.67 mg/ml microtubule-associated proteins (MAPs). The amount of GDP bound to tubulin was determined from reaction mixtures prepared with tubulin in-[8-^{14}C]GDP and nonradiolabeled GTP; the amount of GTP bound to tubulin was determined from reaction mixtures prepared with nonradiolabeled tubulin-GDP and [8-^{14}C]GTP. Maytansine was added at the end of the incubation to stabilize the tubulin-nucleotide complex.[112] See Reference 63 for further experimental details.

to discuss the large amount of work that has been done with GTP and GDP analogues without describing the effects of such compounds on tubulin polymerization reactions.

1. Polyphosphate-Modified Analogues

The first analogues examined for interactions with tubulin were the nonhydrolyzable GTP analogues pCH$_2$ppG and guanosine 5′-(β,γ-imidotriphosphate) (pNHppG). Many laboratories reported that they were able to support microtubule assembly, particularly at high concentrations.[3,29,32,34,35,75,97,99-101] It thus appears inarguable that microtubule assembly *per se* does not require GTP hydrolysis, for these analogues were not altered in the assembly process while being incorporated into the newly formed microtubules. Although not quantitatively evaluated in any laboratory, pCH$_2$ppG and pNHppG undoubtedly have a relatively low affinity for tubulin, since high concentrations are required for assembly reactions to occur, and both analogues have only a limited ability to inhibit the binding of GTP to tubulin and its hydrolysis by the protein (Table 2).[3,34,35,37,102]

A number of interesting properties have been reported for microtubules formed with pCH$_2$ppG and/or pNHppG. Unike microtubules formed with GTP, in which the exchangeable nucleotide (i.e., GDP) becomes inaccessible to nucleotide in the medium,[34,36] pCH$_2$ppG within microtubules has been reported to be readily exchangeable with GTP added following assembly.[101] Some groups have reported that microtubules formed with pNHppG are highly stable to disassembly induced by dilution or Ca^{2+},[3,34,35,99,100,103] and they interpreted this finding as indicating that the function of GTP hydrolysis was to permit microtubule disassembly. However, not only have other groups reported that pCH$_2$ppG and pNHppG microtubules could be readily disassembled,[29,79] but highly stable microtubules are formed with two other analogues of GTP, 2′, 3′-dideoxyguanosine 5′-triphosphate (ddGTP)[94] and guanosine 5′-(α,β-methylenetriphosphate) (ppCH$_2$pG),[104] both of which are hydrolyzed during microtubule assembly. These findings suggest that enhanced microtubule stability following pNHppG-supported assembly derives not from the absence of hydrolysis, but rather from subtle effects of the nucleotide analogue on microtubule conformation, perhaps deriving from allosteric factors.

A number of groups also examined the effects of the nonhydrolyzable GTP analogues on subunit flux (or treadmilling) through microtubules.[75,77,78] They all found that treadmilling was much reduced with microtubules formed with the nonhydrolyzable analogues and interpreted this to mean that GTP hydrolsis was required for this phenomenon to occur. Again, however, treadmilling is also negligible with the two hydrolyzable GTP analogues, ppCH$_2$pG[105] and ddGTP.[106] As with the question of microtubule stability, this strongly indicates that it is more likely conformational effects of nucleotide on polymer, rather than nucleotide hydrolysis, which determines whether treadmilling will occur.

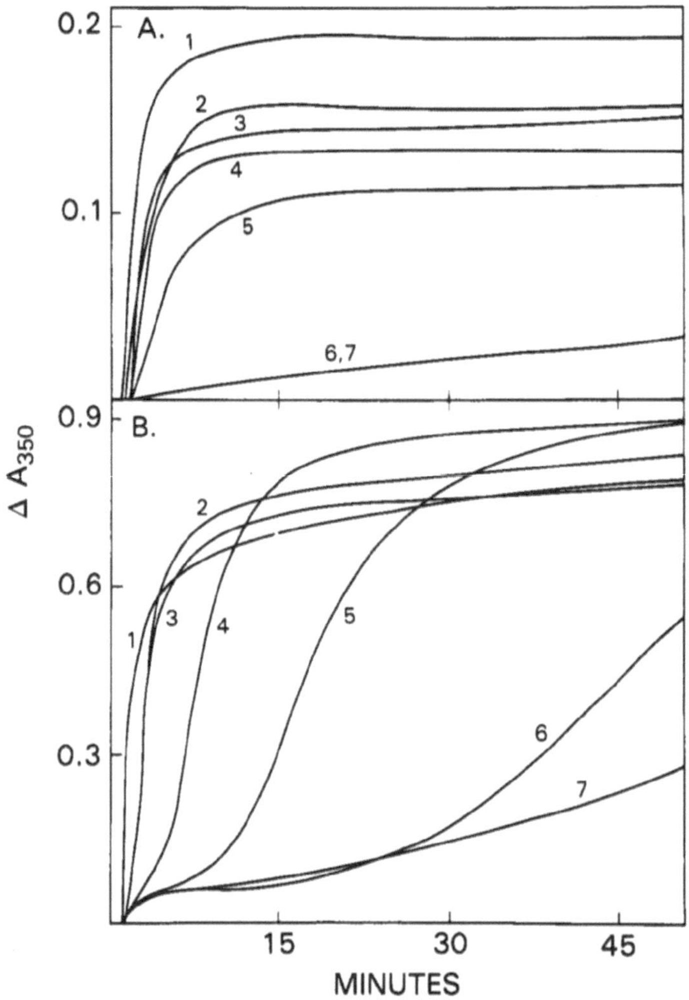

FIGURE 6. Effects of Mg^{2+} on GTP-dependent tubulin polymerization and on inhibition of polymerization by GDP. (A) With microtubule-associated proteins. Each reaction mixture contained 14 μM tubulin-GDP, 0.3 mg/ml heat-treated microtubule-associated proteins, 0.1 mM GTP, 0.1 4-morpholineethanesulfonate (pH 6.4), and further additions as follows: curve 1, 0.5 mM $MgCl_2$; curve 2, none; curves 3 to 7, 0.2 mM GDP and 0.5 mM, 0.1 mM, 50 μM, 25 μM, and no $MgCl_2$, respectively. (B) In glutamate. Each reaction mixture contained 14 μM tubulin-GDP, 0.1 mM GTP, 1.0 M monosodium glutamate, and further additions, as follows: curve 1, 1.0 mM $MgCl_2$; curve 2, none; curves 3 to 7, 0.2 mM GDP and 1.0 mM, 0.25 mM, 0.15 mM, 0.1 mM, and no $MgCl_2$, respectively. All reaction components were repurified to minimize contaminating Mg^{2+}. See Reference 53 for further experimental details.

Sandoval and colleagues have published an interesting series of papers exploring the effects of the GTP analogue ppCH$_2$pG and the GDP analogue guanosine 5′-(α,β-methylenediphosphate) (pCH$_2$pG) on tubulin polymerization.[104,105,107-110] In their first reported observations[104] they noted that ppCH$_2$pG supported a vigorous, extensive microtubule assembly reaction both with and without microtubule-associated proteins, that pCH$_2$pG was formed during the reaction, and that the microtubules formed with the analogue were unusually stable to cold and Ca^{2+}. The presence of calcium in the reaction mixture, however, resulted in altered polymer morphology and subsequent depolymerization if GDP was added. In our own laboratory we have found that ppCH$_2$pG, unlike pCH$_2$ppG and pNHppG, supports

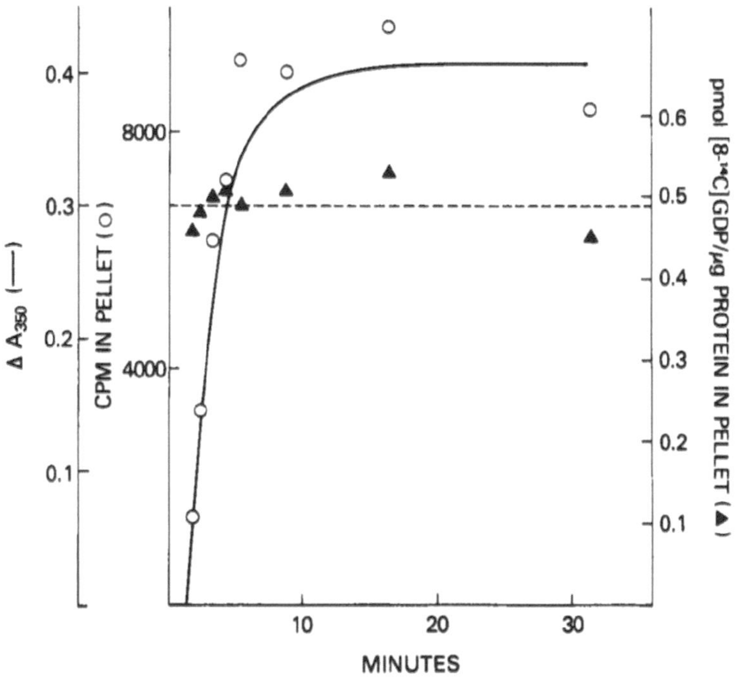

FIGURE 7. Time course of direct incorporation of GDP into microtubules. The reaction mixture contained 20 μM tubulin-[8-^{14}C]GDP, 40 μM nonradiolabeled GTP, and 0.7 mg/ml of heat-treated microtubule-associated proteins. Microtubule assembly was followed turbidimetrically (solid line) with aliquots removed at intervals to harvest assembled microtubules by centrifugation. The pellets were dissolved and aliquots taken for determination of radioactivity and protein. Total CPM (circles) and the ratio of radioactivity to protein (triangles) were determined. See Reference 62 for further experimental details.

tubulin polymerizatin at low concentrations, with an effective minimum that differs little from that of GTP.[49,61] Nevertheless, ppCH$_2$pG, like pCH$_2$ppG and pNHppG, competes poorly with GTP in binding to tubulin and is not an effective inhibitor of GTP hydrolysis (Table 2).[102]

When the GDP analogue pCH$_2$pG was examined, it was found to stimulate a feeble tubulin polymerization reaction; although no change turbidity occurred with the GDP analogue, if a reaction mixture containing the analogue and either microtubule protein or purified tubulin was centrifuged at high speed, a pellet containing very short microtubules was obtained.[107] These microtubules were insensitive to calcium but disassembled in the cold. The nucleotide was unaltered in the reaction.

Even in the presence of antimitotic drugs, ppCH$_2$pG was able to induce aberrant polymerization reactions;[108] with partially purified microtubule-associated proteins (microtubule-associated protein 2 [MAP-2] and tau factor were resolved from each other) polymers with differing morphologies were formed when the analogue was present (as opposed to only microtubules with GTP).[109] When ppCH$_2$pG hydrolysis and GTP hydrolysis were compared in relation to microtubule assembly, both nucleotides appeared to be hydrolyzed significantly slower than tubulin was incorporated into polymer (see References 38, 39, 73, and 74), with the analogue also being hydrolyzed much slower than GTP.[105] It was also concluded that ppCH$_2$pG substantially enhanced microtubule nucleation reactions, and that microtubules formed with the analogue were unable to participate in subunit flux (treadmilling — see above). In addition, microtubules formed with ppCH$_2$pG contained little, if any, nucleotide which had not been hydrolyzed to pCH$_2$pG.[105] Finally, ppCH$_2$pG was injected

TABLE 2
Comparison of Polyphosphate-Modified Analogues to GDP as Inhibitors of GTP Hydrolysis and Binding to Tubulin

	Percent inhibition	
Nucleotide	GTP hydrolysis	GTP binding
GDP	79	36
pCH_2ppG	12	4
pNHppG	23	28
NH_2ppG	4	16
$ppCH_2pG$	50	17
pCH_2pG	9	4
GTPγS	93	36
GDPβS	25	6
GTPβS(A)	20	7
GTPβS(B)	17	4

Note: In the hydrolysis experiment reaction mixtures contained 15 μ*M* tubulin-GDP, 0.5 mg/ml heat-treated microtubule-associated proteins, 0.1 m*M* [γ-^{32}P]GTP, and the inhibiting nucleotide at 0.3 m*M*. Incubation was for 40 min at 37°C. In the binding experiment reaction mixtures contained 15 μ*M* tubulin-GDP, 0.12 m*M* [8-^{14}C]GTP, and the inhibiting nucleotide at 0.5 m*M*. Incubation was for 15 min at 37°C. In the nucleotide binding experiment, maytansine was *not* added at the end of the incubation. See Reference 102 for further experimental details. NH_2ppG, guanylyl phosphoramidate (other abbreviations in text).

into several types of living cells, resulting in severe disruption both of microtubule patterns in these cells and of cellular functions associated with microtubules.[110]

Several laboratories have examined interactions of tubulin with phosphorthioate analogues of GDP and GTP — primarily guanosine 5′-*O*-(3-thiotriphosphate) (GTPγS), but also guanosine 5′-*O*-(2-thiodiphosphate) (GDPβS), guanosine 5′-*O*-(2-thiotriphosphate), S_p diastereoisomer [GTPβS(A)], and guanosine 5′-*O*-(2-thiotriphosphate), R_p diastereoisomer [GTPβS(B)].[30,102,111] Kirsch and Yarbrough[30] first reported that in 25% glycerol microtubule protein was induced to polymerize by high concentrations of GTPγS. The analogue also induced tubulin-O to polymerize in reaction mixtures containing glycerol and dimethylsulfoxide. Both reactions were sluggish compared to GTP-supported polymerization. The polymer, morphologically identical to standard microtubules, was cold stable and contained unhydrolyzed GTPγS. In our own laboratory[102] we first became interested in GTPγS when we observed that it more potently inhibited tubulin-dependent, microtubule-associated protein-induced GTP hydrolysis than did GDP (Table 2 and Figure 8).[102] We confirmed that in glycerol (with microtubule-associated proteins) a sluggish polymerization reaction occurred with GTPγS, but both in glutamate an with microtubule-associated proteins (without glycerol) GTPγS had no ability to support polymerization. With microtubule-associated proteins the analogue was also more potent than GDP as an inhibitor of microtubule assembly. GTPγS was neither broken down by tubulin nor did it form a covalent bond with the protein.

One unusual feature of the interaction of GTPγS with tubulin was that binding of the analogue to the protein, unlike the binding of GTP, seemed to be markedly enhanced at higher reaction temperatures and in the presence of microtubule-associated proteins (Table 3). To a lesser extent this was also observed with GDP (Table 3); but subsequent experiments in which tubulin-nucleotide complexes were stabilized[63] by the antimitotic drug maytansine[112] (see Section IV.B.2) demonstrated that significant amounts of GDP dissociated from tubulin

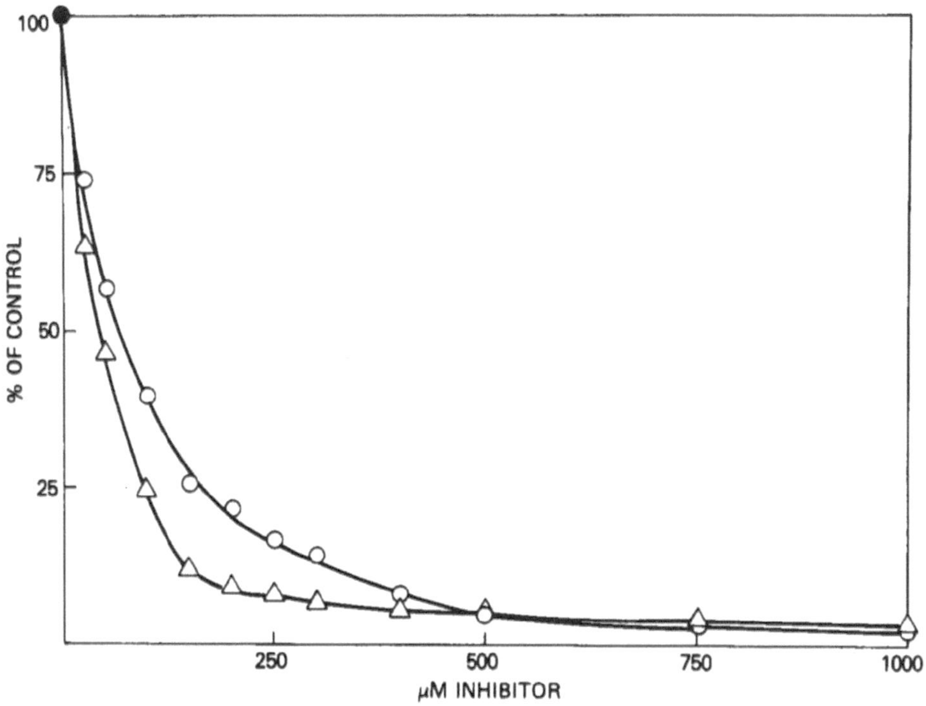

FIGURE 8. Inhibition of tubulin-dependent GTP hydrolysis by GTPγS and GDP. Each reaction mixture contained 15 μ*M* tubulin-GDP, 0.5 mg/ml heat-treated microtubule-associated proteins, 0.1 m*M* [γ-^{32}P]GTP, and the indicated concentrations of GTPγS (triangles) or GDP (circles). See Reference 102 for further experimental details.

TABLE 3
Effect of Reaction Temperature and Microtubule-Associated Proteins on the Binding of GDP, GTP, and GTPγS to Tubulin

	Mole nucleotide bound per mole tubulin			
Nucleotide	**0°C**	**0°C + MAPs**	**37°C**	**37°C + MAPs**
GTP	0.72	0.71	0.76	0.80
GDP	0.38	0.51	0.59	0.72
GTPγs	0.10	0.24	0.53	0.88

Note: Reaction mixtures contained 15 μ*M* tubulin-GDP, the indicated radiolabeled nucleotide, and 0.5 mg/ml heat-treated microtubule-associated proteins (MAPs) if indicated, and they were incubated for 15 min at the indicated temperature. In this experiment maytansine was *not* added at the end of the incubation. For further experimental details, see Reference 102.

during the analytical procedure. This may also have been the case for tubulin-bound GTPγS. Nevertheless, in a number of polymerization systems at lower temperatures or without microtubule-associated proteins, GDP surpassed GTPγS as an inhibitor of polymerization, whereas when polymerization required both 37°C and microtubule-associated proteins GTPγS was the superior inhibitor.[102]

Roychowdhury and Gaskin[111] confirmed that glycerol was required for an assembly reaction with GTPγS, observing that with 30%, instead of 25%, glycerol microtubule assembly was as extensive with the analogue as with GTP, although the reaction was still slower than the GTP-supported reaction. In the absence of glycerol GTPγS enhanced the formation of ring oligomers. The rate and extent of polymer formation, as well as the distribution of tubulin between microtubules and ring oligomers, were markedly affected by the precise concentrations of GTPγS and Mg^{2+} in the reaction mixture. Oligomeric structures, in particular, were favored at higher Mg^{2+} concentrations. Again, there was little if any hydrolysis of GTPγS. These workers suggested that GTPγS probably inhibits microtubule assembly by sequestering tubulin and microtubule-associated proteins in stable oligomeric structures; they concluded that the affinity of GTPγS for tubulin was only slightly less than that of GTP.

In contrast to GTPγS, the two isomers of GTPβS and GDPβS appeared to interact only weakly with tubulin (Table 2).[102] They had only minor effects on the tubulin-dependent binding or hydrolysis of GTP. Nevertheless, the two GTPβS isomers stimulated tubulin polymerization in glycerol, in glutamate, and with microtubule-associated proteins, provided the analogues were present in high concentrations. In glutamate and with microtubule-associated proteins GTPβS(B) was somewhat more active than GTPβS(A). Anomolous polymeric forms were observed much more commonly with the latter isomer. In glycerol the two isomers of GTPβS had equivalent activity, but the polymer formed with GTPβS(B) was cold stable while that formed with GTPβS(A) depolymerized rapidly at low temperatures.[102]

The chromium (III) salt of GTP (crGTP) has been prepared and its interactions with tubulin examined.[113,114] The cation binds so tightly to the nucleotide that it is not readily displaced, and thus the complex serves as an analogue of the magnesium salt of GTP. With crGTP microtubule assembly was much more rapid than with GTP, although the final extent of assembly was similar with the two compounds. Microtubules formed with the analogue were cold-labile but largely stable to calcium depolymerization. At least half of the crGTP that entered microtubules was not hydrolyzed. Although no studies analyzing the binding of crGTP to unpolymerized tubulin have been reported, crGTP-supported microtubule assembly is highly sensitive to inhibition by GDP. This implies that the affinity of the nucleotide for tubulin is low compared to its affinities for GDP and GTP (see also References 41 and 61).

Monasterio and Timasheff[115,116] described interactions of guanosine 5′-(γ-fluorotriphosphate) (GTPγF) with tubulin. In their polymerization studies they used purified tubulin in a reaction system containing 3.4 *M* glycerol and high concentrations of Mg^{2+}. GTPγF was unable to support tubulin polymerization under these conditions; when analogue and GTP were approximately equimolar, the polymerization reaction was almost completely inhibited. GTPγF would thus appear to be the most potent nucleotide inhibitor of tubulin polymerization yet described, except that Monasterio and Timasheff unfortunately did not directly compare it to either GDP or GTPγS under their reaction conditions.[115] This is of importance, since a large, but undefined, excess of GTPγF had only minimal inhibitory effect on the binding of radiolabeled GTP to tubulin (the analogue was estimated to bind to tubulin about 1/50th as avidly as GTP). GTPγF also inhibited colchicine-induced GTP hydrolysis, but again no comparison was made to GDP or GTPγS. Monasterio[116] next prepared a tubulin-colchicine-GTPγF-Mn^{2+} complex and examined its spectral properties to estimate the distance between the exchangeable nucleotide site and the divalent cation site in tubulin. This will be discussed below (Section III.C).

Kirsch and Yarbrough[30,117] evaluated interactions with tubulin of three GTP analogues with bulky substituents at the γ-phosphate position — guanosine 5′-*O*-(3-thio(acetylaminoethyl)-1-naphthylamine-5-sulfonic acid triphosphate) (a derivative of GTPγS); guanosine 5′-triphosphoro-γ-1-(5-sulfonic acid)naphthylamidate; and guanosine 5′-triphosphoro-γ-anilinylamidate. These three compounds were unable to support tubulin polymerization even when tubulin-O was used. The first two compounds, however, have fluorescent derivatives attached to the γ-phosphate. Based on spectroscopic evidence, these two analogues did bind to tubulin-O, but with an affinity which was estimated to be about 100-fold less than that of GTP.

Two groups have reported studies with a 5′-hydroxyl derivatized nucleoside, 5′-*p*-fluorosulfonylbenzoylguanosine (FSBG),[118,119] and these studies are included here since the analogue is considered a nucleotide, rather than a nucleoside, analogue which can form covalent crosslinks to proteins.[120] Steiner[118] reported that FSBG induced the polymerization of microtubule protein derived from human platelets, but that the polymer formed was cold stable. A covalent bond was formed between the analogue and β-tubulin, and up to 0.7 molar equivalent of FSBG was bound by tubulin. Prasad and Luduena[119] examined interactions of FSBG with bovine brain tubulin. They reported that 2 molar equivalents of the analogue alkylated two sulfhydryl groups of tubulin, and that the reaction could be reversed with 2-mercaptoethanol. Radiolabel was completely lost from the protein if the tubulin was denatured, precluding subunit identification. They were unable to demonstrate any polymerization reaction supported by FSBG, but FSBG treatment of tubulin prevented subsequent polymerization supported by GTP.

2. Ribose-Modified Analogues

Penningroth and Kirschner[32] reported that 2′-deoxyguanosine 5′-triphosphate (2′dGTP) could replace GTP in microtubule assembly with both standard microtubule protein and with a tubulin-O preparation containing microtubule-associated proteins. Under both conditions lower concentrations of 2′dGTP than of GTP were required for equivalent reactions.

Kirsch and Yarbrough[30] reported studies with periodate-oxidized GTP (oxGTP, the ribose-cleaved dialdehyde; the analogous compound obtained from GDP will be referred to as oxGDP) and its borohydride-reduced derivative (ox-redGTP, the ribose cleaved dialcohol; the analogous compound obtained from GDP will be referred to as ox-redGDP). Both analogues, at high concentrations, supported polymerization of microtubule protein in glycerol and of tubulin-O in glycerol-dimethylsulfoxide. Covalent interaction of substoichiometric amounts of oxGTP, via Schiff base formation, with both tubulin subunits was demonstrated; but these reactions were not inhibited by excess GTP and were interpreted as representing nonspecific interactions.

Maccioni and Seeds[121] examined interactions of oxGTP and 3′-*p*-azidobenzoylguanosine 5′-triphosphate (3′azbGTP) with tubulin-O (derived from microtubule protein). Both analogues at high concentrations were able to promote polymerization reactions, and covalent bonds (via Schiff base formation or after photoactivation) were formed in nearly equal amounts with both α- and β-tubulin with both oxGTP and 3′azbGTP. The oxGTP was hydrolyzed to oxGDP during the polymerization process. These workers (unlike Kirsch and Yarbrough,[30] but there were substantial differences in reaction conditions) observed that GTP inhibited covalent linkage of oxGTP to tubulin, and that near-stoichiometric amounts of oxGTP reacted with the protein.

In our own laboratory we have performed an extensive analysis of interactions of a reasonably complete series of ribose-modified GDP and GTP analogues with tubulin.[41,89,90,94,106,122-124] We have studied two reaction conditions in detail, the glutamate-induced polymerization of purified tubulin and polymerization dependent on heat-treated microtubule-associated proteins. Both protein preparations have been documented to contain negligible amounts of nucleoside diphosphate kinase and nonspecific nucleoside triphosphatase activities.[4,64,125] Only minor differences were observed in nucleotide behavior in the two reaction systems, and, consequently, only the results with microtubule-associated proteins will be discussed here.

The available ribose-modified analogues were as follows: 2′,3′-dideoxyguanosine 5′-

di- and triphosphate (ddGDP and ddGTP); 3′-deoxyguanosine 5′-di- and triphosphate (3′-dGDP and 3′dGTP); 2′-deoxyguanosine 5′-di- and triphosphate (2′dGDP and 2′ dGTP); 9-β-D-arabinofuranosylguanine 5′-di- and triphosphate (araGDP and araGTP); the di- and triphosphate derivatives at the side-chain hydroxyl of 9-[(2-hydroxyethoxy)methyl]guanine (acycloguanosine) (acycloGDP and acycloGTP); ox-redGDP and ox-redGTP; 2′- and 3′-*O*-methylguanosine 5′-di- and triphosphate (2′MeGDP, 2′MeGTP, 3′MeGDP, and 3′MeGTP); and guanosine 2′- and 3′-monophosphate, 5′-di- and triphosphate (2′PGDP, 2′PGTP, 3′PGDP, and 3′PGTP). From the point of view of their interactions with tubulin, these analogues can be subdivided into two groups. The deoxy analogues consist of 2′d-, 3′d-, dd-, and araGDP and GTP. The second group consists of substituted analogues that bear a methyl or phosphate at the 2′ or 3′ position and of the open ribose ring analogues ox-redGDP, ox-redGTP, acycloGDP, and acycloGTP.

With all four deoxyGTP analogues, microtubule assembly occurred with a lower nucleotide concentration than with GTP, was more extensive, and onset of the reaction was early relative to the GTP-supported reaction. All the ribose-substituted and open ribose ring analogues were deficient relative to GTP. Either polymer morphology was aberrant, a higher nucleotide concentration was required for assembly, less polymer was formed, polymerization was delayed, or a combination of deficiencies occurred. The two tetraphosphates, in particular, were without activity in supporting microtubule assembly.[123]

One analogue, ddGTP, seemed particularly notable in that a polymerization reaction occurred without microtubule-associated proteins with 10 μ*M* tubulin, although the reaction required a higher nucleotide concentration, was delayed, incomplete, and resulted in anomolous structures.[123] Furthermore, alone of all the diphosphates examined, ddGDP at high concentrations was able to support microtubule assembly. This reaction required microtubule-associated proteins. The microtubules formed contained ddGDP in near-stoichiometric amounts, and the nucleotide was unaltered in the assembly reaction.[124]

Radiolabeled analogues were also prepared, and their hydrolysis relative to their polymerization reactions was examined. With 2′dGTP, 3′dGTP, and araGTP (Figure 9) the pattern was qualitatively similar to that obtained with GTP. Under the reaction condition used in the experiment of Figure 9, onset of hydrolysis of the three analogues was earlier and the reactions somewhat brisker than GTP hydrolysis. Similarly, the assembly reactions with the analogues began earlier than the GTP-supported reaction. In the four small panels the individual hydrolytic and polymerization reactions are correlated on a molar basis. In all cases the two reactions were simultaneous in onset and initially stoichiometric. The reactions supported by ddGTP (Figure 10) were somewhat different. With microtubule-associated proteins a sluggish hydrolysis reaction began prior to any change in turbidity. At the onset of polymerization nucleotide hydrolysis markedly accelerated. The two reactions appeared to be stoichiometric, as the slopes of the curves are similar. Without microtubule-associated proteins, however, ddGTP hydrolysis and the associated polymerization reaction were completely uncoupled. Hydrolysis began about 5 min before assembly; ddGTP hydrolysis without microtubule-associated proteins, although delayed in onset, was much faster than the reaction with microtubule-associated proteins. The significance of this rapid uncoupled reaction is not at present understood. Polymerization with the four deoxyGTP analogues resulted in their incorporation into polymer as the diphosphates and in the displacement of [8-^{14}C]GDP originally bound in the exchangeable site, demonstrating that the polymerization these compounds induced was mediated through the exchangeable nucleotide site (Table 4).[41]

It is tempting to explain these polymerization results with a simple model in which the nucleotide binds at the exchangeable site with the ribose moiety directed inward. Thus the deoxy analogues would have no difficulty entering the site, perhaps with an enhanced affinity, while binding of the substituted and open ribose ring analogues would be reduced as a

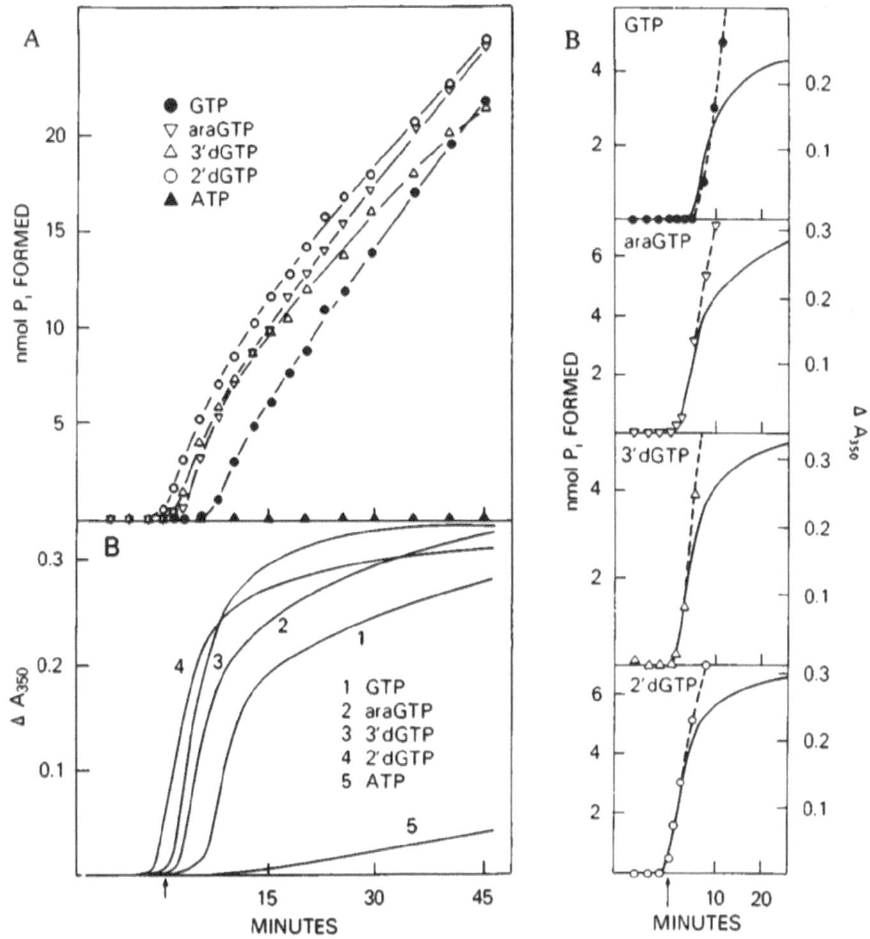

FIGURE 9. Simultaneous nucleotide hydrolysis and microtubule assembly with GTP, 2′dGTP, 3′dGTP, and araGTP. Reaction mixtures contained 10 μ*M* tubulin-GDP, 0.2 mg/ml heat-treated microtubule-associated proteins, and the indicated radiolabeled nucleotide at 0.1 m*M*. Panel A presents the hydrolytic data, panel B the turbidimetric data, and the small panels the initial phase of each reaction, with hydrolysis and polymer represented on the same molar scale. The amount of polymer was estimated on the basis of protein recovered by centrifugation after a 45-min incubation. See Reference 41 for further experimental details.

consequence of steric factors. It is unfortunately not consistent with all the available data. First, the deoxyGDP analogues were all significantly less active than GDP as inhibitors of tubulin polymerization, with ddGDP supporting rather than inhibiting assembly. When all the GDP analogues were evaluated for their inhibitory effects on polymerization, we found that there was absolutely no correlation between relative activities of triphosphates in supporting tubulin polymerization and of diphosphates in inhibiting this reaction. We also found that the polymerization reaction without microtubule-associated proteins induced by ddGTP could be inhibited by GTP and other GTP analogues as well as by GDP and GDP analogues. In this case relative inhibitory activities of triphosphates and the cognate diphosphates was practically identical.[90] Moreover, all four deoxyGTP analogues induce a polymerization reaction at temperatures substantially below that observed with GTP, for even at 20°C microtubule assembly with GTP is extremely slow (see below). Consequently GTP was examined as a potential inhibitor of polymerization at 20°C with all four deoxyGTP ana-

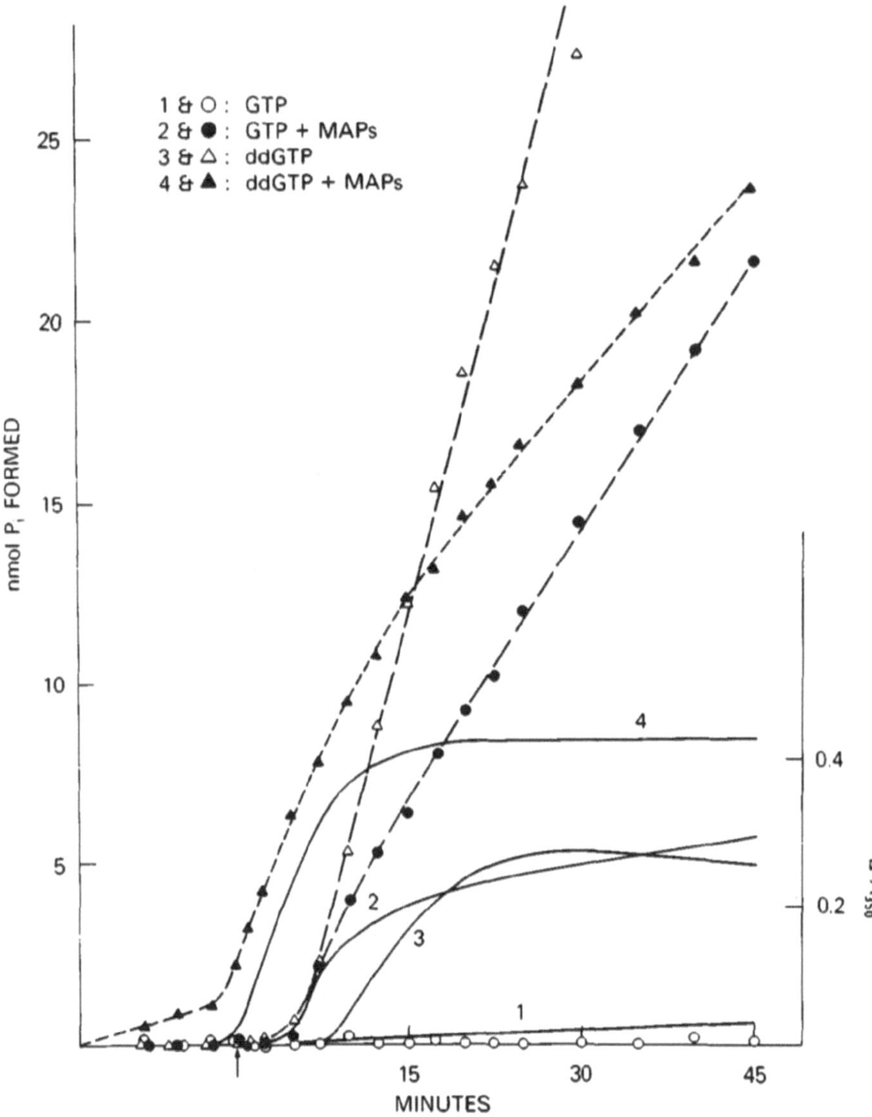

FIGURE 10. Effect of microtubule-associated proteins on nucleotide hydrolysis and polymerization with GTP and ddGTP. Reaction mixtures contained 10 μ*M* tubulin-GDP, the indicated radiolabeled nucleotide at 0.1 m*M*, and 0.2 mg/ml heat-treated microtubule-associated proteins (MAPs), if indicated. The solid lines represent the turbidimetric data, the symbols the hydrolytic data. Hydrolysis and polymer are represented on the same molar scale. The amount of polymer was estimated on the basis of protein recovered by centrifugation after a 45-min incubation. See Reference 41 for further experimental details.

logues. These reactions were substantially inhibited by equimolar GTP, particularly if reaction rates are considered (Figure 11).[41] The reaction with 2′dGTP was inhibited least, those with 3′dGTP and ddGTP most, in agreement with the relative affinities obtained in the inhibition studies with the GDP analogs (see Reference 90).

We also examined the binding of both GTP and GDP analogues to unpolymerized tubulin and their ability to displace [8-^{14}C]GDP from the exchangeable site (Table 5). None of the deoxyGTP and GDP analogues was as effective as GTP and GDP, respectively, in either binding to tubulin or displacing radiolabeled GDP from tubulin. The analogues themselves display the same order of activity as observed in the inhibition of polymerization assays —

TABLE 4
Recovery of Ribose-Modified Analogues in Polymer and their Displacement of [8-^{14}C]GDP from Tubulin during Polymerization

Nucleotide	Mole nucleotide per mole tubulin in polymer pellet	
	Analogue	GDP
UTP + NDK	—	0.84
GTP	1.02	0.01
dGTP	1.06	0.01
araGTP	1.06	0.03
3'dGTP	1.09	0.03
ddGTP	1.16	0.08
ddGDP	1.01	—

Note: For experiments in which analogue incorporation into polymer was measured, reaction mixtures contained 10 μM tubulin-GDP, 0.2 mg/ml heat-treated microtubule-associated proteins, and the indicated nucleotide at 0.1 mM radiolabeled in either the α- or β-phosphate with ^{32}P. Thin-layer chromatographic analysis of the polymer pellet demonstrated that in all cases the recovered radiolabeled compounds were in the form of the diphosphates. For the experiments in which GDP displacement during polymerization was measured, reaction mixtures contained 10 μM tubulin-[8-^{14}C]GDP, 0.2 mg/ml heat-treated microtubule-associated proteins, and the indicated nonradiolabeled nucleotide at 1 mM. For further experimental details, see References 41 and 124. NDK, nucleoside diphosphate kinase (other abbreviations in text).

2'-deoxy analogues most active, 3'-deoxy and dideoxy analogues least active, and ara analogues of intermediate activity.[41] Radiolabeled ox-redGTP, acycloGTP, 2'MeGTP, and 3'MeGTP also bound poorly to unpolymerized tubulin.[49]

Finally we performed a series of cross-inhibition experiments on nucleotide effects on binding and hydrolysis, examining the effects of ddGTP and 3'dGTP on the binding and hydrolysis of GTP (Figure 12A) and of GTP on the hydrolysis of ddGTP and 3'dGTP (Figure 12B). The two analogues were without effect on the binding of GTP to tubulin, and had little inhibitory effect on GTP hydrolysis at 37°C. GTP, on the other hand, potently inhibited hydrolysis of 3'dGTP and ddGTP with microtubule-associated proteins; GTP was even more potent as an inhibitor of ddGTP hydrolysis in the absence of the microtubule-associated proteins.[41]

The ribose moiety is clearly an important recognition feature for binding at the exchangeable nucleotide binding site of tubulin, for every analogue we have examined with a ribose modification is deficient in some aspect of binding at this site. It is important to note that support of polymerization, even at low nucleotide concentrations, is inadequate evidence to conclude that binding at the exchangeable site is efficient, for it appears that the rate-limiting step in polymerization does not involve nucleotide binding. In fact, many studies have indicated that the initiation of assembly, or microtubule nucleation, is the rate-limiting step in the assembly process,[126] and our findings with ribose-modified analogues are consistent with this view. The deoxyGTP analogues, despite reduced affinity for the exchangeable site, have properties predicted for compounds which would enhance microtubule nucleation. First, they enhance polymerization at lower temperatures. Second, microtubule-associated proteins are believed to enhance nucleation more than elongation.[127] Under our experimental conditions, the microtubule-associated proteins were not required for polymerization with ddGTP at 10 μM tubulin. With 20 μM tubulin they were not required with 3'dGTP, 2'dGTP, or araGTP. With GTP 50 μM tubulin was needed to eliminate the

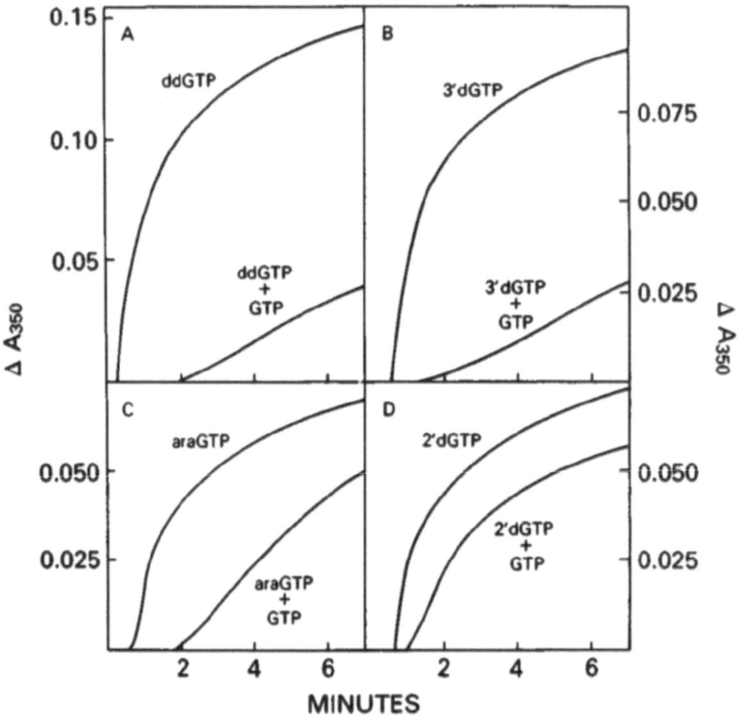

FIGURE 11. Inhibition by equimolar GTP of microtubule assembly supported by ribose-modified GTP analogs at 20°C. Each reaction mixture contained 10 μ*M* tubulin-GDP, 0.3 mg/ml heat-treated microtubule-associated proteins, the indicated analogue at 0.5 m*M*, and, if indicated, 0.5 m*M* GTP. Polymerization was at 20°C. See Reference 41 for further experimental details.

TABLE 5
Binding of Ribose-Modified Analogues and their Displacement of GDP from the Exchangeable Nucleotide Site of Unpolymerized Tubulin

	Mole nucleotide bound per mole tubulin	
Nucleotide	**Analogue bound**	**GDP bound**
None	—	0.40
GTP	0.62	0.01
dGTP	0.42	0.02
araGTP	0.42	0.05
3'dGTP	0.05	0.12
ddGTP	0.12	0.24
GDP	0.34	0.04
dGDP	0.17	0.08
araGDP	0.06	0.18
3'dGDP	0.01	0.20
ddGDP	0.03	0.30

Note: For measurement of the binding of analogues, reaction mixtures contained 10 μ*M* nonradiolabeled tubulin-GDP and the radiolabeled analogue at 0.1 m*M*. For measurement of the displacement of GDP by analogue, reaction mixtures contained 10 μ*M* tubulin-[8-^{14}C]GDP and the nonradiolabeled analogue at 0.1 m*M*. In this experiment maytansine was *not* added at the end of the incubation. For further experimental details, see Reference 41.

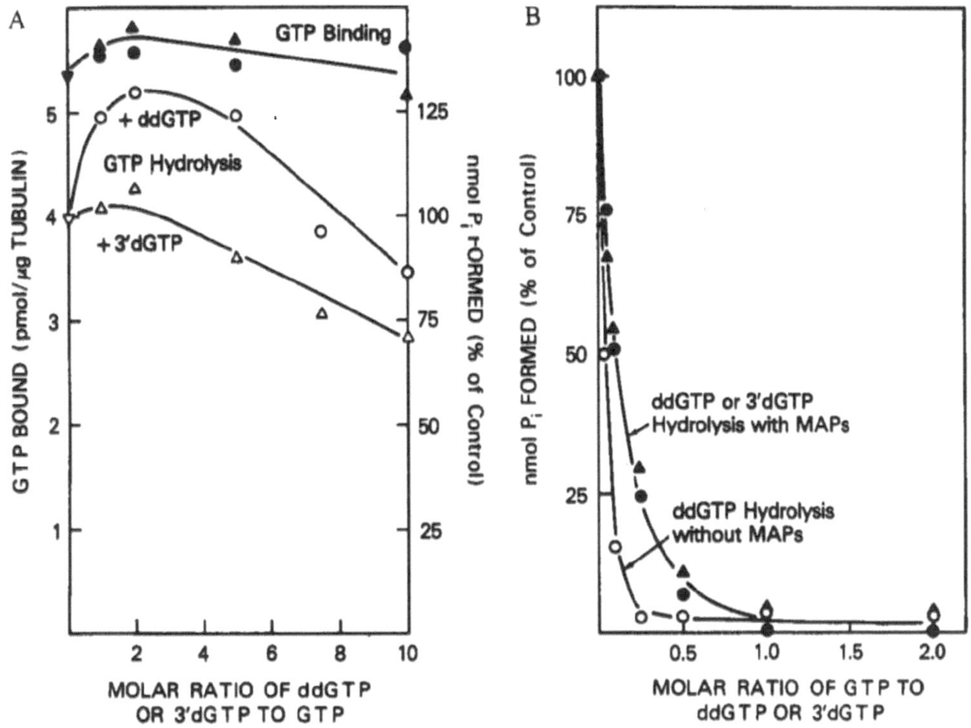

FIGURE 12. Effects of ddGTP and 3′dGTP on GTP binding and hydrolysis (A) and of GTP on ddGTP and 3′dGTP hydrolysis (B). A. In the binding experiment (solid symbols) each reaction mixture contained 10 µM tubulin-GDP, 50 µM radiolabeled GTP, and the indicated concentration of either ddGTP (circles) or 3′dGTP (triangles). Maytansine was *not* used at the end of the incubation in this experiment. In the hydrolysis experiment (open symbols) each reaction mixture contained 10 µ*M* tubulin-GDP, 0.2 mg/ml heat-treated microtubule-associated proteins, 0.1 m*M* radiolabeled GTP, and the indicated concentration of ddGTP or 3′dGTP, as indicated. B. Each reaction mixture contained 10 µ*M* tubulin-GDP, 0.1 m*M* radiolabeled ddGTP (circles) or 3′dGTP (triangles), the indicated concentration of GTP, and 0.2 mg/ml heat-treated microtubule-associated proteins (MAPs), as indicated. See Reference 41 for further experimental details.

absolute requirement for microtubule-associated proteins. Finally, nucleation enhancement is generally invoked when average microtubule length is reduced.[74,105] This was in fact observed with all four deoxyGTP analogues relative to GTP, with the greatest reduction in length observed with ddGTP, the least reduction with 2′dGTP (Figure 13). We therefore conclude that enhanced polymerization with deoxyGTP analogues represents not increased affinity for the exchangeable nucleotide site but enhanced activity in nucleation — the rate-limiting step in polymerization — despite reduced affinity for the exchangeable site.[41]

The polymerization reactions supported by ddGDP and ddGTP permitted us to take a new approach to potential roles for GTP hydrolysis in microtubule assembly. It had been proposed that GTP hydrolysis permitted subsequent microtubule disassembly[3,34,35,99,100,103] and/or was required for the treadmilling reaction.[75,77,78] Microtubules containing ddGDP in the exchangeable site were constructed with either ddGTP (with nucleotide hydrolysis) or ddGDP (without hydrolysis), and their stability[94] and ability to treadmill[106] were examined. Both the ddGDP- and ddGTP-formed microtubules had identical stability upon dilution; but microtubules formed with ddGTP were stable in Ca^{2+}, while those formed with ddGDP were labile when the cation was added to them. A similar result was obtained when temperature effects were examined (Figure 14). With ddGTP and ddGDP microtubules were formed at lower temperatures than with GTP; once formed, the microtubules were stable at

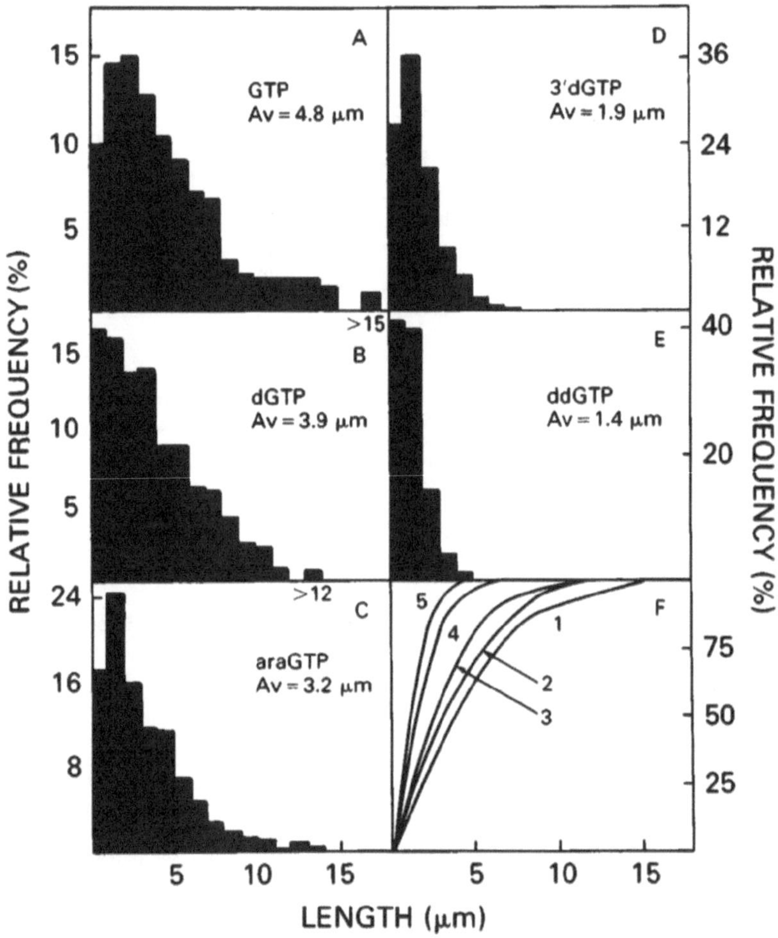

FIGURE 13. Comparison of microtubule lengths in GTP- and ribose-modified analog-supported polymerization. (A to E) Histograms of microtubules formed with the indicated nucleotide. (F) Cumulative length distributions with each nucleotide: curve 1, GTP; curve 2, 2′dGTP; curve 3, araGTP; curve 4, 3′dGTP; and curve 5, ddGTP. See Reference 41 for further experimental details.

lower temperatures than the GTP tubules. However, both the temperature required for formation and for complete disassembly were lower with ddGTP than with ddGDP. These results clearly suggest that GTP hydrolysis enhances, rather than reduces microtubule stability. When treadmilling was examined (Figure 15), we found that both ddGDP- and ddGTP-formed microtubules, as opposed to GTP-formed microtubules, displayed minimal tubulin flux through the polymer. The analogues thus provide no insight into the role of GTP in this process.

3. Guanine-Modified Analogues

Interactions of only a few guanine-modified analogues with tubulin have been described, and their interactions per se with tubulin were only of secondary interest. Geahlen and Haley[42,128] were interested in 8-azidoguanosine 5′-triphosphate (azidoGTP) primarily as a tool to locate the exchangeable nucleotide binding site on tubulin. Using microtubule protein, they found that the analogue, following binding to tubulin and photoactivation, formed a covalent bond primarily to β-tubulin. Photoactivation of the tubulin-analogue complex re-

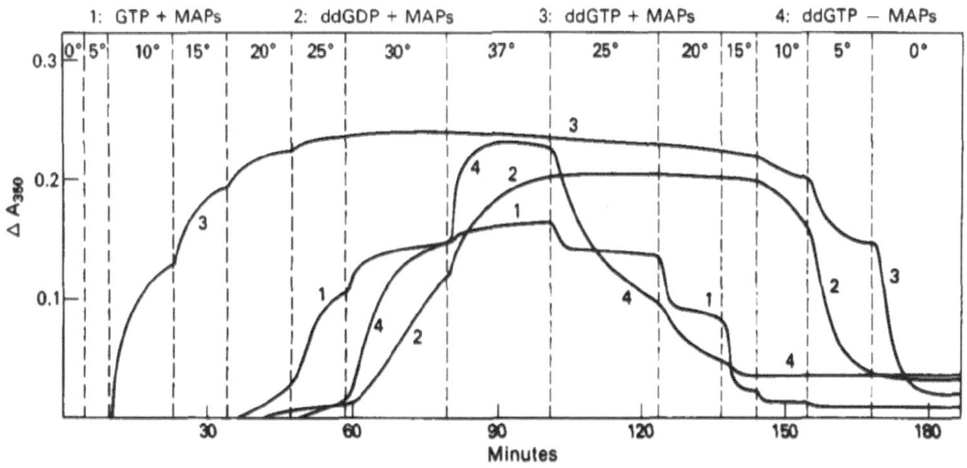

FIGURE 14. Effects of temperature on tubulin polymerization and depolymerization with ddGDP and ddGTP. Each reaction mixture contained 10 μ*M* tubulin-GDP, 0.2 mg/ml heat-treated microtubule-associated proteins (MAPs) if indicated, and the indicated nucleotide at 0.4 m*M*. Temperature changes were made with an electronic temperature controller. See Reference 94 for further experimental details.

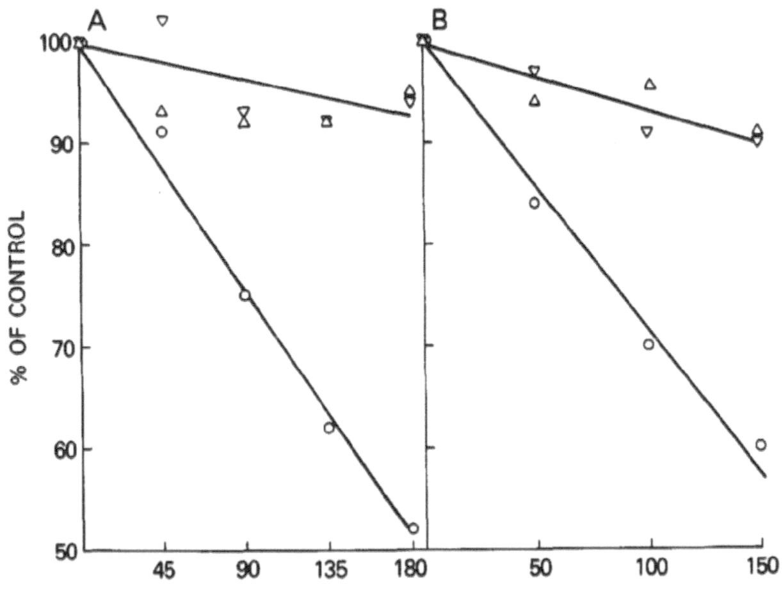

FIGURE 15. Evaluation of flux (treadmilling) of tubulin subunits through microtubules formed with GTP, ddGTP, and ddGDP. Assembly was for 1.5 h with radiolabeled nucleotides, followed by a chase for the indicated times with a tenfold molar excess of nonradiolabeled nucleotide. Microtubules were harvested by centrifugation and radioactivity in the pellets determined. Circles, GTP; triangles, ddGTP; and inverted triangles, ddGDP. (A) Flux in 0.5 m*M* $MgCl_2$. (B) Flux without exogenous $MgCl_2$. See Reference 106 for further experimental details.

sulted in inactivation of the tubulin, and photolabeling was inhibited by GTP. The analogue was able to support tubulin polymerization at concentrations as low as 10 μ*M*, was hydrolyzed to 8-azidoguanosine 5′-diphosphate (azidoGDP) during the reaction, and without photoactivation the azidoGTP-supported polymerization reaction was cold reversible. The dissociation constant of azidoGTP was estimated to be nearly identical to that of GTP (1.4 times higher). Although the binding affinity of azidoGDP was nearly the same, it photolabeled β-tubulin only one fourth as well as azidoGTP. Interestingly, in a preliminary report, Haley et al.[129] described a covalent reaction of azidoGTP with α-tubulin when purified tubulin, rather than microtubule protein, was used to study the exchangeable binding site.

Mann et al.[130] examined interactions with tubulin of a sulfhydryl reactive analogue of GTP, 2-amino-6-(*S*-dinitrophenylthio)purine riboside triphosphate, also with the goal of defining the exchangeable nucleotide site. They found it to react about equally with both subunits, with about 1.3 mol of radiolabeled (and hydrolyzed) analogue covalently linked to each mole of tubulin. *N*-Ethylmaleimide inhibited binding of the GTP analogue to tubulin and vice versa, but *N*-ethylmaleimide did not displace guanine nucleotude bound to tubulin. It is thus unclear whether the analogue was binding primarily as a GTP analogue or as a sulfhydryl reactive agent, but the data presented favors the latter interpretation. No quantitative details were described.

At least three reports have appeared describing interactions of 2-amino-6-mercapto-9-β-ribofuranosylpurine 5′-di- and triphosphate (thioGDP and thioGTP) with tubulin.[59,70,72] These workers wanted to use the fluorescent properties of the analogues to explore nucleotide interactions at the exchangeable site. They observed a decrease in intrinsic tubulin fluorescence when the analogues bound to tubulin and concluded that conformational changes were involved in the binding of nucleotides to the protein. The binding of the analogues was inhibited by GTP, and thioGDP and thioGTP bound to tubulin with affinities comparable to those of GDP and GTP. Using different reaction conditions, Engelborghs and Eccleston[70] obtained dissociation rate constants for GTP ranging from 1.9 to 9.0 min^{-1} and for thioGTP of 2.6 to 11 min^{-1}. Fishback and Yarbrough,[59,72] examining interactions of the analogues with tubulin-O, obtained a dissociation equilibrium constant for GTP of 0.02 μM, for thioGTP of 0.03 μM, and for thioGDP of 0.08 μM, and a dissociation rate constant for GDP of 0.96 min^{-1}, for thioGDP of 1.2 min^{-1}, and for thioGTP of 0.36 min^{-1}. In addition, thioGTP was nearly as effective as GTP in promoting tubulin polymerization, although only studies at high nucleotide concentrations were described. Under some reaction conditions structures with aberrant morphology were formed.[59] About 0.9 mol of analogue per mole of tubulin entered polymer, with analogue incorporation inhibited by both GDP and GTP, and the thioGTP was hydrolyzed to thioGDP during polymerization.

4. Summary

Despite low affinities for tubulin, the nonhydrolyzable GTP analogues pCH_2ppG and pNHppG support microtubule assembly at high analogue concentrations. DeoxyGTP analogues (particularly ddGTP) and $ppCH_2pG$ also have low affinities for tubulin, but they support microtubule assembly at concentrations as low or lower than effective GTP concentrations, apparently through marked enhancement of nucleation reactions. Two GDP analogues, pCH_2pG and ddGTP, support microtubule assembly reactions, with the latter much more effective than the former. Polymers formed with all these analogues have properties which differ significantly from those of microtubules induced with GTP. The only ribose- or polyphosphate-modified analogue which seems to have a high affinity for the exchangeable site is GTPγS, and its affinity appears to be markedly affected by components other than tubulin in the reaction mixture. Although GTPγS acts primarily as an inhibitor of tubulin polymerization reactions, it can support the formation of microtubules, again with altered properties, in reaction mixtures containing glycerol. Two guanine-modified analogues, azidoGTP and thioGTP, have high affinities for the exchangeable site and readily support microtubule assembly. Properties of polymers formed with these two analogues have not yet been described in detail.

C. NONGUANINE NUCLEOTIDES

Although a variety of effects of nonguanine nucleotides (mostly ATP) on microtubules both in cells and in cell-free reaction mixtures have been described, there are relatively few reports in which direct interactions with tubulin have been examined. Penningroth and Kirschner[32] described the preparation of microtubule protein containing primarily tubulin-O by assembly with pCH_2ppG and subsequent gel filtration to remove the analogue from the exchangeable site. Both before and after removal of the exchangeable GDP these workers observed polymerization with ATP, dATP, CTP, dCTP, UTP, and ITP at concentrations much higher than the GTP concentration required for microtubule assembly. After GDP depletion much higher concentrations of these nonguanine nucleotides were required, but the GTP concentration required for assembly decreased. Despite this difference, it remained possible that the nonguanine nucleotides were only active because of the nucleoside diphosphate kinase present in a microtubule protein preparation. Two nonhydrolyzable ATP analogues [adenosine 5′-(β,γ-methylenetriphosphate) and adenosine 5′-(β,γ-imidotriphosphate)], however, were able to support assembly only after removal of exchangeable site GDP, provided high concentrations of the analogues were used. These workers thus interpreted their results to indicate that virtually any nucleoside 5′-triphosphate had some affinity for the exchangeable site and could support tubulin polymerization under appropriate conditions. Similarly, Kirsch and Yarbrough[30] observed that 6-mercaptopurine ribonucleoside 5′-triphosphate supported polymerization of tubulin-O much better than of tubulin containing exchangeable site nucleotide.

Steiner[131] examined interactions of human platelet tubulin with adenosine 3′,5′-cyclic monophosphate (cAMP). He observed enhancement of GTP-dependent microtubule assembly with 1.0 and 10 μ*M* cAMP, formation of a covalent bond between α-tubulin and 8-azidoadenosine 3′,5′-cyclic monophosphate, and a K_D value for the binding of cAMP to tubulin of about 4 n*M*.

Zabrecky and Cole[98,132-134] obtained remarkably similar results with ATP, in a polymerization system using purified tubulin and high concentrations of glycerol. Initially they observed that ATP induced formation primarily of a network of aggregated rings, in contrast to microtubules formed with GTP. They then found that ATP appeared to act synergistically with GTP in promoting microtubule assembly, although both nucleotides were examined only at relatively high concentrations. They observed binding of ATP (but not its hydrolysis) to tubulin with an apparent K_D value of 0.2 m*M*. With 8-azidoadenosine 5′-triphosphate they demonstrated formation of a covalent bond between the analogue and α-tubulin. (See Section B.3, for the contrast in the covalent bond formed between azidoGTP and the tubulin subunits depending on whether microtubule protein or purified tubulin was used.) These workers interpreted their findings as demonstrating a third nucleotide binding site on tubulin, distinct from the nonexchangeable and exchangeable GTP sites.

These observations of Steiner[131] and of Zabrecky and Cole[98,132-134] led us to examine interactions of bovine tubulin-GDP free of nucleoside diphosphate kinase activity with cAMP, ATP, ATP analogues, ITP, XTP, CTP, and UTP.[61] We found no effects under several reaction conditions with cAMP; but in glycerol we obtained polymerization with millimolar concentrations of ATP, micromolar concentrations of GTP, and intermediate concentrations of ITP and XTP, with ITP more effective than XTP. UTP and CTP were inactive at concentrations up to 5 m*M*. Several lines of evidence led us to conclude that the ATP-dependent polymerization reaction was mediated through the exchangeable GTP site rather than a third nucleotide site. First, we found little difference in polymer morphology with different nucleotides, obtaining mixtures of microtubules, rings, and sheets. Second, we found that polymerization patterns obtained with mixtures of ATP and GTP were consistent with additive rather than synergistic effects. Third, while GTP-induced polymerization was highly resistant to inhibition by GDP (complete inhibition of polymerization with 10 μ*M* GTP required 0.2 m*M* GDP), ATP-induced polymerization was exquisitely sensitive to minimal concentrations of GDP (complete inhibition of polymerization with 2.5 m*M* ATP

TABLE 6
Displacement of GDP from the Exchangeable Site during Polymerization with ATP, dATP, and GTP

Nucleotide conc mM	Mole [8-^{14}C]GDP per mole tubulin in polymer pellet			
	UTP + NDK	GTP	ATP	dATP
0.008	—	0.72	—	—
0.01	—	0.46	—	—
0.05	—	0.22	—	—
0.1	—	0.11	—	—
0.4	—	—	—	0.68
0.7	—	—	0.63	0.54
1.0	0.94	0.04	0.59	0.49
2.5	—	—	0.43	0.37
5.0	—	0.03	0.29	0.30

Note: Reaction mixtures contained 30 μ*M* tubulin-[8-^{14}C]GDP and the indicated nucleotide at the indicated concentrations, together with 3.4 *M* glycerol and 5 m*M* $MgCl_2$ (to induce polymerization). See Reference 61 for further experimental details. NDK, nucleoside diphosphate kinase.

required only 50 μ*M* GDP). Fourth, relative activities of a series of ribose- and polyphosphate-modified ATP analogues closely paralleled those of the comparable series of GTP analogues. Fifth, ATP was hydrolyzed by tubulin, and ATP hydrolysis was inhibited by GDP and GTP. Sixth, tubulin-dependent hydrolysis of ATP and GTP were affected identically by antimitotic drugs. Finally, and most convincing, polymerization of tubulin-[8-^{14}C]GDP with ATP, dATP, as well as GTP, led to progressive displacement of radiolabeled GDP in the polymer (Table 6).

III. DIVALENT CATIONS

A. MAGNESIUM

The magnesium cation appears to be an integral component of the tubulin molecule, for elemental analysis has demonstrated that one molar equivalent of Mg^{2+} is tightly bound to the protein.[52,53,135-137] Even dialysis against high concentrations of EDTA does not remove this Mg^{2+} cation.[52,135,138] Several groups have described reaction conditions in which not only this tightly bound Mg^{2+} was needed, but in which there was also a requirement for additional, exogenous Mg^{2+} for tubulin polymerization.[135-143] The Mg^{2+} cation can also be used to precipitate tubulin as a step in its purification.[144] High concentrations of the cation favor formation of ring oligomers from purified tubulin.[65,144-146] Frigon and Timasheff[146] also estimated that there were approximately 50 weak binding sites for Mg^{2+} on tubulin.

Under other reaction conditions, however, there is no absolute requirement for exogenous Mg^{2+} for tubulin polymerization, although exogenous cation does appear to enhance polymerization reactions.[53,62,63] These reaction conditions include tubulin polymerization induced by glutamate and microtubule assembly reactions in sulfonate buffers, provided submillimolar GTP concentrations were used. Under these reaction conditions, in the absence of exogenous Mg^{2+}, EDTA inhibited tubulin polymerization, implying that the tightly bound Mg^{2+} was essential for microtubule assembly.[53]

Much attention has focused on potential interactions of both the tightly bound cation and of exogenously added Mg^{2+} with nucleotides. Arai et al.[3] first described a virtually absolute requirement for exogenous Mg^{2+} for the binding of radiolabeled GTP to tubulin

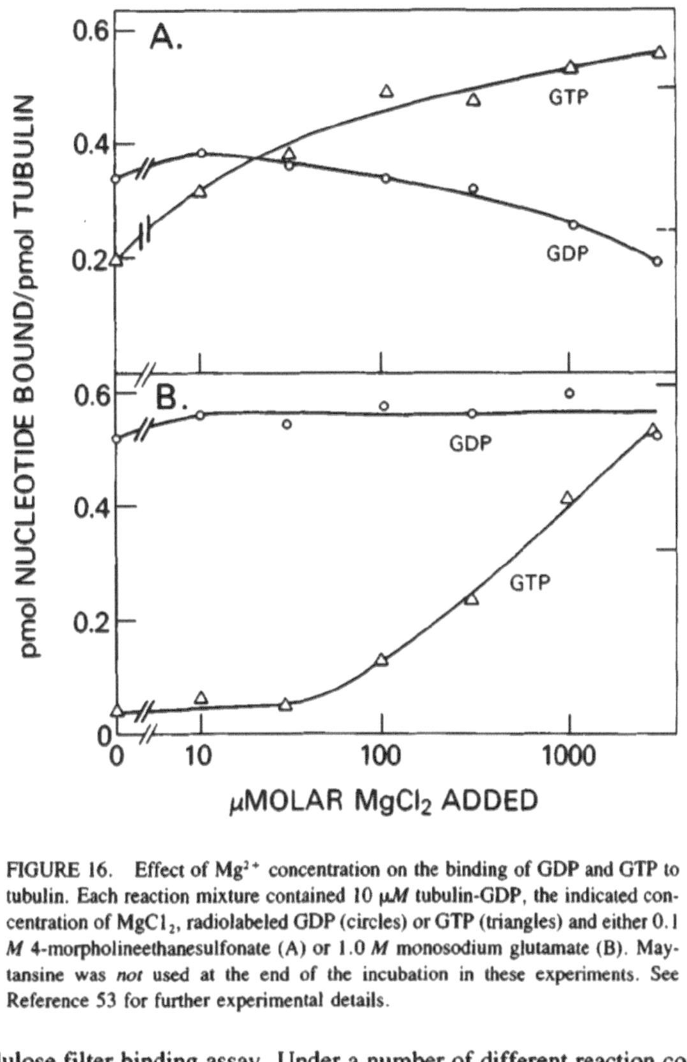

FIGURE 16. Effect of Mg^{2+} concentration on the binding of GDP and GTP to tubulin. Each reaction mixture contained 10 μ*M* tubulin-GDP, the indicated concentration of $MgCl_2$, radiolabeled GDP (circles) or GTP (triangles) and either 0.1 *M* 4-morpholineethanesulfonate (A) or 1.0 *M* monosodium glutamate (B). Maytansine was *not* used at the end of the incubation in these experiments. See Reference 53 for further experimental details.

in a nitrocellulose filter binding assay. Under a number of different reaction conditions, our laboratory observed that omission of exogenous Mg^{2+} from the reaction mixture markedly enhanced the effectiveness of GDP as an inhibitor of tubulin polymerization.[53,90,122] We went on to demonstrate that this arose from a marked difference in the binding properties of GDP and GTP to tubulin in the presence and absence of exogenous Mg^{2+}.[53] The binding of GDP to tubulin was essentially unaffected by the Mg^{2+} concentration of the reaction mixture (Figure 16), or even by addition of EDTA in the absence of exogenous Mg^{2+} (Table 7). The binding of GTP to tubulin was markedly enhanced by addition of exogenous Mg^{2+} to reaction mixtures (beyond the tightly bound molar equivalent of Mg^{2+}) (Figure 16), and addition of EDTA almost eliminated GTP binding to tubulin (Table 7). Similarly, radiolabeled GDP was readily displaced from tubulin by GDP, but not GTP, in the absence of exogenous Mg^{2+} (Table 7). Correia et al.[54] have recently confirmed these observations and extended them in a more quantitative analysis; they found little difference in the affinities of GDP and GDP-Mg^{2+} for tubulin, but they concluded that the binding affinity of GTP was at least 1100-fold less than that of GDP, GDP-Mg^{2+}, and GTP-Mg^{2+}. Nucleotide binding data at several Mg^{2+} concentrations further demonstrated the differing affinities of GDP and GTP for tubulin as the Mg^{2+} varied.[63] Expressed as a dimensionless equilibrium constant for the exchange of GTP for GDP on tubulin, a value of 0.2 was obtained in the

TABLE 7

Effects of Exogenous Mg^{2+} and EDTA on the Binding of radiolabeled GDP and GTP to Tubulin and Displacement of [8-^{14}]GDP from the Exchangeable Site by Exogeneous Nucleotide

Reaction condition	Mole nucleotide per mole tubulin		[8-^{14}C]GDP displaced (percent of control)	
	GDP bound	GTP bound	By GDP	By GTP
0.1 *M* Mes	0.37	0.19	47	25
0.1 *M* Mes + 1 mM Mg^{2+}	0.26	0.53	50	82
0.1 *M* Mes + 0.2 mM EDTA	0.40	0.04	54	9

Note: In the binding experiments reaction mixtures contained 10 μ*M* tubulin-GDP, either [8-^{14}C]GDP or [8-^{14}C]GTP at 100 μ*M*, and other components as indicated. In the displacement experiment reaction mixtures contained 10 μ*M* tubulin-[8-^{14}C]GDP, nonradiolabeled GDP or GTP at 10 μ*M*, and other components as indicated. In these experiments maytansine was *not* added at the end of the incubation. Mes, 4-morpholineethanesulfonate. See Reference 53 for further experimental details

absence of exogenous Mg^{2+}, 2.7 with 0.2 m*M* Mg^{2+}, 5.0 with 0.5 m*M* Mg^{2+}, 10.8 with 2 m*M* Mg^{2+}, and 10.6 with 4 m*M* Mg^{2+}.[63]

Based on studies with Mn^{2+} (see Section C) Jemiolo and Grisham[52] and Monasterio[116] have proposed that the tightly bound Mg^{2+} cation is bound as a complex with nucleotide at the exchangeable GTP site. We feel that this is inconsistent with the requirement for *additional* Mg^{2+} for the binding of GTP to tubulin[9,53,54] and with the inhibition of microtubule assembly in sulfonate buffers by high concentrations of GTP in the absence of exogenous Mg^{2+}.[53] Moreover, we have recently observed release of tubulin-bound Mg^{2+} despite retention of nucleotide at the exchangeable site.[49]

Low concentrations of Mg^{2+} also favor increased direct incorporation of tubulin-[8-^{14}C]GDP into microtubules (without GTP hydrolysis), presumably as a consequence of increased concentrations of tubulin-GDP at the expense of tubulin-GTP in reaction mixtures.[62] Further investigation in our laboratory of Mg^{2+} effects on tubulin-nucleotide interactions[63] have demonstrated that the higher the concentration of exogenous Mg^{2+} in the reaction mixture (excluding the range from 0 to 0.2 m*M*), the higher the concentration of GTP required for the initiation of microtubule assembly (Figure 17A). Even more striking, as a consequence of enhanced binding of GTP to tubulin at higher Mg^{2+} concentrations, was the finding that the minimum concentration of tubulin-GTP required for initiation of assembly rose steadily as the Mg^{2+} concentration rose and was minimal in the absence of exogenous Mg^{2+} (Figure 17B). On the other hand, efficiency of incorporation of tubulin-GDP into elongating microtubules was about half that of incorporation of tubulin-GTP at low Mg^{2+} concentrations; but both tubulin-GTP and tubulin-GDP were incorporated into elongating microtubules with equal efficiency at high Mg^{2+} concentrations. These findings were interpreted in terms of enhanced oligomer formation at higher Mg^{2+} concentrations:[65,144-146] as oligomers contain increasing concentrations of tubulin-GDP they become less effective as nucleating species but retain their ability to elongate microtubules.

Olmsted and Borisy[135] found that exogenous Mg^{2+} concentrations over 10 m*M* inhibited microtubule assembly from microtubule protein, with the extent of assembly unaffected from 0 to 10 m*M* but maximal rates occurring when the cation was present at 0.1 to 1.0 m*M*. Lee and Timasheff[142] interpreted their findings on the effects of Mg^{2+} in the glycerol-dependent polymerization of purified tubulin as indicating incorporation of an additional mole of Mg^{2+} per mole of tubulin into polymer. Martin et al.[147] have examined Mg^{2+} effects on microtubule nucleation, elongation, and cold-induced disassembly using microtubule protein. They found that nucleation was maximal at 0.5 m*M* Mg^{2+} with microtubule length

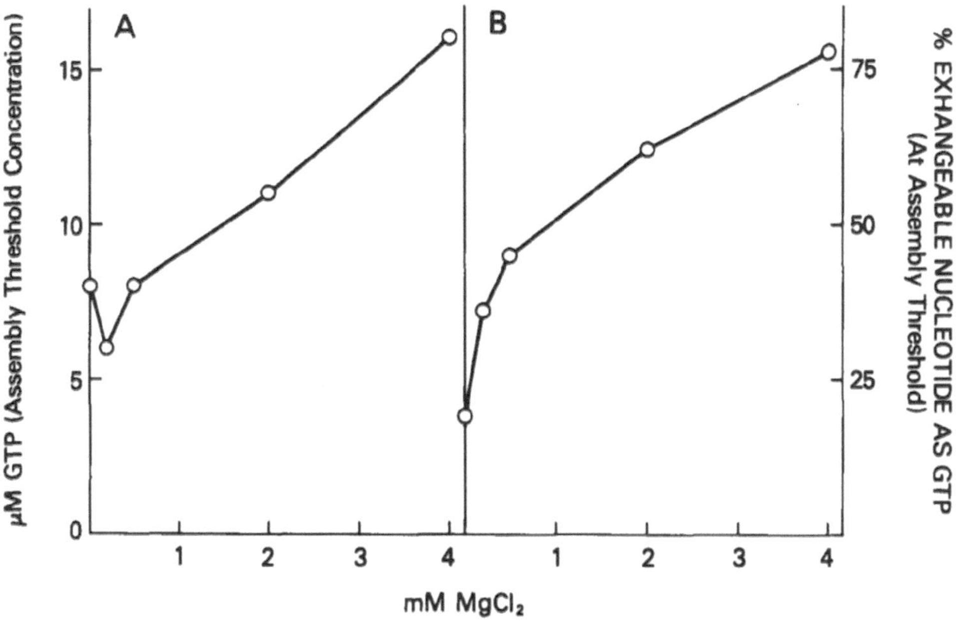

FIGURE 17. Effect of Mg^{2+} concentration on the GTP requirement for the initiation of microtubule assembly. (A) GTP concentration for initiation of assembly as a function of Mg^{2+} concentration. Reaction mixtures contained 20 μ*M* tubulin-GDP, 0.7 mg/ml heat-treated microtubule-associated proteins, and $MgCl_2$ and GTP as indicated. The figure presents the highest GTP concentration at which tubulin polymerization was never observed at each Mg^{2+} concentration (the assembly threshold). (B) The proportion of tubulin bearing GTP in the exchangeable site at the GTP assembly threshold as a function of Mg^{2+} concentration. Reaction mixtures contained either tubulin-GDP (nonradiolabeled) + [8-^{14}C]GTP or tubulin-[8-^{14}C]GDP + nonradiolabeled GTP at appropriate concentrations to determine the proportion of tubulin-bound GDP and GTP. Reactions were stopped by addition of maytansine. See Reference 63 for further experimental details.

increasing at both higher and lower concentrations of the cation. The elongation rate had an identical Mg^{2+} optimum. Disassembly rates, on the other hand, increased as the Mg^{2+} concentration rose. Unger et al.[148] obtained similar results in studies of glycerol-induced polymerization of microtubule protein. Higher Mg^{2+} concentrations delayed microtubule assembly, implying decreased nucleation, and yielded polymer with decreased stability to cold and Ca^{2+}. Finally, Mg^{2+} enhances the stability of tubulin, reducing its rate of denaturation.[5,53,60,137]

B. CALCIUM

Weisenberg[139] first reported that microtubule assembly with a partially purified tubulin preparation required ethylene glycol bis(β-aminoethyl ether)-*N,N,N′,N′*-tetraacetate (EGTA) and was inhibited by $CaCl_2$. Soon after, addition of Ca^{2+} to preformed microtubules was noted to cause their disassembly.[140,149-151] Subsequently, the role of Ca^{2+} as an inhibitor of microtubule assembly and as an agent inducing the depolymerization of preformed microtubules has been investigated in many laboratories.

Hayashi and Matsumara[152] first described the binding of $^{45}Ca^{2+}$ to microtubule protein, localized the site of binding to tubulin, and estimated the K_D as 15 μ*M*. The binding of the isotope to tubulin (in the form of microtubule protein) was then examined carefully by Solomon,[153] who reported one high affinity site for Ca^{2+} with a K_D of 3 μ*M* together with approximately 16 low affinity sites with an average K_D of 0.3 m*M*. Binding to the high affinity sites was inhibited by both Mg^{2+} and KCl, while binding to the low affinity sites was inhibited by KCl only. The high affinity Ca^{2+} site was distinct from the colchicine, vinblastine, and exchangeable nucleotide sites.

The proteolytic enzyme subtilisin cleaves oligopeptides (2000 to 4000 Da molecular weight) from the carboxyl-termini of both α- and β-tubulin, yielding a residual protein almost completely resistant to Ca^{2+} inhibition of polymerization,[154] and with the polymer formed insensitive to Ca^{2+}-induced depolymerization.[155] These observations led Serrano et al.[154] to reevaluate the binding of $^{45}Ca^{2+}$ to tubulin. They found 1.4 high affinity sites and 12 low affinity sites, with dissociation constants similar to those of Solomon.[153] They went on to localize the Ca^{2+}-binding properties of the tubulin subunits and their proteolytic fragments. Even following denaturation and separation on polyacrylamide gels, both subunits bound approximately equal amounts of Ca^{2+}, but subtilisin-digested tubulin only bound Ca^{2+} at the low affinity sites. The subtilisin oligopeptides continued to bind Ca^{2+} with high affinity; using other proteolytic enzymes, it appeared that the carboxyl-termini of both subunits retained the capacity to bind Ca^{2+}. Consequently, Serrano et al.[154] concluded that there are two high affinity Ca^{2+} sites on tubulin, one at the carboxyl-terminus of each tubulin subunit. Similar results were obtained by Fong et al.,[156] although these workers reported only six low affinity sites. In addition Fong et al.[156] noted substantial inhibition of Ca^{2+} binding by Mg^{2+} at both classes of site, with the major effect on the number of sites and little change in their affinity for Ca^{2+}. It is thus possible that only one of the two high affinity Ca^{2+} sites is inhibited by Mg^{2+}.

Ca^{2+} effects on microtubule assembly can be divided into three categories: (1) direct effects of the cation on microtubule components, (2) effects on microtubules mediated through Ca^{2+} binding proteins, and (3) indirect effects mediated through enzymes which alter microtubule components following their activation by Ca^{2+}. It is generally believed that physiologically significant Ca^{2+} effects can only occur in the micromolar range, although often *in vitro* dramatic Ca^{2+} effects require millimolar concentrations of the cation.

Microtubule assembly in tissue homogenates tends to be significantly more sensitive to inhibition by Ca^{2+} than does microtubule assembly from microtubule protein.[135,139,150,157,158] An especially notable example of this was described by Nishida and Sakai.[158] Moreover, almost without exception, inhibition of polymerization of partially purified or electrophoretically homogeneous tubulin requires lower Ca^{2+} concentrations than does inhibition of polymerization of microtubule protein or of reconstituted systems of tubulin + microtubule-associated proteins, often at the micromolar Ca^{2+} concentrations which may be physiologically significant.[136,140-142,159-162] Additional microtubule-associated proteins can reduce Ca^{2+}-sensitivity,[162,163] while microtubule-associated proteins from Ehrlich ascites cells may even eliminate Ca^{2+}-sensitivity.[164] Ca^{2+} sensitivity is generally enhanced at lower temperatures,[159,161,165] lower tubulin or microtubule protein concentrations,[159] and higher potassium[150,166] and magnesium[135,148,167] concentrations. Ca^{2+} inhibition of microtubule assembly is believed to occur primarily at the level of nucleation, with lesser effects on elongation.[159,161,162] Ca^{2+}-induced depolymerization appears to occur primarily from microtubule ends, rather than as a diffuse disintegration along the entire microtubule.[163,168-171] Ca^{2+} also strongly inhibits tubulin-dependent GTP hydrolysis, probably as a consequence of inhibiting polymerization.[172]

The calcium-binding protein studied most extensively in terms of its interactions with microtubule components is calmodulin. Studies with cells and cytoskeletal extracts have demonstrated association of calmodulin with both interphase and spindle microtubules.[171,173,174] Injection of a Ca^{2+}-calmodulin complex into fibroblasts causes a localized depolymerization of microtubules.[175] *In vitro* calmodulin markedly reduces the Ca^{2+} concentration, often to micromolar levels, required for both inhibition of microtubule assembly and for depolymerization of preformed microtubules, although calmodulin must usually be present in superstoichiometric amounts (relative to tubulin) for these effects to be observed (generally, the higher the Ca^{2+} concentration used, the lower the calmodulin concentration required, and vice versa).[159,160,163,171,174,176] An exception to this generalization is the report of Job et al.[177] on the effects of calmodulin + Ca^{2+} on cold-stable microtubules. Normally these structures are completely resistant to depolymerization by Ca^{2+}, but at a calmodulin to tubulin ratio of 1:10 (but a high Ca^{2+} concentration of 0.8 m*M*) they depolymerize rapidly (also see below).[177] The effect of calmodulin + Ca^{2+} in inhibiting microtubule assembly is enhanced at higher KCl concentrations.[176]

The target microtubule protein molecule for calmodulin remains controversial. Kumagai and Nishida[178] first reported that tubulin, and not microtubule-associated proteins, bound calmodulin, and subsequently this binding was carefully quantitated; tubulin was found to bind two molecules of calmodulin, in the presence of Ca^{2+}, with similar affinities (K_D = 3.5 μ*M*).[179] Other workers have not been able to confirm this binding of calmodulin to tubulin, and instead observed binding to microtubule-associated proteins.[180,181] Sobue et al.[180] described binding only to tau factor, while Lee and Wolff[181] found binding to both tau factor and MAP-2. The binding of calmodulin to tau factor and MAP-2 was confirmed by Kumagai et al.[182] The apparent K_D values of calmodulin for both tau factor and MAP-2 are similar (2 to 7 μ*M*).[181,182] The binding site of calmodulin on MAP-2 has been localized to the small assembly promoting region (molecular weight about 40,000 Da).[183] In support of the idea that only microtubule-associated proteins interact with calmodulin are the observations that the polymerization of purified tubulin is not inhibited by calmodulin + Ca^{2+} [160-162] and that microtubule-associated proteins clearly alter calmodulin effects.[159,160,163] Nevertheless, the effects of calmodulin + Ca^{2+} on microtubule assembly cannot be duplicated by merely manipulating addition of microtubule-associated proteins to tubulin,[162,182] leaving open the possibility that interaction of tubulin with microtubule-associated proteins may create a functional calmodulin site on tubulin.

A second Ca^{2+}-binding protein which has received attention as a potential agent in regulating microtubule assembly is the S-100 family of proteins derived from mammalian brain tissue. These are three small, related, dimeric proteins which are soluble in 100% ammonium sulfate. They contain Ca^{2+} binding sequences, but their affinity for Ca^{2+} is at least tenfold less than that of calmodulin.[184] The S-100 proteins enhance microtubule disassembly in the presence of millimolar Ca^{2+} concentrations,[185,186] perhaps more effectively than calmodulin,[185] and they inhibit microtubule assembly in the presence of micromolar Ca^{2+} levels.[186,187] Their effect is enhanced at lower microtubule protein concentrations, lower temperatures, higher KCl concentrations, and higher pH values.[186,188,189] At high pH values, in fact, the S-100 proteins inhibit microtubule assembly without Ca^{2+} in the reaction mixture, although Ca^{2+} does enhance their effect (Ca^{2+} contamination in reaction components was not, however, rigorously excluded).[189] Unlike with calmodulin, the S-100 proteins + Ca^{2+} inhibit the polymerization of purified tubulin as well as that of microtubule protein.[188,189] In addition, binding of S-100 proteins to tubulin[189,190] and to tau factor[191-193] has been described.

Finally, a few reports have described more indirect effects of Ca^{2+} on microtubule assembly. Sandoval and Weber[194] described a Ca^{2+}-activated protease which degraded high molecular weight microtubule-associated proteins, thereby inhibiting microtubule assembly. Three groups have reported a calmodulin + Ca^{2+}-dependent protein kinase which phosphorylated microtubule components (both tubulin and microtubule-associated proteins), resulting in either microtubule disassembly[195] or inhibition of microtubule assembly.[196,197] Similarly, Job et al.[198] described a calmodulin + Ca^{2+}-dependent phosphorylation reaction which resulted in the depolymerization of cold-stable microtubules.

C. OTHER DIVALENT CATIONS

Several groups have described detailed studies on the effects of Zn^{2+} on tubulin polymerization.[114,199-202] Multiple bizarre structures were formed, most particularly extensive sheets of protofilaments. In electron micrographs the protofilaments appear to be parallel, with their axis both parallel and at an angle to the longitudinal axis of the sheets. The latter sheets were formed only in the absence of either GTP or microtubule-associated proteins and often closed into structures which could be described as macrotubules.[200] Image reconstruction, however, indicates that the apparently parallel protofilaments actually alternate in polarity so that they are more accurately described as oriented in an antiparallel fashion, as opposed to the parallel protofilaments of microtubules.[203,204] The Zn^{2+}-induced polymer was resistant to Ca^{2+}- and colchicine-induced depolymerization, but different laboratories have reported both stability and lability to cold.[199,202] The polymer is rapidly dispersed by addition of either sulfhydryl reactive agents or Zn^{2+}-chelating agents.[200,202] The formation of the Zn^{2+}-induced polymer is not inhibited by podophyllotoxin or colchicine (with radiolabeled colchicine recovered in the polymer), while vinblastine inhibits its formation only if high concentrations of Mg^{2+} are also present.[202] At very low Zn^{2+} concentrations microtubules rather than sheets are formed;[201] in the presence of both taxol and Zn^{2+}, purified tubulin forms sheets without GTP.[114] Based on the effects of Zn^{2+} concentration on sheet morphology (sheets are wider and shorter at higher cation levels) de la Torre et al.[205] suggested that Zn^{2+} promotes lateral tubulin-tubulin interactions between protofilaments and found up to 1.3 mol of Zn^{2+} per mole of tubulin in the sheet polymers. Eagle et al.[206] examined interactions of Zn^{2+} with both unpolymerized tubulin and the sheet polymers. They found over 60 binding sites for Zn^{2+} on tubulin, of which about seven to ten were of higher affinity. Depending on reaction conditions, the K_D values for the higher affinity sites were 1 to 3 μM and for the lower affinity sites, 16 to 85 μM. The binding of Zn^{2+} to tubulin did not displace the tightly bound Mg^{2+} cation, as determined by elemental analysis, and the binding of Zn^{2+} was strongly inhibited by Cu^{+} and Cd^{2+}, weakly inhibited by Mg^{2+}, Mn^{2+}, and Co^{2+}, and unaffected by Ca^{2+}. Eagle et al.,[206] too, found up to 1.3 molar equivalent of Zn^{2+}, relative to tubulin, in the sheet polymer. Several groups have concluded that Zn^{2+} exerts its effects by binding directly to tubulin rather than as a Zn^{2+}-GTP complex[114,205,206] and that Zn^{2+} induces formation of tubulin oligomers as well as sheet polymers.[205,206] Zn^{2+} also accelerates the inactivation of tubulin.[206]

Buttlaire et al.[138] described interactions of Mn^{2+} with purified tubulin. The cation could substitute for the Mg^{2+} requirement in dimethylsulfoxide-induced microtubule assembly from purified tubulin. Exploiting the electron paramagnetic resonance (EPR) spectroscopic properties of Mn^{2+}, these workers described its binding to tubulin. There appeared to be one high affinity site with a K_D value of approximately 2 μM and about eight low affinity sites with an average K_D value of 0.4 mM. Mg^{2+}, Co^{2+}, and Zn^{2+}, but not Ca^{2+}, could partially displace bound Mn^{2+} from tubulin. Binding of Mn^{2+} at its high affinity site appeared to displace the tightly bound atom of Mg^{2+}, but this conclusion was not confirmed by elemental analysis. The EPR spectrum of tubulin-bound Mn^{2+} indicated that the cation bound at the high affinity site was readily accessible to the solvent. Gaskin[114] reported that Mn^{2+} could replace Mg^{2+} with microtubule protein in supporting microtubule assembly. Jemiolo and Grisham[52] confirmed by elemental analysis that Mn^{2+} displaced the tightly bound Mg^{2+} from tubulin; they obtained binding data for Mn^{2+} by EPR spectroscopy virtually identical to those of Buttlaire et al.[138] Jemiolo and Grisham[52] prepared nucleotide-depleted tubulin and, exploiting the EPR properties of Mn^{2+}, demonstrated that the binding of the cation to this tubulin was substantially reduced unless its GTP content was increased. They also noted that the EPR spectrum of tubulin-bound Mn^{2+} was very similar to the spectra of Mn(II)-GTP and Mn(II)-GDP (the spectrum of tubulin-bound Mn^{2+} differed significantly in its details, however, from the equivalent spectrum presented by Buttlaire et al.[138]). Jemiolo and Grisham[52] concluded that the tightly bound Mn^{2+} (and by extension, the tightly bound Mg^{2+}) was bound to tubulin by the exchangeable nucleotide. This conclusion was supported by Monasterio,[116] who constructed a tubulin-Mn^{2+}-GTPγF-colchicine complex and examined its spectral properties. These indicated that the divalent cation and the γ-phosphate group of the exchangeable nucleotide were separated by 6 to 8 Å, suggesting that they are bound to tubulin as a complex.

Co^{2+}, like Zn^{2+}, induces both purified tubulin and microtubule protein to form extensive sheet polymers[114,201,207] and, at low concentrations, microtubules.[201] Wallin et al.[207] reported that the Co^{2+}-induced sheets were stable to colchicine- and Ca^{2+}-induced depolymerization. However, Haskins et al.[201] found them to be sensitive to cold, colchicine, GDP, vincristine, and high ionic strength. The proportion of sheets, as opposed to microtubules, formed in Co^{2+}-containing solutions was increased at lower pH values, higher dimethylsulfoxide concentrations, and lower GTP concentrations.[201,208] Himes et al.[208] studied the binding of Co^{2+} to tubulin under a variety of reaction conditions, and they found great variability in the number of binding sites and their affinities. GTP in the reaction mixture particularly reduced the amount of Co^{2+} bound to the protein. In its absence, there were about 8 high affinity sites with a K_D value of about 20 μM and an additional 20 low affinity binding sites. At the opposite extreme, with GTP in the reaction mixture, there were one to two high affinity sites with a K_D value of 0.7 μM. Binding of Co^{2+} to tubulin was associated with loss of about half the tightly bound Mg^{2+} cation and about one third of its guanine nucleotide. Binding of Co^{2+} was strongly inhibited by Cu^{+}, Cd^{2+}, and Zn^{2+}, moderately inhibited by Mn^{2+}, weakly inhibited by Mg^{2+}, and not inhibited by Ca^{2+}. Tubulin containing one to two tightly bound Co^{2+} cations required additional Co^{2+} to polymerize. The polymer generally contained about one molar equivalent each of Co^{2+} and Mg^{2+}. It was highly labile if EDTA was added to it.[208] Himes et al.[208] concluded that direct binding of Co^{2+} at both high and low affinity sites on tubulin was required for the cation to induce tubulin polymerization.

High concentrations of Hg^{2+}, Cu^{2+}, and Cd^{2+} inhibit microtubule assembly and substantially reduce the free sulfhydryl content of tubulin.[207] High concentrations of Sr^{2+} and Ba^{2+}, like Ca^{2+}, both inhibit microtubule assembly and cause the disassembly of preformed microtubules.[148]

Finally, although it is not a divalent cation, a recent report describing the promotion of microtubule assembly by Al^{3+} should be noted here.[209] It was estimated that free Al^{3+} in the picomolar range was sufficient to initiate microtubule assembly in a reaction condition without microtubule-associated proteins. The polymerization reaction occurred without GTP hydrolysis, and the microtubules formed were stable to Ca^{2+}-induced depolymerization but labile to cold. These findings were practically identical to those obtained with CrGTP[113] (see Section II.B), suggesting that the active species is a relatively stable Al(III)-GTP complex.

D. SUMMARY

Tubulin has complex interactions with divalent cations and is able to bind them at both high and low affinity sites, with as many as 60 sites reported. The relationship of high affinity sites for different cations remains largely unclear at the present time. Similarly, specific information is presently unavailable on the role of both high and low affinity sites in such diverse phenomena as enhancement vs. inhibition of assembly, or in formation of microtubules vs. aberrant polymers vs. oligomers. It seems likely that divalent cations, particularly Mg^{2+}, play a role in the binding of GTP at the exchangeable site, although it remains uncertain what part they play in the binding of either GDP at the exchangeable site or of GTP at the nonexchangeable site. At the present time it seems most likely that Ca^{2+} has multiple effects on intracellular microtubules. The physiological effects of micromolar free Ca^{2+} concentrations may result from cumulative effects at multiple loci: direct binding to tubulin; binding to microtubule-associated proteins, and perhaps tubulin, via calmodulin; binding to tubulin and tau factor via S-100 proteins; and indirect effects on microtubule components mediated through Ca^{2+}-modulated effectors.

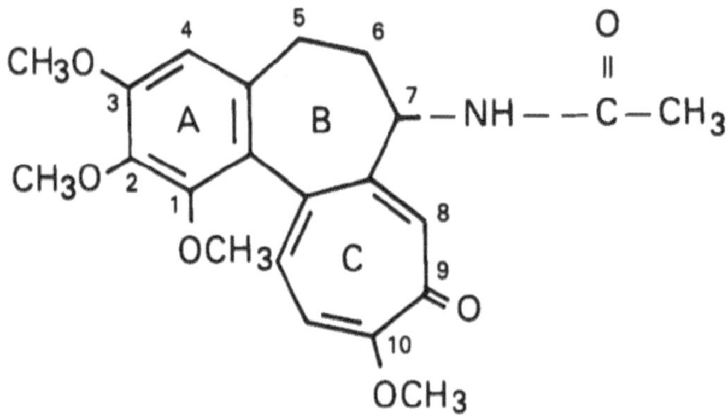

FIGURE 18. Structural formula of colchicine.

IV. ANTIMITOTIC DRUGS

This section of the chapter will be almost entirely restricted to agents which have been documented to cause mitotic arrest in mammalian cells at concentrations of 10 μM or less and to have substantial effects on tubulin polymerization when present in reaction mixtures in concentrations stoichiometric with that of tubulin (although observations with tubulin from other sources will also be summarized).

A. DRUGS BINDING AT THE COLCHICINE SITE

1. Colchicine and Colchicine Analogues

Although its medical uses are now limited, colchicine (obtained from *Colchicum autumnale* and *Gloriosa superba;* structure in Figure 18) has been of major importance in biochemical studies of microtubules and in studies of their intracellular functions. In fact, the initial purification of tubulin exploited its strong interaction with colchicine.[1] Until *in vitro* microtubule assembly conditions were established,[139,210] the tubulin-colchicine interaction was the most intensively examined aspect of microtubule biochemistry.

The colchicine binding activity of tubulin is perhaps the most readily quantifiable characteristic of the lability of the protein, and it enabled various workers to describe reaction components that stabilize tubulin: guanine nucleotides,[1,3,5,211-215] glucose,[211] sucrose,[216-219] glycerol,[3,217,220] glutamate,[5,211,221] organic phosphates and other anions,[5,221] vinblastine,[211,214,215,218,222-225] colchicine itself,[1,225,226] microtubule protein components,[227,228] albumin,[5] and Mg^{2+}.[5,53,212] At the same time, like guanine nucleotide binding (see above), the colchicine binding activity of tubulin is more resistant to inactivation than is the ability of tubulin to polymerize. Workers have reported relative resistance of colchicine binding to time-dependent decay,[219] ionizing radiation,[229] alkylation of cysteine,[230,231] histidine,[51] lysine,[232] or arginine[233] residues, and proteolytic digestion of tubulin.[234]

The interaction of the drug with tubulin is relatively slow, dependent on increased reaction temperature, noncovalent, and so tight that, for most purposes, it can be treated as an irreversible reaction.[1,211,212,214,215,222-224,235-237] Many workers have studied the binding reaction, attempting to define it quantitatively. Reported K_D values have ranged from 0.02 to 5 μM.[212-215,218,222-226,236-251] The lowest K_D values were obtained by workers who attempted to determine the reverse rate constant, separating this slow reaction from the faster denaturation of tubulin. These values ranged from 0.009 to 0.049 h^{-1}.[214,226,241,242,245] Forward rate constants have ranged from 10^5 to 2×10^6 M^{-1} h^{-1}.[213,214,226,236,241,242,245] Much of this variability is probably due to differences in reaction conditions, reaction temperature, and source of tubulin.[224,241,242,244,246] In addition, the binding of colchicine to tubulin has been reported to reduce threefold the rate of dissociation of the α-β tubulin dimer.[252] There have been at least three reports of thermodynamic parameters of the overall binding reaction — one to rat brain tubulin,[223] one to sea urchin tubulin,[224] and one to vinblastine paracrystals of sea urchin tubulin.[238] The enthalpy values reported were 7 to 16 kcal/mol; the entropy values, 47 to 80 entropy units; and the free energy values, −9.2 kcal/mol at 37°C and −6.7 kcal/mol at 13°C. The activation energy of the foward reaction has been calculated as 20 kcal/mol;[253] and estimated for the reverse reaction as about 30 kcal/mol.[254]

Most workers, using Scatchard analysis, have reported only one class of binding site (although reported stoichiometry has ranged from as low as 0.1 to as high as 0.9).[215,222-225,238,242,246] There have, however, been several reports in which Scatchard analysis indicated two classes of binding sites. Williams et al.[255] described a high-performance liquid chromatographic Hummel-Dreyer analysis of colchicine binding to tubulin, with the K_D value of the low affinity sites estimated at 65 μ*M*. Ringel and Sternlicht[256] found ^{13}C nuclear magnetic resonance evidence for low affinity colchicine site(s) with a K_D value of 650 μ*M*. In addition, three analogues derivatized at the amino group of the B ring (bromocolchicine, colcemid, and an allocolchicine spin probe) have been noted to bind to tubulin at both high and low affinity sites with K_D values in the ranges described above for both sites.[257-259]

Many workers have undertaken studies to help localize the colchicine binding site on tubulin. Assuming that inhibition of the covalent reaction by colchicine represents specificity, two affinity analogues modified at the amino group on the B ring form covalent bonds either predominantly[257] or almost entirely[249] with α-tubulin. The resulting conclusion that the colchicine site is on α-tubulin has been supported by findings with protease-digested tubulin:[234,260] radiolabeled colchicine binds to isolated carboxyl terminal 12- and 16-KDa fragments of α-tubulin, with K_D values of 0.2 to 0.4 μ*M*, although the stoichiometry of binding was low (0.1 mole colchicine per mole of fragment). In addition, a mutant Chinese hamster ovary (CHO) cell line resistant to colcemid with an altered α-tubulin has been described.[261] On the other hand, mutant CHO cell lines resistant to colchicine and colcemid with altered β-tubulins have also been isolated,[262] as have *Aspirgillus nidulans* mutants, with altered β-tubulins, which were resistant to benomyl[263] (a carbamate antimitotic, presumably binding at the colchicine site, see Section 5). In addition, Luduena and co-workers[264-267] have extensively documented that colchicine and a number of other agents binding at the colchicine site substantially inhibit the alkylation and intrapolypeptide crosslinking of β-tubulin by sulfhydryl reactive agents (the alkylation of α-tubulin was also inhibited). When high concentrations of single ring analogues of the A and C rings were examined, the following results were obtained:[267] (1) aklylation of sulfhydryl groups by iodoacetamide was enhanced by both types of analogue and (2) the formation of intrapolypeptide β-tubulin crosslinks was not affected by C ring analogues (e.g., tropolone methyl ether), but was extensively inhibited by A ring analogues (e.g., 3,4,5-trimethoxybenzaldehyde). These results were interpreted as indicating that effects on alkylation by iodoacetamide were a consequence of colchicine-induced conformational changes, while the effects of intrapolypeptide crosslink formation represented direct blockage of the involved sulfhydryl group(s) by the trimethoxybenzene A ring of colchicine.[267]

While it is universally agreed that colchicine binds directly to the tubulin dimer — and to paracrystals and other oligomeric structures induced by vinblastine[238,248,268-272] — there have been differences in reports on both ring oligomers and microtubules. Some workers have found little colchicine bound to ring oligomers,[149,245,256] while others find that these structures do bind colchicine.[144,273,274] In this regard, microtubule-associated proteins, which enhance oligomerization, have been reported both to act as apparent competitive inhibitors of the binding of colchicine to tubulin[246] and to reduce the affinity of tubulin for colchicine.[274] Similarly, preformed microtubules have been reported to have little ability to bind colchicine.[140,222,248] Microtubule protein has much reduced colchicine binding ability following

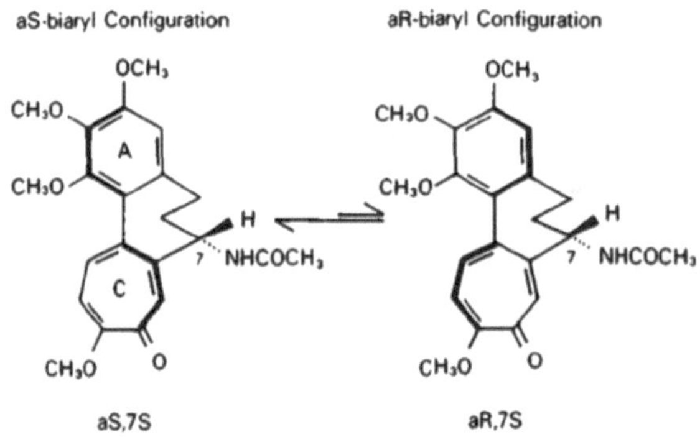

FIGURE 19. Proposed biaryl configuration of colchicine.[250]

polymerization.[76,229] However, there have been two reports describing experiments with B ring amino-derivatized colchicine analogues which indicate that colchicine can bind along the entire length of preformed microtubules.[275,276] In one case, the colchicine derivative had peroxidase activity, permitting visualization of labeled microtubules in the electron microscope.[275] In the other, the colchicine derivative was intensely fluorescent and clearly bound to microtubules at concentrations which did not disrupt them.[276]

The molecular mechanism of the binding of colchicine to tubulin has received considerable attention. It is probable that colchicine itself undergoes some conformational change during the binding process, for the tropolone moiety becomes fluorescent upon binding to the protein (excitation maximum at 362 nm, emission maximum at 435 nm, quantum yield about 0.03)[223,277] and its circular dichroic spectrum is altered (loss of a negative band at 340 nm).[247] Bhattacharyya and Wolff[278] have noted that the fluorescence can be mimicked by increasing solvent viscosity or by otherwise immobilizing the tropolone ring, suggesting that this moiety is immobilized upon binding of the drug to tubulin. This interpretation is supported by the study of Deinum and Lincoln[259] with an allocolchicine spin probe. The electron spin resonance spectrum of the drug-tubulin complex indicated that the probe was tightly bound and highly immobilized by the protein. Detrich et al.[247] suggested that the change in the circular dichroic spectrum represented a flattening of the colchicine molecule upon binding to tubulin, with a marked decrease in the dihedral angle between the A and C rings from about 55° observed in crystals of colchicine and colchicine analogues[279-281] to less than 20°. Yeh et al.[250] recently noted that colchicine can nominally exist in solution in two biaryl configurations, termed aS and aR (Figure 19). From circular dichroic measurements on colchicine and deacetamidocolchicine both in solution and bound to tubulin, they concluded, first, that the drug which bound to tubulin was primarily in the aS configuration and, second, that its circular dichroic properties derived primarily from the biaryl configuration rather than from the asymmetric carbon at position 7.

The development of fluorescence upon the binding of colchicine to tubulin has permitted a detailed kinetic analysis of the reaction.[245,274] These experiments led to the conclusion that the reaction is not a simple bimolecular binding process. There was both a fast and slow phase to fluorescence development. Analysis of the fast phase was most consistent with a model in which more rapid binding of colchicine to tubulin was followed by a slower conformational change in the protein.[245,274] Thermodynamic parameters of the rapid binding step include an enthalpy value of -8 kcal/mol and an entropy value of -15 entropy units.[274,282] (The slow phase of fluorescence development is poorly understood, but it may represent slow conversion of a nonbinding tubulin conformation to one which can bind the drug.[274])

The concept that tubulin undergoes a conformational change upon binding colchicine is supported by changes in its sulfhydryl group reactivity,[259,264-267,283] its immunological properties,[284] and its spectral properties.[282] Andreu and Timasheff[282,285] examined interactions of both colchicine and single ring analogues of the drug with tubulin (also see below) and proposed that the mechanism of colchicine binding involves an initial, rate-limiting interaction of the tropolone C ring with the protein, followed by a rapid interaction of the trimethoxybenzene A ring. They pointed out the near identity of the thermodynamic parameters they had obtained for the binding of tropolone methyl ether and those obtained by Lambier and Engelborghs[274] with colchicine noted above. Andreu and Timasheff[282,285] also proposed that the major contribution to the conformational change in tubulin derived from binding of the C ring, as the spectral changes observed with colchicine occurred to a large extent with tropolone methyl ether, but not with podophyllotoxin.

Upon binding to tubulin, colchicine induces a GTPase reaction which occurs independently of polymerization.[36,55,56,172] Pantaloni's group[55] initially reported that this reaction occurred independently of tubulin concentration, implying that it was a property of the individual tubulin molecule; but subsequently they observed marked cooperativity effects, with GTP hydrolysis a sigmoidal function of tubulin concentration.[286] We have similarly observed a marked effect of tubulin concentration on colchicine-induced GTP hydrolysis.[49] These findings strongly imply that with colchicine nucleotide hydrolysis requires tubulin-tubulin interactions. Andreu and Timasheff,[282] however, found that colchicine-induced GTP hydrolysis was independent of protein concentration, and Arai and Kaziro[35] reported inhibition of hydrolysis by the drug. Not surprisingly, colchicine does not affect the binding of GTP to tubulin.[112,238]

The most dramatic effects of colchicine on cells are mitotic arrest (the appearance of large numbers of cells with condensed chromosomes), the failure of spindle formation, and the disappearance of cytoskeletal microtubules. This latter effect is most impressively demonstrated by immunofluorescent staining of drug-treated cells.[287] A large number of workers have reported that addition of colchicine to preformed microtubules causes their disassembly,[76,139,143,150,151,199,288-292] but this phenomenon has not been examined very quantitatively. Although depolymerization has been reported at 3,[290] 5,[292] and 30 μM[151] colchicine, most workers have used much higher drug concentrations, and on occasion little or no depolymerization has been observed at these higher concentrations.[76,140,293,294] The depolymerization reaction appears to occur in an endwise fashion.[76,289,295] It has been reported that microtubule-associated proteins significantly reduce the rate of depolymerization.[76,289] Deery and Weisenberg[76] argued that depolymerization occurs because the binding of tubulin to colchicine lowers the tubulin concentration below the critical concentration required to maintain polymer, but that the tubulin-colchicine complex also retains some ability to stabilize polymer.

Bergen and Borisy[296] examined a system in which colchicine failed to cause microtubule disassembly. If highly stable axoneme microtubules serve as nucleation centers, brain microtubules will elongate on them in two directions, which can be readily distinguished, presumably with different polarity. When diluted, the newly formed microtubules disassemble, while the axoneme remains intact. Microtubules both grow and shrink faster at the distal (+) anoneme end than at the proximal (−) end. Bergen and Borisy[296] found that colchicine and the tubulin-colchicine complex could completely inhibit dilution-induced disassembly of brain microtubules in this system, with the proximal end more readily stabilized than the distal end (50% stabilization at a three- to fivefold lower concentration of tubulin-colchicine complex).

The key aspect of the effect of colchicine on cells is undoubtedly the potent inhibition of the alkaloid of microtubule assembly. Taylor[297] was the first to point out that colchicine appeared to exert its effect in cells at substoichiometric concentrations: mitotic arrest occurred when less than 5% of the cellular tubulin was complexed with the drug. Quantitative results with *in vitro* tubulin polymerization from different laboratories have, however, shown considerable variability. In the following discussion, and subsequently with other drugs, IC_{50} values for inhibition of polymerization will be followed by either "plateau" or "rate" to indicate which parameter was measured by the investigators. In our own experience with multiple antimitotic agents evaluation of rates, without exception, results in IC_{50} values three- to fivefold lower than evaluation of the extent of assembly.[49,298] In addition, depending on the rate of binding of drug to tubulin, IC_{50} values can be enhanced by a drug-tubulin preincubation prior to assembly.[299,300] Unless noted in the summary presented here, it can be assumed that no preincubation was performed. Finally, lower tubulin concentrations in reaction mixtures results in lower IC_{50} values.[49] I will first summarize studies in which unliganded colchicine was added to reaction mixtures.

In three studies in which viscosity was used to follow microtubule assembly in brain tissue homogenates, IC_{50} (plateau) values of 0.2[301] and 0.5 μM[157,302] were obtained. The IC_{50} values were less than 5% of the tubulin concentrations. In a fourth report describing experiments in which viscometry was used, total inhibition of microtubule assembly was observed with 1 μM colchicine.[303] Virtually all studies in which turbidity was used to follow microtubule assembly yielded higher IC_{50} values: 3 μM (rate);[304] 3 μM (plateau; preincubated);[305] 3 μM (plateau);[306] 4 μM (plateau; preincubated);[307] 8 μM (plateau);[308] 10 μM (rate and plateau);[309] 10 μM (rate; preincubated);[310] 20 μM (rate and plateau);[251] 0.1 mole drug per mole tubulin (plateau; preincubated — i.e., 1 μM with tubulin at 1 mg/ml);[108] and, in our own laboratory, 3 μM (plateau; preincubated)[299] and 6 to 9 μM (plateau).[299,311] In the one exception an IC_{50} value of 0.2 μM (plateau) was obtained.[312] These studies were all performed with either microtubule protein or with tubulin + microtubule-associated proteins in sulfonate buffers. A relatively high IC_{50} value (10 μM, plateau) was reported for inhibition of dimethylsulfoxide-induced polymerization of purified tubulin.[313] Three laboratories have also reported extensive inhibition of polymerization at low colchicine concentrations, although detailed evaluation at multiple drug concentrations was not described. Andreu and Timasheff[282] observed 80% inhibition of the glycerol-Mg^{2+}-induced polymerization (plateau) of purified tubulin with 0.5 μM-colchicine. With tubulin + microtubule-associated proteins Sloboda and Rosenbaum[294] obtained nearly total inhibition with 1 μM colchicine. Margolis and Wilson,[293] using microtubule protein, obtained little inhibition with either 1 or 0.2 μM colchicine without preincubation. With preincubation, 1.0 μM colchicine was totally inhibitory, while 0.2 μM drug was about 60% inhibitory (plateau). It should be noted in this latter study that the preincubated control reaction was much less vigorous than the non-preincubated control, suggesting deterioration of the protein preparation. In addition, Margolis and Wilson[293] performed their analysis at 30°C, as compared to the 37°C used in other studies with microtubule protein.It is possible that colchicine, and other antibutulin agents, are more effective inhibitors at lower temperatures, as is the case with GDP.[49]

In addition, several groups have examined the effect on the polymerization reaction of mixing preformed tubulin-colchicine complex with tubulin. Margolis and Wilson[293] found that 3% tubulin-colchicine complex in microtubule protein inhibited polymerization by 40% (plateau). Sternlicht and Ringel,[314] in a study with microtubule fragments, which limited their analysis to the elongation phase, observed about 50% inhibition of both rate and plateau when the molar ratio of the tubulin-colchicine complex to total tubulin was about 1.5%. Subsequently, they found that both the dimethylsulfoxide-induced polymerization of purified tubulin and polymerization of microtubule protein were somewhat less sensitive to addition of the tubulin-colchicine complex — in the former, about 4% complex was required and, in the latter, over 15% complex to obtain 50% inhibition (plateaus).[315] Keates and Mason[316] observed complete inhibition of microtubule protein polymerization at about 15% tubulin-colchicine complex; Bergen and Borisy[317] found that just over 10% complex totally inhibited elongation onto axoneme microtubules and that 4% complex inhibited growth rates by 50%; Andreu et al.[318] observed 50% inhibition (plateau) with 5% tubulin-colchicine complex of glycerol-induced polymerization of purified tubulin; and Farrell and Wilson[319] reported that 12% complex added to microtubule protein inhibited the assembly rate by 70% and extent by 50%.

The mechanism by which colchicine inhibits microtubule assembly has been examined in a number of laboratories. Generally, these studies have attempted above all to explain the substoichiometric effects of the drug. Margolis and Wilson[293] observed that addition of tubulin-colchicine complex to microtubule fragments did not destablize the polymer but prevented further elongation, and that there was no apparent binding of free colchicine to the fragments. They proposed that the tubulin-colchicine complex bound to microtubule ends and thereby blocked the elongation reaction (the end poisoning model). These workers[320] subsequently found that incorporation of a single tubulin-colchicine complex halted microtubule growth, but up to eight complexes could be added to a single microtubule end. In addition, dilutional depolymerization led to no loss of colchicine from residual microtubules, indicating that the tubulin-colchicine complex was entirely, and tightly, bound to the assembly end of the polymer. Lambeir and Engelborghs[321] and Keates and Mason[316] also concluded that the major effect of colchicine was to stop elongation at microtubule ends, but they observed a rapid equilibrium, rather than irreversible binding, of the tubulin-colchicine complex at microtubule ends. Both groups concluded that the affinities of tubulin and tubulin-colchicine for microtubule ends were comparable in magnitude, but they differed on the effects of colchicine on initiation of assembly. Keates and Mason[316] observed no effect of colchicine on microtubule number, while Lambeir and Engelborghs[321] reported that colchicine significantly inhibited nucleation.

Sternlicht and Ringel[314,315,322] described very different results, leading to a different model of colchicine action. Under a variety of reaction conditions they observed a copolymerization phenomenon in which tubulin-colchicine complex was incorporated along with unliganded tubulin into microtubules. Higher concentrations of the complex in the reaction mixture led to an increasing proportion of tubulin-colchicine entering microtubules, even as the overall polymerization reaction was progressively inhibited. (Note the similarity of these findings to the direct incorporation of GDP into microtubules described above, especially when GDP was used to progressively inhibit microtubule assembly.) They also noted that the amount of complex in polymer was proportional to microtubule length, not microtubule number, a finding inconsistent with complex incorporation into microtubule ends only (microtubule length and number were controlled by the number of seeds). They proposed that inhibition of assembly occurred because increasing amounts of tubulin-colchicine complex in microtubules resulted, as a consequence of conformational changes, in a progressively lower affinity of polymer ends for unpolymerized tubulin.[314,315] In addition they found that the apparent association rate constants of tubulin and complex for microtubules was similar, but that the dissociation rate constant of the tubulin-colchicine complex was at least five times larger than that of tubulin.[322]

Farrell and Wilson[323] noted that both the end poisoning and copolymerization models appeared to be valid, depending on the concentration of tubulin-colchicine complex in the reaction mixture. At high concentrations microtubule assembly abruptly ceased, leading to apparent end poisoning; but at low concentrations, episodic incorporation of complexes into polymer led to a transient decrease in the rate of microtubule elongation, a result comparable to that of Sternlicht and Ringel.[314,315] Farrell and Wilson[323] also reported that binding of the tubulin-colchicine complex to the assembly end was reversible, and that the apparent association rate constant of the complex was about three times that of tubulin. More recently these workers observed that the number of tubulin-colchicine complexes incorporated into polymer was almost constant and independent of microtubule length, with average length decreasing as complex concentration rose (reaction variable was the concentration of tubulin-colchicine complex).[319]

Several groups specifically investigated whether there were differential effects of colchicine on the two ends of microtubules.[76,296,317,319,320,324] Margolis et al.[320] concluded that a tubulin-colchicine cap at the assembly end of microtubules was stable, and Bergen and Borisy[296] observed stabilization of both microtubule ends by colchicine (see above). Deery and Weisenberg[76] found that tubulin-colchicine was in reversible equilibrium with both ends of the microtubule and that colchicine eliminated subunit flux through the polymer. These latter workers also observed little difference in the dilutional stability of normal and colchicine-treated microtubules, noting that depolymerization occurred from both ends with both types of polymer. Deery and Weisenberg[76] proposed that the mechanism of colchicine action could involve interference with cooperative interactions of oligomers during assembly, perhaps as a consequence of tubulin-colchicine entry into such oligomers. Bergen and Borisy[317] also examined the effects of the tubulin-colchicine complex on microtubule elongation, with the reaction initiated with axoneme fragments. They observed nearly equal inhibition at both ends, with the complex progressively reducing the association rate constants without affecting the dissociation rate constants. Farrell and Wilson,[319] on the other hand, noted major differences between the two microtubule ends. An increasing concentration of tubulin-colchicine complex led to a progressive decrease in subunit flux through the microtubule. This resulted primarily from a rapid decrease in the dissociation rate constant from the assembly end, although there was also a slower decrease, as a function of complex concentration, of the dissociation rate constant from the disassembly end. In addition, Jordan and Farrell[324] observed that addition of the tubulin-colchicine complex to microtubules at steady state resulted in a marked decrease in the association rate constant at the assembly end (50% reduction at about 5% complex) and a smaller decrease in the dissociation rate constant at the disassembly end (50% reduction at about 10% complex).

This brief summary of reports in the literature on the mechanism of colchicine inhibition of assembly reveals many contradictions that remain to be reconciled. Moreover, despite the clear inhibition of microtubule assembly by colchicine, there have been a number of reports of nonmicrotubule polymer formation in the presence of colchicine[108,318,325-328] and colchicine analogues.[249,329] Generally, somewhat atypical reaction conditions were required, and this was the case in all the most extensive reports. Sandoval and Weber[108] found that when $ppCH_2pG$ replaced GTP in supporting polymerization, colchicine no longer inhibited turbidity development. Instead of microtubules, the polymer consisted of long, helical ribbons with a decreasing number of protofilaments (minimum, three to four) as the drug concentration increased. The stoichiometry of colchicine to tubulin in the polymer was as high as 0.8. Salterelli and Pantaloni[327] and Andreu and Timasheff[318,328] described an extensive polymerization reaction of the tubulin-colchicine complex occurring at relatively high Mg^{2+} concentrations (at least 5 m*M*[327]). The two groups obtained similar critical concentrations of tubulin (1.0 to 1.2 mg/ml), and in both cases the rection was GTP dependent, cold reversible, and inhibited by GDP. Both agreed that the thermodynamic parameters of the reaction were similar to those obtained for microtubule assembly (with glycerol in the reaction mixture, required for the reaction without colchicine). However, the two groups described different polymer morphologies, calcium sensitivities, and polymer stability following GTP exhaustion.

There has been a great deal of work performed with colchicine analogues, but for reasons of space the following summary will be largely restricted to those with which interactions with tubulin have been demonstrated. Although their *in vitro* interactions with tubulin are feeble, single ring analogues of colchicine are of great interest. The only ones reported to have significant antimitotic activity are the hallucinogen mescaline (structure in Figure 20) and, to a lesser extent, its derivative *N*-acetylmescaline.[330] High concentrations of the A ring analogues 2,3,4- and 3,4,5-trimethoxybenzaldehyde stimulate tubulin-dependent GTP hydrolysis and inhibit the binding of colchicine to tubulin,[56] as well as alter the alkylation

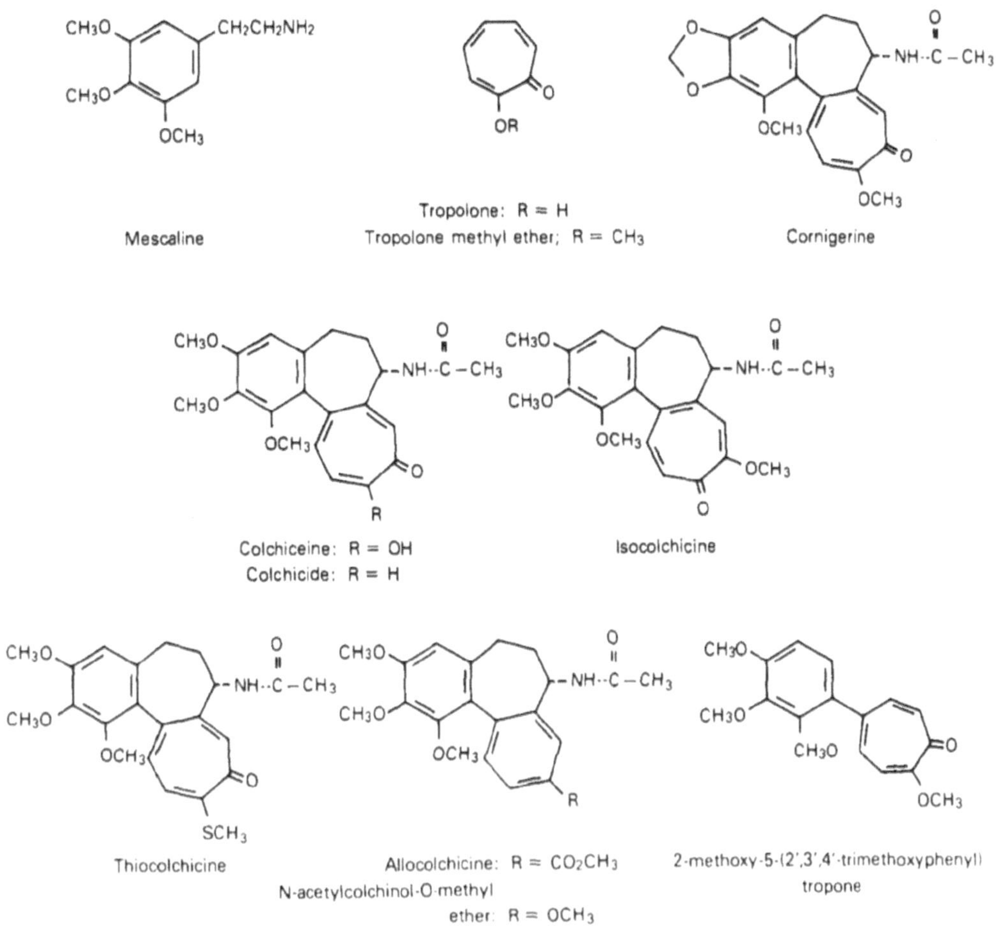

FIGURE 20. Structural formulas of colchicine analogues.

pattern of tubulin sulfhydryl groups (see above).[267] Tropolone and tropolone methyl ether (structures in Figure 20) inhibit colchicine binding to tubulin (K_i values 50 to 60 μM)[223,253] and alter the sulfhydryl alkylation pattern.[267] Tropolone methyl ether also inhibits tubulin-dependent GTP hydrolysis.[172] The most extensive studies of single ring analogues were those of Andreu and Timasheff.[282,285] Although they observed feeble binding of mescaline to tubulin (K_D value of 100 μM), it did not appear to be at the colchicine site, and a more specific binding of *N*-acetylmescaline was even weaker (K_D value of 2.3 m*M*). They also observed feeble binding, inhibited by colchicine, of both tropolone and tropolone methyl ether to tubulin. Two different procedures yielded high K_D values of 0.5 and 3 m*M* for the latter compound. The single ring analogues all inhibited tubulin polymerization at high concentrations, but they did not enhance either the binding of each other to tubulin or their inhibitory effects on polymerization. As noted above, tropolone methyl ether had qualitatively similar effects as colchicine on a number of spectral properties of tubulin.

There have been relatively few reports of interactions of A ring-modified analogues with tubulin. Removal of any of the three methyl groups reduces the inhibitory effects of the analogues on colchicine binding to tubulin,[331,332] on the growth of L1210 leukemia cells,[332] and on tubulin polymerization.[49] (However, contrary to my experience,[49] Dvorackova et al.[307] reported that 2- and 3-demethylcolchicine were comparable to colchicine in their inhibitory effects on microtubule assembly.) Removal of any two of the three methyl groups reduces activity in all assays still further.[49,332,333] The least important of the three methyl groups is that at the 3-position, the most important that at the 1-position.[331,332] Alternate, generally larger, substituents on the A ring have invariably resulted in analogues with reduced or negligible activity.[223,239,243,253,307,331,333] The one A ring analogue with equal, or even enhanced, activity relative to colchicine is one with a rather unusual structure (see Figure 20), the natural product cornigerine obtained from *Colchicum cornigerum.*[334] Overall cornigerine is about 1.5 times as potent as colchicine in binding to tubulin, inhibiting cell growth, and inhibiting microtubule assembly.[299,307,332,333]

Similarly, there have been relatively few reports of C ring modifications (see structures in Figure 20). Removal of the methyl group (to yield colchiceine) or the methoxy group at position 10 (to yield colchicide) and reversal of the carbonyl oxygen and methoxy group at positions 9 and 10 (to yield isocolchicine) result in nearly inactive compounds.[223,239,253,301,331-333,335] On the other hand, replacement of the methoxy group at position 10 with a thiomethyl group (to yield thiocolchicine), both as a single change or in combination with other modifications, almost invariably produces more potent compounds.[239,243,331,332,335,336] This is true whether cytotoxicity, inhibition of colchicine binding to tubulin, or effects on tubulin polymerization is evaluated. Dvorackova et al.[307] reported that *N'*-acetylcolchicamide (acetamide group replacing the methoxy group at position 10) was similar to colchicine in inhibiting microtubule assembly, but that two compounds with bulky substituents at position 10 were nearly inert. Zweig and Chignell[239] found that colchiceinamide (amino group instead of methoxy group at position 10) was a strong inhibitor of the binding of colchicine to tubulin.

A seemingly more profound change in the molecule is conversion of the 7-membered tropolone C ring to a phenyl ring (structures in Figure 20). Zweig and Chignell[239] noted significant inhibition of colchicine binding to tubulin by *N*-acetylcolchinol and *N*-acetyliodocolchinol, although both compounds seemed moderately less active than tubulin. Fitzgerald[301] reported that allocolchicine was equivalent to colchicine in inhibiting mitosis of HeLa cells and microtubule assembly, and Iorio[337] described potent inhibition of the binding of colchicine to tubulin by *N*-acetylcolchinol-*O*-methyl ether. We recently observed with both allocolchicine and *N*-acetylcolchinol-*O*-methyl ether significantly greater inhibition, as compared to colchicine, of the binding of radiolabeled colchicine to tubulin and of mitosis in L1210 cells.[332] Both compounds are also more potent inhibitors of tubulin polymerization than colchicine, binding rapidly to the protein, probably even at 0°C.[49]

Most reported modifications of colchicine have been at the B ring, and numerous active derivatives at the amino nitrogen (e.g., colcemid, in which the acetyl group is replaced by a methyl group) have been described.[223,239,249,253,257-259,275,276,283,307,331-333,335,336,338-345] In addition, deacetylcolchicine,[239,307,331,345,346] deacetylthiocolchicine,[239,243] deacetamidocolchicine,[250,301,332,342,343,345] and 5,6-dehydro-7-deacetamidocolchicine[331] have significant antitubulin activity, although most workers have described these compounds as less active than colchicine itself. A wide range of activities were described for analogues derivatized at the amino group, with no structure-activity relationships apparent. Some of these analogues have had valuable properties for investigating tubulin biochemistry, serving as potential spin probes[259,341,344] or chemical[275] or fluorescent markers;[276,339] or, by virtue of reactive groups, serving as affinity labels.[249,257,283] One important modification in the B ring that tubulin does not tolerate is reversal of configuration at position 7 — synthetic (+)-colchicine (7R configuration) has at most a feeble interaction with the protein.[250,331,332,345]

One last modification in the B ring yielding an active analogue is especially notable — namely, its complete elimination. The compound 2-methoxy-5-(2',3',4'-trimethoxyphenyl)tropone (MTPT, structure in Figure 20) consists of only a trimethoxybenzene ring and a tropolone methyl ether ring directly linked together. MTPT was synthesized by Fitzgerald,[301] who noted its near identity with colchicine as an inhibitor of HeLa cell growth (an observation we have confirmed with L1210 murine leukemia cells[347]) and of *in vitro* microtubule assembly. The crystal structure of MTPT was examined by Rossi et al.,[348] who found that it differed little from that of colchicine and active colchicine analogues.[279-281] The dihedral angles between the trimethoxybenzene and tropolone rings are similar, as are the alternating bond lengths in the tropolone rings. In the MTPT crystal, the bond between the two rings was entirely single bond in character, indicating that free rotation about it should occur. The methoxy group of the tropolone ring in the MTPT crystal, however, occupied the position it would take in isocolchicine rather than in colchicine. Presumably, on binding to tubulin the relative position of this group would be comparable to its position in colchicine, requiring that the two rings rotate about 180° relative to each other.[348]

Flavin and Slaughter[338] first reported that colcemid (methyl instead of acetyl group at the amine of the B ring) bound much faster than colchicine to tubulin. Banerjee and Bhattacharyya[340] noted in addition that this binding reaction was reversible, with a significant reaction occurring even at 0°C, in contrast to the slow, irreversible, temperature-dependent binding of colchicine. Similar properties were observed with other B ring-modified analogues, including MTPT which bound to tubulin almost instantly even at 0°C.[258,342,343,345] In a structure-function analysis, Bhattacharyya et al.[345] demonstrated that the rapid binding rate of MTPT was progressively reduced by adding the B ring (deacetamidocolchicine), the amine group (deacetylcolchicine), and amide derivatives of increasing size. Significant reversibility, however, was not observed with any derivative of deacetylcolchicine except for colcemid, with only MTPT displaying nearly complete reversibility.[345] Similar observations were reported by Banerjee et al.[349] Other groups have confirmed the rapid and reversible binding of MTPT to tubulin,[254,329,347,350] and a similar rapid intracellular recovery of microtubules has been described.[351] Kinetic and thermodynamic parameters of the MTPT binding reaction have been studied in several laboratories (with significant differences obtained) for comparison to analogous results with colchicine.[254,329,349,350,352]

2. Podophyllotoxin and Podophylloxotin Analogues

Like colchicine, podophyllotoxin (obtained from *Podophyllum peltatum* and *Podophyllum emodi;* structure in Figure 21), has limited clinical use. Wilson and Friedkin[211,235] first noted that podophyllotoxin inhibited the binding of radiolabeled colchicine to tubulin. Subsequently, many laboratories reported that podophyllotoxin was a potent competitive inhibitor of colchicine binding, with reported K_i values ranging from 0.3 to 3 μM.[214,215,218,222-225,239,243,251,300,353-356] Bryan[238] also described competitive inhibition by podophyllotoxin of the binding of colchicine to vinblastine-induced tubulin paracrystals, but the reported K_i value was higher (30 μM). However, since the K_D value for colchicine was also high in this system (about 5 μM), it seems likely that these drugs have reduced access to their binding site in the paracrystals.

Using radiolabeled podophyllotoxin, Cortese et al.[253] and Flavin and Slaughter[338] obtained K_D values for the drug of 0.6 to 0.7 μM. With radiolabeled podophyllotoxin these workers observed more rapid and less temperature-dependent binding of the drug than of colchicine to tubulin (also see Reference 357). Cortese et al.[253] observed one class of binding site (stoichiometry 0.8) and calculated the activation energy for podophyllotoxin binding as 15 kcal/mol, the association rate constant (37°C) as $3.8 \times 10^6\ M^{-1}\ h^{-1}$ (tenfold that of colchicine), and the dissociation rate constant (37°C) as $1.9\ h^{-1}$. The binding rate constant and activation energy have been confirmed by indirect measurements.[352] Cortese et al.[253] also noted that radiolabel from podophyllotoxin was more readily lost than that from colchicine following dilution of the drug-protein complex. This evidence for reversibility of binding confirmed reports based on apparent displacement of prebound podophyllotoxin by radiolabeled colchicine.[214,243,357] On the other hand, Andreu and Timasheff[282] did not observe displacement of podophyllotoxin from tubulin by colchicine. Tropolone and tropolone methyl ether did not inhibit the binding of radiolabeled podophyllotoxin to tubulin,[253] but they did

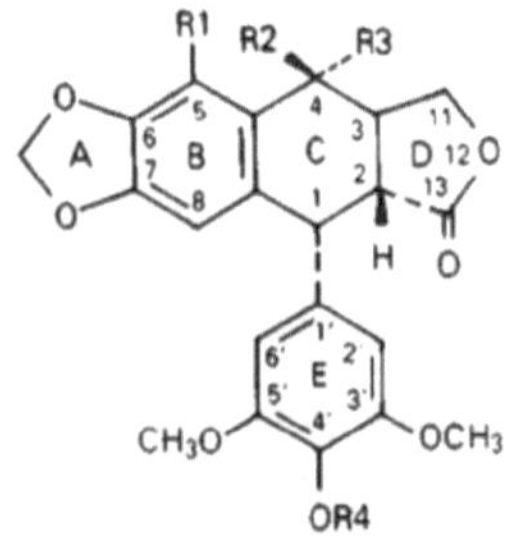

Podophyllotoxin:	R1 = R2 = H R3 = OH; R4 = CH_3
4′-Demethylpodophyllotoxin:	R1 = R2 = R4 = H R3 = OH
Deoxypodophyllotoxin:	R1 = R2 = R3 = H R4 = CH_3
4′-Demethyldeoxypodophyllotoxin:	R1 = R2 = R3 = R4 = H
α-Peltatin:	R1 = OH; R2 = R3 = R4 = H
β-Peltatin:	R1 = OH; R2 = R3 = H R4 = CH_3
β-Peltatin A-methyl ether:	R1 = OCH_3 R2 = R_3 = H R4 = CH_3
Epipodophyllotoxin:	R1 = R3 = H R2 = H R4 = CH_3
4′-Demethylepipodophyllotoxin:	R1 = R3 = R4 = H; R2 = OH

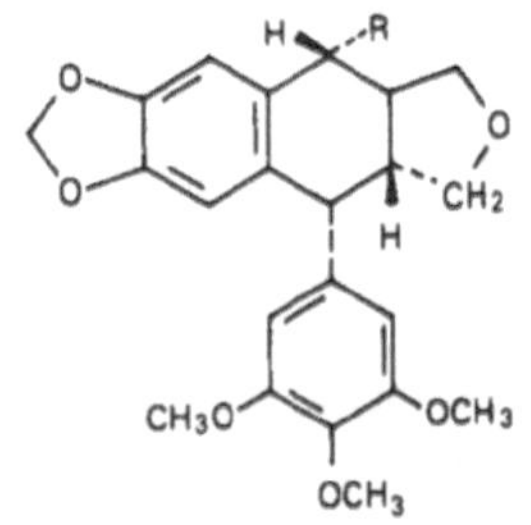

Podophyllotoxin cyclic ether: R = OH
Deoxypodophyllotoxin cyclic ether : R = H

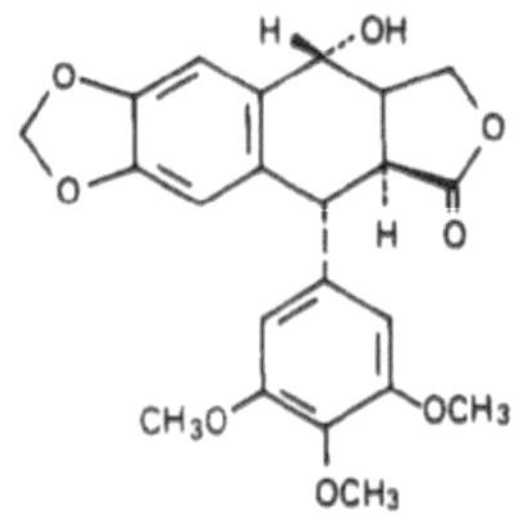

Picropodophyllotoxin

FIGURE 21. Structural formulas of podophyllotoxin and podophyllotoxin analogues.

interfere with the binding of colchicine.[253,282] This had led to the conclusion that the binding sites for the two drugs do not completely overlap.

There were similarities and differences relative to colchicine when several other effects of podophyllotoxin on tubulin were examined. Immunological properties of tubulin are altered in the presence of the drug.[284] Its inhibitory effects on the alkylation of tubulin sulfhydryl groups are similar to those of colchicine.[264-267] However, in contrast to these similarities, Detrich et al.[252] found that, in contrast to colchicine, podophyllotoxin did not affect the dissociation of the α-β dimer, and Andreu and Timasheff[282] reported that podophyllotoxin, unlike colchicine, had at most minor effects on the spectral properties of tubulin. In contrast to the clear-cut stimulation of tubulin-dependent GTP hydrolysis by colchicine, podophyllotoxin generally inhibits this reaction.[55,56,172,299] To some extent, however, this GTPase inhibition depends on reaction conditions. In glutamate, low podophyllotoxin concentrations were weakly stimulatory,[56] while high drug concentrations cause an initial profound depression of the reaction (lasting about 10 to 15 min), followed by vigorous GTP hydrolysis. Thus, at early time points high concentrations of podophyllotoxin are inhibitory and at late time points, stimulatory.[358] Podophyllotoxin, like colchicine, did not affect the binding of GTP to tubulin.[112]

Although not as extensively studied as with colchicine, there are similar discrepancies in the literature regarding polymer stability when podophyllotoxin was added to microtubules. Margolis and Wilson[293] observed no depolymerization when 50 μM podophyllotoxin was added to preformed microtubules, and Margolis[77] subsequently reported only a minimal drop in turbidity under similar reaction conditions. Karr and Purich[359] also used 50 μM podophyllotoxin, but they observed almost complete, but very slow, dissassembly of preformed microtubules. Complete depolymerization took over 4 h, with a half-time of about 40 min. Caplow and Zeeberg[360] observed comparable depolymerization with 50 μM podophyllotoxin, but it seemed less complete (about 60% disassembly in 1 h). Farrell et al.,[361] in contrast, observed more rapid depolymerization. Using a system derived from *Strongylocentrotus purpuratus* (outer doublet tubulin containing few, if any, microtubule-associated proteins), they observed at least 70% depolymerization, with the reaction nearly complete in 10 min. Drug (50 μ*M*) and tubulin (50 μ*M*) were nearly stoichiometric in this study. Kumar[362] described a similar rapid, but complete depolymerization by 50 μ*M* podophyllotoxin of microtubules formed from tubulin + MAP-2. Sackett et al.,[363] however, obtained total and almost instant disassembly of polymer formed from subtilisin-digested tubulin with 50 μ*M* podophyllotoxin.

There have been several studies on the minimum amount of podophyllotoxin required for the inhibition of microtubule assembly. The most potent effect was observed by Wilson et al.[364] in a viscometric study with microtubule protein; they reported an IC_{50} value of 0.04 μ*M* (rate) at 30°C. Other studies were all performed turbidimetrically at 37°C and yielded the following IC_{50} values for microtubule assembly: 1 μ*M* (rate);[304] 2 μ*M* (plateau; preincubated);[305] 2 μ*M* (plateau);[308] 0.8 μ*M* (plateau);[312] 0.5 μ*M* (plateau);[365] 0.6 μ*M* (plateau);[366] 2 μ*M* (rate);[367] 0.1 mol drug per mole tubulin (plateau; preincubated — i.e., 1 μ*M* with tubulin at 1 mg/ml);[108] and 3 μ*M* (plateau).[251] In addition, there are two reported studies with purified tubulin. In glycerol-Mg^{2+} the IC_{50} value was 0.3 μ*M* (plateau),[282] while in glutamate it was 3 μ*M* (plateau; preincubated).[356]

The mechanism by which podophyllotoxin inhibits tubulin polymerization has been explored by several laboratories, primarily from the point of view of the effects of the drug on microtubule ends. Margolis and Wilson[368] reported that podophyllotoxin, presumably as a tubulin-drug complex, capped the assembly end of microtubules, preventing further elongation, without affecting the disassembly end; Margolis[77] noted an apparent requirement for continuing GTP hydrolysis to maintain this cap. Karr and Purich[359] also concluded that the primary effect of podophyllotoxin was to cap the assembly end, but that the process was reversible. Farrell et al.[361] noted an initial acceleration of tubulin loss at the disassembly end of microtubules following addition of podophyllotoxin, in addition to capping of the assembly end. Caplow and Zeeberg[360] concluded that podophyllotoxin capped both ends of microtubules reversibly. Jordan and Farrell[324] found that the drug inhibited tubulin addition at the assembly end (IC_{50} value about 0.5 μ*M* podophyllotoxin) and tubulin loss at the dissembly end (IC_{50} value about 5 μ*M*).

As with colchicine, Sandoval and Weber[108] found that $ppCH_2pG$ markedly reduced the inhibitory effect of podophyllotoxin on tubulin polymerization as compared to the GTP-supported reaction. The polymer formed, however, was not microtubules, but again ribbons containing three to four protofilaments. While colchicine did not prevent formation of this aberrant polymer, podophyllotoxin retained some inhibitory properties; but about 140 times as much drug was required to suppress the reaction with the nucleotide analogue as with GTP. No other podophyllotoxin-induced structures have yet been described.

A number of studies of interactions of tubulin with podophyllotoxin analogues have been reported. Figure 21 summarizes the structures of a number of these compounds. Most have been examined both as inhibitors of the binding of radiolabeled colchicine to tubulin[243,353,355] and as inhibitors of microtubule assembly.[366] Alterations in the podophyllotoxin structure that made little difference or even enhanced its activity were demethylation at the 4′ position, removal of the hydroxyl at the 4 position, and addition of a hydroxy or methoxy group at the 5 position — i.e., good activity was observed with deoxypodophyllotoxin, 4′-demethylpodophyllotoxin, 4′-demethyldeoxypodophyllotoxin, α-peltatin, β-peltatin, and β-peltatin A-methyl ether.[243,338,355,366] About a 10- to 20-fold reduction in activity was observed with the reversal of configuration at position 4 which occurs in epipodophyllotoxin,[338,355,366] but 4′-demethylepipodophyllotoxin was about 2 to 3 times as active as epipodophyllotoxin as an inhibitor of microtubule assembly.[366] Bulky substituents at positions 4, 5, or 4′ resulted in analogues with little antitubulin activity,[243,253,338,355,366] including notably the clinically valuable VP-16-213 (etoposide) and VM-26 (teniposide),[366] which are potent DNA topoisomerase inhibitors.[369] Zavala et al.[305] described very interesting properties for isodeoxypodophyllotoxin (reversal of configuration at position 1 in the structure of deoxypodophyllotoxin, see Figure 21). The analogue was as active as podophyllotoxin as an inhibitor of microtubule assembly, but had little effect on the binding of radiolabeled colchicine to tubulin. In addition, it appeared that only one of its optical isomers was active.

Analogues in which the lactone D ring is altered have invariably been less active than podophyllotoxin, often almost inert.[243,353,355,366] The one partial exception to this generalization are the podophyllotoxin and deoxypodophyllotoxin cyclic ethers (Figure 21) in which the carbonyl group at position 13 is replaced by a methylene group. Although they are one fifth to one tenth as active as podophyllotoxin as inhibitors of colchicine binding,[243,353,355] they have nearly identical activity in inhibiting microtubule assembly.[366] Another D ring analogue which should be specifically noted in picropodophyllotoxin (Figure 21), with its reversal of configuration at position 2, for many workers have demonstrated its greatly reduced activity.[214,235,243,355,366] Conversion of the C ring to an unsaturated ring also yielded an inactive analog (dehydropodophyllotoxin).[366]

Crystal structures for at least two podophyllotoxin analogues have been reported, those of 5′-demethoxy-β-peltatin A methyl ether[370] and of 2′-bromopodophyllotoxin;[371] but there has been no report of interactions of either of these compounds with tubulin. In both cases rings A, B, C, and D were described as being roughly coplanar, and the plane of ring E was virtually perpendicular to the A-D plane. Based on molecular models, Gensler and Gatsonis[372] noted that the E ring in podophyllotoxin had limited mobility, while that of picropodophyllotoxin could rotate freely. A nuclear magnetic resonance study[373] of podophyllotoxin and a number of its analogues indicated that overall drug conformation was similar for active and inactive compounds, except for picropodophyllotoxin. This suggested that steric factors were probably a major component in the failure of analogues modified at the 4 position or in the D ring to inhibit microtubule assembly.[366,373]

Jurd and collaborators have reported the synthesis[374-377] of several extensive series of antineoplastic[378] compounds which, structurally, appear to be analogues of podophyllotoxin (Figure 22). We have confirmed that they act as antimitotic agents and have examined their *in vitro* interactions with tubulin, particularly from the point of view of methoxy substituents on the parts of the molecules analogous to the E ring of podophyllotoxin.[298,356,379]

The first series of agents, represented by the structures on the left-hand side of Figure 22 ("benzyl-benzodioxole compounds"), were the most potent.[298] The most active agents were nearly equivalent to podophyllotoxin as competitive inhibitors of the binding of radiolabeled colchicine to tubulin (K_i values of 0.5, 0.6, and 1.0 μM for podophyllotoxin, NSC 350102, and NSC 321567, respectively) but only about one third as active as inhibitors of tubulin polymerization. Like podophyllotoxin, the benzyl-benzodioxole compounds did not affect the binding of GTP to tubulin,[112] even though they inhibited tubulin-dependent GTP hydrolysis. Their inhibitory effect on GTP hydrolysis was not quite as strong as that of podophyllotoxin.

Structure-activity analysis[298] of available agents demonstrated requirements for an intact dioxole ring and substituents at positions 5 and 4′. The optimum substituent at the 5 position was an ethoxy or methoxy group, but some activity was seen as well with 3-carbon substituents. At the 4′ position a methoxy group was required, but some activity was also obtained with an ethoxy group. No definite substituent pattern at the bridge carbon was observed. The most surprising finding with the benzyl-benzodioxole compounds occurred when the portion of the molecule analogous to the E ring was further modified to more closely resemble that of podophyllotoxin; all compounds with a 3′,4′,5′-trimethoxybenzene ring were inactive. Roach et al.[380] examined the effects of benzyl-benzodioxole compounds on the alkylation of tubulin sulfhydryl groups and found that they had effects similar to those of colchicine and podophyllotoxin.

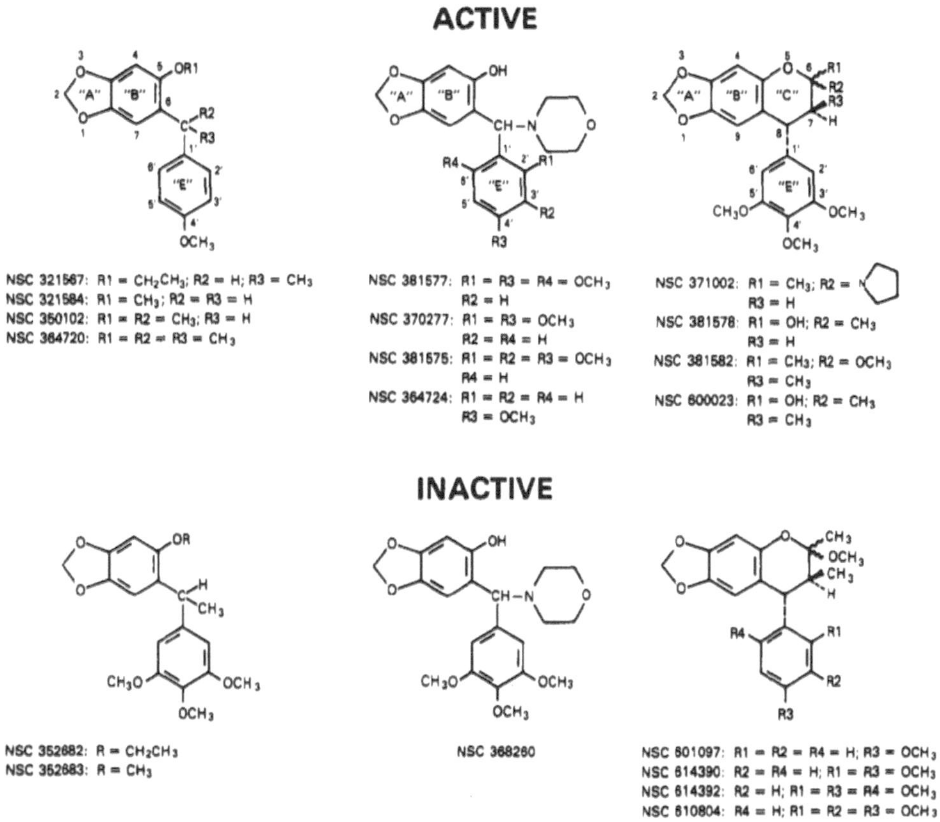

FIGURE 22. Structural formulas of benzyl-benzodioxole, morpholino-benzyl-benzodioxole, and benzopyran compounds — active and inactive analogues.

The most active agents in the second group of compounds synthesized by Jurd[376] (structures in the midportion of Figure 22; ''morpholino compounds'') were about one fourth as active as podophyllotoxin as inhibitors of tubulin polymerization but very feeble as inhibitors of colchicine binding (K_i value for the most inhibitory morpholino compound, NSC 381577, was 8 μM).[379] The morpholino compounds were, however, even more potent than podophyllotoxin in inhibiting tubulin-dependent GTP hydrolysis. Structure-activity studies were performed with morpholino compounds modified in the portion of the molecule analogous to the E ring, for these were readily prepared.[376] As with the benzyl-benzodioxole compounds, the 4′-monomethoxy morpholino compound was much more inhibitory than the 3′,4′,5′-trimethoxy compound. However, the most active agents were those with a 2′,4′-dimethoxy or a 2′,4′,6′-trimethoxybenzene ''E'' ring (the agent with a 2′,3′,4′-trimethoxybenzene ''E'' ring was also active).[379]

Thus in two sets of compounds which seem most comparable to podophyllotoxin in their interactions with tubulin, increasing apparent structural analogy to podophyllotoxin results in loss of affinity for tubulin.

The third class of agents synthesized by Jurd[377] (structures on the right-hand side of Figure 22; ''benzopyran compounds'') were the least active both as inhibitors of tubulin polymerization (about one fifth as active as podophyllotoxin) and of colchicine binding (K_i value for the most active inhibitor, NSC 381582, was 11 μM).[356] A large number of analogues with substituents at positions 6 and 7 were prepared, but no structure-activity correlations were apparent; most benzopyran compounds had little inhibitory effect on tubulin. There

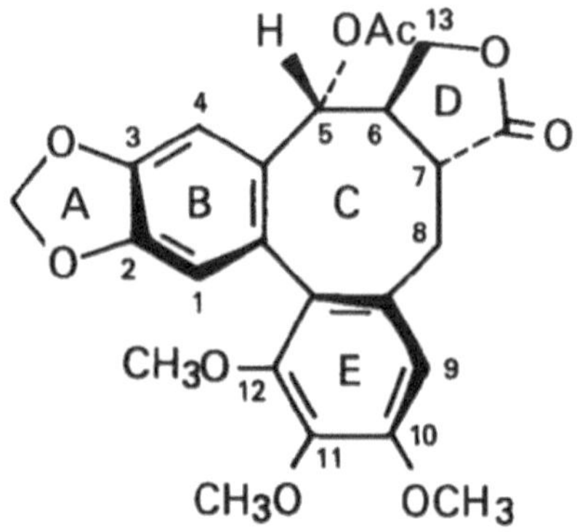

FIGURE 23. Structural formula of steganacin.

were two major differences, however, between the four active benzopyran compounds (Figure 22), on the one hand, and the active benzyl-benzodioxole and morpholino compounds, on the other. First, in active agents the portion of the molecule analogous to the E ring of podophyllotoxin was a 3′,4′,5′-trimethoxybenzene ring. Analogues with other substituent patterns (most notable a 4′-monomethoxybenzene ring) were inactive. Second, the benzopyran compounds stimulated tubulin-dependent GTP hydrolysis, like colchicine, rather than inhibited the reaction, like podophyllotoxin. One factor that may explain the relative inactivity of the benzopyran compounds bearing a substituent at position 7 (the R3 group, see Figure 22), is that the configurations of the hydrogens at positions 7 and 8 are identical to those of the inactive compound picropodophyllotoxin (*trans*) rather than to those of podophyllotoxin (*cis*).

3. Steganacin and Congeners

Steganacin was isolated by Kupchan et al.[381] from the stems and stem bark of *Steganotaenia araliacea*. The presentation of the structure in Figure 23 was adapted from Taafrout et al.[382] Note that the molecule has the same biaryl configuration demonstrated by Yeh et al.[250] to be the active conformation of colchicine (see Figure 19). In addition, this rendition of steganacin emphasizes the analogous configurations of its acetyl group with the acetamido group of colchicine (see Figure 19) and of its lactone D ring with the lactone D ring of podophyllotoxin (see Figure 21). Based on a total synthesis of optically active steganone (probably identical to natural steganone, which has a keto group at position 5), Robin et al.[383] argued that Kupchan et al.[381] had assigned the incorrect configuration to steganacin. The structure presented in Figure 21 follows Robin and collaborators,[382,383] rather than that of earlier work.[308,354,381]

Only a few studies have been published in which interactions of steganacin and its analogues with tubulin have been examined. Wang et al.[308] first described the antimitotic activity of steganacin, and this was confirmed by Schiff et al.[354] The IC_{50} value for inhibition of microtubule assembly was reported to be 2 μM (plateau)[308,354] and 4 μM (plateau; preincubated).[305] The latter study was performed with racemic, synthetic steganacin (see below). High drug concentrations cause slow, partial disassembly of preformed microtubules; inhibition of assembly was partially reversible by removing the drug by gel filtration.[308] Steganacin competitively inhibits the binding of radiolabeled colchicine to tubulin,[308] and K_i values of 3μM for natural steganacin[354] and 10 μM for synthetic, racemic steganacin[305] have been reported. Steganacin resembles colchicine rather than podophyllotoxin in its effect on tubulin-dependent GTP hydrolysis, for it stimulates the reaction rather than inhibiting it.[298,299,379]

Schiff et al.[354] examined effects of several steganacin analogs with tubulin. Steganangin, in which a 5-carbon ester group replaces the acetate group at position 5, had activity similar to that of steganacin (also observed by Wang et al.[308]). Episteganacin (reversal of configuration at position 5) and steganol (hydroxyl instead of acetate at position 5), however, were both inactive. Schiff et al.[354] also compared synthetic, racemic steganacin to the natural product, and, surprisingly, found little difference between them (the natural product was slightly more inhibitory), implying significant activity in the optical isomer of natural steganacin.

Zavala et al.[305] confirmed the poor activity of steganol and episteganacin with tubulin *in vitro*, and they found in addition that steganone, stegane (no acetyl group at position 5), and compounds with a disrupted C ring were inactive. They found one active analogue, isopicrostegane, which had a very different structure from steganacin: no substituent at position 5, a *cis* lactone D ring, and reversal of the biaryl configuration of steganacin.

The crystal structures of the two inactive analogues have been described. Kupchan et al.[381] examined episteganol (hydroxyl instead of acetyl group at position 5, with configuration the reverse of steganacin) and found that the planes of the two aromatic rings were oriented at about 120° to each other. Pascard et al.[384] studied 8α-hydroxyisopicrostegane, since they felt the hydroxyl group at position 8, although it eliminated activity, would not result in much change in the conformation of isopicrostegane. The planes of the two phenyl rings were approximately perpendicular.

4. Combretastatin and Congeners

The bark of the South African tree *Combretum caffrum* is a newly discovered source for multiple antimitotic compounds, and these are the simplest natural products which bind to tubulin. Pettit et al.[385-387] described the isolation and characterization of the first of these compounds and named it combretastatin (structure in Figure 24). Combretastatin competitively inhibited the binding of colchicin to tubulin (K_i value, 1 μM). It inhibited microtubule assembly and the polymerization of purified tubulin.[347] The drug also was similar to colchicine and podophyllotoxin in inhibiting the alkylation of tubulin sulfhydryl groups.[267] It was not, however, the most potent of the *C. caffrum* compounds.

Pettit and collaborators went on to describe the isolation and synthesis of an additional three dozen compounds,[300,311,388-390] ten of which more strongly inhibited colchicine binding and/or tubulin polymerization than the original combretastatin.[300] Three of the four most inhibitory compounds are natural products and have been designated combretastatin A-1, combretastatin A-2, and combretastatin A-4 (structures in Figure 24; unnamed agents are also shown in Figure 24 and will be designated as "compound 24/#"). The fourth agent with particularly potent antitubulin effects *in vitro*, compound 24/1, is synthetic. Combretastatins A-4 and A-2 are in addition highly cytotoxic (IC_{50} values with L1210 murine leukemia cells of 0.007 and 0.1 μM, respectively), while the two agents with vicinal hydroxyl groups are chemically labile and, probably as a consequence, have reduced cytotoxicity (IC_{50} values of 0.6 and 4 μM for combretastatin A-1 and compound 24/1, respectively). It is worth noting that the analogy between combretastatins A-4 and A-2 is identical to that which occurs with the *Colchicum*-derived drugs colchicine and cornigerine (see above).

Combretastatins A-1 and A-4 inhibit colchicine binding to tubulin (see Table 8) more potently than any other drug we have examined or than any other reported in the literature.[300] With inhibitor and radiolabeled colchicine in equimolar concentrations, these agents inhibit binding of the latter to tubulin by 97 to 99%, with combretastatin A-1 slightly more inhibitory than combretastatin A-4. (For comparison, equimolar podophyllotoxin inhibited colchicine binding by 85 to 90%. Combretastatin A-2 and compound 24/1 were also more potent than podophyllotoxin in this reaction.) The K_i value obtained with combretastatin A-4 was 0.14 μM, while the lability of combretastatin A-1 prevented accurate determination of a K_i value.

The only modification to the combretastatin A-4 structure which resulted in increased

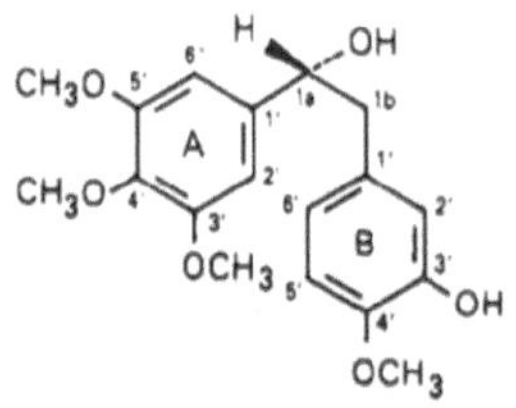

Combretastatin

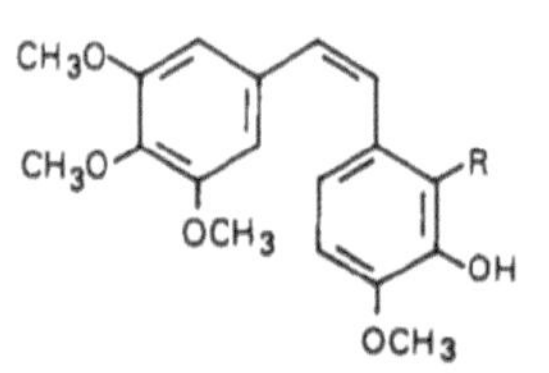

Combretastatin A-4: R=H
Combretastatin A-1: R = OH

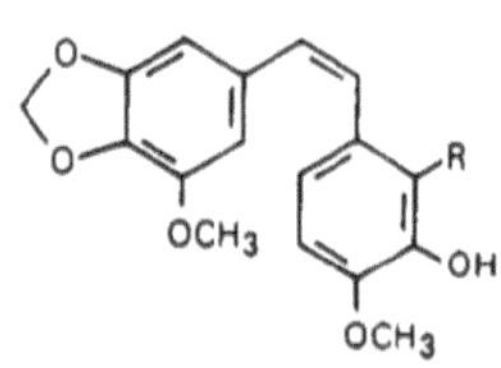

Combretastatin A-2: R=H
Compound 24/1: R=OH

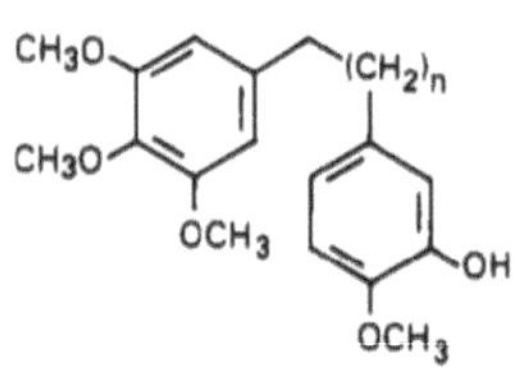

Compound 24/2: n=1
Compound 24/3: n=2

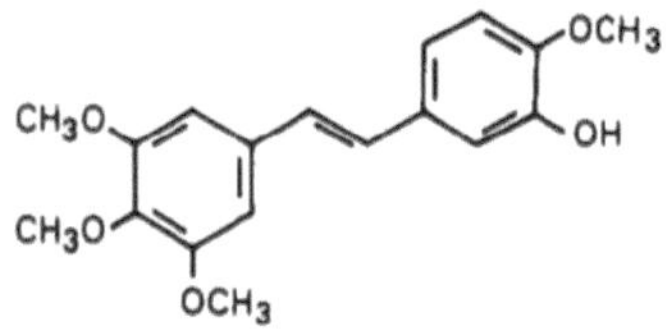

Compound 24/4

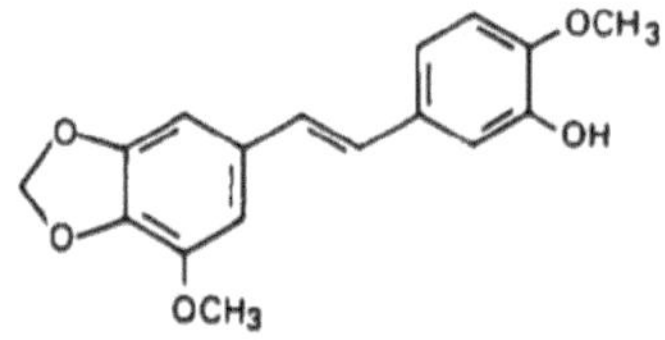

Compound 24/5

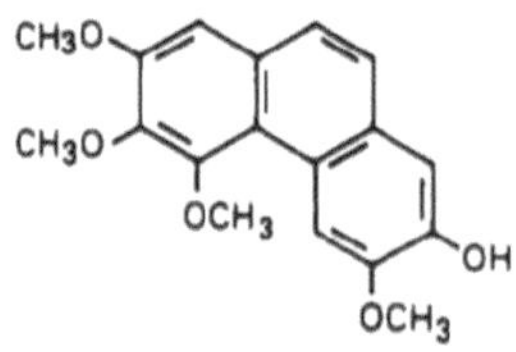

Compound 24/6

FIGURE 24. Structural formulas of natural products derived from *Combretum caffrum* and synthetic analogues.

in vitro activity against tubulin was addition of a second hydroxyl group at position 2′. Significant inhibition of both tubulin polymerization and colchicine binding was nevertheless observed with many other single modifications:[300] reduction of the bridge double bond (compound 24/2); acetate group at position 3′; demethylation at position 5 (equivalent to position 2); methoxy group at position 3′; reduction of the bridge double bond with hydroxyl group at position 1a (R configuration, the original combretastatin); racemic combretastatin (some, but considerably less than 50%, reduction in activity compared to the natural product); reduction of the bridge double bond with acetate group at position 1a; and a 3-carbon saturated bridge (compound 24/3). In addition, good inhibitory activity remained after reduction of the bridge double bond in both combretastatin A-1 and compound 24/1. Two unexpectedly active agents were the *trans* variants of combretastatins A-4 and A-2 (compounds 24/4 and 24/5, respectively).

Generally, when two modifications which each produced relatively minor reductions in activity were combined, agents with little inhibitory effect were obtained (for example, in a natural product, methoxy group at position 3′ combined with demethylation at position 5). In compounds with a reduced bridge double bond, little inhibition occurred with a hydroxyl group at position 1b, or with a hydroxyl at position 4′ and a methoxy group at position 3′. Several natural products were isolated with fused A and B rings, and all exhibited major loss of activity relative to combretastatin A-4 (e.g., compound 24/6, a true phenanthrene structure).[300]

The active *C. caffrum* compounds all stimulate tubulin-dependent GTP hydrolysis, although their precise quantitative effects are highly variable and do not appear to correlate well either with inhibitory activities in the other assays or with their structural features.[49]

TABLE 8
Comparison of Natural and Synthetic Analogues of Combretastatin as Inhibitors of the Growth of L1210 Murine Leukemia Cells, Tubulin Polymerization, and Colchicine Binding

Compound	IC_{50} (μ*M*)		Colchicine binding, inhibitor:colchicine (percent inhibition)		
	L1210 growth	Tubulin polymerization	0.2:1	1:1	10:1
Combretastatin	0.05	5—7.5	—	66	94
Combretastatin A-4	0.007	2—3	79	96	—
Combretastatin A-1	0.6	2—3	73	99	—
Combrestastatin A-2	0.1	4—5	65	90	—
Compound 24/1	4.0	2—3	50	94	—
Compound 24/2	0.2	4—5	—	74	97
Compound 24/3	1.0	10—15	—	42	87
Compound 24/4	0.02	7.5—10	—	75	94
Compound 24/5	0.4	20—25	49	85	—
Compound 24/6	1.0	100	—	38	80

Note: Assays to evaluate inhibitory effects on the growth of L1210 murine leukemia cells, tubulin polymerization, and the binding of radiolabeled colchicine to tubulin were described in detail in Reference 300. In the colchicine binding assay, the concentrations of tubulin and [^{3}H]colchicine were 1 and 5 μ*M*, respectively.

Combretastatin does not affect the binding of GTP to tubulin.[112] Combretastatins A-2 and A-4 (the most potent stable compounds), on the basis of indirect studies, appear to bind to tubulin rapidly and probably in the cold.[49]

The crystals of several *C. caffrum* compounds have been examined in detail.[311,385,389] In the highly active combretastatin A-1 the two *cis* phenyl ring planes were at a 66° angle to each other.[311] In contrast, in a weakly inhibitory analogue with a reduced bridge the two phenyl rings were nearly coplanar.[389]

5. Alkyl Carbamates of Aromatic Amines

The aniline derivative isopropyl *N*-phenylcarbamate (IPC) has long been known to cause mitotic arrest in plant tissues. It has been described as causing disruption of microtubule organization rather than inhibition of polymerization;[391] but several groups have described cellular changes consistent with inhibition of microtubule assembly by IPC.[392-394] In addition, IPC has negligible effect on microtubule assembly with mammalian tubulin.[49,304]

a. Derivatives of Benzimidazole

A number of derivatives of benzimidazole have well-established uses in agriculture, veterinary medicine, and clinical medicine as antifungal and anthelmintic agents. Almost all of them are derivatives of the methyl carbamate of 2-aminobenzimidazole, as well as methyl benzimidazol-2-yl carbamate (carbendazim) itself (structures in Figure 25).

Davidse[396] found that carbendazim inhibited mitosis in the fungus *Aspirgillus nidulans* and proposed microtubules as the target of the drug. Subsequently Davidse and Flach[397,398] examined binding of radiolabeled carbendazim to fungal cell extracts and noted binding of the drug to a protein with the molecular weight of tubulin. The reaction occurred in the cold and was competitively inhibited by colchicine, nocodazole, and thiabendazole (apparent K_i values of 0.4, 0.08, and 35 μ*M*, respectively), but not by podophyllotoxin. Carbendazim, however, has only minimal effects on mammalian cells and tubulin.[56,304,309,397,399-401]

Borgers et al.[402] reported that when parasite-infested animals were treated with meben-

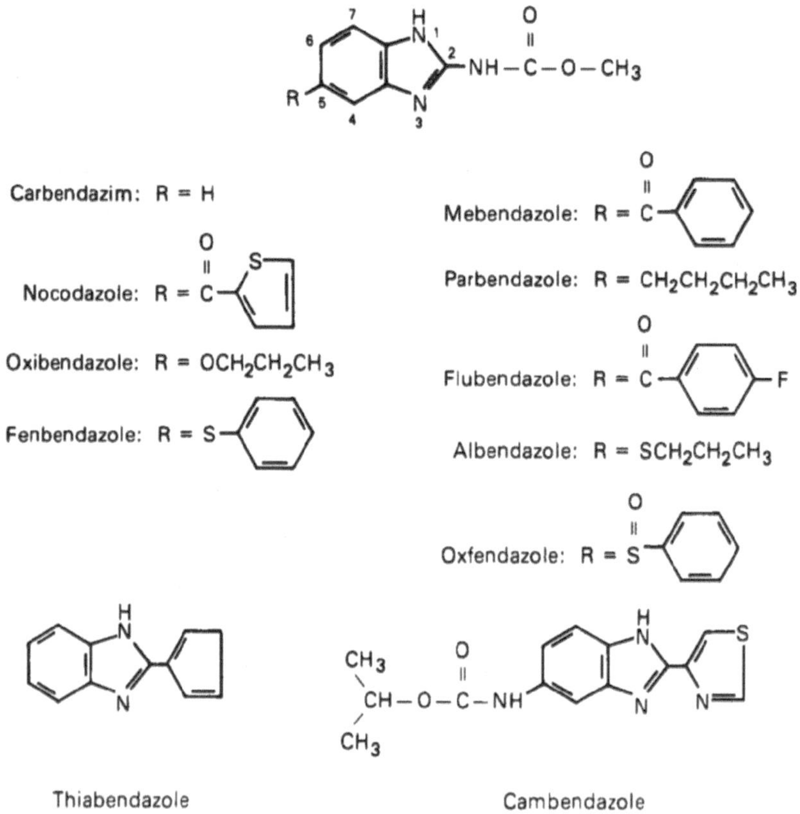

FIGURE 25. Structural formulas of methyl carbamates of aminobenzimidazole and related compounds.

dazole there was extensive disruption of the microtubules of the parasite with little effect of the drug on the microtubules of the host animal. Delatour and Richard,[399] however, noted that many benzimidazole derivatives have teratogenic effects in mammals and that mitotic arrest occurred with oxibendazole, fenbendazole, mebendazole, and cambendazole. Similarly, Holden et al.[403] reported antimitotic effects with several anthelmentic benzimidazole derivatives. Lacey and Watson[404] found most of these agents to be cytotoxic towards L1210 murine leukemia cells. De Brabander et al.[405] reported that nocodazole (NSC 238159), which had specific antineoplastic effects,[406] caused mitotic arrest in cultured cells, and that its cellular effects were rapidly reversible. This property was successfully used by Zieve et al.[407] to produce synchronous cultures of several mammalian cell lines. The methylcarbamate group at position 2 in nocodazole is required for its antimitotic activity.[310]

The first direct study of interactions of an aminobenzimidazole carbamate with tubulin was that of Hoebeke et al.[408] with nocodazole. These workers noted an IC_{50} value (plateau) of about 1 μM for microtubule assembly, with no depolymerization of preformed microtubules. Nocodazole competitively inhibited the binding of radiolabeled colchicine to tubulin (apparent K_i value, 10 μM). Radiolabeled nocodazole bound to tubulin in the cold at one site on the protein, but maximal binding occurred at 25°C.

Several groups have subsequently examined the inhibition of microtubule assembly (from mammalian microtubule protein) by nocodazole, reporting IC_{50} values which have ranged from 1 to 5 μM.[309,310,400,401] Nocodazole also inhibits the polymerization of purified tubulin under several reaction conditions.[409-411] Samson et al.[409] found that inhibition of microtubule assembly by nocodazole was largely reversible by dialysis of the drug-protein complex. In addition, they found that addition of relatively high concentrations of nocodazole to preformed microtubules caused polymer disassembly, in contrast to the report of Hoebeke et al.[408]

Lee et al.[410] found that nocodazole binds both to tubulin α-β-dimers and to oligomers without altering the dimer-oligomer equilibrium. Based on alterations in the spectral properties of the protein, they concluded that nocodazole induced a conformational change in tubulin. Head et al.[412] performed a kinetic analysis of the binding of nocodazole to tubulin, basing their quantitation on spectrophotometric analysis of bound and free drug. They found that two molar equivalents of nocodazole bound to each tubulin molecule with equal affinity (apparent K_D, 3 μM). The kinetics of the reaction was biphasic, suggesting the possibility of two binding species of tubulin, and was most consistent with a rapid binding of the drug to tubulin followed by a slower conformation change in the protein (see Sections 1 and 2 for similar proposed mechanisms for colchicine and podophyllotoxin binding).

Despite this report of two equivalent nocodazole binding sites,[412] workers using radiolabeled nocodazole have found less than one molar equivalent of drug bound to tubulin.[408,413] In addition, only one site for radiolabeled mebendazole (apparent K_D, 4 μM)[414] and one for radiolabeled parbendazole[415] was found. (All these studies were with mammalian brain tubulin.)

Lee et al.[410] observed that the tubulin-nocodazole complex reacted faster than unliganded tubulin with the sulfhydryl reactive agents *p*-mercuric benzoate and 5,5′-dithiobis(2-nitrobenzoate). In contrast, Luduena and colleagues[264-266] found that nocodazole, like colchicine and other colchicine analogues, inhibited the alkylation of tubulin sulfhydryl groups by iodoacetamide and the formation of intra-β-tubulin sulfhydryl crosslinks. Nocodazole stimulates tubulin-dependent GTP hydrolysis[56,411] and has no effect on the binding of GTP to tubulin.[112] Recently Ladd et al.[413] described the synthesis of a nocodazole photoaffinity analogue, 4′-(1-azi-2,2,2-trifluoroethyl)nocodazole. This agent and nonradiolabeled nocodazole were nearly identical in inhibiting the binding of radiolabeled nocodazole to tubulin, indicating that the analogue will be of value in localizing the drug binding site on tubulin.

Several workers have investigated the effects of anthelmintic benzimidazole derivatives on the polymerization of mammalian brain microtubule protein.[309,313,400,401,404,414,415] Although each group evaluated different agents, IC_{50} values in a relatively narrow range from 2 to 9 μM (plateau) have been reported for mebendazole, parbendazole, oxibendazole, fenbendazole, and flubendazole. Two somewhat different plateau values have been reported for albendazole (7 μM[401] and 20 μM[400]).

Lacey and Watson[401,404] described a particularly extensive structure-activity analysis with 5-substituted derivatives of methyl carbamates of 2-aminobenzimidazole. In addition to the anthelmintics discussed above, they evaluated compounds with noncarbon, alkyl, or alkoxy substituents at the 5-position. With a halogen substituent, activity increased as the substituent increased in size, and the 5-bromo-derivative was as active as the anthelmintic drugs. In both the alkyl and alkoxy series (up to eight carbon atoms in size), only those with a one-carbon substituent or with branch point adjacent to the benzimidazole nucleus had reduced activity as inhibitors of microtubule assembly (the *tert*-butyl derivative was particularly inactive).

Several anthelmintics have been demonstrated to competitively inhibit the binding of colchicine to mammalian tubulin, specifically mebendazole (K_i values of 7 and 8 μM),[400,416,417] fenbendazole (K_i value, 20 μM),[309] and oxibendazole (K_i value, 30 μM).[309] The binding of tubulin of radiolabeled mebendazole, which was inhibited to colchicine,[414] and of radiolabeled parbendazole[415] was discussed above.

Two anthelmintics, oxfendazole and thiabendazole, have little or no ability to cause mitotic arrest in mammalian cells or inhibit microtubule assembly with microtubule protein.[309,313,399-401,403] While the latter is not a carbamate derivative, oxfendazole is the only anthelmintic in this class thus far with no activity against mammalian tubulin. Interestingly, however, partial inhibitory activity was observed with an analogue of oxfendazole in which the methyl carbamate function was replaced by an isopropyl carbamate group.[404]

Cambendazole is another anthelmintic that requires special mention. In this compound the carbamate group (an isopropyl carbamate) is at position 2 and a thiazolyl group is at position 5, the reverse of the pattern in the agents discussed above. Nonetheless, cambendazole causes mitotic arrest in mammalian cells.[358,399,403] Although it is a feeble inhibitor of microtubule assembly from microtubule protein,[309,400] it is reasonably active as an inhibitor of the polymerization of purified tubulin.[358] It is possible that an isopropyl carbamate group, as opposed to a methyl carbamate, enhances drug interactions with tubulin. Friedman and Platzer[309] reported that cambendazole was a noncompetitive inhibitor of the binding of colchicine to tubulin, with a K_i value of 100 μM.

Attempts have also been made to examine interactions of the anthelmintic benzimidazoles with tubulin preparations obtained from various parasitic organisms. Friedman and Platzer[416] prepared a tissue homogenate from *Ascaris suum* embryos and examined inhibition of radiolabeled colchicine binding by mebendazole and fenbendazole. Rather than the competitive inhibition they observed with bovine microtubule protein (with K_i values of 7 and 20 μM), the anthelmintics potently, but noncompetitively, inhibited colchicine binding in the embryonic extracts (K_i values, 0.02 and 0.07 μM). In contrast, Kohler and Bachmann[417] purified tubulin about sevenfold from the intestine of adult *A. suum* and found little difference in the inhibitory effects of mebendazole on colchicine binding with this preparation as compared to porcine microtubule protein (K_i values of 4 and 8 μM, respectively). Sangster et al.[418] examined drug binding in tissue homogenates of *Trichostrongylus colubriformis* and observed inhibition of colchicine binding by mebendazole and oxfendazole (one of the drugs inactive with mammalian tubulin); but, curiously, they were unable to demonstrate inhibition by colchicine of the binding by the extracts of radiolabeled parbendazole, oxibendazole, or thiabendazole. Barrowman et al.,[419] however, observed inhibition of colchicine binding to extracts of *A. suum* by albendazole, mebendazole, fenbendazole, oxfendazole, and thiabendazole. Several radiolabeled benzimidazole derivatives have been demonstrated to bind to a component in tissue homogenates and partially purified tubulin from *Haemonchus contortus* and *Ascaridia galli*.[420,421] Dawson et al.[313] described the inhibition of polymerization of purified *A. galli* tubulin by several anthelmintics with IC_{50} values (plateau) from 5 to 8 μM. These included thiabendazole and oxfendazole, the agents inactive with mammalian tubulin. Barrowman et al.[419] similarly demonstrated inhibition of polymerization by several anthelmintics in tissue homogenates of *A. suum*.

b. Derivatives of 1,2-Dihydropyrido[3,4-b]Pyrazine

Wheeler et al.[422,423] reported that ethyl 5-amino-1,2-dihydro-3-[(*N*-methylanilino)methyl]pyrido[3,4-*b*]pyrazin-7-ylcarbamate (NSC 181928; structure in Figure 26) was a potent antimitotic agent. An extensive series of analogues was then synthesized and examined for cytotoxic and antimitotic effects.[424-426] A methyl substituent at position 3 resulted in a major loss in activity, while direct attachment of the phenyl ring at position 3 (as in NSC 330770) resulted in enhanced activity. Substituents on the aromatic residue at position 3 led to enhanced or reduced activity, but this appeared to be unpredictable. Increasing or decreasing the aromaticity of the A ring or its disruption led to major reductions in cytotoxic activity. A methyl group at position 2 (as in NSC 370147) enhanced activity, while a methyl group at position 1 greatly reduced activity. Replacement of the nitrogen atom at position 1 with either oxygen or sulfur or of the carbon atom at position 8 with nitrogen led to major losses of cytotoxicity. Alteration of the amino group at position 5 usually led to large reductions in activity. The requirement for the carbamate function at position 7 could not be examined critically, as compounds with an amino group at position

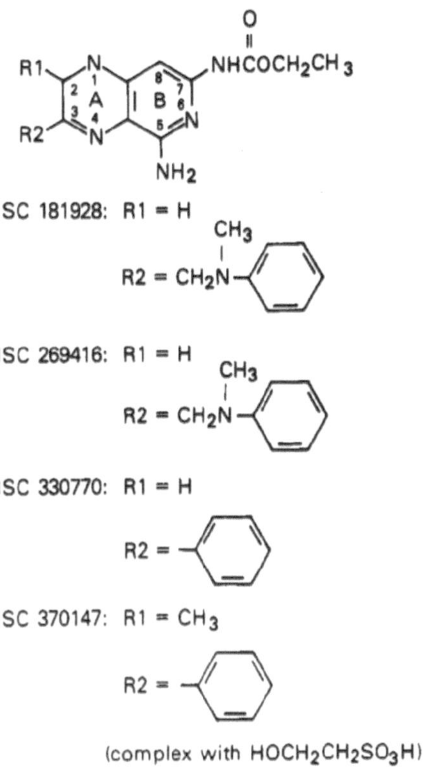

FIGURE 26. Structural formulas of derivatives of 1,2-dihydropyrido[3,4-*b*]pyrazine.

7 were labile.[426] An analogue of NSC 330770 with no substituent at position 7, however, had no antimitotic activity.

In our laboratory we examined interactions of NSC 181928 with purified tubulin.[112,411] We found that it inhibited microtubule assembly and the binding of colchicine to tubulin. It appeared to be somewhat more active than nocodazole in the latter reaction. NSC 181928, like colchicine and nocodazole, stimulated tubulin-dependent GTP hydrolysis and did not interfere with the binding of GTP to tubulin. Bowdon et al.[365] compared the interactions with tubulin of NSC 181928 with those of NSC 269416, NSC 330770, and NSC 370147. IC_{50} values (plateau) with microtubule protein were about 0.3 μM with NSC 181928 and NSC 370147, but were higher (probably about 2 to 3 μM) with the other two compounds. Their order of activity as inhibitors of colchicine binding were NSC 330770 (most active) > NSC 370147 > NSC 181928 > NSC 269416. The inhibition of colchicine binding by NSC 370147 was examined in detail, and the agent was demonstrated to be a competitive inhibitor with a K_i value of 0.6 μM. The most active dihydropyridopyrazine derivatives were more active than nocodazole both as cytotoxic agents[424,426] and in their *in vitro* interactions with tubulin.[365]

c. *Derivatives of Aniline*

The structures presented in Figure 27, tubulozole (NSC 376450),[310,358] NSC 251635,[358,427] and NSC 215914,[358,428] all cause cells to accumulate in mitosis, and they inhibit microtubule assembly and the binding of radiolabeled colchicine to tubulin. In addition, they stimulate tubulin-dependent GTP hydrolysis. Although none has yet been demonstrated to be a competitive inhibitor of colchicine, this similarity in their properties to those of the carbamates discussed above suggests a common mode of action. In all three compounds, a carbamate substituent is attached to a phenyl ring.

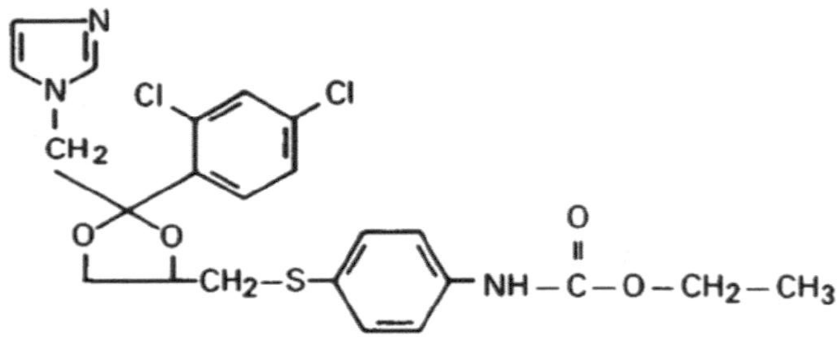

Tubulozole
(NSC 376450)

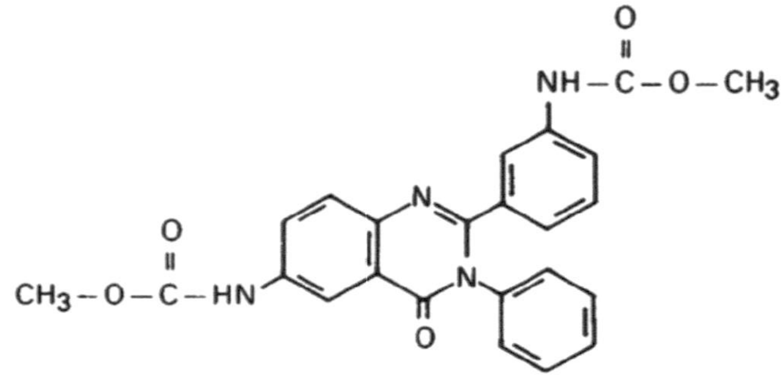

NSC 251635

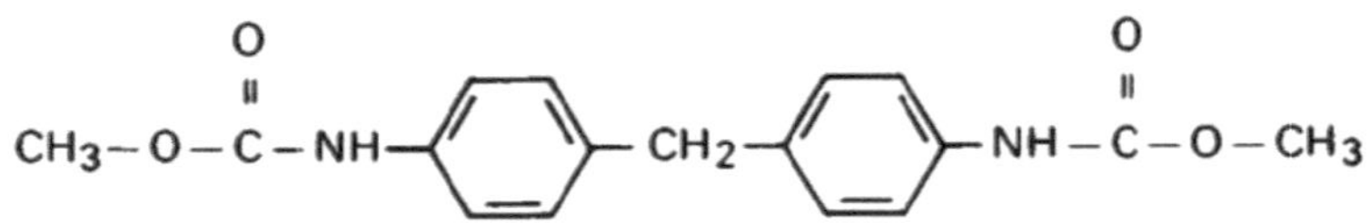

NSC 215914

FIGURE 27. Structural formulas of tubulozole, NSC 215914, and NSC 251635.

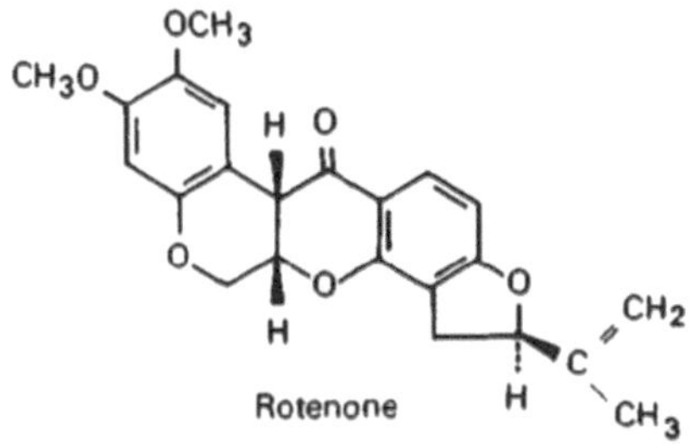

Rotenone

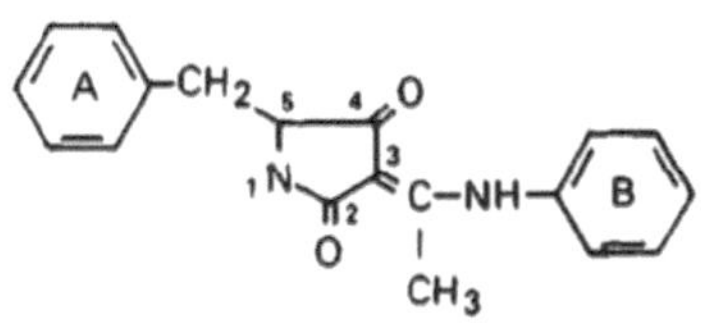

TN-16
(NSC 100445)

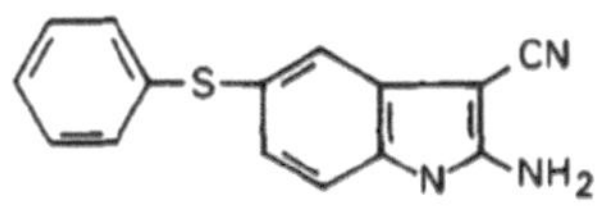

Amphethinile
(NSC 383336)

FIGURE 28. Structural formulas of rotenone, TN-16, and amphethinile.

In the case of tubulozole, only the *cis* isomer at the asymmetric carbon is active, and the carbamate function has been found to be essential for activity.[310] It has an IC_{50} value (plateau) of 0.5 μM as an inhibitor of microtubule assembly. It was more inhibitory than nocodazole, but it caused no disassembly of preformed microtubules.[310]

Both NSC 215914 and NSC 251635 are actually bis-carbamates, and the former is of interest because of its structural symmetry. Both are significantly less effective, both as antimitotic agents and as inhibitors of tubulin polymerization and colchicine binding, than nocodazole, NSC 181928, NSC 330770, and tubulozole, to which they have been directly compared.[358]

A few additional carbamates with effects on tubulin-dependent reactions have been identified, based on their stimulation or inhibition of GTP hydrolysis; but they have not been studied in detail.[358]

6. Other Agents

a. Rotenone

Derived from *Lonchocarpus nicou*, rotenone (structure in Figure 28) is believed to have major inhibitory effects on mitochondrial respiration;[429] but in addition it causes cells to accumulate in metaphase arrest.[430,431] Two groups reported that approximately 10 μM rotenone completely inhibited microtubule assembly from microtubule protein and the polymerization of purified tubulin[432,433] and that this inhibition was largely reversible by charcoal treatment of the drug-tubulin complex;[433] but an IC_{50} value (rate) of 0.5 μM has also been reported.[304] Rotenone inhibits the binding of radiolabeled colchicine to tubulin,[431,433] probably competitively,[433] although its K_i value has not been determined.

b. 3-(1-Anilinoethylidene)-5-Benzylpyrrolidine-2,4-Dione (TN-16)

TN-16 (NSC 100445; structure in Figure 28) was the most cytotoxic agent in a series

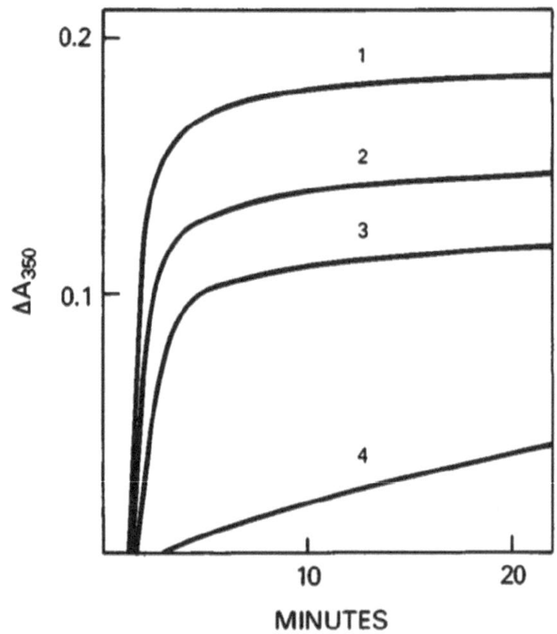

FIGURE 29. Inhibition of microtubule assembly by amphethinile. Each reaction mixture contained 0.1 *M* 4-morpholineethanesulfonate (pH 7.0), 0.5 m*M* $MgCl_2$, 1 m*M* GTP, 1.5 mg/ml tubulin, 0.75 mg/ml heat-treated microtubule-associated proteins, and the following concentrations of amphethinile: curve 1, none; curve 2, 5 μ*M*; curve 3, 15 μ*M*; and curve 4, 30 μ*M*.

of similar compounds synthesized by Yuki et al.[434] With the exception of *para* hydroxy and fluoro substituents in the A and B phenyl rings, respectively, substituents in either phenyl ring resulted in significant losses in cytotoxic activity. The A phenyl could not be replaced by either an alkyl or indole group, nor the B phenyl ring by a pyridine group, with retention of activity. TN-16 caused rapidly reversible mitotic arrest in cultured cells.[435] Interactions of the agent with microtubule protein were studied by Arai.[436] He found that TN-16 inhibited microtubule assembly with an IC_{50} value of 2 μ*M* (plateau) and inhibited colchicine binding to the protein. Inhibition of the latter reaction was comparable to that of nocodazole. TN-16 bound to tubulin in the cold, and the reaction was reversible as bound TN-16 could be subsequently displaced by colchicine. TN-16, like colchicine, inhibits the alkylation of tubulin sulfhydryl groups by iodoacetamide and formation of an intra-β-tubulin crosslink.[437]

*c. **Amphethinile***

Amphethinile, or 2-amino-3-cyano-5-(phenylthio)-indole (NSC 383336; structure in Figure 28) is a recently described antimitotic agent[438] which is in early clinical trials in Europe. It has been described as an antimicrotubule agent, although no data was presented.[438] This finding was confirmed in our laboratory. Figure 29 demonstrates the inhibitory effect of this new agent on microtubule assembly, and Table 9 its moderate inhibitory effect on the binding of radiolabeled colchicine to tubulin.

B. DRUGS BINDING AT THE VINBLASTINE SITE

1. The *Vinca* Alkaloids

From a clinical point of view, the *Vinca* alkaloids (structures in Figure 30) are without question the most important of the antimitotic agents. Vincristine and vinblastine, derived from *Catharanthus roseus* (formerly *Vinca rosea*), and the semisynthetic vindesine have

TABLE 9
Inhibition of Colchicine Binding to Tubulin by Amphethinile

Drug:colchicine	Percent inhibition
Amphethinile:colchicine = 1:1	29
Amphethinile:colchicine = 10:1	89
Podophyllotoxin:colchicine = 1:1	90
Combretastatin A-4:colchicine = 1:1	98

Note: Each reaction mixture contained 1 μM tubulin, 5 μM [^{3}H]colchicine, and the indicated concentration of inhibitor. Incubation was for 10 min at 37°C. See Reference 411 for further experimental details.

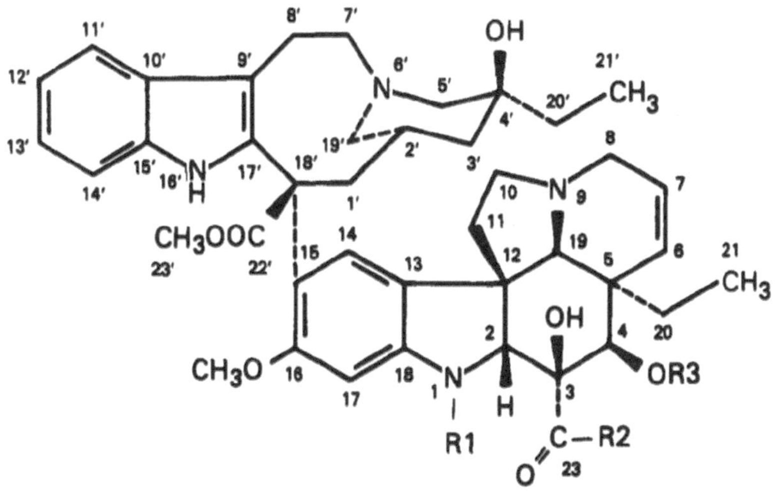

Vinblastine: R1 = CH_3; R2 = OCH_3; R3 = $COCH_3$
Vincristine: R1 = CHO; R2 = OCH_3; R3 = $COCH_3$
Vindesine: R1 = CH_3; R2 = NH_2; R3 = H

FIGURE 30. Structural formulas of vinblastine, vincristine, and vindesine.

important roles in the chemotherapy of malignant diseases.[439] Vincristine, in particular, is a mainstay in the treatment of the neoplasms of childhood, an area in which remarkable progress has been made in the last two decades.[440]

The antimitotic activity of these agents was established shortly after their discovery[441-443] and was subsequently confirmed in many laboratories. Cessation of growth and mitotic arrest with mammalian cells in culture occurs at submicromolar concentrations of the most active *Vinca* compounds, and this is associated with disappearance of cellular microtubules.[444-449]

At higher drug concentrations (at least 10 μM) multiple unique inclusion bodies termed paracrystals appear in cells.[269,395,450-459] One of the first reports of these *Vinca* alkaloid-induced structures resulted from a therapeutic tragedy. A child with meningeal leukemia received an intrathecal injection of vincristine instead of methotrexate, and multiple intraneuronal paracrystals were observed in the nervous tissue examined at autopsy.[452] In experimental studies, paracrystal formation was enhanced if other antimitotic drugs were included in the culture medium.[395,459,460] Since paracrystal formation occurs subsequent to the disappearance of microtubules, it was initially speculated that they represented microtubules with an altered morphology.[451,453] This idea was confirmed by the findings that colchicine bound intracellularly to paracrystals,[461] that antibody raised against isolated paracrystals reacted with the spindle of mitotic cells,[456] that antitubulin antibody reacted with paracrystals,[462] and that isolated paracrystals were composed of virtually pure tubulin.[463]

Bryan[238] examined the interactions of paracrystals isolated from sea urchin eggs with small ligands. Relative to tubulin they contained one molar equivalent of vinblastine, which would not exchange with free vinblastine, and two molar equivalents of guanine nucleotide, which also was not exchangeable. The paracrystals were able to bind one molar equivalent of colchicine and, presumably, podophyllotoxin, since the latter drug inhibited binding of colchicine. Na and Timasheff[464] succeeded in preparing paracrystals (but see below) from purified tubulin *in vitro*. They, too, found about one molar equivalent of vinblastine in the crystals. Both paracrystals isolated from sea urchin eggs[465] and paracrystals formed *in vitro*[464] could bind an additional molar equivalent of vinblastine. Wilson et al.[465] found, in contrast to the earlier report of Bryan,[238] that most of the vinblastine isolated with the sea urchin paracrystals could be displaced by exogenously added vinblastine. They concluded that the two vinblastine binding sites in the paracrystals had similar affinities for the drug (K_D value of 4 μM) and binding was little affected by reaction temperature.

The structure of the vinblastine-induced paracrystals was elegantly analyzed by Fujiwara and Tilney.[268] The isolated paracrystals can be either discoid or elongated, with each end a convex hexagonal face. In the electron microscope the predominant appearance in cross-section is of a hexagonal array of 320 Å tubules, each unit having 6 neighbors. Fujiwara and Tilney[268] concluded that each tubular unit consisted of two tightly spiraled fibrils 180° out of phase, but of the same handedness. They assumed that the fibrils were equivalent to the protofilaments of normal microtubules. Amos et al.[460] reexamined the ultrastructure of sea urchin egg paracrystals and compared it to that of paracrystals formed from purified tubulin *in vitro*. Their conclusion was similar to that of Fujiwara and Tilney,[268] except that the *in vitro* paracrystals were less well organized and their individual units were composed of only one tightly coiled protofilament.

Perhaps in no area of the microtuble/tubulin literature are there such reported differences between laboratories as in the characteristics of the binding of the *Vinca* alkaloids, primarily vinblastine, to tubulin. Although some of these differences may reflect use of different techniques and organisms[244] and varying Mg^{2+} concentrations,[466] the K_D values and number of binding sites should probably be considered as open questions.

The vinblastine and colchicine sites are clearly different, but there appear to be mutual interactions that have generally been considered to represent stabilization of tubulin by both ligands. Most workers have found that including vinblastine in the reaction mixture moderately enhanced the binding of colchicine to tubulin.[211,214,215,218,222-225,235,251,467,468] There have, however, been several reports in which vincristine or vinblastine either had negligible effects on colchicine binding[338,469-471] or acted as a noncompetitive inhibitor.[239,354] Similarly, colchicine has been reported to not alter the binding of radiolabeled *Vinca* alkaloids to tubulin[465,472,473] or to enhance their binding moderately.[237,312,469-471] In comparison to the binding of colchicine, the binding of radiolabeled vinblastine and vincristine to tubulin occurs rapidly and even at low temperatures.[49,237,240,312,359,449,469,473,474]

Most workers have used radiolabeled agents to obtain binding data for the *Vinca* alkaloids with tubulin. Owellen's group[237,240,244] obtained K_D values for vinblastine ranging from 0.2 to 10 μM depending on the source of tubulin and the methodology used, with the latter variable producing greater differences in the binding constants than species variation. They also found that the affinity of vincristine for tubulin was nearly identical to that of vinblastine under the same reaction condition.[237] These workers observed one class of binding site and substoichiometric amounts of drug bound to tubulin. Their studies were performed at 37°C and generally without Mg^{2+}. Wilson et al.[473] reported one class of binding site, a stoichiometry of two vinblastine molecules bound to each tubulin molecule, and a K_D value of 2 to 3 μM. These studies were performed at 4°C with 5 mM Mg^{2+}. Bhattacharyya and Wolff[218] obtained a K_D value for vinblastine of 0.1 μM at 37°C with 10 mM Mg^{2+}. Conrad et al.[475] reported a K_D value of 1 μM for vinblastine, but no further details were presented. Mandelbaum-Shavit et al.[469] reported a K_D value of 2 μM for vincristine obtained at 37°C without Mg^{2+}. Lacey et al.[251] observed substoichiometric binding and a K_D value for vinblastine of 40 μM at 37°C with 0.5 mM Mg^{2+}. Takahashi et al.[476] observed stoichiometric binding of vinblastine at a single site, with a K_D value of 0.2 μM. Their study was performed at 37°C with 0.5 mM Mg^{2+}.

Bhattacharyya and Wolff[312] obtained results indicating two distinct classes of vinblastine binding sites, each with a stoichiometry of one. The high affinity site had a K_D value of 0.2 μM, the low affinity site of 10 μM. This study was performed at 37°C with 10 mM Mg^{2+}. They also observed that the low affinity binding site was much more stable than the high affinity site. Bhattacharyya and Wolff[312] proposed that the high affinity site was responsible for inhibition of microtubule assembly by the *Vinca* alkaloids, while occupancy of the low affinity site led to aggregation and paracrystal formation (see below). We, too, have found two classes of vinblastine binding site on tubulin.[471] In our study at 20°C with 0.5 mM Mg^{2+}, the K_D values were 0.5 μM for the high affinity site and 10 μM for the low affinity site.

Timasheff's group initially evaluated binding of vinblastine and vincristine to tubulin by fluorimetric titration without Mg^{2+} at 25°C.[474,477] Their results indicated two equivalent binding sites with a relatively high dissociation constant of about 30 μM for vincristine and 40 μM for vinblastine. In a subsequent study with radiolabeled vinblastine,[466,478] this group found that the apparent affinity of vinblastine for tubulin varied substantially with the tubulin concentration: the affinity of the drug for the protein rose as the protein concentration increased. Na and Timasheff[478] argued that the binding of vinblastine to tubulin could not be evaluated properly without simultaneously determining the effect of vinblastine-induced tubulin oligomerization on drug binding (see below). Mathematical analysis of their data led them to conclude that rather than two binding sites for vinblastine of equal affinity, there are actually two classes of sites — (1) a single high affinity site with a K_D value of about 20 μM (altered by protein concentration) and (2) perhaps two low affinity sites with a K_D value of about 200 μM (not altered by protein concentration).[478] They also found that in their system 1 mM Mg^{2+} substantially enhanced the binding of vinblastine to tubulin, the K_D value of the high affinity site falling to about 3 μM.[466]

Wilson[357] also described multiple low affinity sites for vinblastine on tubulin (about 25) and suggested they were relatively nonspecific and involved in the precipitation of tubulin observed at millimolar drug concentrations (see below).

Lee et al.[474] reported that spectroscopic studies indicated only minor changes in conformation of the drug-tubulin complex and suggested that the binding of vinblastine occurred in a hydrophobic area of the protein. Upon binding vinblastine, however, tubulin reactivity with antibodies[284] and sulfhydryl reactive agents[264-266,479] changes considerably, suggesting conformational alterations.

Localization of the *Vinca* binding site on tubulin has been examined in a few laboratories. A CHO cell line resistant to maytansine (a drug binding at the *Vinca* site, see Section 2), which produced an altered α-tubulin,[480] and a rhabdomyosarcoma cell mutant resistant to vincristine, which produced an altered β-tubulin,[481] have been reported. Safa and Felsted[482] synthesized a photoaffinity analogue of vinblastine modified at position 23 (see Figure 30) and found it to be nearly identical to vinblastine in its cytotoxicity and its ability to induce paracrystal formation. Safa et al.[471] went on to demonstrate similar properties of the analogue and vinblastine in their effects on microtubule assembly and tubulin aggregation and in binding to tubulin. Photoactivation of the analogue-tubulin complex resulted in specific covalent labeling of the α- and β-subunits in a 3:2 ratio. This was not altered over a wide range of drug concentrations, suggesting that a single site was being photolabeled. Luduena and colleagues[264-266,479] have examined the effects of vinblastine on the alkylation of tubulin sulfhydryl groups. Reaction of iodoacetamide with both subunits was strongly inhibited by vinblastine, but the drug had distinct effects on two intra-β crosslinks. One crosslink, between cys 239 and cys 354[483] (CL-1) occurs with all tubulin preparations, and its formation was stimulated by vinblastine. A second crosslink, between cys 12 and cys 201 or 211 (CL-2),[484] occurs only if nucleotide-depleted tubulin is used, and its formation was weakly inhibited by the drug.

Several laboratories have reported on different aspects of the effects of the *Vinca* alkaloids on tubulin-GTP interactions. The nucleotide enhances the binding of vincristine to tubulin,[485] but the binding of vinblastine to tubulin results in profound inhibition of the GTPase activity of the protein, perhaps directly as a consequence of its inhibition of microtubule assembly (see below).[36,55,56] Both vinblastine and vincristine have been reported to enhance the binding of radiolabeled GTP to tubulin both at 0 and at 37°C,[468] to have no effect on nucleotide exchange in the tubulin-colchicine complex at 37°C,[55] and to weakly inhibit exchange of nucleotide, without displacing bound nucleotide, with a greater effect observed at 0 than at 37°C.[112] As noted above, Bryan[238] found that isolated paracrystals contained two nonexchangeable molar equivalents of guanine nucleotide, but vinblastine-induced aggregation of tubulin *in vitro* has been reported to cause partial release of GTP from the protein.[23,486]

Like the colchicine class of antimitotic drugs, the *Vinca* alkaloids inhibit both microtubule assembly and the polymerization of purified tubulin. There is also a wide variation in reported IC_{50} values for vinblastine and vincristine. The lowest values, 0.02 to 0.07 μM (rate), were obtained in tissue homogenates with viscosity changes measured at 30°C.[364,487] Most workers have obtained IC_{50} values ranging from 0.1 to 0.8 μM, regardless of technique used to measure polymerization, reaction temperature, whether rate or plateau was evaluated, or whether tissue homogenate, microtubule protein, or purified tubulin was used.[55,112,251,304,312,365,448,477,487-492] Three groups, including ours, have reported the IC_{50} (plateau) for vincristine or vinblastine to be 1 to 2 μM.[112,471,493-495] When directly compared, vincristine has always been somewhat more inhibitory than vinblastine.[302,304,488,493]

Although Sajo[489] described nearly complete depolymerization of microtubules (measured viscometrically) following addition of 10 μM vinblastine to polymer, most workers have found that solution turbidity persists when the *Vinca* alkaloids are added to preformed microtubules. The inhibition of polymerization by these agents is unique in that the *Vinca* alkaloids do not simply prevent microtubule assembly, but, instead, induce the formation of alternate oligomeric or polymeric structures. This process has been examined from differing points of view by many workers.

Initially, it was observed that high concentrations of vinblastine (0.1 to 2.0 m*M*) caused a nearly quantitative precipitation of tubulin independent of temperature.[269-271,496,497] Usually the precipitate had an ultrastructural appearance reminiscent of that of paracrystals,[270,496,497] but other proteins besides tubulin could also be precipitated.[271]

Subsequent work in many laboratories[55,272,470,471,476,487,488,491,492,498-509] has demonstrated that lower drug concentrations (from 5 to 100 μM) yielded oligomeric to polymeric rings and tightly coiled spirals with a one and/or two filament substructure, with these filaments often assumed to be equivalent to the protofilaments of microtubules. The coiled aggregates form from both unpolymerized protein[272,470,471,476,492,498-504,506-509] and from polymer[272,487,488,491,499,503,505,507,508] at low temperatures[272,471,488,499,502,508] as well as at 37°C in the absence of GTP,[49,55] and in the presence of colchicine.[55,272,476,499,501] Although spiral aggregate formation occurs with purified tubulin in glutamate[470] and at higher Mg^{2+} concentrations both with glycerol[55] and with dimethylsulfoxide,[499] the reaction in the absence of these components in strongly enhanced by microtubule-associated proteins,[491,499-502,506] particularly tau factor.[500,501,506] Removal of the carboxyl-termini of both tubulin subunits with subtilisin also resulted in marked enhancement of vinblastine-induced spiral aggregation.[506] Radiolabeled *Vinca* alkaloids in pelleted aggregate was measured by two groups. Haskins et al.[499] noted 0.3 mol of drug per mole of tubulin with 10 μM vincristine. Wilson et al.[487] found increasing amounts of vinblastine in the pellet as the drug concentration rose, with a maximum of about 0.7 molar equivalent at 50 μM vinblastine.

It is also worth noting that at least one instance of spiral aggregate, as opposed to paracrystal, formation has been noted *in vivo*.[510] Spermatids in crane fly pupae have a row of cold-stable microtubules attached to a structure known as the perimitochondrial dense body. When treated with vinblastine the microtubules in these cells are replaced by spiral aggregates still attached to the dense body.

Timasheff's group examined vinblastine and vincristine-induced tubulin oligomerization intensively with purified tubulin at relatively low drug:protein ratios.[144,466,474,477,478,511,512] Their primary investigational tool was analytical ultracentrifugation, and they demonstrated increasingly large oligomers as a function of Mg^{2+}, tubulin, and *Vinca* alkaloid concentrations, each taken independently, until ultimately large precipitates were formed. Analysis indicated that the data obtained was best explained by an isodesmic self-association model (multiple steps with identical association constants) with ligand binding stabilizing the oligomers. The best fit required a model in which binding of vinblastine could occur to both single tubulin molecules and tubulin oligomers, and in which both tubulin and tubulin-drug complexes could self-associate. It was concluded that drug:tubulin stoichiometry in the oligomers was approximately 1:1, with the drug binding in a high affinity site. The K_A for oligomerization was estimated as 20,000 M^{-1} without Mg^{2+} and as 50,000 M^{-1} in 1.0 mM Mg^{2+}.

Wilson et al.,[487] as noted above, observed particularly potent inhibition of microtubule assembly by vinblastine in brain tissue homogenates (IC_{50} value of 0.07 μM), but inhibition of polymerization of purified microtubule protein was still highly sensitive to the drug (IC_{50} value of 0.6 μM, with a tubulin concentration of 16 μM). These workers examined tubulin flux through the microtubule and found that 50% inhibition occurred at 0.14 μM vinblastine. In parallel experiments in which radiolabeled vinblastine was added to microtubules, they observed a limited amount of binding of the drug to the polymer with an apparent K_D of 2 μM. Correlating these data with the average microtubule length, they calculated that tubulin flux was inhibited by 50% when the average vinblastine to microtubule ratio was about 1.2. They concluded that substoichiometric inhibition of microtubule assembly could be explained by the binding of a small number of vinblastine molecules or vinblastine-tubulin complexes to one or both microtubule ends. Wilson et al.[487] also noted that as the vinblastine concentration was further increased microtubules started to disassemble into spiral aggregates and rings, and this was associated with increasing ratios of vinblastine to tubulin in pellets harvested by centrifugation. This latter reaction was then studied more extensively.[505] Formation of aggregates increased with incubation time. Careful preparation of specimens for electron microscopy demonstrated an average of five to six spiral aggregates, with occasionally up to eight, attached to each microtubule end. Both ends of individual microtubules were involved in this process. Although no aggregates were observed along the body of the microtubule, its structural integrity seemed to have deteriorated. Despite shearing the aggregates from them, the residual microtubules had an increasing amount of radiolabeled vinblastine bound to them as a function of incubation time. Both as the vinblastine concentration rose and as the incubation time increased, microtubules became decreasingly sensitive to decoration by DEAE-dextran. These workers argued that this vinblastine-induced depolymerization reaction occurred by a mechanism in which increasing amounts of vinblastine bound to microtubules led to progressive breakdown of their organization.

Maytansine[49,490,504] and rhizoxin,[476] two other antimitotic agents which bind at the *Vinca* site on tubulin (see Sections 2 and 4), inhibit the spiral aggregation reaction, even at concentrations substoichiometric to that of tubulin and at concentrations well below that of vinblastine in the reaction mixture. Fellous et al.[504] found that aggregation with 10 μM tubulin and 40 μM vinblastine was totally inhibited by 2 μM maytansine, and Takahashi et al.[476] observed complete inhibition of aggregation with 10 μM tubulin, 100 μM vinblastine, and 2 μM rhizoxin.

Most of the observations described above indicate that the major *Vinca* alkaloid-induced phenomena — inhibition of microtubule assembly, paracrystal and spiral aggregate formation, substoichiometric inhibition of aggregation by maytansine and rhizoxin — probably result from interaction of these agents at a single site on tubulin. The rapid precipitation of tubulin observed at high drug concentrations may involve either low affinity sites or ionic interactions with tubulin of the agents in their cationic form.

There have been relatively few systematic studies of interactions of *Vinca* alkaloid analogues with tubulin (see structure in Figure 30, numbered as described by Moncrief and Lipscomb[513]). Analogues of the tetracyclic (catharanthine) and pentacyclic (vindoline) portions of the vinblastine molecule have had little or no inhibitory effect on tubulin.[302,464,472,486,514] Of the three agents used clinically, vinblastine and vincristine differ minimally in the substituent at position 1, while vindesine differs from vinblastine in the substituents at positions 4 and 23. No major differences have been described between these three compounds in their interactions with tubulin.[302,465,472,488]

Tubulin appears to be relatively tolerant to modifications in the vindoline moiety. Many, but not all, alterations at position 23 yield compounds with significant inhibitory activity.[302,465,471,472,475,482] Deacetylation or alternate esters at position 4 have little effect on vinblastine interactions with tubulin.[302,357,464,465,472,475,514] Removal of the substituent at position 1[465,472] or saturation of the 6-7 bond,[472] however, does lead to a reduction in activity. Molecules with more substantial alterations in the vindoline ring are inert (reducing it to an indole derivative[493] or introducing a bridging group between positions 2 and 5, together with removal of most vindoline substituents[493,494]).

A number of analogues modified in the catharanthine moiety have also been examined. Reversal of configuration at position 4′ reduces inhibition of microtubule assembly less than twofold.[493,494] Most compounds with modifications at position 4′, including an epoxide or a double bond between positions 3′ and 4′, have comparable or only moderately (up to tenfold) reduced inhibitory activity as compared to vinblastine;[302,472,492-494,503,509] but their antimitotic effects are much more variable.[503,509] Addition of an hydroxyl group at position 3′ has only minor effect on *in vitro* interactions with tubulin,[302,472] but this compound is 100-fold less active than vinblastine as an antimitotic agent.[514] Reversal of configuration at positions 2′ and/or 18′ resulted in compounds with little or no activity.[493,494] Deacetylation at position 18′ resulted in an inactive analogue; but a bulky modification in the 18′ acetyl substituent only reduced inhibition of microtubule assembly 20-fold.[493,494] A compound with a direct bond between positions 3′ and 9′ (with loss of the hydrogen nominally at position 16′), together with reversed configuration at position 4′, was almost inactive.[493] "Reduced vinblastine" (hydroxymethyl groups at positions 3 and 18′; hydroxyl at position 4) also had little effect on microtubule assembly.[493]

Finally, three structurally similar compounds with major modifications in the catharanthine moiety, with very divergent activities, were described by Potier et al.[494] (structures in Figure 31). In all three compounds the bridge between positions 6′ and 9′ was disrupted, and the 3′-4′ bond was double with elimination of the 4′-hydroxyl group. Compound 31/1 was half as active as vinblastine as an inhibitor of microtubule assembly, compound 31/2, one tenth as active, and compound 31/3, almost inactive.

The crystal structure of at least one *Vinca* alkaloid derivative has been described in

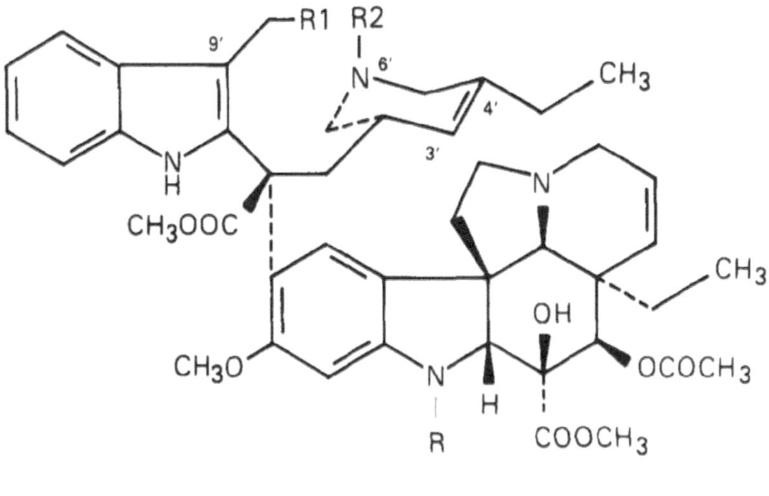

31/1: R1 = OCH_3: R2 = CH_2CN
31/2: R1 = OCH_3; R2 = CH_3
31/3: R1 = H; R2 = CH_3

FIGURE 31. Structural formulas of vinblastine analogues with a disrupted catharanthine moiety.

detail, vincristine methiodide dihydrate, in which a quarternary amine was formed by adding a methyl group at position 6′.[513] The mean planes of the vindoline and cathantharine moieties are oriented in the crystal at about an 80° angle to each other.

2. The Maytansinoids

Kupchan and collaborators[515-519] described the isolation and chemical characterization of a series of ansa macrolide compounds from *Maytenus ovatus, M. buchananii, M. serrata,* and the related plant *Putterlickia verrucosa.* The most extensively studied of these compounds is known as maytansine (NSC 153858; structure in Figure 32). but all of them possess a 19-member macrolide ring. Similar compounds were also isolated from *Colubrina texensis,* which belongs to a different plant family, by Wani et al.;[520] and from a microorganism, a *Nocardia* species, by Asai et al.[521,522] The most carefully evaluated compounds from the latter organism are known as ansamitocin P-3 and ansamitocin P-4 (structures in Figure 32). Both maytansine[519,523-525] and the ansamitocins[448,522] are potent antimitotic agents.

The binding of radiolabeled maytansine to tubulin is relatively fast, reversible, and occurs in the cold,[49,469] with a K_D value at 37°C of 0.7 μM.[469] This binding reaction is inhibited by vincristine[469] and vinblastine.[470] Vincristine was demonstrated to act as a competitive inhibitor of the binding of radiolabeled maytansine with an apparent K_i value of 10 μM.[469] Neither vincristine[469] nor vinblastine[49,470] completely displaces bound radiolabeled maytansine from tubulin, but the reason for this has never been established. Nonradiolabeled maytansine, however, displaces virtually all radiolabeled vincristine or vinblastine bound to tubulin.[469-471] The k_i value obtained for maytansine vs. vincristine binding was 0.4 μM,[469,526] and vs. vinblastine binding, 0.5 μM.[490] Lacey et al.[251] reported that 1.6 μM maytansine inhibited the binding of radiolabeled vinblastine by 50%. These reports have all indicated that maytansine binds to tubulin with a higher affinity than either vinblastine or vincristine. In addition, ansamitocin P-3 was found to inhibit the binding of radiolabeled vinblastine to tubulin, but the data suggested the inhibition might not be competitive.[476] Maytansine had no effect or mildly stimulated the binding of radiolabeled colchicine to tubulin.[251,469,470]

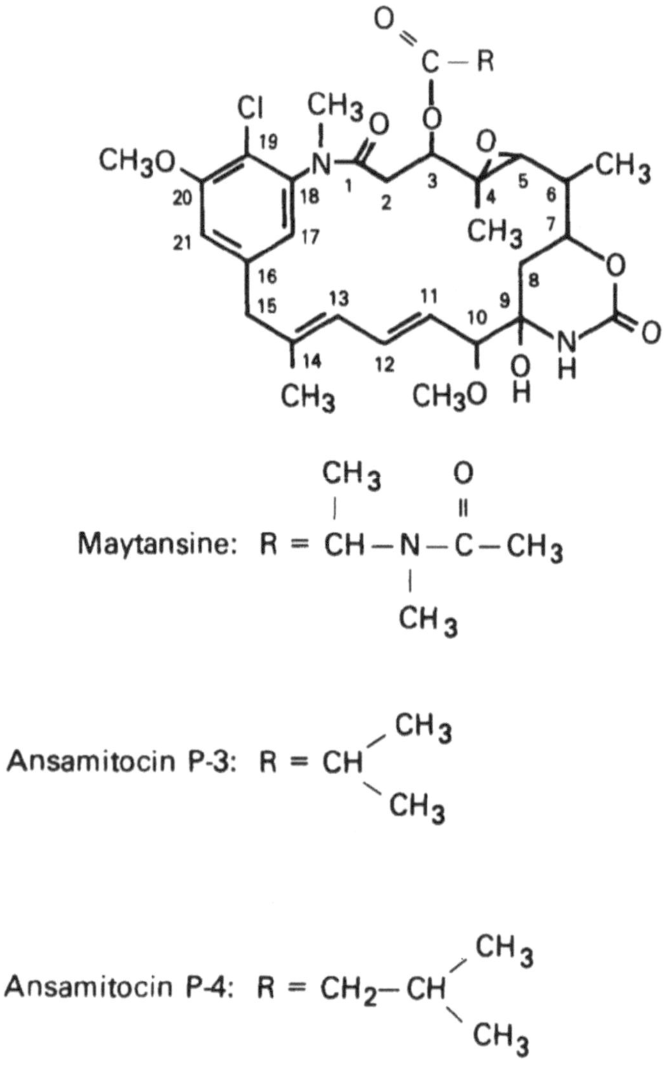

FIGURE 32. Structural formulas of maytansine, ansamitocin P-3, and ansamitocin P-4.

Luduena and colleagues have noted distinctive effects of maytansine on the alkylation of tubulin sulfhydryls.[264-266,479] Alkylation was least affected by maytansine of all antimitotic agents examined, nor did the drug alter the relative alkylation of the two subunits. When both vinblastine and maytansine were included in the reaction mixture, the maytansine effect predominated. The effects of maytansine on cysteine-cysteine crosslinks were also unique. Although maytansine was similar to vinblastine in enhancing formation of CL-1, its effects on CL-2 (requiring nucleotide-depleted tubulin) were remarkable. The drug almost totally inhibited formation of this crosslink, as did readdition of guanine nucleotide to the tubulin. Especially in view of the probable location of the exchangeable nucleotide site on the β-tubulin subunit (see Section A), this implies that the maytansine binding site may also be on the β-subunit.

Our own studies lead to a similar conclusion.[56,63,112] Maytansine potently inhibits tubulin-dependent GTP hydrolysis[56] and nucleotide exchange on tubulin, particularly at low temperatures (Figure 33).[112] It does not, however, displace nucleotide bound in the exchangeable site and, in fact, prevents its loss from tubulin during gel filtration chromatography.[63,112]

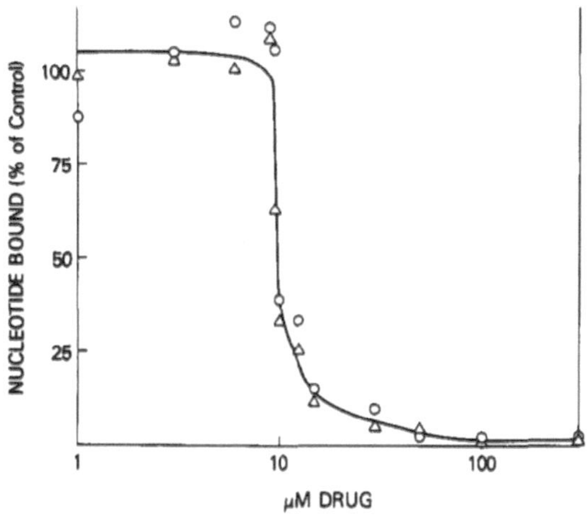

FIGURE 33. Inhibition of the binding of radiolabeled GDP and GTP to tubulin by maytansine. Each reaction mixture contained 10 *μM* tubulin-GDP, the indicated concentration of maytansine, and, added last, either [8-^{14}C]GDP (circles) or [8-^{14}C]GTP (triangles) at 50 *μM*. See Reference 112 for further experimental details.

The simplest explanation for these findings is that maytansine binds on the β-subunit in a manner that prevents entry and exit of nucleotides at the exchangeable site.

Maytansine and the ansamitocins inhibit microtubule assembly requiring microtubule-associated proteins[112,251,448,490,495,504,519,524,526] and the glutamate-induced polymerization of purified tubulin.[49] Fellous et al.[504] reported that a given concentration of maytansine was substantially more inhibitory for polymerization with tau factor than for polymerization with MAP-2. Bhattacharyya and Wolff[490] reported an IC_{50} value of 0.4 *μM* (plateau) for maytansine, but most workers have obtained values from 1 to 5 *μM* for maytansine and the ansamitocins (both rate and plateau).[112,251,448,495,524] Despite its tighter binding to tubulin than vinblastine (see above), maytansine has always been found to be somewhat less effective than vinblastine as an inhibitor of microtubule assembly.[112,251,490] Similarly, in direct comparisons the ansamitocins are less potent assembly inhibitors than vinblastine and vincristine.[448,495]

Relatively low concentrations (less than 20 μM) of both maytansine and the ansamitocins cause extensive depolymerization of preformed microtubules.[448,524] The ansamitocins have been demonstrated to cause disappearance of intracellular microtubules.[448]

As noted above (Section 1), maytansine and the ansamitocins do not cause spiral aggregate formation in a manner similar to the *Vinca* alkaloids.[49,448,490,495,504] Rather, maytansine inhibits *Vinca* alkaloid-induced aggregate formation.[49,490,504] This inhibition is quite potent, for maytansine at concentrations substoichiometric to those of both tubulin and vinblastine totally inhibited aggregate formation (e.g., 2 *μM* maytansine, 10 *μM* tubulin, 140 *μM* vinblastine).[49,504] Maytansine also caused dissolution of vinblastine-induced aggregates.[504]

Little is known about tubulin interactions of maytansine analogues, other than ansamitocin P-3 and ansamitocin P-4,[448,476,495] which differ only in the ester substituent at position 3 (see Figure 32). York et al.[526] compared four analogues to maytansine as inhibitors of vincristine binding to tubulin. A compound with an altered ester at position 3 was nearly as effective as maytansine, while three compounds lacking a substituent at position 3 together with introduction of a 2-3 double bond were significantly less inhibitory. One of them had an apparent noncompetitive effect on vincristine binding, but the calculated K_i value was 5 *μM* as compared to the competitive K_i value of 0.4 *μM* for maytansine. Kupchan et al.[519]

described preliminary results comparing the inhibitory effects of a large series of analogues on microtubule assembly, fertilized sea urchin egg mitosis, and growth of KB cells in culture. There was little correlation in the three assays, and they only described compounds with significant inhibitory effects on microtubule assembly. The following modifications did not appear to alter greatly the inhibitory effect of maytansine on polymerization (see Figure 32): altered ester at position 3, hydroxyl group at position 3, etherification of the hydroxyl at position 9, thioether instead of hydroxyl group at position 9, no substituent at position 3 together with introduction of a 2-3 double bond (in contrast to the finding of York et al.[526]), no *N*-methyl group, and acetyl group at position 15.

Kupchan et al.[515,516] and Bryan et al.[527] described the crystal structure of maytansine 3-bromopropyl ether. They found that its substituents are oriented in a manner that minimizes intramolecular repulsions. The two longer sides of the 19-member ring are approximately parallel and about 5.4 Å apart, with an apparent opening in its center. The face of the ring with the ester group is relatively hydrophilic, while the opposite face is relatively hydrophobic.

3. Phomopsin A

Lupinosis is a severe liver disease of livestock grazing on lupins. It is caused by fungal toxins produced by *Phomopsis leptostomiformis*,[528] and one of the characteristic pathologic changes observed in the livers of diseased animals is hepatocytes arrested in mitosis.[529] Culvenor and colleagues purified and established the structures of the active components in fungal extracts.[251,530-532] The compound present in the largest amounts was named phomopsin A (NCS 381839; structure in Figure 34). It is a hexapeptide composed of modified amino acids, with a 13-membered ring including an ether linkage. Phomopsin A caused experimental lupinosis, including the characteristic mitotic liver cells.[530] It also caused mitotic arrest of cells in culture,[533,534] leading to the prediction that it would interfere with microtubule assembly.[533]

Phomopsin A inhibited the binding of radiolabeled vinblastine to tubulin, although competitive inhibition was not rigorously demonstrated.[251] It was about twice as effective as maytansine and three times as effective as nonradiolabeled vinblastine as an inhibitor of this binding reaction. Phomopsin A moderately enhanced the binding of colchicine to tubulin.[251]

The drug potently inhibited microtubule assembly from microtubule protein[251,534] and glycerol-induced polymerization of purified tubulin,[534] with IC_{50} values (rate and plateau) of 0.4 to 0.6 μM. In a direct comparison to the other drugs, phomopsin A was equivalent to vinblastine and more potent than maytansine as an inhibitor of microtubule assembly.[251] The mycotoxin caused the disappearance of microtubules in cultured cells and their complete depolymerization *in vitro*.[534] This latter reaction was not characterized by total loss of turbidity, but by formation of rings and spiral aggregates similar to but not identical in morphology to those induced by vinblastine. Rings and spirals formed from glycerol-induced microtubules as well as from those composed of microtubule protein.[534]

Lacey et al.[251] found that three analogues of phomopsin A had almost identical effects on microtubule assembly and vinblastine binding. These were phomopsin B, which lacks the chloride atom at position 14 and is produced by the fungus in smaller quantities; octahydrophomopsin A, in which the four carbon-carbon double bonds are reduced; and phomopsinamine A, which lacks the substituent at position 24.

Mackay et al.[532] described the crystal structure of phomopsin A, permitting accurate determination of its structure. There are two molecules per unit in the crystal, with the phenyl ring having somewhat different orientations in them. The side chain projects out from the 13-membered ring in both molecules.

4. Rhizoxin

Rice seedling blight, caused by the fungus *Rhizopus chinensis*, is characterized by an abnormal swelling of the plant roots secondary to failure of cell division.[535] Several com-

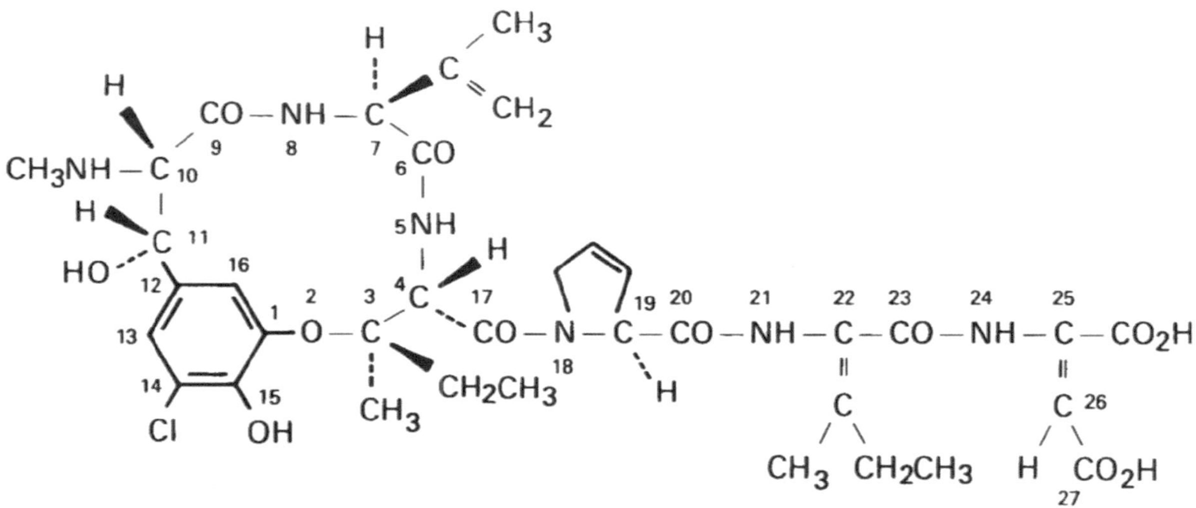

Phomopsin A (NSC 381839)

FIGURE 34. Structural formulas of phomopsin A (NSC 381839).

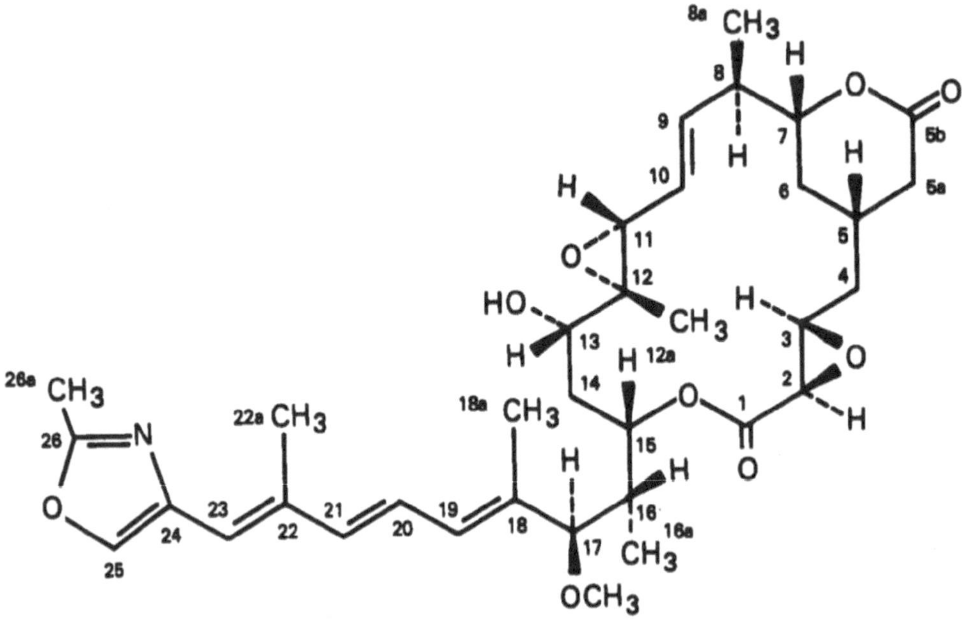

FIGURE 35. Structural formula of rhizoxin (NSC 332598).

ponents able to reproduce the disease were isolated and their structures determined by Iwasaki and collaborators.[536-539] The most important of these agents was named rhizoxin (NSC 332598; structure in Figure 35). Its most prominent structural feature is a 16-membered macrolide ring. Rhizoxin has antifungal[536] and antitumor[540] properties and caused the accumulation of cells in metaphase arrest.[540]

The binding of radiolabeled rhizoxin to tubulin at 37°C occurs rapidly and is reversible. The reaction was evaluated by Scatchard analysis, and this indicated one high affinity binding site with a K_D value of 0.2 μM.[476] Binding of the drug was inhibited both by vinblastine and ansamitocin P-3, with the latter being the more potent inhibitor, but not by colchicine.[476] Double reciprocal plots indicated competitive inhibition by both agents (K_i values of 0.1 and 3 μM for ansamitocin P-3 and vinblastine, respectively). Inhibition of the binding of radiolabeled vinblastine to tubulin by rhizoxin, however, did not appear to be purely competitive, and the apparent K_i value for rhizoxin was very low (about 0.01 μM).[476]

Rhizoxin inhibited microtubule assembly with an IC_{50} value of 5 μM (plateau), with simulataneous experiments indicating that its inhibitory effect was less than that of vinblastine and identical to that of ansamitocin P-3.[495] Rhizoxin also caused extensive microtubule depolymerization as measured by loss of turbidity.[495]

At high concentrations there was no evidence for spiral aggregate formation with rhizoxin.[476,495] In fact, low concentrations of rhizoxin inhibited the vinblastine-induced reaction; with 10 μM tubulin and 100 μM vinblastine, 2 μM rhizoxin eliminated aggregate formation.[476]

Several related compounds were compared to rhizoxin for inhibitory effects on microtubule assembly.[495] Little change in activity occurred if the methoxy substituent at position 17 was replaced with an hydroxyl group or if either the 2-3 epoxide alone or both epoxides were replaced with double bonds. A fourfold reduction in activity occurred when the lactone group was disrupted. Major loss of activity occurred if either epoxide group was hydrolyzed and replaced by an hydroxyl group (at either position 3 or 12) or if the hydroxyl group at position 13 was esterified.

The crystal structure of one relatively inactive analogue was evaluated.[537] The side chain projects out from the 16-membered ring.

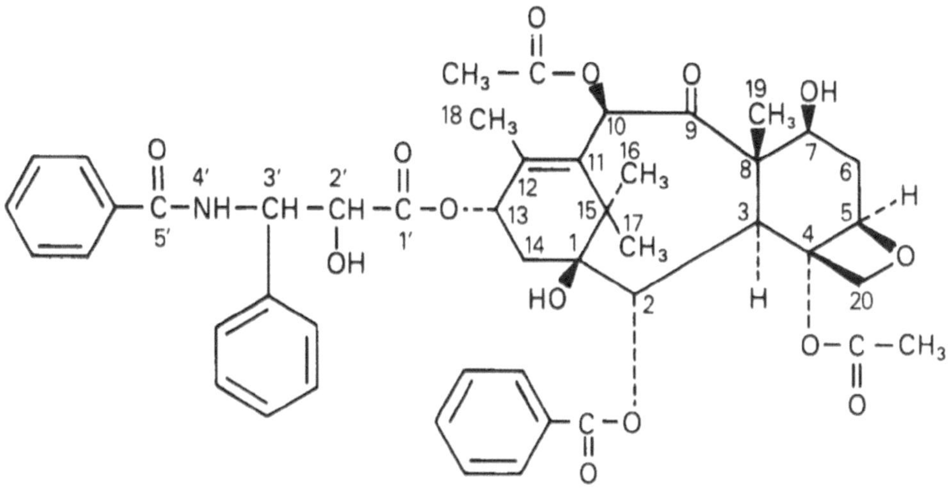

FIGURE 36. Structural formula of taxol.

Finally, rhizoxin is notable as one of the few antimitotic agents which has potent effects in both plant and animal tissues, considering the plant disease which led to its isolation[535] and its antineoplastic properties in animals.[540]

C. TAXOL

Isolated first from *Taxus brevifolia* and other *Taxus* species by Wani et al.[541] and subsequently from *T. wallichiana* by Miller et al.[542] and from *T. baccata* by Senilh et al.,[543] taxol (NSC 125973; structure in Figure 36) is an antimitotic agent[544-546] with a unique mode of action. Rather than causing depolymerization of microtubules, it stabilizes them, enhancing their formation and preventing their disassembly (see below). As a consequence, it has remarkable effects on microtubules in cells and tissues distinct from those of other antimitotic drugs, including increased microtubule formation, generation of microtubule bundles and abnormal arrays of microtubules with other cellular organelles, and abnormal persistence of microtubules. Such effects have been evaluated by many workers, and a summary of this literature is beyond the scope of this chapter. The reader is referred to an excellent review by Manfredi and Horwitz.[547]

Instead of inhibiting microtubule assembly, taxol markedly enhances all aspects of tubulin polymerization:[114,125,362,548-562] initiation and elongation are more rapid; the extent of assembly is greater; the critical concentration of tubulin required for polymerization is substantially reduced; microtubule assembly will occur without microtubule-associated proteins or GTP and at low temperatures; and the microtubules formed in the presence of taxol are more stable than normal microtubules to depolymerization by dilution, calcium, cold, and inhibitory drugs (see below). These effects of taxol have been observed under a wide variety of reaction conditions, utilizing purified tubulin, microtubule protein, and tubulin + microtubule-associated proteins.

In addition to its failure to depolymerize taxol-stabilized microtubules, Ca^{2+} did not inhibit taxol-induced microtubule assembly.[362,548] With Zn^{2+} + taxol, even in the absence of GTP, extensive sheet polymers formed.[114]

Taxol causes a marked reduction in the rate of tubulin subunit flux through microtubules, whether subunit uptake or release was measured.[362,553,558] Combined with the resistance of taxol-stabilized microtubules to dilutional disassembly,[549,558] this led to the conclusion that the taxol mechanism of action must involve a marked reduction of the dissociation rate constant of tubulin α-β dimers from the polymer.[548,549,553,558] This occurs at both microtubule ends, with about a four- to fivefold reduction in the dissociation rate constants (measured by dilution into taxol-free buffer).[553,558]

Although maximum effects of taxol *in vitro* are obtained when the drug and tubulin are approximately stoichiometric in their concentrations, the drug has significant effects on polymerization reactions and polymer at lower concentrations. Schiff et al.[548] detected enhanced polymerization of microtubule protein at 1 mg/ml (about 8 μM tubulin) with as little as 0.05 μM taxol. Sedimentation of 2 mg/ml of microtubule protein at 0°C was half maximal with 2 μM taxol, and at 37°C with salt (representing selective tubulin polymerization) with 1 μM taxol.[551] With purified tubulin at 15 μM, partial assembly occurred with 1 μM drug, and 80% of maximal assembly with 5 μM taxol.[549] In another study with purified tubulin,[362] half-maximal stimulation of polymerization occurred with taxol in a molar ratio to tubulin of about 25%. The rate of cold-induced disassembly of microtubules formed with 2 mg/ml microtubule protein was reduced by 50% with 0.5 μM taxol.[556] Approximately 6 μM taxol inhibited subunit flux by 50% in reaction mixtures containing 5.3 mg/ml microtubule protein (about 40 μM tubulin).[558]

Besides enhancing their stability, taxol alters physical properties of microtubules. Because of enhanced nucleation of assembly, microtubules formed in the presence of the drug are significantly shorter than those formed in its absence.[548,559] Moreover, as the incubation time increases, polymer morphology[548,550] and microtubule length distribution[550,554,559] can change substantially with only minor changes in turbidity or protein harvested by centrifugation. Wallin et al.[559] compared taxol effects on linear dichroic and viscometric properties of microtubule solutions under conditions in which these properties were stable in control microtubules, and in which there was little difference between control and taxol microtubules in their mass and turbidimetric properties. They found that the linear dichroism of the taxol-induced microtubules increased with incubation time and was much lower than that of the control microtubules, indicating that the latter were more readily oriented in parallel than the taxol tubules. At roughly equal turbidity readings, the reaction mixture containing taxol-induced microtubules was initially about four times as viscous as the control microtubule solution. While the viscosity of the control solution was stable, that of the taxol solution fell over about 45 min to the control value. In addition, electron micrographs suggested that the taxol-induced microtubules were less rigid than control microtubules.

Taxol does not appear to significantly alter the distribution of microtubule-associated proteins which copolymerize with tubulin,[292,362,549,551,553,559,563] nor prevent their binding to polymer initially formed from purified tubulin.[362] This has permitted purification of microtubule protein from cells and tissues with low microtubule content.[551]

In the presence of microtubule-associated proteins taxol induced microtubules to form in clusters of two to five tubules, and it was suggested that this may represent a stage in bundle formation.[564,565] The microtubule-associated proteins had a globular as opposed to filamentous appearance in the electron microscope. Taxol also caused formation of filamentous aggregates of purified microtubule-associated proteins.[565] Extensive microtubule bundles have been induced in rat brain tissue homogenates by taxol. A factor was extracted from the bundles by a high concentration of NaCl which reproduced bundle formation when added to microtubule protein in the presence of the drug.[566]

The binding of radiolabeled taxol,[552,555,567,568] 7-epitaxol,[561] and 7-acetyltaxol[569] to microtubule protein and tubulin has been examined. Parness and Horwitz[567] documented the binding of [^{3}H]taxol to tubulin polymers, harvested by centrifugation, both with and without microtubule-associated proteins or GTP. Near stoichiometric amounts of drug, relative to tubulin, were bound by the polymer. K_D values of 0.8 and 1.2 μM were obtained in the presence and absence of microtubule-associated proteins, respectively. The [^{3}H]taxol also bound to preformed microtubules and could be displaced by nonradiolabeled taxol. Ca^{2+} did not affect binding of radiolabeled taxol to polymer. [^{3}H]taxol had a reduced ability to bind to sea urchin sperm flagellar outer doublet microtubules.[555] Fourfold enhancement of binding occurred if these microtubules were disassembled by sonication and reassembled in the presence of taxol.

Takoudju et al.[569] examined the binding of 7-[^{3}H]acetyltaxol (see below) to purified tubulin by equilibrium dialysis. They observed binding only when a polymerization reaction occurred. Under their conditions there was neither tubulin polymerization nor drug binding at 0°C or in the presence of several other antimitotic agents (see below). They observed a stoichiometric binding of 7-acetyltaxol to the polymer, and a K_D value of 0.3 μM was obtained.

There is no information currently available on the binding site of taxol on tubulin. Taxol-resistant and taxol-dependent mutant CHO cell lines have been described with alterations in both α- and β-tubulin.[570,571]

Taxol does not interfere with nucleotide binding to tubulin;[112,549] GTP hydrolysis occurs in the presence of taxol,[125,548,549,554] with exchangeable site GDP recovered in the taxol-induced polymer.[553] Further, taxol was able to induce polymerization of charcoal-[548] or alkaline phosphatase-treated[554] tubulin in the absence of exogenous nucleotide, implying that an occupied exchangeable site was not required for the reaction. Nevertheless, GTP in the reaction mixture does enhance taxol-induced polymerization.[125,562,567]

GTP hydrolysis and microtubule assembly can occur as either independent or closely linked events in the presence of taxol.[125,554] In our laboratory[125] we examined GDP, GTP, and temperature effects in a reaction system containing purified tubulin with and without heat-treated microtubule-associated proteins. No exogenous Mg^{2+} was included in the reaction mixtures, which also contained 0.1 M glutamate. The net effect of taxol in this reaction system was significant enhancement of tubulin-dependent GTP hydrolysis. At 0°C taxol-induced microtubule assembly required *both* microtubule-associated proteins and GTP. This reaction was associated with simultaneous and initially stoichiometric GTP hydrolysis and could be completely inhibited by GDP. Without microtubule-associated proteins, tubulin polymerization (sheets were the reaction product) was induced by taxol at 37°C, and GTP was again required. As at 0°C with microtubule-associated proteins, hydrolysis and polymerization were simultaneous in onset and initially stoichiometric, with GDP a potent inhibitor of both reactions. With microtubule-associated proteins at 37°C, microtubule assembly and GTP hydrolysis were completely uncoupled, with hydrolysis either preceding or lagging behind polymerization, depending on precise reaction conditions. In this case, exogenous GDP strongly inhibited the hydrolysis reaction but not polymerization.

We also found that GTP hydrolysis at the turbidity plateau was almost identical when tubulin + microtubule-associated proteins was compared to tubulin + taxol with and without microtubule-associated proteins.[125] This makes it difficult to ascribe the plateau hydrolysis reaction to microtubule ends,[83,554] since nearly identical GTPase reactions occurred with perhaps a fourfold increase in ends[548] (more and shorter microtubules with than without taxol when microtubule-associated proteins were in the reaction) or when sheets instead of microtubules were formed (the reaction product formed with tubulin + taxol).

Carlier and Pantaloni[554] examined taxol effects on GTP hydrolysis using purified tubulin in reaction mixtures containing high concentrations of glycerol and Mg^{2+}. Taxol accelerated the onset of polymerization, with some enhancement of total polymer formed, but in this system taxol did not significantly alter the GTP hydrolysis reaction, especially its early, rapid phase. Carlier and Pantaloni[554] also examined nucleotide interactions with taxol-induced microtubules. When tubulin-[^{3}H]GTP was used to form taxol-induced polymer, up to half the radiolabel could be displaced if nonradiolabeled GTP was added after a 2-min incubation, but less than 10% was displaced after a 60-min incubation. The radiolabel in polymer formed with tubulin-[^{3}H]GDP was only minimally displaced by nonradiolabeled GTP even after short incubation times. Taxol-induced polymer formed with tubulin-O did not bind radiolabeled GTP.

Carlier and Pantaloni[554] noted that GTP hydrolysis at the turbidity plateau was progressively reduced with increasing concentrations of taxol relative to the control reaction in the glycerol system (as opposed to the result we obtained with microtubule-associated proteins, see above). This inhibitory effect could also be obtained by adding taxol to preformed microtubules. These workers suggested that GTP hydrolysis at microtubule ends was inhibited only at higher taxol concentrations as a consequence of a reduced affinity of the ends, as opposed to the microtubule core, for the drug. A similar inhibition by taxol of GTP hydrolysis at the turbidity plateau, however, also occurs with glutamate-induced tubulin sheets.[411,572]

There have been several studies on the effects of taxol analogs on microtubule systems *in vitro* (see Figure 36).[552,556,557,561,569] For activity with mammalian tubulin, an ester group at position 13 is essential.[552,556] A second active natural product named cephalomannine has activity comparable to that of taxol.[552,556] Instead of a benzoate residue at position 4′, cephalomannine has an α-methylcrotonyl moiety [$HC(CH_3)$=$C(CH_3)$–, the two methyl groups in *cis* configuration] at position 4′.[542] Several alterations have been made at position 7 with little effect on the drug activity: acetylation,[556,557,569] epimerization,[561] and formation of a xylose derivative.[556] Deacetylation at position 10 does not greatly affect interactions with tubulin,[552,556] but acetylation at position 2′ causes a substantial loss in activity.[556,557]

Lataste et al.[556] also examined the ability of taxol and taxol analogues to stabilize microtubules derived from *Physarum polycephalum* amoebae. Structure-activity correlations with mammalian brain tubulin were excellent, except that analogues without the C13 ester group were also highly active in stabilizing polymer (although they were not cytotoxic). This permitted a further structure-activity analysis of modifications in the taxane ring system, assuming that conclusions reached with the *P. polycephalum* microtubule protein could be extended to mammalian microtubule protein. The only change that caused a major loss of activity was reduction and acetylation of the ketone group at position 9. The following changes caused minor (less than tenfold) activity losses: epimerization of the hydroxyl group at position 7 (contrary to the negligible loss of activity with mammalian tubulin[561]), epimerization of the methyl group at position 8, oxidation of the hydroxyl at either position 10 or position 13 to a ketone group, and addition of water across the 11-12 double bond (hydroxyl at position 11).

Combinations of taxol with other antimitotic agents have been examined in many laboratories, both in biochemical and in cellular studies. Most reports with cells have described persistance of microtubules if the culture medium contained taxol as well as a second drug which otherwise usually caused disappearance of microtubules. Such results have been obtained with steganacin,[545] nocodazole,[573,574] colchicine,[574,575] colcemid,[576-579] and vincristine.[580,581] An exception to this generalization was observed with taxol + colcemid in sea urchin eggs — the latter agent caused disappearance of taxol-induced microtubules.[582] Manfredi et al.[568] examined the binding of [^{3}H]taxol to cells, which correlated well with the amount of polymerized tubulin they contained. When cells were incubated with a 100-fold molar excess of colchicine, podophyllotoxin, vinblastine, or nocodazole over radiolabeled taxol, microtubule assembly was inhibited and radiolabeled taxol binding to the cells was much reduced. Since the vinblastine-treated cells contained paracrystals, it appears that these structures do not bind taxol.[568]

Cell-free microtubule assembly in the presence of taxol plus a second antimitotic agent has been examined by many workers. Podophyllotoxin,[362,553] colchicine,[362,560] and vinblastine[567] have been reported to totally inhibit taxol-induced microtubule assembly, but incomplete inhibition, with formation of short microtubules, was also described with podophyllotoxin.[567] The inhibition of the taxol reaction by vinblastine was accompanied by spiral aggregate formation.[567] Vinblastine in the presence of taxol continued to inhibit tubulin-dependent GTP hydrolysis.[554] Taxol-induced assembly was only partially inhibited by nocodazole or maytansine and not affected at all by griseofulvin.[362] Preformed taxol-induced microtubules were resistant to depolymerization by podophyllotoxin[362,567] and vinblastine,[567] although some disassembly occurred, with partial formation of ribbons or spiral aggregates with podophyllotoxin or vinblastine, respectively.[567]

The binding of radiolabeled taxol to polymer was inhibited by podophyllotoxin and vinblastine as consequence of their inhibition of the assembly reaction.[567] The binding of radiolabeled 7-acetyltaxol to purified tubulin, as measured by equilibrium dialysis, was inhibited by colchicine, podophyllotoxin, β-peltatin, nocodazole, vinblastine, and maytansine.[569] This was attributed to inhibition of taxol-induced polymerization under the experimental conditions.

Taxol does not inhibit the binding of radiolabeled colchicine, podophyllotoxin, or vinblastine to tubulin,[362,549] unless taxol and microtubule protein are preincubated prior to the addition of radiolabeled colchicine.[549] It is difficult to interpret these findings in terms of binding sites, however, since taxol appears to bind preferentially to polymer, while the other agents have higher affinities for unassembled tubulin (see above).

Choudhury et al.[560] reported that MTPT and deacetamidocolchicine only partially inhibited taxol-induced assembly of purified tubulin, and they found that both compounds were incorporated into a taxol-induced polymer and bound to a preformed taxol-induced polymer which was cold-labile. They assumed that microtubule assembly had occurred, with the colchicine site therefore accessible on taxol-induced microtubules, but no morphological examination of the polymer was described. Somewhat analogously, Carlier and Pantaloni[554] found that taxol, particularly at high concentrations, enhanced GTP hydrolysis by the tubulin-colchicine complex. Both groups studied reactions at a high glycerol concentration, but at relatively low Mg^{2+} concentrations where polymerization of the tubulin-colchicine complex had not been observed.[327] Nevertheless, MTPT, like colchicine, can induce a GTP-dependent aggregation reaction.[329] We have found MTPT more effective than colchicine in stimulating this reaction.[49] Moreover, Howard and Timasheff[562] have found that taxol enhances aggregation of the tubulin-colchicine complex (examined at 16 mM Mg^{2+} in glycerol). The stimulatory effect of taxol was less potent than in the microtubule assembly reaction, and the tubulin-colchicine aggregation reaction retained its requirement for GTP and its temperature sensitivity even in the presence of taxol. Obviously these competing reactions hamper straightforward interpretation of results obtained in the presence of multiple ligands.

D. OTHER DRUGS

1. Derivatives of 5,6-Diphenylpyridazin-3-One

A series of derivatives of 5,6-diphenylpyridazin-3-one (DPP derivatives; structural formula in Figure 37; the two phenyl rings bear identical substituents in all cases) was synthesized by Powers and collaborators as potential antihypertensive agents.[583,584] Several compounds displayed significant herbicidal activity,[585] and initial toxicological studies of herbicidal compounds included the observation of bizarre mitotic figures in renal tissue.[586,587] A number of DPP derivatives were submitted to the National Cancer Institute for evaluation of antineoplastic activity, and a significant increase in survival of mice given simultaneous injections of P388 murine leukemia cells and selected DPP derivatives was observed.

Interactions of these agents with tubulin were evaluated in our laboratory.[470] Several derivatives were cytotoxic at micromolar concentrations for L1210 murine leukemia cells in culture, with accumulation of cells arrested in mitosis at cytotoxic drug levels. Six agents have so far been identified which inhibit microtubule assembly by at least 50% at concentrations below 10 μM (see below for structural details). The DPP derivatives also inhibit glutamate- and glycerol-induced polymerization reactions. Higher concentrations of most,

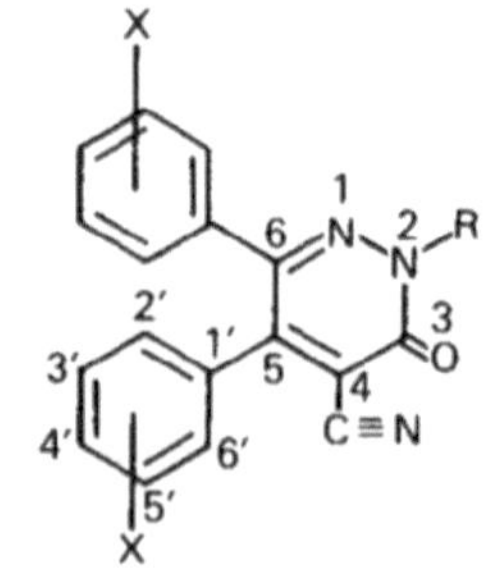

FIGURE 37. Structural formula of the derivatives of 5,6-diphenylpyridazin-3-one.

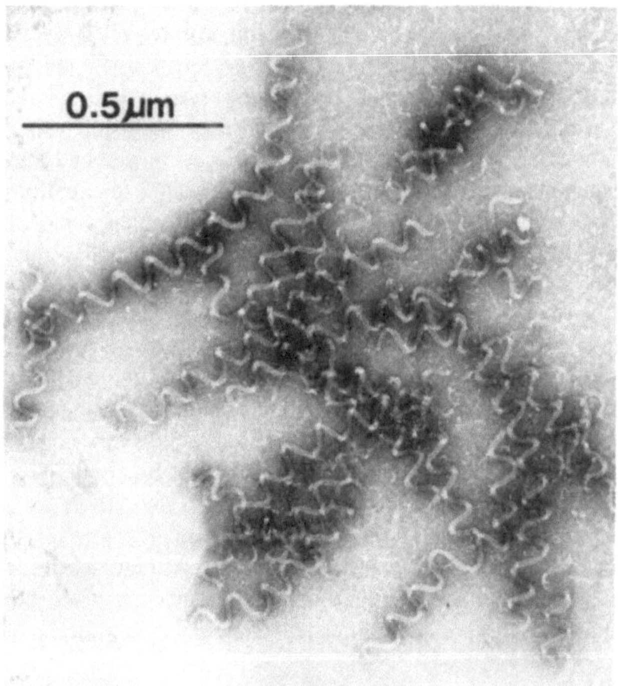

FIGURE 38. Spiral tubulin oligomers induced by the DPP derivative NSC 362453 (see text for structural details) in 1 *M* glutamate.

if not all, active compounds cause an aggregation reaction in which loosely coiled helices are formed (Figure 38). These helices appear to consist of two to three parallel filaments. The aggregation reaction did not require GTP, and the aggregates were cold stable. Thus far, they appear to be more well defined when formed from purified tubulin in glutamate, than when formed in reaction mixtures containing both tubulin and microtubule-associated proteins.

The DPP derivatives strongly stimulated tubulin-dependent GTP hydrolysis (about five times as much nucleotide was degraded with these agents as with colchicine). The DPP derivatives do not interfere with the binding of radiolabeled colchicine, vinblastine, maytansine, or GTP to tubulin.

Structure-activity studies were performed by Hess and Powers[588] on the effects of DPP derivatives on flagellar regeneration in *Chlamydomonas reinhardtii* as a plant model system and in our laboratory[470] on microtubule assembly with bovine tubulin and microtubule-associated proteins as an animal model system. The most active agents in the two systems were entirely different.

For inhibition of microtubule assembly we could discern no obvious structure-activity relationship for the substituent at position 2. Activity was only observed if both phenyl rings were present, and maximal activity required a nitrile group at position 4. No activity was observed in compounds with a 2′-2′ carbon-carbon bond, to yield a tetracyclic structure. A clear structure-activity pattern was observed with the symmetrical substituents on the phenyl rings. Activity was progressively enhanced with a fluorine atom at position 4′, a bromine at 4′, a chlorine at 4′, and two chlorines at 3′ and 4′. Available compounds indicated that a methyl group at position 4′ is equivalent to a bromine at that position and that a compound with a chlorine substituent at position 4′ is about three to four times more active than one with a chlorine at position 2′, which is in turn two to three times more inhibitory than an unsubstituted compound.

As noted above, six DPP derivatives are particularly potent as inhibitors of microtubule assembly. Their structures are as follows (see Figure 37): NSC 362449 (X = H; R = CH_2COOH); NSC 362453 (X = Cl at 4′; R = CH_2CH_2-O-$COCH_2CH_2COOH$); NSC 362455 (X = Cl at 3′ and 4′; R = CH_2CH_2OH); NSC 602746 (X = Cl at 3′ and 4′; R = H); NSC 608593 (X = H; R = CH_2CN_4H [tetrazole-5-yl-methyl]); and NSC 613241 (X = F at 4′; R = CH_2CN_4H [tetrazole-5-yl-methyl]).

2. Agents which Attack Tubulin Sulfhydryl Groups

A role for sulfhydryl groups in mitosis has long been postulated,[589] and the studies of Nath and Rebhun[590] and of Oliver et al.[591] established that reduced glutathione was essential for optimal microtubule assembly in cells. Moreover, *in vitro* microtubule assembly can be completely inhibited by alkylation of a limited number of tubulin sulfhydryl groups.[50,230,231,259,592-595] A number of agents appear to inhibit microtubule assembly by reacting specifically at tubulin sulfhydryl groups,[449,589,594-604] but only organic mercurials[589,599-603] and 2,4-dichlorobenzyl thiocyanate and its analogues[449,604] have been documented to cause both mitotic arrest and inhibition of tubulin polymerization at relatively low concentrations.

a. Methylmercuric Chloride

Organic mercurials were demonstrated to cause mitotic arrest in plant[589] and mammalian[599,601] cells many years ago, but documentation of a specific effect on microtubules has only been reported with methylmercuric chloride (CH_3HgCl).[600,602,603] The inhibitory effects of these compounds are believed to derive from their interaction with protein sulfhydryl groups.[589,599-603] Methylmercuric chloride completely inhibited growth of a glioma cell line at 5 μM, with accumulation of mitotic cells and disappearance of cellular microtubules, but no inhibition was observed at 1 μM.[601] Fibroblast microtubules disappeared when cells were treated with 10 μM methylmercuric chloride, and this effect was prevented by including cysteine or a chelating agent (dimercaptosuccinate) in the culture medium.[603] Methylmercuric chloride strongly inhibited *in vitro* microtubule assembly at 20 to 30 μM,[600,602,603] but it had little effect on the reaction at concentrations below 10 μM.[600,602]

b. 2,4-Dichlorobenzyl Thiocyanate

In our laboratory we are examining effects of 2,4-dichlorobenzyl thiocyanate (DCBT; NSC 145813; structure in Figure 39) on cells and tubulin.[449,604] Initial studies[449] demonstrated that the agent inhibited the growth of L1210 murine leukemia cells (IC_{50} value, 0.2 μM), with accumulation of cells in mitotic arrest, with similar cytotoxicity in other cell lines as well. CHO cells treated with the drug and immunofluorescently stained with antibody to α-

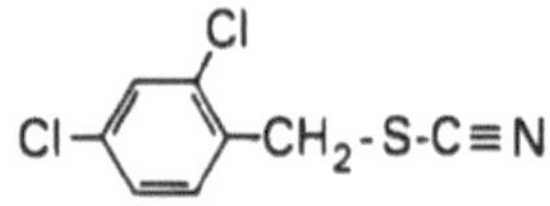

FIGURE 39. Structural formula of 2,4-dichlorobenzyl thiocyanate.

tubulin displayed no cellular microtubules. They were replaced by structures in the form of aggregated spirals. Resistance to DCBT was observed in a mutant CHO cell line with an altered β-tubulin, and resistance was still further enhanced in a cell line with multiple copies of the mutant gene. Tubule polymerization was not inhibited by DCBT unless drug and protein were preincubated prior to the addition of GTP. Effects of the drug were then dependent on the length and temperature of the preincubation, as well as the drug concentration. Under the conditions chosen for study, the IC_{50} value (plateau; preincubated) was 4 μ*M*, with complete inhibition observed at 7 μ*M* DCBT. Structure-activity studies with DCBT analogues demonstrated that both chloride atoms were required for maximal activity, with 4-chlorobenzyl thiocyanate two to four times as active as 2-chlorobenzyl thiocyante. A third chloride atom at position 5 in the phenyl ring did not significantly alter the effects of the agent on either cell growth or inhibition of tubulin polymerization *in vitro*.

In more recent studies[604] with radiolabeled DCBT (three variants — [^{35}S]DCBT, [^{3}H-benzyl]DCBT, and [^{14}C-nitrile]DCBT), we found that multiple molecules of the agent react with tubulin, with a covalent bond formed between the 2,4-dichlorobenzylmercaptan moiety and the protein and release of cyanate anion to the medium. Since dithiothreitol protected tubulin from DCBT and could reverse its effects on the protein (complete removal of covalently bound drug; partial restoration of activity in polymerization), we believe that DCBT forms mixed disulfides with cysteine residues of tubulin. When tubulin was incubated with a superstoichiometric amount of either the ^{35}S- or ^{3}H-labeled DCBT (identical results were obtained with the two drug preparations), both tubulin subunits reacted with the DCBT. With equal amounts of protein and drug, the β-tubulin subunit was predominantly labeled (Figure 40). With 5 μ*M* DCBT and 25 μ*M* tubulin, a level of drug which results in nearly complete inhibition of polymerization, virtually all radiolabel was in the β-subunit. In current studies,[49] this β-tubulin was digested with cyanogen bromide and virtually all radiolabel was recovered in a single peptide. This peptide is presently being sequenced to identify the cysteine residue responsible for the potent inhibition by DCBT of tubulin polymerization.

3. Griseofulvin

Griseofulvin (structure in Figure 41) has relatively feeble effects on mammalian cell growth and microtubule assembly and is included here for the sake of completeness, as its properties have been examined by a number of workers. Derived from *Penicillium griseofulvin*, the drug is used clinically, with oral administration, in the treatment of fungal diseases of the skin and nails. Even though it has almost no clinical toxicity, it has been documented to cause mitotic arrest in rat tissues[605] and in mammalian cells in culture.[606-608] One report described persistant microtubules in the arrested cells.[607] Similarly, in the plasmodial phase of *Physarum polycephalum* griseofulvin caused mitotic arrest, but the cells did not lose their microtubules. Instead, they contained inclusion bodies consisting of large numbers of microtubules.[609] An immunofluorescence study of both HeLa and murine 3T3 cells, however, demonstrated that griseofulvin can cause complete disappearance of cellular microtubules.[608] The drug also caused dissolution of meiotic spindle microtubules in unfertilized annelid eggs.[610]

A number of workers have examined effects of griseofulvin on microtubule assembly or the polymerization of purified tubulin. In most cases, significant inhibition of polymerization required high concentrations of the drug (over 50 μ*M*).[55,304,306,608,611,612] Keates,[613]

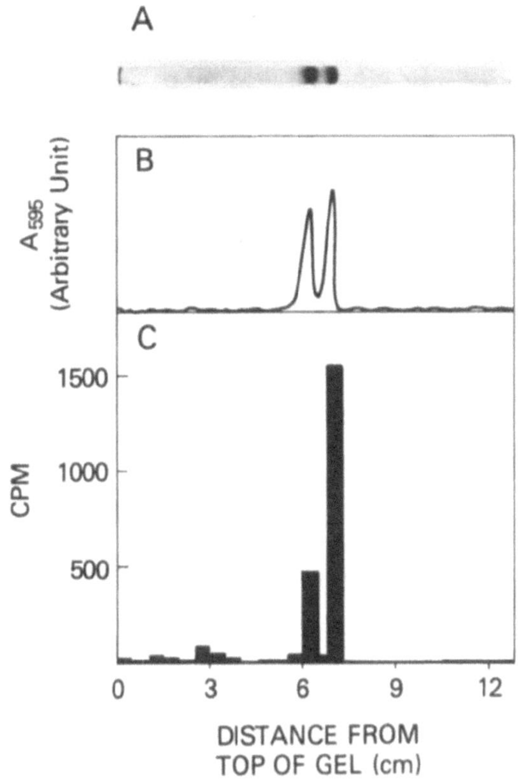

FIGURE 40. Alkylation of β-tubulin by DCBT. The reaction mixture contained 10 μ*M* tubulin and 10 μ*M* [^{35}S]DCBT in 1 *M* glutamate. Incubation was for 2 h at 37°C. The sample was processed for electrophoresis as described in Reference 604.

however, reported that just over 10 μ*M* griseofulvin inhibited the rate of microtubule assembly by 50%, although significant effects on the assembly plateau only occurred at drug concentrations over 60 μ*M*. Wehland et al.[611] also found that griseofulvin (100 μ*M*) caused a slow, but nearly complete, disassembly of preformed microtubules.

Besides inhibiting microtubule assembly, high concentrations of griseofulvin cause the formation of a cold-stable aggregate from either microtubule protein[306,608] or from tubulin + microtubule-associated proteins.[611] Formation of this aggregate requires both components and is inhibited by colchicine. The aggregate contains tubulin and microtubule-associated proteins and is disrupted by colchicine.[611]

Interactions of microtubule components with radiolabeled griseofulvin have been reported by two groups. Roobol et al.[614] found that the drug bound to microtubule-associated proteins, while Sloboda et al.[612] observed stoichiometric binding of griseofulvin to tubulin.

Other reported effects of high concentrations of griseofulvin on tubulin have included stimulation of tubulin-dependent GTP hydrolysis[55] and enhancement of vinblastine-induced paracrystal formation in sea urchin eggs.[395] Creasey and Chou[467] found that griseofulvin inhibited the binding of radiolabeled colchicine to a cell homogenate, but at least three other groups[55,215,338] reported that griseofulvin had no effect on colchicine binding.

4. Estramustine

Since estramustine (structure in Figure 41) and its phosphate derivative (at position 17) interact with microtubule-associated proteins and not tubulin (see below), it, too, is included

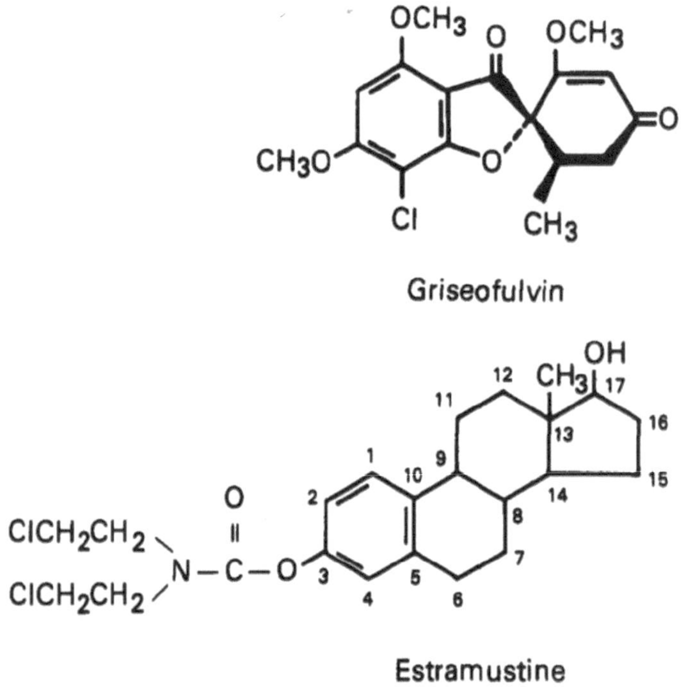

FIGURE 41. Structural formulas of griseofulvin and estramustine.

here for the sake of completeness. Used clinically in the treatment of prostatic carcinoma,[615] estramustine caused prostatic carcinoma cell lines to accumulate in mitotic arrest.[616-618] An immunofluorescence study demonstrated that microtubules disappeared from drug-treated cells.[619]

Kanje et al.[620] found that high concentrations of estramustine phosphate inhibited microtubule assembly, with an IC_{50} value (plateau) of about 150 μM, caused rapid depolymerization, and increased the critical concentration of microtubule protein required for assembly. Wallin et al.[621] demonstrated binding of radiolabeled estramustine phosphate to microtubule-associated proteins rather than tubulin, and they observed that adding microtubule-associated proteins to a reaction mixture overcame the inhibitory effects of the drug. These workers subsequently found that estramustine phosphate had nearly equivalent inhibitory effects (plateau IC_{50} values of about 100 μM) on microtubule assembly dependent on either tau factor, MAP-2, or protease-digested MAP-2 and had no inhibitory effect at 100 μM on dimethylsulfoxide-induced polymerization of purified tubulin.[622]

Somewhat different *in vitro* results were recently reported by Stearns and Tew,[292] who used estramustine. They found strong binding of radiolabeled drug to MAP-2, and perhaps to MAP-1, but not to tau factor. Scatchard analysis of their data yielded a K_D value of about 15 μM, with one class of binding site, and a stoichiometry of about 20 drug molecules per protein molecule. When polymer was harvested by centrifugation at differing drug concentrations, MAP-2 disappeared from the pellet more rapidly than tubulin. By turbidimetry and electron microscopy, 20 μM estramustine totally inhibited microtubule assembly, but lower drug concentrations were not evaluated. As opposed to the rapid disassembly observed with estramustine phosphate,[620,622] depolymerization with estramustine was relatively slow.[292]

Neither estramustine[292] nor estramustine phosphate[620,622] prevented formation of taxol-induced microtubules. These agents, however, caused a marked reduction in the amount of microtubule-associated proteins bound to the taxol-induced polymer.

V. CONCLUSION

Tubulin interacts with a large number of small ligands at multiple diverse binding sites. Despite the enormous progress documented in this review, little is known about the locations of these sites, their relationships to each other, or the specific molecular features of the ligands that determine their high affinities for the protein. Greater understanding of these aspects of tubulin-ligand interactions should permit rational design of agents which will interfere with microtubule structure and function. This may allow greater control of neoplastic and inflammatory diseases.

ACKNOWLEDGMENT

The author would like to thank Dr. David G. Johns for his continuing interest and support.

REFERENCES

1. **Weisenberg, R. C., Borisy, G. G., and Taylor, E. W.,** The colchicine-binding protein of mammalian brain and its relation to microtubules, *Biochemistry*, 7, 4466, 1968.
2. **Detrich, H. W., III, Berkowitz, S. A., Kim, H., and Williams, R. C., Jr.,** Binding of glycerol by microtubule protein, *Biochem. Biophys. Res. Commun.*, 68, 961, 1976.
3. **Arai, T., Ihara, Y., Arai, K., and Kaziro, Y.,** Purification of tubulin from bovine brain and its interaction with guanine nucleotides, *J. Biochem. (Tokyo)*, 77, 647, 1975.
4. **Hamel, E. and Lin, C. M.,** Glutamate-induced polymerization of tubulin: characteristics of the reaction and application to the large-scale purification of tubulin, *Arch. Biochem. Biophys.*, 209, 29, 1981.
5. **Hamel, E. and Lin, C. M.,** Stabilization of the colchicine-binding activity of tubulin by organic acids, *Biochim. Biophys. Acta*, 675, 226, 1981.
6. **Waxman, P. G., del Campo, A. A., Lowe, M. C., and Hamel, E.,** Induction of polymerization of purified tubulin by sulfonate buffers: marked differences between 4-morpholineethane sulfonate (Mes) and 1,4-piperazineethanesulfonate (Pipes), *Eur. J. Biochem.*, 120, 129, 1981.
7. **Hamel, E., del Campo, A. A., Lowe, M. C., Waxman, P. G., and Lin, C. M.,** Effects of organic acids on tubulin polymerization and associated guanosine 5′-triphosphate hydrolysis, *Biochemistry*, 21, 503, 1982.
8. **Stephens, R. E., Renaud, F. L., and Gibbons, I. R.,** Guanine nucleotide associated with protein of the outer fibers of flagella and cilia, *Science*, 156, 1606, 1967.
9. **Ponstingl, H., Krauhs, E., Little, M., and Kempf, T.,** Complete amino acid sequence of α-tubulin from porcine brain, *Proc. Natl. Acad. Sci. U.S.A.*, 78, 2757, 1981.
10. **Krauhs, E., Little, M., Kempf, T., Hofer-Warbinek, R., Ade, W., and Ponstingl, H.,** Complete amino acid sequence of β-tubulin from porcine brain, *Proc. Natl. Acad. Sci. U.S.A.*, 78, 4156, 1981.
11. **Bryan, J. and Wilson, L.,** Are cytoplasmic microtubules heteropolymers?, *Proc. Natl. Acad. Sci. U.S.A.*, 68, 1762, 1971.
12. **Feit, H., Slusarek, L., and Shelanski, M. L.,** Heterogeneity of tubulin subunits, *Proc. Natl. Acad. Sci. U.S.A.*, 68, 2028, 1971.
13. **Feit, H., Neudeck, U., and Baskin, F.,** Comparison of the isoelectric and molecular weight properties of tubulin subunits, *J. Neurochem.*, 28, 697, 1977.
14. **Berkowitz, S. A., Katagiri, J., Binder, H. K., and Williams, R. C., Jr.,** Separation and characterization of microtubule proteins from calf brain, *Biochemistry*, 16, 5610, 1977.
15. **Lu, R. C. and Elzinga, M.,** Chromatographic resolution of the subunits of calf brain tubulin, *Anal. Biochem.*, 77, 243, 1977.
16. **Stephens, R. E.,** High resolution preparative SDS-polyacrylamide gel electrophoresis: fluorescent visualization and electrophoretic elution-concentration of protein bands, *Anal. Biochem.*, 65, 369, 1975.

17. **Little, M.**, Identification of a second β chain in pig brain tubulin, *FEBS Lett.*, 108, 283, 1979.
18. **Eipper, B. A.**, Rat brain microtubule protein: purification and determination of covalently bound phosphate and carbohydrate, *Proc. Natl. Acad. Sci. U.S.A.*, 69, 2283, 1972.
19. **Arce, C. A., Rodriguez, J. A., Barra, H. S., and Caputto, R.**, Incorporation of L-tyrosine, L-phenylalanine and L-3,4-dihydroxyphenylalanine as single units into rat brain tubulin, *Eur. J. Biochem.*, 59, 145, 1975.
20. **Maruta, H., Greer, K., and Rosenbaum, J. L.**, The acetylation of alpha-tubulin and its relationship to the assembly and disassembly of microtubules, *J. Cell Biol.*, 103, 571, 1986.
21. **Lee, J. C., Frigon, R. P., and Timasheff, S. N.**, The chemical characterization of calf brain microtubule protein subunits, *J. Biol. Chem.*, 248, 7253, 1973.
22. **Luduena, R. F., Shooter, E. M., and Wilson, L.**, Structure of the tubulin dimer, *J. Biol. Chem.*, 252, 7006, 1977.
23. **Berry, R. W. and Shelanski, M. L.**, Interactions of tubulin with vinblastine and guanosine triphosphate, *J. Mol. Biol.*, 71, 71, 1972.
24. **Jacobs, M., Smith, H., and Taylor, E. W.**, Tubulin: nucleotide binding and enzymic activity, *J. Mol. Biol.*, 89, 455, 1974.
25. **Kobayashi, T.**, Nucleotides bound to brain tubulin and reconstituted microtubules, *J. Biochem. (Tokyo)*, 76, 201, 1974.
26. **Levi, A., Cimino, M., Mercanti, D., and Calissano, P.**, Studies on the binding of GTP to the microtubule protein, *Biochim. Biophys. Acta*, 365, 450, 1974.
27. **Spiegelman, B. M., Penningroth, S. M., and Kirschner, M. W.**, Turnover of tubulin and the N site GTP in Chinese hamster ovary cells, *Cell*, 12, 587, 1977.
28. **Caplow, M. and Zeeberg, B.**, Stoichiometry for guanine nucleotide binding to tubulin under polymerizing and nonpolymerizing conditions, *Arch. Biochem. Biophys.*, 203, 404, 1980.
29. **Penningroth, S. M. and Kirschner, M. W.**, Nucleotide binding and phosphorylation in microtubule assembly *in vitro*, *J. Mol. Biol.*, 115, 643, 1977.
30. **Kirsch, M. and Yarbrough, L. R.**, Assembly of tubulin with nucleotide analogs, *J. Biol. Chem.*, 256, 106, 1981.
31. **Purich, D. L. and MacNeal, R. K.**, Properties of tubulin treated with alkaline phosphatase to remove guanine nucleotides from the exchangeable binding site, *FEBS Lett.*, 96, 83, 1978.
32. **Penningroth, S. M. and Kirschner, M. W.**, Nucleotide specificity in microtubule assembly *in vitro*, *Biochemistry*, 17, 734, 1978.
33. **Kobayashi, T.**, Dephosphorylation of tubulin-bound guanosine triphosphate during microtubule assembly, *J. Biochem. (Tokyo)*, 77, 1193, 1975.
34. **Weisenberg, R. C., Deery, W. J., and Dickinson, P. J.**, Tubulin-nucleotide interactions during the polymerization and depolymerization of microtubules, *Biochemistry*, 15, 4248, 1976.
35. **Arai, T. and Kaziro, Y.**, Role of GTP in the assembly of microtubules, *J. Biochem. (Tokyo)*, 82, 1063, 1977.
36. **David-Pfeuty, T., Erickson, H. P., and Pantaloni, D.**, Guanosinetriphosphatase activity of tubulin associated with microtubule assembly, *Proc. Natl. Acad. Sci. U.S.A.*, 74, 5372, 1977.
37. **MacNeal, R. K. and Purich, D. L.**, Stoichiometry and role of GTP hydrolysis in bovine neurotubule assembly, *J. Biol. Chem.*, 253, 4683, 1978.
38. **Hamel, E., Batra, J. K., Huang, A. B., and Lin, C. M.**, Effects of pH on tubulin-nucleotide interactions, *Arch. Biochem. Biophys.*, 245, 316, 1986.
39. **O'Brien, E. T., Voter, W. A., and Erickson, H. P.**, GTP hydrolysis during microtubule assembly, *Biochemistry*, 26, 4148, 1987.
40. **Schilstra, M. J., Martin, S. A., and Bayley, P. M.**, On the relationship between nucleotide hydrolysis and microtubule assembly: studies with a GTP-regenerating system, *Biochem. Biophys. Res. Commun.*, 147, 588, 1987.
41. **Hamel, E., Lustbader, J., and Lin, C. M.**, Deoxyguanosine nucleotide analogues: potent stimulators of microtubule nucleation with reduced affinity for the exchangeable nucleotide site of tubulin, *Biochemistry*, 23, 5314, 1984.
42. **Geahlen, R. L. and Haley, B. E.**, Interactions of a photoaffinity analog of GTP with the proteins of microtubules, *Proc. Natl. Acad. Sci. U.S.A.*, 74, 4375, 1977.
43. **Nath, J. P., Eagle, G. R., and Himes, R. H.**, Direct photoaffinity labeling of tubulin with guanosine 5'-triphosphate, *Biochemistry*, 24, 1555, 1985.
44. **Hesse, J., Maruta, H., and Isenberg, G.**, Monoclonal antibodies localize the exchangeable GTP-binding site in β- and not α-tubulins, *FEBS Lett.*, 179, 91, 1985.
45. **Nath, J. P. and Himes, R. H.**, Localization of the exchangeable nucleotide binding domain in β-tubulin, *Biochem. Biophys. Res. Commun.*, 135, 1135, 1986.
46. **Hesse, J., Thierauf, M., and Ponstingl, H.**, Tubulin sequence region β155-174 is involved in binding exchangeable guanosine triphosphate, *J. Biol. Chem.*, 262, 15472, 1987.

47. **Mandelkow, E. -M., Hermann, M., and Ruhl, U.,** Tubulin domains probed by limited proteolysis and subunit-specific antibodies, *J. Mol. Biol.*, 185, 311, 1985.
48. **Sternlicht, H., Yaffe, M. B., and Farr, G. W.,** A model of the nucleotide-binding site in tubulin, *FEBS Lett.*, 214, 226, 1987.
49. **Hamel, E.,** unpublished observations.
50. **Nishida, E. and Kobayashi, T.,** Relationship between tubulin SH groups and bound guanine nucleotides, *J. Biochem. (Tokyo)*, 81, 343, 1977.
51. **Lee, Y. C., Houston, L. L., and Himes, R. H.,** Inhibition of the self-assembly of tubulin by diethylpyrocarbonate and photooxidation, *Biochem. Biophys. Res. Commun.*, 70, 50, 1976.
52. **Jemiolo, D. K. and Grisham, C. M.,** Divalent cation-nucleotide complex at the exchangeable nucleotide binding site of tubulin, *J. Biol. Chem.*, 257, 8148, 1982.
53. **Huang, A. B., Lin, C. M., and Hamel, E.,** Differential effects of magnesium on tubulin-nucleotide interactions, *Biochim. Biophys. Acta*, 832, 22, 1985.
54. **Correia, J. J., Baty, L. T., and Williams, R. C., Jr.,** Mg^{2+} dependence of guanine nucleotide binding to tubulin, *J. Biol. Chem.*, 262, 17278, 1987.
55. **David-Pfeuty, T., Simon, C., and Pantaloni, D.,** Effect of antimitotic drugs on tubulin GTPase activity and self-assembly, *J. Biol. Chem.*, 254, 11696, 1979.
56. **Lin, C. M. and Hamel, E.,** Effects of inhibitors of tubulin polymerization on GTP hydrolysis, *J. Biol. Chem.*, 256, 9242, 1981.
57. **Zeeberg, B. and Caplow, M.,** Determination of free and bound microtubular protein and guanine nucleotide under equilibrium conditions, *Biochemistry*, 18, 3880, 1979.
58. **Karr, T. L. and Purich, D. L.,** Examination of tubulin-nucleotide interactions by protein fluorescence quenching measurements, *Biochem. Biophys. Res. Commun.*, 84, 957, 1978.
59. **Fishback, J. L. and Yarbrough, L. R.,** Interaction of 6-mercapto-GTP with bovine brain tubulin: equilibrium aspects, *J. Biol. Chem.*, 259, 1968, 1984.
60. **Croom, H. B., Correia, J. J., Baty, L. T., and Williams, R. C., Jr.,** Release of exchangeably bound guanine nucleotides from tubulin in a magnesium-free buffer, *Biochemistry*, 24, 768, 1985.
61. **Duanmu, C., Lin, C. M., and Hamel, E.,** Tubulin polymerization with ATP is mediated through the exchangeable GTP site, *Biochim. Biophys. Acta*, 881, 113, 1986.
62. **Hamel, E., Batra, J. K., and Lin, C. M.,** Direct incorporation of guanosine 5′-diphosphate into microtubules without guanosine 5′-triphosphate hydrolysis, *Biochemistry*, 25, 7054, 1986.
63. **Lin, C. M. and Hamel, E.,** Interrelationships of tubulin-GDP and tubulin-GTP in microtubule assembly, *Biochemistry*, 26, 7173, 1987.
64. **Hamel, E. and Lin, C. M.,** Separation of active tubulin and microtubule-associated proteins by ultracentrifugation and isolation of a component causing the formation of microtubule bundles, *Biochemistry*, 23, 4173, 1984.
65. **Howard, W. D. and Timasheff, S. N.,** GDP state of tubulin: stabilization of double rings, *Biochemistry*, 25, 8992, 1986.
66. **Spann, U., Renner, W., Mandelkow, E. -M., Bordas, J., and Mandelkow, E.,** Tubulin oligomers and microtubule assembly studied by time-resolved X-ray scattering: separation of prenucleation and nucleation events, *Biochemistry*, 26, 1123, 1987.
67. **Kravit, N. G., Regula, C. S., and Berlin, R. D.,** A reevaluation of the structure of purified tubulin in solution: evidence for the prevalence of oligomers over dimers at room temperature, *J. Cell Biol.*, 99, 108, 1984.
68. **Correia, J. J. and Williams, R. C., Jr.,** Characterization of oligomers of tubulin by two-dimensional native electrophoresis, *Arch. Biochem. Biophys.*, 239, 120, 1985.
69. **Terry, B. J. and Purich, D. L.,** Nucleotide release from tubulin and nucleoside 5′-diphosphate kinase action in microtubule assembly, *J. Biol. Chem.*, 254, 66, 1979.
70. **Engelborghs, Y. and Eccleston, J.,** Fluorescence stopped-flow study of the binding of S6-GTP to tubulin, *FEBS Lett.*, 141, 78, 1982.
71. **Brylawski, B. P. and Caplow, M.,** Rate for nucleotide release from tubulin, *J. Biol. Chem.*, 258, 760, 1983.
72. **Yarbrough, L. R. and Fishback, J. L.,** Kinetics of interaction of 2-amino-6-mercapto-9-β-ribofuranosylpurine 5′-triphosphate with bovine brain tubulin, *Bicohemistry*, 24, 1708, 1985.
73. **Carlier, M. -F. and Pantaloni, D.,** Kinetic analysis of guanosine 5′-triphosphate hydrolysis associated with tubulin polymerization, *Biochemistry*, 20, 1918, 1981.
74. **Carlier, M. -F. and Pantaloni, D.,** Assembly of microtubule protein: role of guanosine di- and triphosphate nucleotides, *Biochemistry*, 21, 1215, 1982.
75. **Terry, B. J. and Purich, D. L.,** Assembly and disassembly properties of microtubules formed in the presence of GTP, 5′-guanylyl imidodiphosphate, and 5′-guanylyl methylenediphosphate, *J. Biol. Chem.*, 255, 10532, 1980.

76. **Deery, W. J. and Weisenberg, R. C.,** Kinetic and steady-state analysis of microtubules in the presence of colchicine, *Biochemistry,* 20, 2316, 1981.
77. **Margolis, R. L.,** Role of GTP hydrolysis in microtubule treadmilling and assembly, *Proc. Natl. Acad. Sci. U.S.A.,* 78, 1586, 1981.
78. **Cote, R. H. and Borisy, G. G.,** Head-to-tail polymerization of microtubules *in vitro, J. Mol. Biol.,* 150, 577, 1981.
79. **Karr, T. L., Podrasky, A. E., and Purich, D. L.,** Participation of guanine nucleotides in nucleation and elongation steps of microtubule assembly, *Proc. Natl. Acad. Sci. U.S.A.,* 76, 5475, 1979.
80. **Carlier, M.-F., Hill, T. L., and Chen, Y.-D.,** Interference of GTP hydrolysis in the mechanism of microtubule assembly: an experimental study, *Proc. Natl. Acad. Sci. U.S.A.,* 81, 771, 1984.
81. **Mitchison, T. and Kirschner, M. W.,** Dynamic instability of microtubule growth, *Nature (London),* 312, 232, 1984.
82. **Caplow, M., Shanks, J., and Brylawski, B. P.,** Concerning the anomalous kinetic behavior of microtubules, *J. Biol. Chem.,* 260, 12675, 1985.
83. **David-Pfeuty, T.,** Mechanism of GTP hydrolysis at microtubule ends, *Biophys. Chem.,* 12, 121, 1980.
84. **Carlier, M.-F. and Pantaloni, D.,** Kinetic analysis of cooperativity in tubulin polymerization in the presence of guanosine di- or triphosphate nucleotides, *Biochemistry,* 17, 1908, 1978.
85. **Zackroff, R. V., Weisenberg, R. C., and Deery, W. J.,** Equilibrium and kinetic analysis of microtubule assembly in the presence of guanosine diphosphate, *J. Mol. Biol.,* 139, 641, 1980.
86. **Jameson, L. and Caplow, M.,** Effect of guanosine diphosphate on microtubule assembly and stability, *J. Biol. Chem.,* 255, 2284, 1980.
87. **Bayley, P. M. and Martin, S. R.,** Inhibition of microtubule elongation by GDP, *Biochem. Biophys. Res. Commun.,* 137, 351, 1986.
88. **Bayley, P. M., Butler, F. M. M., and Manser, E. J.,** Control of nucleation in microtubule self-assembly, *FEBS Lett.,* 205, 230, 1986.
89. **Hamel, E. and Lin, C. M.,** Interaction of tubulin with ribose-modified analogs of GDP and GTP: evidence for two mutually exclusive exchangeable nucleotide binding sites, *Proc. Natl. Acad. Sci. U.S.A.,* 78, 3368, 1981.
90. **Hamel, E. and Lin, C. M.,** Inhibition of tubulin polymerization with ribose-modified analogs of GDP and GTP: reduced inhibition with microtubule-associated proteins and magnesium, *Biochim. Biophys. Acta,* 797, 117, 1984.
91. **Engelborghs, Y. and Van Houtte, A.,** Temperature jump relaxation study of microtubule elongation in the presence of GTP/GDP mixtures, *Biophys. Chem.,* 14, 195, 1981.
92. **Martin, S. R. and Bayley, P. M.,** Effects of GDP on microtubules at steady state, *Biophys. Chem.,* 27, 67, 1987.
93. **Zeeberg, B. and Caplow, M.,** An isoenergetic exchange mechanism which accounts for tubulin-GDP stabilization of microtubules, *J. Biol. Chem.,* 256, 12051, 1981.
94. **Hamel, E., del Campo, A. A., and Lin, C. M.,** Stability of tubulin polymers formed with dideoxyguanosine nucleotides in the presence and absence of microtubule-associated proteins, *J. Biol. Chem.,* 259, 2501, 1984.
95. **Lee, S. -H., Kristofferson, D., and Purich, D. L.,** Microtubule interactions with GDP provide evidence that assembly-disassembly properties depend on the method of brain microtubule protein isolation, *Biochem. Biophys. Res. Commun.,* 105, 1605, 1982.
96. **Manser, E. J. and Bayley, P. M.,** Incorporation of GDP-tubulin during elongation of microtubules *in vitro, Biochem. Biophys. Res. Commun.,* 131, 386, 1985.
97. **Maccioni, R. B. and Seeds, N. W.,** Residual nucleotide and tubulin's ability to polymerize with nucleotide analogs, *J. Biol. Chem.,* 257, 3334, 1982.
98. **Zabrecky, J. R. and Cole, R. D.,** ATP-induced aggregates of tubulin rings, *J. Biol. Chem.,* 255, 11981, 1980.
99. **Arai, T. and Kaziro, Y.,** Effect of guanine nucleotides on the assembly of brain microtubules: ability of 5′-guanylyl imidodiphosphate to replace GTP in promoting the polymerization of microtubules *in vitro, Biochem. Biophys. Res. Commun.,* 69, 369, 1976.
100. **Sutherland, J. W. H.,** Comparison of the effects of different adenosine and guanosine nucleotides on the assembly of bovine neurotubules, *Biochem. Biophys. Res. Commun.,* 72, 933, 1976.
101. **David-Pfeuty, T. and Huitorel, P.,** Tubulin polymerization in the presence of GMP-PCP, *Biochem. Biophys. Res. Commun.,* 95, 535, 1980.
102. **Hamel, E. and Lin, C. M.,** Guanosine 5′-*O*-(3-thiotriphosphate), a potent nucleotide inhibitor of microtubule assembly, *J. Biol. Chem.,* 259, 11060, 1984.
103. **Weisenberg, R. C. and Deery, W. J.,** Role of nucleotide hydrolysis in microtubule assembly, *Nature (London),* 263, 792, 1976.
104. **Sandoval, I. V., MacDonald, E., Jameson, J. L., and Cuatrecasas, P.,** Role of nucleotides in tubulin polymerization: effect of guanylyl 5′-methylenediphosphonate, *Proc. Natl. Acad. Sci. U.S.A.,* 74, 4881, 1977.

105. **Sandoval, I. V. and Weber, K.,** Guanosine 5′-(α,β-methylene)triphosphate enhances specifically microtubule nucleation and stops the treadmill of tubulin protomers, *J. Biol. Chem.*, 255, 6966, 1980.
106. **Hamel, E. and Lin, C. M.,** Dideoxyguanosine nucleotides and microtubule assembly, *Ann. N.Y. Acad. Sci.*, 466, 635, 1986.
107. **Sandoval, I. V., Jameson, J. L., Niedel, J., MacDonald, E., and Cuatrecasas, P.,** Role of nucleotides in tubulin polymerization: effect of guanosine 5′-methylene diphosphonate, *Proc. Natl. Acad. Sci. U.S.A.*, 75, 3178, 1978.
108. **Sandoval, I. V. and Weber, K.,** Polymerization of tubulin in the presence of colchicine or podophyllotoxin: formation of a ribbon structure induced by guanylyl-5′-methylene diphosphonate, *J. Mol. Biol.*, 134, 159, 1979.
109. **Sandoval, I. V. and Weber, K.,** Different tubulin polymers are produced by microtubule-associated proteins MAP_2 and τ in the presence of guanosine 5′-(α,β-methylene) triphosphate, *J. Biol. Chem.*, 255, 8952, 1980.
110. **Wehland, J. and Sandoval, I. V.,** Cells injected with guanosine 5′-(α,β-methylene)triphosphate, an α,β-nonhydrolyzable analog of GTP, show anomalous patterns of tubulin polymerization affecting cell translocation, intracellular movement, and the organization of Golgi elements, *Proc. Natl. Acad. Sci. U.S.A.*, 80, 1938, 1983.
111. **Roychowdhury, S. and Gaskin, F.,** Magnesium requirements for guanosine 5′-*O*-(3-thiotriphosphate) induced assembly of microtubule protein and tubulin, *Biochemistry*, 25, 7847, 1986.
112. **Huang, A. B., Lin, C. M., and Hamel, E.,** Maytansine inhibits nucleotide binding at the exchangeable site of tubulin, *Biochem. Biophys. Res. Commun.*, 128, 1239, 1985.
113. **MacNeal, R. K. and Purich, D. L.,** Chromium (III)-nucleotide complexes as probes of the guanosine 5′-triphosphate-induced microtubule assembly, *Arch. Biochem. Biophys*, 191, 233, 1978.
114. **Gaskin, F.,** *In vitro* microtubule assembly regulation by divalent cations and nucleotides, *Biochemistry*, 20, 1318, 1981.
115. **Monasterio, O. and Timasheff, S. N.,** Inhibition of tubulin self-assembly and tubulin-colchicine GTPase activity by guanosine 5′-(γ-fluorotriphosphate), *Biochemistry*, 26, 6091, 1987.
116. **Monasterio, O.,** ^{19}F nuclear magnetic resonance measurement of the distance between the E-site GTP and the high affinity Mg^{2+} in tubulin, *Biochemistry*, 26, 6099, 1987.
117. **Yarbrough, L. R. and Kirsch, M.,** Binding of fluorescent analogs of GTP to the exchangeable nucleotide binding site of tubulin, *J. Biol. Chem.*, 256, 112, 1981.
118. **Steiner, M.,** Affinity labeling of the exchangeable nucleotide binding site in platelet tubulin, *Biochem. Biophys. Res. Commun.*, 123, 92, 1984.
119. **Prasad, A. R. S. and Luduena, R. F.,** Chemical modification of bovine brain tubulin with the guanine nucleotide affinity analogue 5′-*p*-fluorosulfonylbenzoylguanosine, *Biochem. Int.*, 14, 85, 1987.
120. **Pal, P. K., Reischer, R. J., Wechter, W. J., and Colman, R. F.,** A new affinity label for guanosine nucleotide sites in proteins, *J. Biol. Chem.*, 253, 6644, 1978.
121. **Maccioni, R. B. and Seeds, N. W.,** Affinity labeling of tubulin's exchangeable guanosine 5′-triphosphate binding site, *Biochemistry*, 22, 1572, 1983.
122. **Lustbader, J. and Hamel, E.,** Di- and triphosphate derivatives of acyclo- and arabinosylguanine: effects on the polymerization of purified tubulin, *Biochim. Biophys. Acta*, 719, 215, 1982.
123. **Hamel, E., del Campo, A. A., Lustbader, J., and Lin, C. M.,** Modulation of tubulin-nucleotide interactions by microtubule-associated proteins: polymerization with ribose-modified analogues of guanosine 5′-triphosphate, *Biochemistry*, 22, 1271, 1983.
124. **Hamel, E., del Campo, A. A., and Lin, C. M.,** Microtubule assembly with the guanosine 5′-diphosphate analogue 2′,3′-dideoxyguanosine 5′-diphosphate, *Biochemistry*, 22, 3664, 1983.
125. **Hamel, E., del Campo, A. A., Lowe, M. C., and Lin, C. M.,** Interactions of taxol, microtubule-associated proteins, and guanine nucleotides in tubulin polymerization, *J. Biol. Chem.*, 256, 11887, 1981.
126. **Voter, W. A. and Erickson, H. A.,** The kinetics of microtubule assembly: evidence for a two-stage nucleation reaction, *J. Biol. Chem.*, 259, 10430, 1984.
127. **Sloboda, R. D., Dentler, W. L., and Rosenbaum, J. L.,** Microtubule-associated proteins and the stimulation of tubulin assembly *in vitro*, *Biochemistry*, 15, 4497, 1976.
128. **Geahlen, R. L. and Haley, B. E.,** Use of a GTP photoaffinity probe to resolve aspects of the mechanism of tubulin polymerization, *J. Biol. Chem.*, 254, 11982, 1979.
129. **Haley, B. E., Ponstingl, H., and Doenges, K.-H.,** Photoaffinity labelling of pure tubulin using 8-azidoguanosine triphosphate at the α-subunit, *Hoppe-Seyler's Z. Physiol. Chem.*, 364, 1137, 1983.
130. **Mann, K., Giesel, M., and Fasold, H.,** Isolation of native microtubules from porcine brain and characterization of SH groups essential for polymerization at the GTP binding sites, *FEBS Lett.*, 92, 45, 1978.
131. **Steiner, M.,** 3′,5′-cyclic AMP binds to and promotes polymerisation on platelet tubulin, *Nature (London)*, 272, 834, 1978.
132. **Zabrecky, J. R. and Cole, R. D.,** Effect of ATP on the kinetics of microtubule assembly, *J. Biol. Chem.*, 257, 4633, 1982.

133. **Zabrecky, J. R. and Cole, R. D.,** Binding of ATP to tubulin, *Nature (London)*, 296, 775, 1982.
134. **Zabrecky, J. R. and Cole, R. D.,** Localization of the ATP binding site on α-tubulin, *Arch. Biochem. Biophys.*, 225, 475, 1983.
135. **Olmsted, J. B. and Borisy, G. G.,** Ionic and nucleotide requirements for microtubule polymerization *in vitro*, *Biochemistry*, 14, 2996, 1975.
136. **Himes, R. H., Burton, P. R., and Gaito, J. M.,** Dimethyl sulfoxide-induced self-assembly of tubulin lacking associated proteins, *J. Biol. Chem.*, 252, 6222, 1977.
137. **Williams, R. C., Jr. and Detrich, H. W., III,** Separation of tubulin from microtubule-associated proteins on phosphocellulose. Accompanying alterations in concentrations of buffer components, *Biochemistry*, 18, 2499, 1979.
138. **Buttlaire, D. H., Czuba, B. A., Stevens, T. H., Lee, Y. C., and Himes, R. H.,** Manganous ion binding to tubulin, *J. Biol. Chem.*, 255, 2164, 1980.
139. **Weisenberg, R. C.,** Microtubule formation *in vitro* in solutions containing low calcium concentrations, *Science*, 177, 1104, 1972.
140. **Lee, Y. C., Samson, F. E., Jr., Houston, L. L., and Himes, R. H.,** The *in vitro* polymerization of tubulin from beef brain, *J. Neurobiol.*, 5, 317, 1974.
141. **Lee, J. C. and Timasheff, S. N.,** The reconstitution of microtubules from purified calf brain tubulin, *Biochemistry*, 14, 5183, 1975.
142. **Lee, J. C. and Timasheff, S. N.,** *In vitro* reconstitution of calf brain microtubules: effects of solution variables, *Biochemistry*, 16, 1754, 1977.
143. **Herzog, W. and Weber, K.,** *In vitro* assembly of pure tubulin into microtubules in the absence of microtubule-associated proteins and glycerol, *Proc. Natl. Acad. Sci. U.S.A.*, 74, 1860, 1977.
144. **Weisenberg, R. C. and Timasheff, S. N.,** Aggregation of microtubule subunit protein. Effects of divalent cations, colchicine and vinblastine, *Biochemistry*, 9, 4110, 1970.
145. **Frigon, R. P. and Timasheff, S. N.,** Magnesium-induced self-association of calf brain tubulin. I. Stoichiometry, *Biochemistry*, 14, 4559, 1975.
146. **Frigon, R. P. and Timasheff, S. N.,** Magnesium-induced self-association of calf brain tubulin. II. Thermodynamics, *Biochemistry*, 14, 4567, 1975.
147. **Martin, S. R., Butler, F. M. M., Clark, D. C., Zhou, J.-M., and Bayley, P. M.,** Magnesium ion effects on microtubule nucleation *in vitro*, *Biochim. Biophys. Acta*, 914, 96, 1987.
148. **Unger, E., Bohm, K., Wondraczek, H., and Vater, M.,** Properties of microtubules stabilized by glycerol: cold resistance, influence of Mg^{2+} and heavier alkaline-earth ions, *Adv. Biosci.*, 3, 25, 1984.
149. **Kirschner, M. W., Williams, R. C., Weingarten, M., and Gerhart, J. C.,** Microtubules from mammalian brain: some properties of their depolymerization products and a proposed mechanism of assembly and disassembly, *Proc. Natl. Acad. Sci. U.S.A.*, 71, 1159, 1974.
150. **Haga, T., Abe, T., and Kurokawa, M.,** Polymerization and depolymerization of microtubules *in vitro* as studied by flow birefringence, *FEBS Lett.*, 39, 291, 1974.
151. **Kuriyama, R. and Sakai, H.,** Viscometric demonstration of tubulin polymerization, *J. Biochem. (Tokyo)*, 75, 463, 1974.
152. **Hayashi, M. and Matsumura, F.,** Calcium binding to bovine brain tubulin, *FEBS Lett.*, 58, 222, 1975.
153. **Solomon, F.,** Binding sites for calcium on tubulin, *Biochemistry*, 16, 358, 1977.
154. **Serrano, L., Valencia, A., Caballero, R., and Avila, J.,** Localization of the high affinity calcium-binding site on tubulin molecule, *J. Biol. Chem.*, 261, 7076, 1986.
155. **Bhattacharyya, B., Sackett, D. L., and Wolff, J.,** Tubulin, hybrid dimers and tubulin S: stepwise charge reduction and polymerization, *J. Biol. Chem.*, 260, 10208, 1985.
156. **Fong, K. C., Babitch, J. A., and Anthony, F. A.,** Calcium binding to tubulin, *Biochem. Biophys. Acta*, 952, 13, 1988.
157. **Olmsted, J. B. and Borisy, G. G.,** Characterization of microtubule assembly in porcine brain extracts by viscometry, *Biochemistry*, 12, 4282, 1973.
158. **Nishida, E. and Sakai, H.,** Calcium-sensitivity of the microtubule reassembly system: difference between crude brain extract and purified microtubular protein, *J. Biochem. (Tokyo)*, 82, 303, 1977.
159. **Berkowitz, S. A. and Wolff, J.,** Intrinsic calcium sensitivity of tubulin polymerization: the contributions of temperature, tubulin concentration, and associated proteins, *J. Biol. Chem.*, 256, 11216, 1981.
160. **Lee, Y. C. and Wolff, J.,** Two opposing effects of calmodulin on microtubule assembly depend on the presence of microtubule-associated proteins, *J. Biol. Chem.*, 257, 6306, 1982.
161. **Keller, T. C. S., III, Jemiolo, D. K., Burgess, W. H., and Rebhun, L. I.,** *Strongylocentrotus purpuratus* spindle tubulin. II. Characteristics of its sensitivity to Ca^{++} and the effects of calmodulin isolated from bovine brain and *S. purpuratus* eggs, *J. Cell Biol.*, 93, 797, 1982.
162. **Nishida, E. and Kumagai, H.,** Calcium-sensitivity of sea urchin tubulin in *in vitro* assembly and the effects of calcium-dependent regulator (CDR) proteins isolated from sea urchin eggs and porcine brains, *J. Biochem. (Tokyo)*, 87, 143, 1980.

163. **Schliwa, M., Euteneuer, U., Bulinski, J. C., and Izant, J. G.,** Calcium lability of cytoplasmic microtubules and its modulation by microtubule-associated proteins, *Proc. Natl. Acad. Sci. U.S.A.*, 78, 1037, 1981.
164. **Doenges, K. H.,** Assembly of non-neural microtubules in the presence of calcium ions, *FEBS Lett.*, 89, 157, 1978.
165. **Pirollet, F., Job, D., Fischer, E. H., and Margolis, R. L.,** Purification and characterization of sheep brain cold-stable microtubules, *Proc. Natl. Acad. Sci. U.S.A.*, 80, 1560, 1983.
166. **Nishida, E.,** Effects of solution variables on the calcium sensitivity of the microtubule assembly system, *J. Biochem. (Tokyo)*, 84, 507, 1978.
167. **Rosenfeld, A., Zackroff, R. V., and Weisenberg, R. C.,** Magnesium stimulation of calcium binding to tubulin and calcium induced depolymerization of microtubules, *FEBS Lett.*, 65, 144, 1976.
168. **Summers, K. and Kirschner, M. W.,** Characteristics of the polar assembly and disassembly of microtubules observed in vitro by darkfield light microscopy, *J. Cell Biol.*, 83, 205, 1979.
169. **Karr, T. L., Kristofferson, D., and Purich, D. L.,** Calcium ion induces endwise depolymerization of bovine brain microtubules, *J. Biol. Chem.*, 255, 11853, 1980.
170. **Weisenberg, R. C. and Deery, W. J.,** The mechanism of calcium-induced microtubule disassembly, *Biochem. Biophys. Res. Commun.*, 102, 924, 1981.
171. **Deery, W. J., Means, A. R., and Brinkley, B. R.,** Calmodulin-microtubule association in cultured mammalian cells, *J. Cell Biol.*, 98, 904, 1984.
172. **Andreu, J. M. and Timasheff, S. N.,** The ligand- and microtubule assembly-induced GTPase activity of purified calf brain tubulin, *Arch. Biochem. Biophys.*, 211, 151, 1981.
173. **Welsh, M. J., Dedman, J. R., Brinkley, B. R., and Means, A. R.,** Calcium-dependent regulator protein: localization in mitotic apparatus of eucaryotic cells, *Proc. Natl. Acad. Sci. U.S.A.*, 75, 1867, 1978.
174. **Marcum, J. M., Dedman, J. R., Brinkley, B. R., and Means, A. R.,** Control of microtubule assembly-disassembly by calcium-dependent regulator protein, *Proc. Natl. Acad. Sci. U.S.A.*, 75, 3771, 1978.
175. **Keith, C., DiPaola, M., Maxfield, F. R., and Shelanski, M. L.,** Microinjection of Ca^{2+}-calmodulin causes a localized depolymerization of microtubules, *J. Cell Biol.*, 97, 1918, 1983.
176. **Nishida, E., Kumagai, H., Ohtsuki, I., and Sakai, H.,** The interactions between calcium-dependent regulator protein of cyclic nucleotide phosphodiesterase and microtubule proteins. I. Effect of calcium-dependent regulator protein on the calcium sensitivity of microtubule assembly, *J. Biochem. (Tokyo)*, 85, 1257, 1979.
177. **Job, D., Fischer, E. H., and Margolis, R. L.,** Rapid disassembly of cold-stable microtubules by calmodulin, *Proc. Natl. Acad. Sci. U.S.A.*, 78, 4679, 1981.
178. **Kumagai, H. and Nishida, E.,** The interactions between calcium-dependent regulator protein of cyclic nucleotide phosphodiesterase and microtubule proteins. II. Association of calcium-dependent regulator protein with tubulin dimer, *J. Biochem. (Tokyo)*, 85, 1267, 1979.
179. **Kumagai, H., Nishida, E., and Sakai, H.,** The interaction between calmodulin and microtubule proteins. IV. Quantitative analysis of the binding between calmodulin and tubulin dimer, *J. Biochem. (Tokyo)*, 91, 1329, 1982.
180. **Sobue, K., Fujita, M., Muramoto, Y., and Kakiuchi, S.,** The calmodulin-binding protein in microtubules is tau factor, *FEBS Lett.*, 132, 137, 1981.
181. **Lee, Y. C. and Wolff, J.,** Calmodulin binds to both microtubule-associated protein 2 and τ proteins, *J. Biol. Chem.*, 259, 1226, 1983.
182. **Kumagai, H., Nishida, E., Kotani, S., and Sakai, H.,** On the mechanism of calmodulin-induced inhibition of microtubule assembly *in vitro*, *J. Biochem. (Tokyo)*, 99, 521, 1986.
183. **Lee, Y. C. and Wolff, J.,** The calmodulin-binding domain of microtubule-associated protein 2, *J. Biol. Chem.*, 259, 8041, 1984.
184. **Donato, R.,** S-100 proteins, *Cell Calcium*, 7, 123, 1986.
185. **Baudier, J., Briving, C., Deinum, J., Haglid, K., Sorskog, L., and Wallin, M.,** Effect of S-100 proteins and calmodulin on Ca^{2+}-induced disassembly of brain microtubule proteins, *in vitro*, *FEBS Lett.*, 147, 165, 1982.
186. **Donato, R.,** Effect of S-100 protein on assembly of brain microtubule proteins *in vitro*, *FEBS Lett.*, 162, 310, 1983.
187. **Endo, T. and Hidaka, H.,** Effect of S-100 protein on microtubule assembly-disassembly, *FEBS Lett.*, 161, 235, 1983.
188. **Donato, R.,** Mechanism of action of S-100 protein(s) on brain microtubule protein assembly, *Biochem. Biophys. Res. Commun.*, 124, 850, 1984.
189. **Donato, R.,** Calcium-independent, pH-regulated effects of S-100 proteins on assembly-disassembly of brain microtubule protein *in vitro*, *J. Biol. Chem.*, 263, 106, 1988.
190. **Donato, R.,** Quantitative analysis of the interaction between S-100 proteins and brain tubulin, *Cell Calcium*, 8, 283, 1987.

191. **Baudier, J., Mochly-Rosen, D., Newton, A., Lee, S.-H., Koshland, D. E., Jr., and Cole, R. D.,** Comparison of S100b protein with calmodulin: interactions with melittin and microtubule-associated τ proteins and inhibition of phosphorylation of τ proteins by protein kinase C, *Biochemistry*, 26, 2886, 1987.
192. **Baudier, J. and Cole, R. D.,** Reinvestigation of the sulfhydryl reactivity in bovine brain S100b (ββ) protein and the microtubule-associated τ proteins. Ca^{2+} stimulates disulfide cross-linking between the S100b β-subunit and the microtubule-associated τ(2) protein, *Biochemistry*, 27, 2728, 1988.
193. **Baudier, J. and Cole, R. D.,** Interactions between the microtubule-associated τ proteins and S100b regulate τ phosphorylation by the Ca^{2+} calmodulin-dependent protein kinase II, *J. Biol. Chem.*, 263, 5876, 1988.
194. **Sandoval, I. V. and Weber, K.,** Calcium-induced inactivation of microtubule formation in brain extracts: presence of a calcium-dependent protease acting on polymerization-stimulating microtubule-associated proteins, *Eur. J. Biochem.*, 92, 463, 1978.
195. **Yamauchi, T. and Fujisawa, H.,** Disassembly of microtubules by the action of calmodulin-dependent protein kinase (kinase II) which occurs only in the brain tissues, *Biochem. Biophys. Res. Commun.*, 110, 287, 1983.
196. **Yamamoto, H., Fukunaga, K., Tanaka, E., and Miyamoto, E.,** Ca^{2+}- and calmodulin-dependent phosphorylation of microtubule-associated protein 2 and τ factor, and inhibition of microtubule assembly, *J. Neurochem.*, 41, 1119, 1983.
197. **Wandosell, F., Serrano, L., Hernandez, M. A., and Avila, J.,** Phosphorylation of tubulin by a calmodulin-dependent protein kinase, *J. Biol. Chem.*, 261, 10332, 1986.
198. **Job, D., Rauch, C. T., Fischer, E. H., and Margolis, R. L.,** Regulation of microtubule cold stability by calmodulin-dependent and -independent phosphorylation, *Proc. Natl. Acad. Sci. U.S.A.*, 80, 3894, 1983.
199. **Larsson, H., Wallin, M., and Edstrom, A.,** Induction of a sheet polymer of tubulin by Zn^{2+}, *Exp. Cell Res.*, 100, 104, 1976.
200. **Gaskin, F. and Kress, Y.,** Zinc ion-induced assembly of tubulin, *J. Biol. Chem.*, 252, 6918, 1977.
201. **Haskins, K. M., Zombola, R. R., Boling, J. M., Lee, Y. C., and Himes, R. H.,** Tubulin assembly induced by cobalt and zinc, *Biochem. Biophys. Res. Commun.*, 95, 1703, 1980.
202. **Banerjee, A., Roychowdhury, S., and Bhattacharyya, B.,** Zinc-induced self assembly of goat brain tubulin: some novel aspects, *Biochem. Biophys. Res. Commun.*, 105, 1503, 1982.
203. **Crepeau, R. H., McEwen, B., and Edelstein, S. J.,** Differences in α and β polypeptide chains of tubulin resolved by electron microscopy with image reconstruction, *Proc. Natl. Acad. Sci. U.S.A.*, 75, 5006, 1978.
204. **Baker, T. S. and Amos, L. A.,** Structure of the tubulin dimer in zinc-induced sheets, *J. Mol. Biol.*, 123, 89, 1978.
205. **de la Torre, J., Villasante, A., Corral, J., and Avila, J.,** Factors implicated in determining the structure of zinc tubulin-sheets: lateral tubulin-tubulin interaction is promoted by the presence of zinc, *J. Supramol. Struct. Cell. Biochem.*, 17, 183, 1981.
206. **Eagle, G. R., Zombola, R. R., and Himes, R. H.,** Tubulin-zinc interactions: binding and polymerization studies, *Biochemistry*, 22, 221, 1983.
207. **Wallin, M., Larsson, H., and Edstrom, A.,** Tubulin sulfhydryl groups and polymerization *in vitro:* effects of di- and trivalent cations, *Exp. Cell Res.*, 107, 219, 1977.
208. **Himes, R. H., Lee, Y. C., Eagle, G. R., Haskins, K. M., Babler, S. D., and Ellermeier,** The relationship between cobalt binding to tubulin and the stimulation of assembly, *J. Biol. Chem.*, 257, 5839, 1982.
209. **Macdonald, T. L., Humphreys, W. G., and Martin, R. B.,** Promotion of tubulin assembly by aluminum ion *in vitro*, *Science*, 236, 183, 1987.
210. **Shelanski, M. L., Gaskin, F., and Cantor, C. R.,** Microtubule assembly in the absence of added nucleotides, *Proc. Natl. Acad. Sci. U.S.A.*, 70, 765, 1973.
211. **Wilson, L.,** Properties of colchicine binding protein from chick embryo brain. Interactions with vinca alkaloids and podophyllotoxin, *Biochemistry*, 9, 4999, 1970.
212. **Williams, J. A. and Wolff, J.,** Colchicine-binding protein and the secretion of thyroid hormone, *J. Cell Biol.*, 54, 157, 1972.
213. **Rappaport, L., Leterrier, J. F., and Nunez, J.,** Protein-kinase activity, *in vitro* phosphorylation and polymerization of purified tubulin, *Ann. N.Y. Acad. Sci.*, 253, 611, 1975.
214. **McClure, W. O. and Paulson, J. C.,** The interaction of colchicine and some related alkaloids with rat brain tubulin, *Mol. Pharmacol.*, 13, 560, 1977.
215. **Hadjian, A. J., Guidicelli, C., and Chambaz, E. M.,** Characterization of tubulin in the bovine adrenal cortex cytosol, *FEBS Lett.*, 77, 233, 1977.
216. **Frigon, R. P. and Lee, J. C.,** The stabilization of calf-brain microtubule protein by sucrose, *Arch. Biochem. Biophys.*, 153, 587, 1972.
217. **Solomon, F., Monard, D., and Rentsch, M.,** Stabilization of colchicine-binding activity of neuroblastoma, *J. Mol. Biol.*, 78, 569, 1973.
218. **Bhattacharyya, B. and Wolff, J.,** Membrane-bound tubulin in brain and thyroid tissue, *J. Biol. Chem.*, 250, 7639, 1975.

219. **Barton, J. S.,** Polymerization and colchicine binding: two independent properties of tubulin, *Biochim. Biophys. Acta*, 532, 155, 1978.
220. **Keates, R. A. B.,** Stabilization of microtubule protein in glycerol solutions, *Can. J. Biochem.*, 59, 353, 1981.
221. **Jennett, R. B., Tuma, D. J., Sorrell, W. T., and Sorrell, M. F.,** Stabilization of hepatic colchicine-binding activity by organic acids, *Arch. Biochem. Biophys.*, 236, 304, 1985.
222. **Wilson, L. and Meza, I.,** The mechanism of action of colchicine: colchicine binding properties of sea urchin sperm tail outer doublet tubulin, *J. Cell Biol.*, 58, 709, 1973.
223. **Bhattacharyya, B. and Wolff, J.,** Promotion of fluorescence upon binding of colchicine to tubulin, *Proc. Natl. Acad. Sci. U.S.A.*, 71, 2627, 1974.
224. **Pfeffer, T. A., Asnes, C. F., and Wilson, L.,** Properties of tubulin in unfertilized sea urchin eggs, *J. Cell Biol.*, 69, 599, 1976.
225. **Barnes, L. D. and Roberson, G. M.,** Tubulin and microtubules from bovine kidney: purification, properties, and characterization of ligand binding, *Arch. Biochem. Biophys.*, 196, 511, 1979.
226. **Sherline, P., Leung, J. T., and Kipnis, D. M.,** Binding of colchicine to purified microtubule protein, *J. Biol. Chem.*, 250, 5481, 1975.
227. **Bamburg, J. R., Shooter, E. M., and Wilson, L.,** Assay of microtubule protein in embryonic chick dorsal root ganglia, *Neurobiology*, 3, 162, 1973.
228. **Wiche, G. and Furtner, R.,** Microtubule associated proteins stabilize the colchicine binding activity of tubulin, *FEBS Lett.*, 116, 247, 1980.
229. **Zaremba, T. G. and Irwin, R. D.,** Effect of ^{60}Co gamma irradiation on the polymerization of microtubules *in vitro*, *Radiat. Res.*, 71, 300, 1977.
230. **Kuriyama, R. and Sakai, H.,** Role of tubulin -SH groups in polymerization to microtubules, *J. Biochem. (Tokyo)*, 76, 651, 1974.
231. **Ikeda, Y. and Steiner, M.,** Sulfhydryls of platelet tubulin: their role in polymerization and colchicine binding, *Biochemistry*, 17, 3454, 1978.
232. **Mellado, W., Slebe, J. C., and Maccioni, R. B.,** Tubulin carbamoylation: functional amino groups in microtubule assembly, *Biochem. J.*, 203, 675, 1982.
233. **Maccioni, R. B., Vera, J. C., and Slebe, J. C.,** Arginyl residues involvement in microtubule assembly, *Arch. Biochem. Biophys.*, 207, 248, 1981.
234. **Serrano, L., Avila, J., and Maccioni, R. B.,** Limited proteolysis of tubulin and the localization of the binding site for colchicine, *J. Biol. Chem.*, 259, 6607, 1984.
235. **Wilson, L. and Friedkin, M.,** The biochemical events of mitosis. II. The *in vivo* and *in vitro* binding of colchicine in grasshopper embryos and its possible relation to inhibition of mitosis, *Biochemistry*, 6, 3126, 1967.
236. **Borisy, G. G. and Taylor, E. W.,** The mechanism of action of colchicine: binding of colchicine-^{3}H to cellular protein, *J. Cell Biol.*, 34, 525, 1967.
237. **Owellen, R. J., Owens, A. H., Jr., and Donigian, D. W.,** The binding of vincristine, vinblastine and colchicine to tubulin, *Biochem. Biophys. Res. Commun.*, 47, 685, 1972.
238. **Bryan, J.,** Definition of three classes of binding sites in isolated microtubule crystals, *Biochemistry*, 11, 2611, 1972.
239. **Zweig, M. H. and Chignell, C. F.,** Interaction of some colchicine analogs, vinblastine and podophyllotoxin with rat brain microtubule protein, *Biochem. Pharmacol.*, 22, 2141, 1973.
240. **Owellen, R. J., Donigian, D. W., Hartke, C. A., Dickerson, R. M., and Kuhar, M. J.,** The binding of vinblastine to tubulin and to particulate fractions of mammalian brain, *Cancer Res.*, 34, 3180, 1974.
241. **Garland, D. and Teller, D. C.,** A reexamination of the reaction between colchicine and tubulin, *Ann. N.Y. Acad. Sci.*, 253, 232, 1975.
242. **Bhattacharyya, B. and Wolff, J.,** Anion-induced increases in the rate of colchicine binding to tubulin, *Biochemistry*, 15, 2283, 1976.
243. **Kelleher, J. K.,** Tubulin binding affinities of podophyllotoxin and colchicine analogues, *Mol. Pharmacol.*, 13, 232, 1977.
244. **Hains, F. O., Dickerson, R. M., Wilson, L., and Owellen, R. J.,** Differences in the binding properties of vinca alkaloids and colchicine in tubulin by varying protein sources and methodology, *Biochem. Pharmacol.*, 27, 71, 1978.
245. **Garland, D. L.,** Kinetics and mechanism of colchicine binding to tubulin: evidence for ligand-induced conformational change, *Biochemistry*, 17, 4266, 1978.
246. **Nunez, J., Fellous, A., Francon, J., and Lennon, A. M.,** Competitive inhibition of colchicine binding to tubulin by microtubule-associated proteins, *Proc. Natl. Acad. Sci. U.S.A.*, 76, 86, 1979.
247. **Detrich, H. W., III, Williams, R. C., Jr., MacDonald, T. L., Wilson, L., and Puett, D.,** Changes in the circular dichroic spectrum of colchicine associated with its binding to tubulin, *Biochemistry*, 20, 5999, 1981.

248. **Palanivelu, P. and Luduena, R. F.,** Interactions of the τ-tubulin-vinblastine complex with colchicine, podophyllotoxin, and *N,N'*-ethylenebis(iodoacetamide), *J. Biol. Chem.*, 257, 6311, 1982.
249. **Williams, R. F., Mumford, C. L., Williams, G. A., Floyd, L. J., Aivaliotis, M. J., Martinez, R. A., Robinson, A. K., and Barnes, L. D.,** A photoaffinity derivative of colchicine: 6'-(4'-azido-2'-nitrophenylamino)hexanoyldeacetylcolchicine: photolabeling and location of the colchicine-binding site on the α-subunit of tubulin, *J. Biol. Chem.*, 260, 13794, 1985.
250. **Yeh, H. J. C., Chrzanowska, M., and Brossi, A.,** The importance of the phenyl-tropolone 'aS' configuration in colchicine's binding to tubulin, *FEBS Lett.*, 229, 82, 1988.
251. **Lacey, E., Edgar, J. A., and Culvenor, C. C. J.,** Interaction of phomopsin A and related compounds with purified sheep brain tubulin, *Biochem. Pharmacol.*, 36, 2133, 1987.
252. **Detrich, H. W., III, Wiliams, R. C., Jr., and Wilson, L.,** Effect of colchicine binding on the reversible dissociation of the tubulin dimer, *Biochemistry*, 21, 2392, 1982.
253. **Cortese, F., Bhattacharyya, B., and Wolff, J.,** Podophyllotoxin as a probe for the colchicine binding site of tubulin, *J. Biol. Chem.*, 252, 1134, 1977.
254. **Bane, S., Puett, D., Macdonald, T., and Williams, R. C., Jr.,** Binding to tubulin of the colchicine analog 2-methoxy-5-(2',3',4'-trimethyoxyphenyl)tropone: thermodynamic and kinetic aspects, *J. Biol. Chem.*, 259, 7391, 1984.
255. **Williams, R. F., Aivaliotis, M. J., Barnes, L. D., and Robinson, A. K.,** High-performance liquid chromatographic application of the Hummel and Dreyer method for the determination of colchicine-tubulin binding parameters, *J. Chromatogr.*, 266, 141, 1983.
256. **Ringel, I. and Sternlicht, H.,** Carbon-13 nuclear magnetic resonance study of microtubule protein: evidence for a second colchicine site involved in the inhibition of microtubule assembly, *Biochemistry*, 23, 5644, 1984.
257. **Schmitt, H. and Atlas, D.,** Specific affinity labelling of tubulin with bromocolchicine, *J. Mol. Biol.*, 102, 743, 1976.
258. **Ray, K., Bhattacharyya, B., and Biswas, B. B.,** Anion-induced increases in the affinity of colcemid binding to tubulin, *Eur. J. Biochem.*, 142, 577, 1984.
259. **Deinum, J. and Lincoln, P.,** Binding to tubulin of an allocolchicine spin probe: interaction with the essential SH groups and other active sites, *Biochim. Biophys. Acta*, 870, 226, 1986.
260. **Avila, J., Serrano, L., and Maccioni, R. B.,** Regulatory aspects of the colchicine interactions with tubulin, *Mol. Cell. Biochem.*, 73, 29, 1987.
261. **Keates, R. A. B., Sarangi, F., and Ling, V.,** Structural and functional alterations in microtubule protein from Chinese hamster ovary cell mutants, *Proc. Natl. Acad. Sci. U.S.A.*, 78, 5638, 1981.
262. **Cabral, F., Sobel, M. E., and Gottesman, M. M.,** CHO mutants resistant to colchicine, colcemid or griseofulvin have an altered β-tubulin, *Cell*, 20, 29, 1980.
263. **Sheir-Neiss, G., Lai, M. H., and Morris, N. R.,** Identification of a gene for β-tubulin in *Aspergillus nidulans*, *Cell*, 15, 639, 1978.
264. **Luduena, R. F. and Roach, M. C.,** Interaction of tubulin with drugs and alkylating agents. II. Effects of colchicine, podophyllotoxin, and vinblastine on the alkylation of tubulin, *Biochemistry*, 20, 4444, 1981.
265. **Luduena, R. F., Roach, M. C., Trcka, P. P., Little, M., Palanivelu, R., Binkley, P., and Prasad, V.,** β_2-Tubulin, a form of chordate brain tubulin with lesser reactivity toward an assembly-inhibiting sulfhydryl-directed cross-linking reagent, *Biochemistry*, 21, 4787, 1982.
266. **Roach, M. C. and Luduena, R. F.,** Different effects of tubulin ligands on the intrachain cross-linking of β_1-tubulin, *J. Biol. Chem.*, 259, 12063, 1984.
267. **Roach, M. C., Bane, S., and Luduena, R. F.,** The effects of colchicine analogues on the reaction of tubulin with iodo[^{14}C]acetamide and *N,N'*-ethylenebis(iodoacetamide), *J. Biol. Chem.*, 260, 3015, 1985.
268. **Fujiwara, K. and Tilney, L. G.,** Substructural analysis of the microtubule and its polymorphic forms, *Ann. N.Y. Acad. Sci.*, 253, 27, 1975.
269. **Olmsted, J. B., Carlson, K., Klebe, R., Ruddle, F., and Rosenbaum, J.,** Isolation of microtubule protein from cultured neuroblastoma cells, *Proc. Natl. Acad. Sci. U.S.A.*, 65, 129, 1970.
270. **Marantz, R., Ventilla, M., and Shelanski, M.,** Vinblastine-induced precipitation of microtubule protein, *Science*, 165, 498, 1969.
271. **Wilson, L., Bryan, J., Ruby, A., and Mazia, D.,** Precipitation of proteins by vinblastine and calcium ions, *Proc. Natl. Acad. Sci. U.S.A.*, 66, 807, 1970.
272. **Warfield, R. K. N. and Bouck, G. B.,** Microtubule-microtubule transitions: intermediates after exposure to the mitotic inhibitor vinblastine, *Science*, 186, 1219, 1974.
273. **Penningroth, S. M.,** Colchicine binding to an oligomer of tubulin, *Biochem. Biophys. Res. Commun.*, 92, 183, 1980.
274. **Lambeir, A. and Engelborghs, Y.,** A fluorescence stopped flow study of colchicine binding to tubulin, *J. Biol. Chem.*, 256, 3279, 1981.
275. **Zimmermann, H.-P. and Doenges, K.-H.,** Decoration of microtubules by a colchicine derivative, *Cell Biol. Int. Rep.*, 5, 699, 1981.

276. **Hiratsuka, T. and Kato, T.**, A fluorescent analog of colcemid, *N*-(7-nitrobenz-2-oxa-1,3-diazol-4-yl)-colcemid, as a probe for the colcemid binding sites of tubulin and microtubules, *J. Biol. Chem.*, 262, 6318, 1987.
277. **Arai, T. and Okuyama, T.**, Fluorometric assay of tubulin-colchicine complex, *Anal. Biochem.*, 69, 443, 1975.
278. **Bhattacharyya, B. and Wolff, J.**, Immobilization-dependent fluorescence of colchicine, *J. Biol. Chem.*, 259, 11836, 1984.
279. **Margulis, T. N.**, Structure of the mitotic spindle inhibitor colcemid. *N*-desacetyl-*N*-methylcolchicine, *J. Am. Chem. Soc.*, 96, 899, 1974.
280. **Koerntgen, C. and Margulis, T. N.**, X-ray analysis of sulfur-containing colchicine derivatives, *J. Pharm. Sci.*, 66, 1127, 1977.
281. **Lessinger, L. and Margulis, T. N.**, The crystal structure of colchicine. A new application of magic integers to multiple-solution direct methods, *Acta Cryst.*, B34, 578, 1978.
282. **Andreu, J. M. and Timasheff, S. N.**, Conformational states of tubulin liganded to colchicine, tropolone methyl ether, and podophyllotoxin, *Biochemistry*, 21, 6465, 1982.
283. **Schmitt, H. and Kram, R.**, Binding of antimitotic drugs around cysteine residues of tubulin, *Exp. Cell Res.*, 115, 408, 1978.
284. **Morgan, J. L. and Spooner, B. S.**, Immunological detection of microtubule poison-induced conformational changes in tubulin, *J. Biol. Chem.*, 258, 13127, 1983.
285. **Andreu, J. M. and Timasheff, S. N.**, Interaction of tubulin with single ring analogues of colchicine, *Biochemistry*, 21, 534, 1982.
286. **Heusele, C. and Carlier, M.-F.**, GTPase activity of the tubulin-colchicine in relation with tubulin-tubulin interactions, *Biochem. Biophys. Res. Commun.*, 103, 332, 1981.
287. **Weber, K., Pollack, R., and Bibring, T.**, Antibody against tubulin: the specific visualization of cytoplasmic microtubules in tissue culture cells, *Proc. Natl. Acad. Sci. U.S.A.*, 72, 459, 1975.
288. **Borisy, G. G. and Olmsted, J. B.**, Nucleated assembly of microtubules in porcine brain extracts, *Science*, 177, 1196, 1972.
289. **Allen, C. and Borisy, G. G.**, Structural polarity and directional growth of microtubules of *Chlamydomonas* flagella, *J. Mol. Biol.*, 90, 381, 1974.
290. **Haga, T. and Kurokawa, M.**, Microtubule formation from two components separated by gel filtration of a tubulin preparation, *Biochim. Biophys. Acta*, 392, 335, 1975.
291. **Maccioni, R. B. and Seeds, N. W.**, Enhancement of tubulin assembly as monitored by a rapid filtration assay, *Arch. Biochem. Biophys.*, 185, 262, 1978.
292. **Stearns, M. E. and Tew, K. D.**, Estramustine binds MAP-2 to inhibit microtubule assembly *in vitro*, *J. Cell Sci.*, 89, 331, 1988.
293. **Margolis, R. L. and Wilson, L.**, Addition of tubulin-colchicine complex to microtubule ends: the mechanism of substoichiometric colchicine poisoning, *Proc. Natl. Acad. Sci. U.S.A.*, 74, 3466, 1977.
294. **Sloboda, R. D. and Rosenbaum, J. L.**, Decoration and stabilization of intact, smooth-walled microtubules with microtubule-associated proteins, *Biochemistry*, 18, 48, 1979.
295. **Gaskin, F., Cantor, C. R., and Shelanski, M. L.**, Biochemical studies on the *in vitro* assembly and disassembly of microtubules, *Ann. N.Y. Acad. Sci.*, 253, 133, 1975.
296. **Bergen, L. G. and Borisy, G. G.**, Tubulin-colchicine complex (TC) inhibits microtubule depolymerization by a capping reaction exerted preferentially at the minus end, *J. Cell. Biochem.*, 30, 11, 1986.
297. **Taylor, E. W.**, The mechanism of colchicine inhibition of mitosis. I. Kinetics of inhibition and the binding of ^{3}H-colchicine, *J. Cell Biol.*, 25, 145, 1965.
298. **Batra, J. K., Jurd, L., and Hamel, E.**, Structure-function studies with derivatives of 6-benzyl-1,3-benzodioxole, a new class of synthetic compounds which inhibit tubulin polymerization and mitosis, *Mol. Pharmacol.*, 27, 94, 1985.
299. **Hamel, E., Ho, H. H., Kang, G.-J., and Lin, C. M.**, Cornigerine, a potent antimitotic *Colchicum* alkaloid of unusual structure: interactions with tubulin, *Biochem. Pharmacol.*, 37, 2445, 1988.
300. **Lin, C. M., Singh, S. B., Chu, P. S., Dempcy, R. O., Schmidt, J. M., Pettit, G. R., and Hamel, E.**, Interactions of tubulin with potent natural and synthetic analogs of the antimitotic agent combretastatin, a structure-activity study, *Mol. Pharmacol.*, 34, 200, 1988.
301. **Fitzgerald, T. J.**, Molecular features of colchicine associated with antimitotic activity and inhibition of tubulin polymerization, *Biochem. Pharmacol.*, 25, 1383, 1976.
302. **Owellen, R. J., Hartke, C. A., Dickerson, R. M., and Hains, F. O.**, Inhibition of tubulin-microtubule polymerization by drugs of the *Vinca* alkaloid class, *Cancer Res.*, 36, 1499, 1976.
303. **MacKinney, A. A., Jr., Vyas, R., and Mueller, C.**, 1-Acetyl-3-acetoxy-5′5-diphenylhydantoin has colchicine-like activity against microtubular protein, *Res. Comm. Chem. Path. Pharmacol.*, 44, 251, 1984.
304. **Hoebeke, J. and Van Nijen, G.**, Quantitative turbidimetric assay for potency evaluation of colchicine-like drugs, *Life Sci.*, 17, 591, 1975.

305. **Zavala, F., Guenard, D., Robin, J.-P., and Brown, E.,** Structure-antitubulin activity relationships in steganacin congeners and analogues. Inhibition of tubulin polymerization *in vitro* by (±)-isodeoxypodophyllotoxin, *J. Med. Chem.*, 23, 546, 1980.
306. **Roobol, A., Gull, K., and Pogson, C. I.,** Inhibition by griseofulvin of microtubule assembly *in vitro*, *FEBS Lett.*, 67, 248, 1976.
307. **Dvorackova, S., Guenard, D., Picot, F., Simanek, V., and Waisser, K.,** Relationships between chemical structures and antimitotic activities of a group of colchicine alkaloids, *Acta Univ. Palacki. Olomouc.*, 111, 13, 1985.
308. **Wang, R. W. -J., Rebhun, L. I., and Kupchan, S. M.,** Antimitotic and antitubulin activity of the tumor inhibitor steganacin, *Cancer Res.*, 37, 3071, 1977.
309. **Friedman, P. A. and Platzer, E. G.,** Interaction of anthelmintic benzimidazoles and benzimidazole derivatives with bovine brain tubulin, *Biochim. Biophys. Acta*, 544, 605, 1978.
310. **Geuens, G. M. A., Nuydens, R. M., Willebrords, R. E., Van de Veire, R. M. L., Goossens, F., Dragonetti, C. H., Mareel, M. M. K., and De Brabander, M. J.,** Effects of tubulozole on the microtubule system of cells in culture and *in vivo*, *Cancer Res.*, 45, 733, 1985.
311. **Pettit, G. R., Singh, S. B., Niven, M. L., Hamel, E., and Schmidt, J. M.,** Isolation, structure, and synthesis of combretastatins A-1 and B-1, potent new inhibitors of microtubule assembly, derived from *Combretum caffrum*, *J. Nat. Prod.*, 50, 119, 1987.
312. **Bhattacharyya, B. and Wolff, J.,** Tubulin aggregation and disaggregation: mediation by two distinct vinblastine-binding sites, *Proc. Natl. Acad. Sci. U.S.A.*, 73, 2375, 1976.
313. **Dawson, P. J., Gutteridge, W. E., and Gull, K.,** A comparison of the interaction of antihelmintic benzimidazoles with tubulin isolated from mammalian tissue and the parasitic nematode *Ascaridia galli*, *Biochem. Pharmacol.*, 33, 1069, 1984.
314. **Sternlicht, H. and Ringel, I.,** Colchicine inhibition of microtubule assembly via copolymer formation, *J. Biol. Chem.*, 254, 10540, 1979.
315. **Sternlicht, H., Ringel, I., and Szasz, J.,** The co-polymerization of tubulin and tubulin-colchicine complex in the absence and presence of associated proteins, *J. Biol. Chem.*, 255, 9138, 1980.
316. **Keates, R. A. B. and Mason, G. B.,** Inhibition of microtubule polymerization by the tubulin-colchicine complex: inhibition of spontaneous assembly, *Can. J. Biochem.*, 59, 361, 1981.
317. **Bergen, L. G. and Borisy, G. G.,** Tubulin-colchicine complex inhibits microtubule elongation at both plus and minus ends, *J. Biol. Chem.*, 258, 4190, 1983.
318. **Andreu, J. M., Wagenknecht, T., and Timasheff, S. N.,** Polymerization of the tubulin-colchicine complex: relation to microtubule assembly, *Biochemistry*, 22, 1556, 1983.
319. **Farrell, K. W. and Wilson, L.,** Tubulin-colchicine complexes differentially poison opposite microtubule ends, *Biochemistry*, 23, 3741, 1984.
320. **Margolis, R. L., Rauch, C. T., and Wilson, L.,** Mechanism of colchicine-dimer addition to microtubule ends: implications for the microtubule polymerization mechanism, *Biochemistry*, 19, 5550, 1980.
321. **Lambeir, A. and Engelborghs, Y.,** A quantitative analysis of tubulin-colchicine binding to microtubules, *Eur. J. Biochem.*, 109, 619, 1980.
322. **Sternlicht, H., Ringel, I., and Szasz, J.,** Theory for modeling the copolymerization of tubulin and tubulin-colchicine complex, *Biophys. J.*, 42, 255, 1983.
323. **Farrell, K. W. and Wilson, L.,** Proposed mechanism for colchicine poisoning of microtubules reassembled *in vitro* from *Strongylocentrotus purpuratus* sperm tail outer doublet tubulin, *Biochemistry*, 19, 3048, 1980.
324. **Jordan, M. A. and Farrell, K. W.,** Differential radiolabeling of opposite microtubule ends: methodology, equilibrium exchange-flux analysis, and drug poisoning, *Anal. Biochem.*, 130, 41, 1983.
325. **Matsumura, F. and Hayashi, M.,** Polymorphism of tubulin assembly: *in vitro* formation of sheet, twisted ribbon and microtubule, *Biochim. Biophys. Acta*, 453, 162, 1976.
326. **Wiche, G., Honig, L. S., and Cole, R. D.,** Microtubule protein preparations from C_6 glial cells and their spontaneous polymer formation, *J. Cell Biol.*, 80, 553, 1979.
327. **Saltarelli, D. and Pantaloni, D.,** Polymerization of the tubulin-colchicine complex and guanosine 5′-triphosphate hydrolysis, *Biochemistry*, 21, 2996, 1982.
328. **Andreu, J. M. and Timasheff, S. N.,** Tubulin bound to colchicine forms polymers different from microtubules, *Proc. Natl. Acad. Sci. U.S.A.*, 79, 6753, 1982.
329. **Andreu, J. M., Gorbunoff, M. J., Lee, J. C., and Timasheff, S. N.,** Interaction of tubulin with bifunctional colchicine analogues: an equilibrium study, *Biochemistry*, 23, 1742, 1984.
330. **Harrisson, C. M. H., Page, B. M., and Keir, H. M.,** Mescaline as a mitotic spindle inhibitor, *Nature (London)*, 260, 138, 1976.
331. **Rosner, M., Capraro, H.-G., Jacobson, A. E., Atwell, L., Brossi, A., Iorio, M. A., Williams, T. H., Sik, R. H., and Chignell, C. F.,** Biological effects of modified colchicine. Improved preparation of 2-demethylcolchicine, 3-demethylcolchicine, and (+)-colchicine and reassignment of the position of the double bond in dehydro-7-deacetamidocolchicines, *J. Med. Chem.*, 24, 257, 1981.

332. **Brossi, A., Yeh, H. J. C., Chrzanowska, M., Wolff, J., Hamel, E., Lin, C. M., Quinn, F., Suffness, M., and Silverton, J.**, Colchicine and its analogues: recent findings, *Med. Res. Rev.*, 8, 77, 1988.
333. **Brossi, A., Sharma, P. N., Atwell, L., Jacobson, A. E., Iorio, M. A., Molinari, M., and Chignell, C. F.**, Biological effects of modified colchicines. II. Evaluation of catecholic colchicines, colchifolines, colchicide, and novel *N*-acyl- and *N*-aroyldeacetylcolchicines, *J. Med. Chem.*, 26, 1365, 1983.
334. **Rosner, M., Hsu, F.-L., and Brossi, A.**, Synthesis of two aromatic methylenedioxy-substituted colchicine congeners: elucidation of the structure of cornigerine as 2,3-(methylenedioxy)-2,3-didemethoxycolchicine, *J. Org. Chem.*, 46, 3686, 1981.
335. **Dumont, R., Brossi, A., Chignell, C. F., Quinn, F. R., and Suffness, M.**, A novel synthesis of colchicide and analogues from thiocolchicine and congeners: reevaluation of colchicide as a potential antitumor agent, *J. Med. Chem.*, 30, 732, 1987.
336. **Kerekes, P., Sharma, P. N., Brossi, A., Chignell, C. F., and Quinn, F. R.**, Synthesis and biological effects of novel thiocolchicines. III. Evaluation of *N*-acyldeacetylthiocolchicines, *N*-(alkoxycarbonyl)deacetylthiocolchicines, and *O*-ethyldemethylthiocolchicines. New synthesis of thiodemecolcine and antileukemic effects of 2-demethyl- and 3-demethylthiocolchicine, *J. Med. Chem.*, 28, 1204, 1985.
337. **Iorio, M. A.**, Contraction of the tropolonic ring of colchicine by hydrogen peroxide oxidation, *Heterocycles*, 22, 2207, 1984.
338. **Flavin, M. A. and Slaughter, C.**, Microtubule assembly and function in *Chlamydomonas:* inhibition of growth and flagellar regeneration by antitubulins and other drugs and isolation of resistant mutants, *J. Bacteriol.*, 118, 59, 1974.
339. **Clark, J. I. and Garland, D.**, Fluorescein colchicine: synthesis, purification, and biological activity, *J. Cell Biol.*, 76, 619, 1978.
340. **Banerjee, A. C. and Bhattacharyya, B.**, Colcemid and colchicine binding to tubulin: similarity and dissimilarity, *FEBS Lett.*, 99, 333, 1979.
341. **Deinum, J., Lincoln, P., Larsson, T., Lagercrantz, C., and Erkell, L.-J.**, Synthesis and binding to tubulin of an allocolchicine spin probe, *Acta Chem. Scand. B*, 35, 677, 1981.
342. **Ray, K., Bhattacharyya, B., and Biswas, B. B.**, Role of B-ring of colchicine in its binding to tubulin, *J. Biol. Chem.*, 256, 6241, 1981.
343. **Choudhury, G. G., Banerjee, A., Bhattacharyya, B., and Biswas, B. B.**, Interaction of colchicine analogues with purified tubulin, *FEBS Lett.*, 161, 55, 1983.
344. **Sharma, P. N., Brossi, A., Silverton, J. V., and Chignell, C. F.**, Synthesis and binding to tubulin of colchicine spin probes, *J. Med. Chem.*, 27, 1729, 1984.
345. **Bhattacharyya, B., Howard, R., Maity, S. N., Brossi, A., Sharma, P. N., and Wolff, J.**, B ring regulation of colchicine binding kinetics and fluorescence, *Proc. Natl. Acad. Sci. U.S.A.*, 83, 2052, 1986.
346. **Kalra, S. K. and Fragata, M.**, Effect of tropolone on the binding of [^{3}H]colchicine to brain tubulin, *FEBS Lett.*, 42, 112, 1974.
347. **Lin, C. M. and Hamel, E.**, Interactions of combretastatin, a new plant-derived antimitotic agent, with tubulin, *Biochem. Pharmacol.*, 32, 3864, 1983.
348. **Rossi, M., Link, J., and Lee, J. C.**, The structure of 2-methoxy-5-(2′,3′,4′-trimethoxyphenyl)tropone, an effective analog of colchicine, *Arch. Biochem. Biophys.*, 231, 470, 1984.
349. **Banerjee, A., Barnes, L. D., and Luduena, R. F.**, The role of the B-ring of colchicine in the stability of the colchicine-tubulin complex, *Biochim. Biophys. Acta*, 913, 138, 1987.
350. **Engelborghs, Y. and Fitzgerald, T. J.**, Kinetic and thermodynamic aspects of tubulin-ligand interactions: binding of the colchicine analog 2-methoxy-5-(2′,3′,4′-trimethoxyphenyl)tropone, *Ann. N.Y. Acad. Sci.*, 466, 709, 1986.
351. **Diez, J. C., Avila, J., Nieto, J. M., and Andreu, J. A.**, Reversible inhibition of microtubules and cell growth by the bicyclic colchicine analogue MTC, *Cell Motil. Cytoskel.*, 7, 178, 1987.
352. **Engelborghs, Y. and Fitzgerald, T. J.**, A fluorescence stopped flow study of the competition and displacement kinetics of podophyllotoxin and the colchicine analog 2-methoxy-5-(2′,3′,4′-trimethoxyphenyl)tropone on tubulin, *J. Biol. Chem.*, 262, 5204, 1987.
353. **Gensler, W. J., Murthy, C. D., and Trammell, M. H.**, Nonenolizable podophyllotoxin derivatives, *J. Med. Chem.*, 20, 635, 1977.
354. **Schiff, P. B., Kende, A. S., and Horwitz, S. B.**, Steganacin: an inhibitor of HeLa cell growth and microtubule assembly *in vitro*, *Biochem. Biophys. Res. Commun.*, 85, 737, 1978.
355. **Kelleher, J. K.**, Correlation of tubulin-binding and antitumor activities of podophyllotoxin analogs, *Cancer Treat. Rep.*, 62, 1443, 1978.
356. **Batra, J. K., Kang, G. J., Jurd, L., and Hamel, E.**, Methylenedioxy-benzopyran analogs of podophyllotoxin, a new synthetic class of antimitotic agents that inhibit tubulin polymerization, *Biochem. Pharmacol.*, 37, 2595, 1988.
357. **Wilson, L.**, Microtubules as drug receptors: pharmacological properties of microtubule protein, *Ann. N.Y. Acad. Sci.*, 253, 213, 1975.

358. **Duanmu, C., Shahrik, L. K., Ho, H. H., and Hamel, E.,** Tubulin-dependent hydrolysis of guanosine triphosphate as a screening test to identify new antitubulin compounds with potential as antimitotic agents: application to carbamates of aromatic amines, *Cancer Res.*, 49, 1344, 1989.
359. **Karr, T. L. and Purich, D. L.,** A microtubule assembly/disassembly model based on drug effects and depolymerization kinetics after rapid dilution, *J. Biol. Chem.*, 254, 10885, 1979.
360. **Caplow, M. and Zeeberg, B.,** Incorporation of radioactive tubulin into microtubules at steady state: experimental and theoretical analysis of the effect of podophyllotoxin, *J. Biol. Chem.*, 256, 5608, 1981.
361. **Farrell, K. W., Kassis, J. A., and Wilson, L.,** Outer doublet tubulin reassembly: evidence for opposite end assembly-disassembly at steady state and a disassembly end equilibrium, *Biochemistry*, 18, 2642, 1979.
362. **Kumar, N.,** Taxol-induced polymerization of purified tubulin: mechanism of action, *J. Biol. Chem.*, 256, 10435, 1981.
363. **Sackett, D. L., Bhattacharyya, B., and Wolff, J.,** Tubulin subunit carboxyl termini determine polymerization efficiency, *J. Biol. Chem.*, 260, 43, 1985.
364. **Wilson, L., Anderson, K., Grisham, L., and Chin, D.,** Biochemical mechanisms of action of microtubule inhibitors, in *Microtubules and Microtubules Inhibitors: Proceedings of the International Symposium on Microtubules and Microtubule Inhibitors*, Borgers, M. and De Brabander, M., Eds., North-Holland, Amsterdam, 1975, 103.
365. **Bowdon, B. J., Waud, W. R., Wheeler, G. P., Hain, R., Dansby, L., and Temple, C., Jr.,** Comparison of 1,2-dihydropyrido[3,4-*b*]pyrazines (1-deaza-7,8-dihydropteridines) with several other inhibitors of mitosis, *Cancer Res.*, 47, 1621, 1987.
366. **Loike, J. D., Brewer, C. F., Sternlicht, H., Gensler, W. J., and Horwitz, S. B.,** Structure-activity study of the inhibition of microtubule assembly *in vitro* by podophyllotoxin and its congeners, *Cancer Res.*, 38, 2688, 1978.
367. **Farrell, K. W., Morse, A., and Wilson, L.,** Characterization of the *in vitro* reassembly of tubulin derived from stable *Strongylocentrotus purpuratus* outer doublet microtubules, *Biochemistry*, 18, 905, 1979.
368. **Margolis, R. L. and Wilson, L.,** Opposite end assembly and disassembly of microtubules at steady state *in vitro*, *Cell*, 13, 1, 1978.
369. **Chen, G. L., Yang, L., Rowe, T. C., Halligan, B. D., Tewey, K. M., and Liu, L. F.,** Nonintercalative antitumor drugs interfere with the breakage-reunion reaction of mammalian DNA topoisomerase II, *J. Biol. Chem.*, 259, 13560, 1984.
370. **Bates, R. B. and Wood, J. B., III,** The crystal and molecular structure of 5′-demethoxy-β-peltatin A methyl ether, *J. Org. Chem.*, 37, 562, 1972.
371. **Petcher, T. J., Weber, H. P., Kuhn, M., and von Wartburg, A.,** Crystal structure and absolute configuration of 2′-bromopodophyllotoxin, *J. Chem. Soc. Perkin Trans. 2*, 288, 1973.
372. **Gensler, W. J. and Gatsonis, C. D.,** The podophyllotoxin-picropodophyllotoxin equilibrium, *J. Org. Chem.*, 31, 3224, 1966.
373. **Brewer, C. F., Loike, J. D., Horwitz, S. B., Sternlicht, H., and Gensler, W. J.,** Conformational analysis of podophyllotoxin and its congeners. Structure-activity relationship in microtubule assembly, *J. Med. Chem.*, 22, 215, 1979.
374. **Jurd, L., Fye, R. L., and Morgan, J., Jr.,** New types of insect chemosterilants. Benzylphenols and benzyl-1,3-benzodioxole derivatives as additives to housefly diet, *J. Agric. Food Chem.*, 27, 1007, 1979.
375. **Rawlins, S. C., Jurd, L., and Snow, J. W.,** Antifertility effects of benzylphenols and benzyl-1,3-benzodioxoles on screwworm flies, *J. Econ. Entomol.*, 72, 674, 1979.
376. **Jurd, L.,** Mannich base derivatives of bioactive benzyl-1,3-benzodioxo-5-ols, *J. Heterocyclic Chem.*, 22, 993, 1985.
377. **Jurd, L.,** Bioactive heterocyclic analogs of the antitumor drug, podophyllotoxin, *J. Heterocyclic Chem.*, 25, 89, 1988.
378. **Jurd, L., Narayanan, V. L., and Paull, K. D.,** *In vivo* antitumor activity of 6-benzyl-1,3-benzodioxole derivatives against the P388, L1210, B16, and M5076 murine models, *J. Med. Chem.*, 30, 1752, 1987.
379. **Batra, J. K., Jurd, L., and Hamel, E.,** Morpholino derivatives of benzyl-benzodioxole, a study of structural requirements for drug interactions at the colchicine-podophyllotoxin binding site of tubulin, *Biochem. Pharmacol.*, 35, 4013, 1986.
380. **Roach, M. C., Trcka, P. P., Jurd, L., and Luduena, R. F.,** The effects of 6-benzyl-1,3-benzodioxole derivatives on the alkylation of tubulin, *Mol. Pharmacol.*, 32, 432, 1987.
381. **Kupchan, M. S., Britton, R. W., Ziegler, M. F., Gilmore, C. J., Restivo, R. J., and Bryan, R. F.,** Steganacin and steganangin, novel antileukemic lignan lactones from *Steganotaenia araliacea*. *J. Am. Chem. Soc.*, 95, 1335, 1973.
382. **Taafrout, M., Rouessac, F., and Robin, J.-P.,** L'araliangine, nouveau lignane bisbenzocyclooctadienolactonique de *Steganotaenia araliacea*, Hochst., *Tetrahedron Lett.*, 24, 197, 1983.
383. **Robin, J. P., Gringore, O., and Brown, E.,** Asymmetric total synthesis of the antileukemic lignan precursor (−)steganone and revision of its absolute configuration, *Tetrahedron Lett.*, 21, 2709, 1980.
384. **Pascard, C., Robin, J.-P., and Brown, E.,** 8α-Hydroxyisopicrostegane, *Acta Cryst.*, B36, 198, 1980.

385. **Pettit, G. R., Cragg, G. M., Herald, D. L., Schmidt, J. M., and Lohavanijaya, P.**, Isolation and structure of combretastatin, *Can. J. Chem.*, 60, 1374, 1982.
386. **Pettit, G. R., Singh, S. B., and Cragg, G. M.**, Synthesis of natural (−)-combrestastatin, *J. Org. Chem.*, 50, 3404, 1985.
387. **Pettit, G. R., Cragg, G. M., and Singh, S. B.**, Antineoplastic agents, 122. Constituents of *Combretum caffrum*, *J. Nat. Prod.*, 50, 386, 1987.
388. **Pettit, G. R., and Singh, S. B.**, Isolation, structure, and synthesis of combrestastatins A-2, A-3, and B-2, *Can. J. Chem.*, 65, 2390, 1987.
389. **Pettit, G. R., Singh, S. B., Schmidt, J. M., Niven, M. L., Hamel, E., and Lin, C. M.**, Isolation, structure, synthesis and antimitotic properties of combretastatins B-3 and B-4 from *Combretum caffrum*, *J. Natl. Prod.*, 51, 517, 1988.
390. **Pettit, G. R., Singh, S. B., Hamel, E., Lin, C. M., Alberts, D. S., and Garcia-Kendall, D.**, Isolation and structure of the strong cell growth and tubulin inhibitor combretastatin A-4, *Experientia*, 45, 209, 1989.
391. **Hepler, P. K. and Jackson, W. T.**, Isopropyl *N*-phenylcarbamate affects spindle microtubule orientation in dividing endosperm cells of *Haemanthus katherinae* Baker, *J. Cell Sci.*, 5, 727, 1969.
392. **Banerjee, S. and Margulis, L.**, Reversible inhibition of cilia regeneration in *Stentor coeruleus* by isopropyl-*n*-phenyl carbamate, *Nature (London)*, 224, 180, 1969.
393. **Brown, D. L. and Bouck, G. B.**, Microtubule biogenesis and cell shape in *Ochromonas*. III. Effects of the herbicidal mitotic inhibitor isopropyl *N*-phenyl-carbamate on shape and flagellum regeneration, *J. Cell Biol.*, 61, 514, 1974.
394. **Coss, R. A. and Pickett-Heaps, J. D.**, The effects of isopropyl *N*-phenyl carbamate on the green alga *Oedogonium cardiacum*. I. Cell division, *J. Cell Biol.*, 63, 84, 1974.
395. **Starling, D.**, The effects of mitotic inhibitors on the structure of vinblastine-induced tubulin paracrystals from sea-urchin eggs, *J. Cell Sci.*, 20, 91, 1976.
396. **Davidse, L. C.**, Antimitotic activity of methyl benzimidazol-2-yl carbamate (MBC) in *Aspergillus nidulans*, *Pest. Biochem. Physiol.*, 3, 317, 1973.
397. **Davidse, L. C. and Flach, W.**, Differential binding of benzimidazol-2-yl carbamate to fungal tubulin as a mechanism of resistance to this antimitotic agent in mutant strains of *Aspergillus nidulans*, *J. Cell Biol.*, 72, 174, 1977.
398. **Davidse, L. C. and Flach, W.**, Interaction of thiabendazole with fungal tubulin, *Biochim. Biophys. Acta*, 543, 82, 1978.
399. **Delatour, P. and Richard, Y.**, Proprietes embryotoxiques et antimitotiques en serie benzimidazole, *Therapie*, 31, 505, 1976.
400. **Ireland, C. M., Gull, K., Gutteridge, W. E., and Pogson, C. I.**, The interaction of benzimidazole carbamates with mammalian microtubule protein, *Biochem. Pharmacol.*, 28, 2680, 1979.
401. **Lacey, E. and Watson, T. R.**, Structure-activity relationships of benzimidazole carbamates as inhibitors of mammalian tubulin, *in vitro*, *Biochem. Pharmacol.*, 34, 1073, 1985.
402. **Borgers, M., De Nollin, S., Verheyen, A., De Brabander, M., and Thienpont, D.**, Effects of new anthelmintics on the microtubular system of parasites, in *Microtubules and Microtubule Inhibitors: Proceedings of the International Symposium on Microtubules and Microtubule Inhibitors*, Borgers, M. and De Brabander, M., Eds., North-Holland, Amsterdam, 1975, 497.
403. **Holden, H. E., Crider, P. A., and Wahrenburg, M. G.**, Mitotic arrest by benzimidazole analogs in human lymphocyte cultures, *Environ. Mutagen.*, 2, 67, 1980.
404. **Lacey, E. and Watson, T. R.**, Activity of benzimidazole carbamates against L1210 mouse leukemia cells: correlation with *in vitro* tubulin polymerization assay, *Biochem. Pharmacol.*, 34, 3603.
405. **De Brabander, M. J., Van de Veire, R. M. L., Aerts, F. E. M., Borgers, M., and Janssen, P. A. J.**, The effects of methyl [5-(2-thienylcarbonyl)-1H-benzimidazol-2-yl]-carbamate, (R 17934; NSC 238159), a new synthetic antitumoral drug interfering with microtubules, on mammalian cells cultured *in vitro*, *Cancer Res.*, 36, 905, 1976.
406. **Atassi, G. and Tagnon, H. J.**, R 17934-NSC 238159; a new antitumor drug. I. Effect on experimental tumors and factors influencing effectiveness, *Eur. J. Cancer*, 11, 599, 1975.
407. **Zieve, G. W., Turnbull, D., Mullins, J. M., and McIntosh, J. R.**, Production of large numbers of mitotic mammalian cells by use of the reversible microtubule inhibitor nocodazole, *Exp. Cell Res.*, 126, 397, 1980.
408. **Hoebeke, J., Van Nijen, G., and De Brabander, M.**, Interaction of oncodazole (R 17934), a new antitumoral drug, with rat brain tubulin, *Biochem. Biophys. Res. Commun.*, 69, 319, 1976.
409. **Samson, F., Donoso, A. A., Heller-Bettinger, I., Watson, D., and Himes, R. H.**, Nocodazole action of tubulin assembly, axon ultrastructure and fast axoplasmic transport, *J. Pharmacol. Exp. Ther.*, 208, 411, 1979.
410. **Lee, J. C., Field, D. J., and Lee, L. L. Y.**, Effects of nocodazole on structures of calf brain tubulin, *Biochemistry*, 19, 6209, 1980.

411. **Hamel, E. and Lin, C. M.,** Interactions of a new antimitotic agent, NSC-181928, with purified tubulin, *Biochem. Biophys. Res. Commun.*, 104, 929, 1982.
412. **Head, J., Lee, L. L.-Y., Field, D. J., and Lee, J. C.,** Equilibrium and rapid kinetic studies on nocodazole-tubulin interaction, *J. Biol. Chem.*, 260, 11060, 1985.
413. **Ladd, D. L., Harrsch, P. B., and Kruse, L. I.,** Synthesis and tubulin binding of 4′-(1-azi-2,2,2-trifluoroethyl)oncodazole, a photolabile analogue of oncodazole, *J. Org. Chem.*, 53, 417, 1988.
414. **Laclette, J. P., Guerra, G., and Zetina, C.,** Inhibition of tubulin polymerization by mebendazole, *Biochem. Biophys. Res. Commun.*, 92, 417, 1980.
415. **Havercroft, J. C., Quinlan, R. A., and Gull, K.,** Binding of parbendazole to tubulin and its influence on microtubules in tissue-culture cells as revealed by immunofluorescence microscopy, *J. Cell Sci.*, 49, 195, 1981.
416. **Friedman, P. A. and Platzer, E. G.,** Interaction of anthelmintic benzimidazoles with *Ascaris suum* embryonic tubulin, *Biochim. Biophys. Acta*, 630, 271, 1980.
417. **Kohler, P. and Bachmann, R.,** Intestinal tubulin as possible target for the chemotherapeutic action of mebendazole in parasitic nematodes, *Mol. Biochem. Parasit.*, 4, 325, 1981.
418. **Sangster, N. C., Prichard, R. K., and Lacey, E.,** Tubulin and benzimidazole-resistance in *Trichostrongylus colubriformis* (Nematoda), *J. Parasit.*, 71, 645, 1985.
419. **Barrowman, M. M., Marriner, S. E., and Bogan, J. A.,** The binding and subsequent inhibition of tubulin polymerization in *Ascaris suum (in vitro)* by benzimidazole anthelmintics, *Biochem. Pharmacol.*, 33, 3037, 1984.
420. **Lacey, E. and Prichard, R. K.,** Interactions of benzimidazoles (BZ) with tubulin from BZ-sensitive and BZ-resistant isolates of *Haemonchus contortus, Mol. Biochem. Parasit.*, 19, 171, 1986.
421. **Ireland, C. M., Clayton, L., Gutteridge, W. E., Pogson, C. I., and Gull, K.,** Identification and drug binding capabilities of tubulin in the nematode *Ascaridia galli, Mol. Biochem. Parasit.*, 6, 45, 1982.
422. **Wheeler, G. P., Bowdon, B. J., Werling, J. A., and Temple, C., Jr.,** 1-Deaza-7,8-dihydropteridines, a new class of mitotic inhibitors with anticancer activity, *Biochem. Pharmacol.*, 30, 2381, 1981.
423. **Wheeler, G. P., Bowdon, B. J., Werline, J. A., Adamson, D. J., and Temple, C. G., Jr.,** Inhibition of mitosis and anticancer activity against experimental neoplasms by ethyl 5-amino-1,2-dihydro-3-[(*N*-methylanilino)methyl]pyrido[3,4-*b*]pyrazin-7-ylcarbamate (NSC 181928), *Cancer Res.*, 42, 791, 1982.
424. **Temple, C., Jr., Wheeler, G. P., Elliott, R. D., Rose, J. D., Kussner, C. L., Comber, R. N., and Montgomery, J. A.,** New anticancer agents: synthesis of 1,2-dihydropyrido[3,4-*b*]pyrazines (1-deaza-7,8-dihydropteridines), *J. Med. Chem.*, 25, 1045, 1982.
425. **Temple, C., Jr., Wheeler, G. P., Elliott, R. D., Rose, J. D., Comber, R. N., and Montgomery, J. A.,** 1,2-Dihydropyrido[3,4-*b*]pyrazines: structure-activity relationships, *J. Med. Chem.*, 26, 91, 1983.
426. **Wheeler, G. P., Bowdon, B. J., Temple, C., Jr., Adamson, D. J., and Webster, J.,** Biological effects and structure-activity relationships of 1,2-dihydropyrido[3,4-*b*]-pyrazines, *Cancer Res.*, 43, 3567, 1983.
427. **Brassinne, C., Atassi, G., Fruhling, J., Penasse, W., Coune, A., Hildebrand, J., Ruysschaert, J.-M., and Laduron, C.,** Antitumor activity of a water-insoluble compound entrapped in liposomes in L1210 leukemia in mice, *J. Natl. Cancer Inst.*, 70, 1081, 1983.
428. **Hamel, E., Abraham, I., Batra, J. K., Dion, R. L., Duanmu, C., Jurd, L., Lin, C. M., and Powers, L. J.,** New antineoplastic agents which interact with tubulin, in *Cell Membranes and Cancer: Proceedings of the Second International Workshop on Membranes in Tumour Growth,* Galeotti, T., Cittadini, A., Neri, G., Papa, S., and Smets, L. A., Eds., Elsevier, Amsterdam, 1985, 305.
429. **Burgos, J. and Redfearn, E.,** The inhibition of mitochondrial reduced nicotinamide-adenine dinucleotide oxidation by rotenoids, *Biochim. Biophys. Acta*, 110, 475, 1965.
430. **Meisner, H. M. and Sorenson, L.,** Metaphase arrest of Chinese hamster cells with rotenone, *Exp. Cell Res.*, 42, 291, 1966.
431. **Brinkley, B. R., Barham, S. S., Barranco, S. C., and Fuller, G. M.,** Rotenone inhibition of spindle microtubule assembly in mammalian cells, *Exp. Cell Res.*, 85, 41, 1974.
432. **Barham, S. S. and Brinkley, B. R.,** Action of rotenone and related respiratory inhibitors on mammalian cell division. I. Cell kinetics and biochemical aspects, *Cytobios*, 15, 85, 1976.
433. **Marshall, L. E. and Himes, R. H.,** Rotenone inhibition of tubulin self-assembly, *Biochim. Biophys. Acta*, 543, 590, 1978.
434. **Yuki, H., Kitanaka, E., Yamao, A., Kariya, K., and Hashimoto, Y.,** Antitumor activity of 5-substituted 3-(1′-anilinoethylidene)pyrrolidine-2,4-dione, *Gann*, 62, 199, 1971.
435. **Hashimoto, Y., Oshima, H., and Yuki, H.,** Metaphase-arresting action of carcinostatic tenuazonic acid, *Gann*, 63, 79, 1972.
436. **Arai, T.,** Inhibition of microtubule assembly in vitro by TN-16, a synthetic antitumor drug, *FEBS Lett.*, 155, 273, 1983.
437. **Roach, M. C. and Luduena, R. F.,** The effect of TN-16 on the alkylation of tubulin, *Biochem. Biophys. Res. Commun.*, 129, 200, 1985.

438. **McGown, A. T., Ewen, C., Smith, D. B., and Fox, B. W.,** Pre-clinical studies of a novel anti-mitotic agent, amphethinile, *Br. J. Cancer*, 57, 157, 1988.
439. **Lane, M.,** Chemotherapy of cancer, in *Cancer: Diagnosis, Treatment, and Prognosis*, del Regato, J. A., Spjut, H. J., and Cox, J. D., Eds., C.V. Mosby Company, St. Louis, 1985, 93.
440. **Levine, A. S., Ed.,** *Cancer in the Young*, Masson Publishing, New York, 1982.
441. **Palmer, C. G., Livengood, D., Warren, A. K., Simpson, P. J., and Johnson, I. S.,** The action of vincaleukoblastine on mitosis, *in vitro*, *Exp. Cell Res.*, 20, 198, 1960.
442. **Cutts, J. H.,** The effect of vincaleukoblastine on dividing cells *in vivo*, *Cancer Res.*, 21, 168, 1961.
443. **Cardinali, G., Cardinali, G., and Enein, M. A.,** Studies on the antimitotic activity of leurocristine (vincristine), *Blood*, 21, 102, 1963.
444. **George, P., Journey, L. J., and Goldstein, M. N.,** Effect of vincristine on the fine structure of HeLa cells during mitosis, *J. Natl. Cancer Inst.*, 35, 355, 1965.
445. **Krishan, A.,** Time-lapse and ultrastructure studies on the reversal of mitotic arrest induced by vinblastine sulfate in Earle's cells, *J. Natl. Cancer Inst.*, 41, 581, 1969.
446. **Tucker, R. W., Owellen, R. J., and Harris, S. B.,** Correlation of cytotoxicity and mitotic spindle dissolution by vinblastine in mammalian cells, *Cancer Res.*, 37, 4336, 1977.
447. **Sweeney, M. J., Boder, G. B., Cullinan, G. J., Culp, H. W., Daniels, W. D., Dyke, R. W., Gerzon, K., McMahon, R. E., Nelson, R. L., Poore, G. A., and Todd, G. C.,** Antitumor activity of deacetyl vinblastine amide sulfate (vindesine) in rodents and mitotic accumulation studies in culture, *Cancer Res.*, 38, 2886, 1978.
448. **Ootsu, K., Kozai, Y., Takeuchi, M., Ikeyama, S., Igarashi, K., Tsukamoto, K., Sugino, Y., Tashiro, T., Tsukagoshi, S., and Sakurai, Y.,** Effects of new antimitotic antibiotics, ansamitocins, on the growth of murine tumors in vivo and on the assembly of microtubules *in vitro*, *Cancer Res.*, 40, 1707, 1980.
449. **Abraham, I., Dion, R. L., Duanmu, C., Gottesman, M. M., and Hamel, E.,** 2,4-Dichlorobenzyl thiocyanate, an antimitotic agent that alters microtubule morphology, *Proc. Natl. Acad. Sci. U.S.A.*, 83, 6839, 1986.
450. **White, J. G.,** Effects of colchicine and Vinca alkaloids on human platelets. I. Influence on platelet microtubules and contractile function, *Am. J. Pathol.*, 53, 281, 1968.
451. **White, J. G.,** Effects of colchicine and Vinca alkaloids on human platelets. II. Changes in the dense tubular system and formation of an unusual inclusion in incubated cells, *Am. J. Pathol.*, 53, 447, 1968.
452. **Schochet, S. S., Jr., Lampert, P. W., and Earle, K. M.,** Neuronal changes induced by intrathecal vincristine sulfate, *J. Neuropathol. Exp. Neurol.*, 27, 645, 1968.
453. **Bensch, K. G. and Malawista, S. E.,** Microtubule crystals: a new biophysical phenomenon induced by *Vinca* alkaloids, *Nature (London)*, 218, 1176, 1968.
454. **Bensch, K. G. and Malawista, S. E.,** Microtubular crystals in mammalian cells, *J. Cell Biol.*, 40, 95, 1969.
455. **Krishan, A. and Hsu, D.,** Observations on the association of helical polyribosomes and filaments with vincristine-induced crystals in Earle's L-cell fibroblasts, *J. Cell Biol.*, 43, 553, 1969.
456. **Nagayama, A. and Dales, S.,** Rapid purification and the immunological specificity of mammalian microtubular paracrystals possessing an ATPase activity, *Proc. Natl. Acad. Sci. U.S.A.*, 66, 464, 1970.
457. **Bryan, J.,** Vinblastine and microtubules. I. Induction and isolation of crystals from sea urchin oocytes, *Exp. Cell Res.*, 66, 129, 1971.
458. **Stebbings, H.,** Influence of vinblastine sulphate on the deployment of microtubules and ribosomes in telotrophic ovarioles, *J. Cell Sci.*, 8, 111, 1971.
459. **Strahs, K. R. and Sato, H.,** Potentiation of vinblastine crystal formation *in vivo* by puromycin and colcemid, *Exp. Cell Res.*, 80, 10, 1973.
460. **Amos, L. A., Jubb, J. S., Henderson, R., and Vigers, G.,** Arrangement of protofilaments in two forms of tubulin crystal induced by vinblastine, *J. Mol. Biol.*, 178, 711, 1984.
461. **Krishan, A. and Hsu, D.,** Binding of colchicine-^{3}H to vinblastine- and vincristine- induced crystals in mammalian tissue culture cells, *J. Cell Biol.*, 48, 407, 1971.
462. **Weber, K., Bibring, T., and Osborn, M.,** Specific visualization of tubulin-containing structures in tissue culture cells by immunofluorescence: cytoplasmic microtubules, vinblastine-induced paracrystals, and mitotic figures, *Exp. Cell Res.*, 95, 111, 1975.
463. **Bryan, J.,** Vinblastine and microtubules. II. Characterization of two protein subunits from the isolated crystals, *J. Mol. Biol.*, 66, 157, 1972.
464. **Na, G. C. and Timasheff, S. N.,** *In vitro* vinblastine-induced tubulin paracrystals, *J. Biol. Chem.*, 257, 10387, 1982.
465. **Wilson, L., Morse, A. N. C., and Bryan, J.,** Characterization of acetyl-^{3}H-labeled vinblastine binding to vinblastine-tubulin crystals, *J. Mol. Biol.*, 121, 255, 1978.
466. **Na, G. C. and Timasheff, S. N.,** Interaction of vinblastine with calf brain tubulin: effects of magnesium ions, *Biochemistry*, 25, 6222, 1986.

467. **Creasy, W. A. and Chou, T. C.,** The binding of colchicine by sarcoma 180 cells, *Biochem. Pharmacol.*, 17, 477, 1968.
468. **Tan, P. T. and Lagnado, J. R.,** Effects of indole alkaloids and related compounds on the properties of brain microtubular protein, *Biochem. Soc. Trans.*, 3, 121, 1975.
469. **Mandelbaum-Shavit, F., Wolpert-DeFilippes, M. K., and Johns, D. G.,** Binding of maytansine to rat brain tubulin, *Biochem. Biophys. Res. Commun.*, 72, 47, 1976.
470. **Batra, J. K., Powers, L. J., Hess, F. D., and Hamel, E.,** Derivatives of 5,6-diphenylpyridazin-3-one: synthetic antimitotic agents which interact with plant and mammalian tubulin at a new drug-binding site, *Cancer Res.*, 46, 1889, 1986.
471. **Safa, A. R., Hamel, E., and Felsted, R. L.,** Photoaffinity labeling of tubulin subunits with a photoactive analogue of vinblastine, *Biochemistry*, 26, 97, 1987.
472. **Owellen, R. J., Donigan, D. W., Hartke, C. A., and Hains, F. O.,** Correlation of biologic data with physicochemical properties among the *Vinca* alkaloids and their congeners, *Biochem. Pharmacol.*, 26, 1213, 1977.
473. **Wilson, L., Creswell, K. M., and Chin, D.,** The mechanism of action of vinblastine. Binding of [acetyl-^{3}H]vinblastine to embryonic chick brain tubulin and tubulin from sea urchin sperm tail outer doublet microtubules, *Biochemistry*, 14, 5586, 1975.
474. **Lee, J. C., Harrison, D., and Timasheff, S. N.,** Interaction of vinblastine with calf brain microtubule protein, *J. Biol. Chem.*, 250, 9276, 1975.
475. **Conrad, R. A., Cullinan, G. J., Gerzon, K., and Poore, G. A.,** Structure-activity relationships of dimeric *Catharanthus* alkaloids. II. Experimental antitumor activities of *N*-substituted deacetylvinblastine amide (vindesine) sulfates, *J. Med. Chem.*, 22, 391, 1979.
476. **Takahashi, M., Iwasaki, S., Kobayashi, H., Okuda, S., Murai, T., and Sato, Y.,** Rhizoxin binding to tubulin at the maytansine-binding site, *Biochim. Biophys. Acta*, 926, 215, 1987.
477. **Prakash, V. and Timasheff, S. N.,** The interaction of vincristine with calf brain tubulin, *J. Biol.Chem.*, 258, 1689, 1983.
478. **Na, G. C. and Timasheff, S. B.,** Interaction of vinblastine with calf brain tubulin: multiple equilibria, *Biochemistry*, 25, 6214, 1986.
479. **Luduena, R. F. and Roach, M. C.,** Contrasting effects of maytansine and vinblastine on the alkylation of tubulin sulfhydryls, *Arch. Biochem. Biophys.*, 210, 498, 1981.
480. **Schibler, M. J. and Cabral, F. R.,** Maytansine-resistant mutants of Chinese hamster ovary cells with an alteration in α-tubulin, *Can. J. Biochem. Cell Biol.*, 63, 503, 1985.
481. **Houghton, J. A., Houghton, P. J., Hazelton, B. J., and Douglass, E. C.,** *In situ* selection of a human rhabdomyosarcoma resistant to vincristine with altered β-tubulins, *Cancer Res.*, 45, 2706, 1985.
482. **Safa, A. R. and Felsted, R. L.,** Specific *Vinca* alkaloid-binding polypeptides identified in calf brain by photoaffinity labeling, *J. Biol. Chem.*, 262, 1261, 1987.
483. **Little, M. and Luduena, R. F.,** Structural differences between brain β1- and β2-tubulins: implications for microtubule assembly and colchicine binding, *EMBO J.*, 4, 51, 1985.
484. **Little, M. and Luduena, R. F.,** Location of two cysteines in brain β_1-tubulin that can be cross-linked after removal of exchangeable GTP, *Biochim. Biophys. Acta*, 912, 28, 1987.
485. **Bowman, L. C., Houghton, J. A., and Houghton, P. J.,** GTP influences the binding of vincristine in human tumor cytosols, *Biochem. Biophys. Res. Commun.*, 135, 695, 1986.
486. **Ventilla, M., Cantor, C. R., and Shelanski, M. L.,** Some features of the vinblastine-induced assembly of porcine tubulin, *Arch. Biochem. Biophys.*, 171, 154, 1975.
487. **Wilson, L., Jordan, M. A., Morse, A., and Margolis, R. L.,** Interaction of vinblastine with steady-state microtubules *in vitro*, *J. Mol. Biol.*, 159, 125, 1982.
488. **Himes, R. H., Kersey, R. N., Heller-Bettinger, I., and Samson, F. E.,** Action of the *Vinca* alkaloids vincristine, vinblastine, and desacetyl vinblastine amide on microtubules *in vitro*, *Cancer Res.*, 36, 3798, 1976.
489. **Sajo, I.,** Vinblastine inhibition of microtubule assembly *in vitro*, *Acta Biochim. Biophys. Acad. Sci. Hung.*, 12, 259, 1977.
490. **Bhattacharyya, B. and Wolff, J.,** Maytansine binding to the vinblastine sites on tubulin, *FEBS Lett.*, 75, 159, 1977.
491. **Donoso, J. A., Haskins, K. M., and Himes, R. H.,** Effect of microtubule-associated proteins on the interaction of vincristine with microtubules and tubulin, *Cancer Res.*, 39, 1604, 1979.
492. **Borman, L. S., Kuehne, M. E., Matson, P. A., Marko, I., and Zebovitz, T. C.,** Single site-modified congeners of vinblastine dissociate its various anti-microtubule actions, *J. Biol. Chem.*, 263, 6945, 1988.
493. **Zavala, F., Guenard, D., and Potier, P.,** Interaction of vinblastine analogues with tubulin, *Experientia*, 34, 1497, 1978.
494. **Potier, P., Guenard, D., and Zavala, F.,** Resultats recents dans le domaine des alcaloides antitumoraux du groupe de la vinblastine. Etudes biochimiques, *C.R. Soc. Biol.*, 173, 414, 1979.

495. **Takahashi, M., Iwasaki, S., Kobayashi, H., and Okuda, S.**, Studies on macrocyclic lactone antibiotics. XI. Anti-mitotic and anti-tubulin activity of new antitumor antibiotics, rhizoxin and its homologues, *J. Antibiotics*, 40, 66, 1987.

496. **Bensch, K. G., Marantz, R., Wisniewski, H., and Shelanski, M.**, Induction of microtubular crystals by *Vinca* alkaloids, *Science*, 165, 495, 1969.

497. **Marantz, R. and Shelanski, M. L.**, Structure of microtubular crystals induced by vinblastine *in vitro*, *J. Cell Biol.*, 44, 234, 1970.

498. **Erickson, H. P.**, Negatively stained vinblastine aggregates, *Ann. N.Y. Acad. Sci.*, 253, 51, 1975.

499. **Haskins, K. M., Donoso, J. A., and Himes, R. H.**, Spirals and paracrystals induced by *Vinca* alkaloids: evidence that microtubule-associated proteins act as polycations, *J. Cell Sci.*, 47, 237, 1981.

500. **Luduena, R. F., Fellous, A., Francon, J., Nunez, J., and McManus, L.**, Effect of tau on the vinblastine-induced aggregation of tubulin, *J. Cell Biol.*, 89, 680, 1981.

501. **Luduena, R. F., Fellous, A., McManus, L., Jordan, M. A., and Nunez, J.**, Contrasting role of tau and microtubule-associated protein 2 in the vinblastine-induced aggregation of brain tubulin, *J. Biol. Chem.*, 259, 12890, 1984.

502. **Luduena, R. F., Roach, M. C., Jordan, M. A., and Murphy, D. B.**, Different reactivities of brain and erythrocyte tubulins toward a sulfhydryl group-directed reagent that inhibits microtubule assembly, *J. Biol. Chem.*, 260, 1257, 1985.

503. **Jordan, M.A., Himes, R. H., and Wilson, L.**, Comparison of the effects of vinblastine, vincristine, vindesine, and vinepidine on microtubule dynamics and cell proliferation *in vitro*, *Cancer Res.*, 45, 2741, 1985.

504. **Fellous, A., Luduena, R. F., Prasad, V., Jordan, M. A., Anderson, W., Ohayon, R., and Smith, P. T.**, Effects of tau and MAP_2 on the interaction of maytansine with tubulin: inhibitory effect of maytansine on vinblastine-induced aggregation of tubulin, *Cancer Res.*, 45, 5004, 1985.

505. **Jordan, M. A., Margolis, R. L., Himes, R. H., and Wilson, L.**, Identification of a distinct class of vinblastine binding sites on microtubules, *J. Mol. Biol.*, 187, 61, 1986.

506. **Serrano, L., de la Torre, J., Luduena, R. F., and Avila, J.**, The removal of the carboxy-terminal region of tubulin favors its vinblastine-induced aggregation into spiral-like structures, *Arch. Biochem. Biophys.*, 249, 611, 1986.

507. **Mackinlay, S. A., Luduena, R. F., and MacRae, T. H.**, Vinblastine-induced aggregation of brine shrimp *(Artemia)* tubulin, *Biochim. Biophys. Acta*, 882, 419, 1986.

508. **Mackinlay, S. A., Luduena, R. F., and MacRae, T. H.**, Temperature effects on solutions of vinblastine-induced polymers assembled from brine shrimp, *FEBS Lett.*, 203, 301, 1986.

509. **Borman, L. S. and Kuehne, M. E.**, Specific alterations in the biological activities of C-20′-modified vinblastine congeners, *Biochem. Pharmacol.*, 38, 715, 1989.

510. **Behnke, O. and Forer, A.**, Vinblastine as a cause of direct transformation of some microtubules into helical structures, *Exp. Cell Res.*, 73, 506, 1972.

511. **Na, G. C. and Timasheff, S. N.**, Stoichiometry of the vinblastine-induced self-association of calf brain tubulin, *Biochemistry*, 19, 1347, 1980.

512. **Na, G. C. and Timasheff, S. N.**, Thermodynamic linkage between tubulin self-association and the binding of vinblastine, *Biochemistry*, 19, 1355, 1980.

513. **Moncrief, J. W. and Lipscomb, W. N.**, Structure of leurocristine methiodide dihydrate by anomalous scattering methods; relation to leurocristine (vincristine) and vincaleukoblastine (vinblastine), *Acta Cryst.*, 21, 322, 1966.

514. **Wilson, L., Bamburg, J. R., Mizel, S. B., Grisham, L. M., and Creswell, K. M.**, Interaction of drugs with microtubule proteins, *Fed. Proc., Fed. Am. Soc. Exp. Biol.*, 33, 158, 1974.

515. **Kupchan, S. M., Komodo, Y., Court, W. A., Thomas, G. J., Smith, R. M., Karim, A., Gilmore, C. J., Haltiwanger, R. C., and Bryan, R. F.**, Maytansine, a novel antileukemic ansa macrolide from *Maytenus ovatus*, *J. Am. Chem. Soc.*, 94, 1354, 1972.

516. **Kupchan, S. M., Komodo, Y., Branfman, A. R., Dailey, R. G., Jr., and Zimmerly, V. A.**, Novel maytansinoids. Structural interrelations and requirements for antileukemic activity, *J. Am. Chem. Soc.*, 96, 3706, 1974.

517. **Kupchan, S. M., Branfman, A. R., Sneden, A. T., Verma, A. K., Dailey, R. G., Jr., Komoda, Y., and Nagao, Y.**, Novel maytansinoids. Naturally occurring and synthetic antileukemic esters of maytansinol, *J. Am. Chem. Soc.*, 97, 5294, 1975.

518. **Kupchan, S. M., Komodo, Y., Branfman, A. R., Sneden, A. T., Court, W. A., Thomas, G. J., Hintz, H. P. J., Smith, R. M., Karim, A., Howie, G. A., Verma, A. K., Nagao, Y., Dailey, R. G., Jr., Zimmerly, V. A., and Sumner, W. C., Jr.**, The maytansinoids. Isolation, structural elucidation, and chemical interrelation of novel ansa macrolides, *J. Org. Chem.*, 42, 2349, 1977.

519. **Kupchan, S. M., Sneden, A. T., Branfman, A. R., Howie, G. A., Rebhun, L. I., McIvor, W. E., Wang, R. W., and Schnaitman, T. C.**, Structural requirements for antileukemic activity among the naturally occurring and semisynthetic maytansinoids, *J. Med. Chem.*, 21, 31, 1978.

520. **Wani, M. C., Taylor, H. L., and Wall, M. E.,** Plant antitumour agents: colubrinol acetate and colubrinol, antileukaemic ansa macrolides from *Coulubrina texensis, J. Chem. Soc. Chem. Commun.*, 390, 1973.
521. **Asai, M., Mizuta, E., Izawa, M., Haibara, K., and Kishi, T.,** Isolation, chemical characterization and structure of ansamitocin, a new antitumor ansamycin antibiotic, *Tetrahedron*, 35, 1079, 1979.
522. **Higashide, E., Asai, M., Ootsu, K., Tanida, S., Kozai, Y., Hasegawa, T., Kishi, T., Sugino, Y., and Yoneda, M.,** Ansamitocin, a group of novel maytansinoid antibiotics with antitumour properties from *Nocardia, Nature (London)*, 270, 1977.
523. **Wolpert-DeFilippes, M. K., Adamson, R. H., Cysyk, R. L., and Johns, D. G.,** Initial studies on the cytotoxic action of maytansine, a novel ansa macrolide, *Biochem. Pharmacol.*, 24, 751, 1975.
524. **Remillard, S., Rebhun, L. I., Howie, G. A., and Kupchan, S. M.,** Antimitotic activity of the potent tumor inhibitor maytansine, *Science*, 189, 1002, 1975.
525. **Wolpert-DeFilippes, M. K., Bono, V. H., Jr., Dion, R. L., and Johns, D. G.,** Initial studies on maytansine-induced metaphase arrest in L1210 murine leukemia cells, *Biochem. Pharmacol.*, 24, 1735, 1975.
526. **York, J., Wolpert-DeFilippes, M. K., Johns, D. G., and Sethi, V. S.,** Binding of maytansinoids to tubulin, *Biochem. Pharmacol.*, 30, 3239, 1981.
527. **Bryan, R. F., Gilmore, C. J., and Haltiwanger, R. C.,** Crystal and molecular structure and absolute configuration of maytansine (3-bromopropyl) ether, *J. Chem. Soc. Perkin Trans. 2*, 897, 1973.
528. **Van Warmelo, K. T., Marasas, W. F. O., Adelaar, T. F., Kellerman, T. S., Van Rensbrug, I. B. J., and Minne, J. A.,** Experimental evidence that lupinosis of sheep is a mycotoxicosis caused by the fungus *Phomopsis leptostromiformis, J. S. Afr. Med. Assoc.*, 41, 235, 1970.
529. **Gardiner, M. R. and Petterson, D. S.,** Pathogenesis of mouse lupinosis induced by a fungus growing on dead lupins, *J. Comp. Pathol.*, 82, 5, 1972.
530. **Culvenor, C. C. J., Beck, A. B., Clarke, A., Cockrum, P. A., Edgar, J. A., Frahn, J. L., Jago, M. V., Lanigan, G. W., Payne, A. L., Peterson, J. E., Petterson, D. S., Smith, L. W., and White, R. R.,** Isolation of toxic metabolites of *Phomopsis leptostromiformis* responsible for lupinosis, *Aust. J. Biol. Sci.*, 30, 269, 1977.
531. **Culvenor, C. C. J., Cockrum, P. A., Edgar, J. A., Frahn, J. L., Gorst-Allman, C. P., Jones, A. J., Marasas, W. F. O., Murray, K. E., Smith, L. W., Steyn, P. S., Vleggaar, R., and Wessels, P. L.,** Structure elucidation of phomopsin A, a novel cyclic hexapeptide mycotoxin produced by *Phomopsis leptostromiformis, J. Chem. Soc. Chem. Commun.*, 1259, 1983.
532. **Mackay, M. F., Van Donkelaar, A., and Culvenor, C. J.,** The X-ray structure of phomopsin A, a hexapeptide mycotoxin, *J. Chem. Soc. Chem. Commun.*, 1219, 1986.
533. **Petterson, D. S., Howlett, R. M., Robertson, T. A., and Papadimitriou, J. M.,** Alteration in cell division, morphology and motility induced by the toxic principle of lupinosis, *Austr. J. Exp. Biol. Med. Sci.*, 57, 211, 1979.
534. **Tonsing, E. M., Steyn, P. S., Osborn, M., and Weber, K.,** Phomopsin A, the causative agent of lupinosis, interacts with microtubules *in vivo* and *in vitro, Eur. J. Cell Biol.*, 35, 156, 1984.
535. **Ibaragi, T.,** Studies on rice seedling blight. I. Growth injury caused by *Rhizopus* sp. under high temperature, *Ann. Phytopathol. Soc. Jpn.*, 39, 141, 1973.
536. **Iwasaki, S., Kobayashi, H., Furukawa, J., Namikoshi, M., Okuda, S., Sato, Z., Matsuda, I., and Noda, T.,** Studies on macrocyclic lactone antibiotics. VII. Structure of a phytotoxin "rhizoxin" produced by *Rhizopus chinensis, J. Antibiotics*, 37, 354, 1984.
537. **Iwasaki, S., Namikoshi, M., Kobayashi, H., Furukawa, J., Okuda, S., Itai, A., Kasuya, A., Iitaka, Y., and Sato, Z.,** Studies on macrocyclic lactone antibiotics. VIII. Absolute structures of rhizoxin and a related compound, *J. Antibiotics*, 39, 424, 1986.
538. **Iwasaki, S., Namikoshi, M., Kobayashi, H., Furukawa, J., and Okuda, S.,** Studies on macrocyclic lactone antibiotics. IX. Novel macrolides from the fungus *Rhizopus chinensis:* precursors of rhizoxin, *Chem. Pharm. Bull.*, 34, 1387, 1986.
539. **Kobayashi, H., Iwasaki, S., Yamada, E., and Okuda, S.,** Biosynthesis of the antimitotic antitumour antibiotic rhizoxin by *Rhizopus chinensis;* origins of the carbon atoms, *J. Chem. Soc. Chem. Commun.*, 1702, 1986.
540. **Tsuruo, T., Oh-hara, T., Iida, H., Tsukagoshi, S., Sato, Z., Matsuda, I., Iwasaki, S., Okuda, S., Shimizu, F., Sasagawa, K., Fukami, M., Fukuda, K., and Arakawa, M.,** Rhizoxin, a macrocyclic lactone antibiotic, as a new antitumor agent against human and murine tumor cells and their vincristine-resistant sublines, *Cancer Res.*, 46, 381, 1986.
541. **Wani, M. C., Taylor, H. L., Wall, M. E., Coggon, P., and McPhail, A. T.,** Plant antitumor agents. VI. The isolation and structure of taxol, a novel antileukemic and antitumor agent from *Taxus brevifolia, J. Am. Chem. Soc.*, 93, 2325, 1971.
542. **Miller, R. W., Powell, R. G., and Smith, C. R., Jr.,** Antileukemic alkaloids from *Taxus wallichiana* Zucc., *J. Org. Chem.*, 46, 1469, 1981.

543. **Senilh, V., Blechert, S., Colin, M., Guenard, D., Picot, F., Potier, P., and Varenne, P.,** Mise en evidence de nouveaux analogues du taxol extraits de *Taxus baccata*, *J. Nat. Prod.*, 47, 131, 1984.
544. **Fuchs, D. A. and Johnson, R. K.,** Cytologic evidence that taxol, an antineoplastic agent from *Taxus brevifolia*, acts as a mitotic spindle poison, *Cancer Treat. Rep.*, 62, 1219, 1978.
545. **Schiff, P. B. and Horwitz, S. B.,** Taxol stabilizes microtubules in mouse fibroblast cells, *Proc. Natl. Acad. Sci. U.S.A.*, 77, 1561, 1980.
546. **Heidemann, S. R. and Gallas, P. T.,** The effect of taxol on living eggs of *Xenopus laevis*, *Dev. Biol.*, 80, 489, 1980.
547. **Manfredi, J. J. and Horwitz, S. B.,** Taxol: an antimitotic agent with a new mechanism of action, *Pharmacol. Ther.*, 25, 83, 1984.
548. **Schiff, P. B., Fant, J., and Horwitz, S. B.,** Promotion of microtubule assembly *in vitro* by taxol, *Nature (London)*, 277, 665, 1979.
549. **Schiff, P. B. and Horwitz, S. B.,** Taxol assembles tubulin in the absence of exogenous guanosine 5'-triphosphate or microtubule-associated proteins, *Biochemistry*, 20, 3247, 1981.
550. **Thompson, W. C., Wilson, L., and Purich, D. L.,** Taxol induces microtubule assembly at low temperature, *Cell Motil.*, 1, 445, 1981.
551. **Vallee, R. B.,** A taxol-dependent procedure for the isolation of microtubules and microtubule-associated proteins, *J. Cell Biol.*, 92, 435, 1982.
552. **Parness, J., Kingston, D. G. I., Powell, R. G., Harracksingh, C., and Horwitz, S. B.,** Structure-activity study of cytotoxicity and microtubule assembly in vitro by taxol and related taxanes, *Biochem. Biophys. Res. Commun.*, 105, 1082, 1982.
553. **Caplow, M. and Zeeberg, B.,** Dynamic properties of microtubules at steady state in the presence of taxol, *Eur. J. Biochem.*, 127, 319, 1982.
554. **Carlier, M.-F. and Pantaloni, D.,** Taxol effect on tubulin polymerization and associated guanosine 5'-triphosphate hydrolysis, *Biochemistry*, 22, 4814, 1983.
555. **Parness, J., Asnes, C. F., and Horwitz, S. B.,** Taxol binds differentially to flagellar outer doublets and their reassembled microtubules, *Cell Motil.*, 3, 123, 1983.
556. **Lataste, H., Senilh, V., Wright, M., Guenard, D., and Potier, P.,** Relationships between the structures of taxol and baccatine III derivatives and their *in vitro* action on the disassembly of mammalian brain and *Physarum* amoebal microtubules, *Proc. Natl. Acad. Sci. U.S.A.*, 81, 4090, 1984.
557. **Mellado, W., Magri, N. F., Kingston, D. G. I., Garcia-Arenas, R., Orr, G. A., and Horwitz, S. B.,** Preparation and biological activity of taxol acetates, *Biochem. Biophys. Res. Commun.*, 124, 329, 1984.
558. **Wilson, L., Miller, H. P., Farrell, K. W., Snyder, K. B., Thompson, W. C., and Purich, D. L.,** Taxol stabilization of microtubules *in vitro:* dynamics of tubulin addition and loss at opposite microtubule ends, *Biochemistry*, 24, 5254, 1985.
559. **Wallin, M., Nordh, J., and Deinum, J.,** Changes in the hydrodynamic properties of microtubules induced by taxol, *Biochim. Biophys. Acta*, 880, 189, 1986.
560. **Choudhury, G. G., Maity, S., Bhattacharyya, B., and Biswas, B. B.,** B-ring of colchicine and its role in taxol-induced tubulin polymerization, *FEBS Lett.*, 197, 31, 1986.
561. **Ringel, I. and Horwitz, S. B.,** Taxol is converted to 7-epitaxol, a biologically active isomer, in cell culture medium, *J. Pharmacol. Exp. Ther.*, 242, 692, 1987.
562. **Howard, W. D. and Timasheff, S. N.,** Linkages between the effects of taxol, colchicine, and GTP on tubulin polymerization, *J. Biol. Chem.*, 263, 1342, 1988.
563. **Deinum, J., Sorskog, L., Wallin, M., and Dahlback, J.,** Molecular weight dependency of heparin inhibition of microtubule assembly *in vitro*, *Biochim. Biophys. Acta*, 802, 41, 1984.
564. **Albertini, D. F., Herman, B., and Sherline, P.,** *In vivo* and *in vitro* studies on the role of HMW-MAPs in taxol-induced microtubule bundling, *Eur. J. Cell Biol.*, 33, 134, 1984.
565. **Foisner, R. and Wiche, G.,** Promotion of MAP/MAP interaction by taxol, *J. Ultrastruc. Res.*, 93, 33, 1985.
566. **Turner, P. F. and Margolis, R. L.,** Taxol-induced bundling of brain-derived microtubules, *J. Cell Biol.*, 99, 940, 1984.
567. **Parness, J. and Horwitz, S. B.,** Taxol binds to polymerized tubulin *in vitro*, *J. Cell Biol.*, 91, 479, 1981.
568. **Manfredi, J. J., Parness, J., and Horwitz, S. B.,** Taxol binds to cellular microtubules, *J. Cell Biol.*, 94, 688, 1982.
569. **Takoudju, M., Wright, M., Chenu, J., Gueritte-Voegelein, F., and Guenard, D.,** Absence of 7-acetyl taxol binding to unassembled brain tubulin, *FEBS Lett.*, 227, 96, 1988.
570. **Cabral, F., Abraham, I., and Gottesman, M. M.,** Isolation of a taxol-resistant Chinese hamster ovary cell mutant that has an alteration in α-tubulin, *Proc. Natl. Acad. Sci. U.S.A.*, 78, 4388, 1981.
571. **Schibler, M. J. and Cabral, F.,** Taxol-dependent mutants of Chinese hamster ovary cells with alterations in α- and β-tubulin, *J. Cell Biol.*, 102, 1522, 1986.
572. **Hamel, E.,** Antimitotic drugs and tubulin-nucleotide interactions, in *Developments in Cancer Chemotherapy*, Glazer, R. I., Ed., CRC Press, Boca Raton, FL, 1984, 131.

573. **De Brabander, M., Geuens, G., Nuydens, R., Willebrords, R., and De Mey, J.,** Taxol induces the assembly of free microtubules in living cells and blocks the organizing capacity of the centrosomes and kinetochores, *Proc. Natl. Acad. Sci. U.S.A.*, 78, 5608, 1981.
574. **Herman, B., Langevin, M. A., and Albertini, D. F.,** The effects of taxol on the organization of the cytoskeleton in cultured ovarian granulosa cells, *Eur. J. Cell Biol.*, 31, 34, 1983.
575. **Crossin, K. L. and Carney, D. H.,** Evidence that microtubule depolymerization early in the cell cycle is sufficient to initiate DNA synthesis, *Cell*, 23, 61, 1981.
576. **Tokunaka, S., Friedman, T. M., Toyama, Y., Pacifici, M., and Holtzer, H.,** Taxol induces microtubule-rough endoplasmic reticulum complexes and microtubule-bundles in culture chrondroblasts, *Differentiation*, 24, 39, 1983.
577. **Wehland, J., Henkart, M., Klausner, R., and Sandoval, I. V.,** Role of microtubules in the distribution of the Golgi apparatus: effect of taxol and microinjected anti-α-tubulin antibodies, *Proc. Natl. Acad. Sci. U.S.A.*, 80, 4286, 1983.
578. **Forry-Schaudies, S., Murray, J. M., Toyama, Y., and Holtzer, H.,** Effects of colcemid and taxol on microtubules and intermediate filaments in chick embryo fibroblasts, *Cell Motil. Cytoskel.*, 6, 324, 1986.
579. **Manfredi, J. J., Fant, J., and Horwitz, S. B.,** Taxol induces the formation of unusual arrays of cellular microtubules in colchicine-pretreated J774.2 cells, *Eur. J. Cell Biol.*, 42, 126, 1986.
580. **White, J. G. and Rao, G. H. R.,** Effects of a microtubule stabilizing agent on the response of platelets to vincristine, *Blood*, 60, 474, 1982.
581. **White, J. G. and Rao, G. H. R.,** Influence of a microtubule stabilizing agent on platelet structural physiology, *Am. J. Pathol.*, 112, 207, 1983.
582. **Schatten, G., Schatten, H., Bestor, T. H., and Balczon, R.,** Taxol inhibits the nuclear movements during fertilization and induces asters in unfertilized sea urchin eggs, *J. Cell Biol.*, 94, 455, 1982.
583. **Buchman, R., Scozzie, J. A., Ariyan, Z. S., Heilman, R. D., Rippin, D. J., Pyne, W. J., and Powers, L. J.,** Antihypertensive 5,6-diarylpyridazin-3-ones, *J. Med. Chem.*, 23, 1398, 1980.
584. **Fogt, S. W., Scozzie, J. A., Heilman, R. D., and Powers, L. J.,** Lipophilic and hydrophilic esters of 4-acetyl-2-(2-hydroxyethyl)-5,6-bis(4-chlorophenyl)-2*H*-pyridazin-3-one as hypertensive agents, *J. Med. Chem.*, 23, 1445, 1980.
585. **Pyne, W. J., Lowbridge, J., Chang, I. K., Knotz, F., and Powers, L. J.,** European Patent EP46069, 1982.
586. **Mizens, M., Brown, W. R., Laveglia, J., Killeen, J. C., Jr., and Ignatoski, J. A.,** Combined 30-day feeding study and one-generation reproduction study in rats with ethyl 5-cyano-3,4-diphenyl-6-oxo-1[6*H*]pyridazineacetate (DS-29399), *Toxicologist*, 4, 17, 1984.
587. **Mizens, M., Brown, W. R., Killeen, J. C., Jr., and Ignatoski, J. A.,** Subchronic feeding study in rats with five pyridazinone compounds structurally related to DS-29399, *Toxicologist*, 5, 60, 1985.
588. **Hess, F. D. and Powers, L. J.,** unpublished observations.
589. **Ramel, C.,** Genetic effects of organic mercury compounds. I. Cytological investigations on *Allium* roots, *Hereditas (Lund)*, 61, 208, 1969.
590. **Nath, J. and Rebhun, L. I.,** Effects of caffeine and other methylxanthines on the development and metabolism of sea urchin eggs: involvement of $NADP^+$ and glutathione, *J. Cell Biol.*, 68, 440, 1976.
591. **Oliver, J. M., Albertini, D. F., and Berlin, R. D.,** Effects of glutathione-oxidizing agents on microtubule assembly and microtubule-dependent surface properties of human neutrophils, *J. Cell Biol.*, 71, 921, 1976.
592. **Deinum, J., Wallin, M., and Lagercrantz, C.,** Spatial separation of the two essential thiol groups and the binding site of the exchangeable GTP in brain tubulin, *Biochim. Biophys. Acta*, 671, 1, 1981.
593. **Lee, Y. C., Yaple, R. A., Baldridge, R., Kirsch, M., and Himes, R. H.,** Inhibition of tubulin self-assembly *in vitro* by fluorodinitrobenzene, *Biochim. Biophys. Acta*, 671, 71, 1981.
594. **Peyrot, V., Briand, C., Momburg, R., and Sari, J. C.,** *In vitro* mechanism study of microtubule assembly inhibition by *cis*-dichlorodiammine-platinum (II), *Biochem. Pharmacol.*, 35, 371, 1986.
595. **Potchoo, Y., Braguer, D., Peyrot, V., Chauvet-Monges, A. M., Sari, J. C., and Crevat, A.,** *In vitro* inhibition of microtubule assembly by disulfiram, *Int. J. Clin. Pharmacol. Ther. Tox.*, 24, 499, 1986.
596. **Faulstich, H., Stournaras, C., Doenges, K. H., and Zimmermann, H.-P.,** The molecular mechanism of interaction of Et_3Pb^+ with tubulin, *FEBS Lett.*, 174, 128, 1984.
597. **Gabriel, L., Miglietta, A., and Dianzani, M. U.,** 4-Hydroxy-alkenals interaction with purified microtubular protein, *Chem.-Biol. Interactions*, 56, 201, 1985.
598. **Luduena, R. F., Roach, M. C., Trcka, P. P., Mallevais, M. L., and MacRae, T.,** The effect of 2-(4-methyl-1-piperazinylmethyl) acrylophenone dihydrochloride on the alkylation of tubulin, *Arch. Biochem. Biophys.*, 255, 453, 1987.
599. **Umeda, M., Saito, K., Hirose, K., and Saito, M.,** Cytotoxic effect of inorganic, phenyl, and alkyl mercuric compounds on HeLa cells, *Jpn. J. Exp. Med.*, 39, 47, 1969.
600. **Abe, T., Haga, T., and Kurokawa, M.,** Blockage of axoplasmic transport and depolymerization of reassembled microtubules by methyl mercury, *Brain Res.*, 86, 504, 1975.

601. **Miura, K., Suzuki, K., and Imura, N.,** Effects of methylmercury on mitotic mouse glioma cells, *Environ. Res.*, 17, 453, 1978.
602. **Imura, N., Miura, K., Inokawa, M., and Nakada, S.,** Mechanism of methylmercury cytotoxicity: by biochemial and morphological experiments using cultured cells, *Toxicology*, 17, 241, 1980.
603. **Sager, P. R., Doherty, R. A., and Olmsted, J. B.,** Interaction of methylmercury with microtubules in cultured cells and *in vitro*, *Exp. Cell Res.*, 146, 127, 1983.
604. **Bai, R., Duanmu, C., and Hamel, E.,** Mechanism of action of the antimitotic drug 2,4-dichlorobenzyl thiocyanate: alkylation of sulfhydryl groups of β-tubulin, *Biochim. Biophys. Acta*, 5, 994, 12, 1989.
605. **Paget, G. E. and Walpole, A. L.,** Some cytological effects of griseofulvin, *Nature (London)*, 182, 1320, 1958.
606. **Deysson, G.,** Sur les proprietes antimitotiques de la griseofulvine, *Ann. Pharm. Franc.*, 22, 2, 1964.
607. **Grisham, L. M., Wilson, L., and Bensch, K. G.,** Antimitotic action of griseofulvin does not involve disruption of microtubules, *Nature (London)*, 244, 294, 1973.
608. **Weber, K., Wehland, J., and Herzog, W.,** Griseofulvin interacts with microtubules both *in vivo* and *in vitro*, *J. Mol. Biol.*, 102, 817, 1976.
609. **Gull, K. and Trinci, A. P. J.,** Ultrastructural effects of griseofulvin on the myxomycete *Physarum polycephalum:* inhibition of mitosis and the production of microtubule crystals, *Protoplasma*, 81, 37, 1974.
610. **Malawista, S. E., Sato, H., and Bensch, K. G.,** Vinblastine and griseofulvin reversibly disrupt the living mitotic spindle, *Science*, 160, 770, 1968.
611. **Wehland, J., Herzog, W., and Weber, K.,** Interaction of griseofulvin with microtubules, microtubule protein and tubulin, *J. Mol. Biol.*, 111, 329, 1977.
612. **Sloboda, R. D., Van Blaricom, G., Creasey, W. A., Rosenbaum, J. L., and Malawista, S. E.,** Griseofulvin: association with tubulin and inhibition of *in vitro* microtubule assembly, *Biochem. Biophys. Res. Commun.*, 105, 882, 1982.
613. **Keates, R. A. B.,** Griseofulvin at low concentration inhibits the rate of microtubule polymerization *in vitro*, *Biochem. Biophys. Res. Commun.*, 102, 746, 1981.
614. **Roobol, A., Gull, K., and Pogson, C. I.,** Evidence that griseofulvin binds to a microtubule-associated protein, *FEBS Lett.*, 75, 149, 1977.
615. **Perez, C. A., Fair, W. R., Ihde, D. C., and Labrie, F.,** Cancer of the prostate, in *Cancer: Principles and Practice of Oncology*, DeVita, V. T., Jr., Hellman, S., and Rosenberg, S. A., Eds., J. B. Lippincott, Philadelphia, 1985, 929.
616. **Hartley-Asp, B.,** Estramustine-induced mitotic arrest in two human prostatic carcinoma cell lines Du145 and PC-3, *Prostate*, 5, 93, 1984.
617. **Tew, K. D. and Hartley-Asp, B.,** Cytotoxic properties of estramustine are unrelated to its alkylating and steroid constituents, *Urology*, 23, 28, 1984.
618. **Tew, K. D., Woodworth, A., and Stearns, M. E.,** Relationship of glutathione depletion and inhibition of glutathione-*S*-transferase activity to the antimitotic properties of estramustine, *Cancer Treat. Rep.*, 70, 715, 1986.
619. **Stearns, M. E. and Tew, K. D.,** Antimicrotubule effects of estramustine, an antiprostatic tumor drug, *Cancer Res.*, 45, 3891, 1985.
620. **Kanje, M., Deinum, J., Wallin, M., Ekstrom, P., Edstrom, A., and Hartley-Asp, B.,** Effect of estramustine phosphate on the assembly of isolated bovine brain microtubules and fast axonal transport in the frog sciatic nerve, *Cancer Res.*, 45, 2234, 1985.
621. **Wallin, M., Deinum, J., and Friden, B.,** Interaction of estramustine phosphate with microtubule-associated proteins, *FEBS Lett.*, 179, 289, 1985.
622. **Friden, B., Wallin, M., Deinum, J., Prasad, V., and Luduena, R.,** Effect of estramustine phosphate on the assembly of trypsin-treated microtubules and microtubules reconstituted from purified tubulin with either tau, MAP2, or the tubulin-binding fragment of MAP2, *Arch. Biochem. Biophys.*, 257, 123, 1987.

Chapter 5

MICROTUBULE PROTEINS IN NEURONAL CELLS

Javier Díaz-Nido, María A. Hernández, and Jesús Avila

TABLE OF CONTENTS

I. INTRODUCTION

The complex morphological and functional wiring of the nervous system is built of billions of nerve cells, each of which projects numerous long processes or neurites (axons and dendrites) that intertwine and interconnect with remarkable specificity. Schematically, a neuron can be divided into three domains: the dendrites, the cell body (also called soma or perikaryon), and the axon. Dendrites, axons, and cell bodies establish specialized functional contacts (synapses) with one another, determining the pattern of signal transmission in the brain. In this way, as pointed out by Brady and Lasek,[1] the highly asymmetrical shape of neuronal cells is as essential to the functioning of the brain as are the electrical properties of these cells.

The information which determines the basic form of a neuron seems to be partly heritable, since the geometries of equivalent neurons from isogenic animals are extremely similar.[2] Moreover, neurons which are dissociated and cultured *in vitro* tend to retain their original form. For example, sensory ganglion cells become largely unipolar,[3] whereas pyramidal cells from hippocampus elaborate their typical processes.[4-6] Thus, the morphology of a nerve cell appears to arise from endogenous determinants,[7] although its expression is controlled by extracellular cues such as intercellular contacts, extracellular matrix composition, and diffusible signals.[8-10]

At the beginning of the 20th century, neuroanatomists established that an endogenous determinant involved in the generation and maintenance of neuronal morphology was a "neurofibril network" which arose in the cell body and extended into the axon and dendrites.[11-14]

Electron microscopy later showed that "neurofibrils" correspond to bundles of cytoskeletal fibers similar to those found in all eukaryotic cells, mainly neurofilaments (the neuronal form of intermediate filaments) and neurotubules (or microtubules).[15-18]

The functional importance of neurofilaments remains unclear,[19,20] although some evidence favors the view that neurofilaments determine the diameter of large myelinated nerve fibers and, consequently, the speed of nervous impulse conduction.[21-23] Microtubules, on the other hand, are necessary for the extension and maintenance of neuronal processes and serve as tracks for bidirectional transport of various organelles between cell bodies and process tips.

Here we focus on the contribution of microtubules to neuronal cytoskeleton structure, with particular emphasis on the heterogeneity of microtubule proteins and the functional roles of microtubules in neuronal morphogenesis and axonal transport.

II. MICROTUBULES AND NEURONAL CYTOSKELETON STRUCTURE

A. MICROTUBULES IN THE AXONAL CYTOSKELETON

Electron microscopy studies have provided the first real images of the cytoskeletal structure within the different domains of a nerve cell. Most studies have been focused on the axonal cytoskeleton in comparison with the cell body of dendritic cytoskeleton.[18]

Axons contain microtubules, neurofilaments, and actin microfilaments. Whereas microfilaments are mainly found underlying the axonal plasma membrane (axolemma), microtubules and neurofilaments parallel the long axis of the axon and appear to be highly interconnected.[24-27]

Although microtubules are essential elements in all axons, neurofilaments are the predominant cytoskeletal polymers in the largest axons, such as most myelinated axons in vertebrates.[18] Another extreme example is the giant axon of *Myxicola* which contains a large number of neurofilaments and only a few microtubules.[28] In contrast, there are many axons which contain only microtubules, such as the parallel fibers of the cerebellum and the unmyelinated axons of sensory ganglion cells in vertebrates.[21] Likewise, in the crayfish giant axon and all arthropod axons, microtubules are the sole cytoskeletal elements.[21]

There are five features of axonal microtubules which deserve specific attention, namely, their length, definite polarity, assembly, three-dimensional organization within the axon, and resistance to depolymerization.

The length of axonal microtubules has been studied by careful analyses of serial sections of different axons. In this way, it has been shown that most, if not all, microtubules extend less than the entire length of the axon. Their average length was calculated to be about 0.37 to 0.76 mm for mouse saphenous nerve,[29] about 0.1 mm for cultured rat dorsal root ganglion cell axons,[30] and about 0.01 mm for unmyelinated axons of the nematode *Caenorhabditis elegans*.[31] Computer-assisted three-dimensional electron microscopic reconstruction of images obtained from axon-like processes in cultured PC12 cells have also shown that microtubules are discontinuous.[32]

The fact that axonal microtubules are discontinuous implies that they do not grow from a single microtubule-organizing center (MTOC) analogous to that found in nonneural cells. Rather, it appears that axonal microtubules contain sufficient local structural information to allow microtubule reassembly upon rewarming after a previous cold-induced microtubule depolymerization.[33-36] It has been hypothesized that such information could be contained in stable microtubule fragments, as will be discussed below.[35-37]

Interestingly, on rewarming previously cooled axons, microtubules reassemble with the same uniform polar orientation as exhibited before manipulation. Microtubules are intrinsically polar structures and all axonal microtubules have their fast-growing or plus ends pointing toward the axon terminal.[38-41]

Microtubule polarity has been experimentally manipulated in cultures of embryonic chicken sensory neurons.[42] Microtubules within neurites of cultured neurons have the same polarity as described for microtubules in the axons of mature neurons. After neurite amputation, the fragment separated from the cell body collapses to an axoplasm bead which rapidly regrows a neurite. In these neurites, the plus ends of microtubules face the newly formed growth cones and the minus ends are located at the original distal terminal of the neurite. Thus, microtubule reorganization seems to occur concomitantly with neurite regrowth to form a polar microtubule array with reversed polarity. This suggests that microtubule polarity may be the result of local interactions within the extending process, rather than being determined by a MTOC. In this context, it has been proposed that the plus ends of cold-stable microtubule fragments must be recruited or capped near the cell periphery to allow neurite outgrowth. These oriented microtubule fragments would then act as nucleation centers (seeds) generating microtubule polarity.[42]

The absence of a single MTOC also poses a question on the spatial organization of microtubule assembly within axons. In nonneuronal cells, microtubules are nucleated by the centrosome, and polymerization/depolymerization occurs through incorporation/release of tubulin dimers at the growing/shrinking free plus microtubule ends. In neurons, tubulin dimers are generated in the cell body, where protein biosynthesis takes place, but it is not known where they are assembled into axonal microtubules.

One hypothesis holds that axonal microtubules are assembled and become cross-linked with neurofilaments in the neuron cell body, in close proximity to their synthesis sites. The microtubule-neurofilament complex would then be conveyed along the axon, giving rise to the slow component of axonal transport[43-45] (see Section V.B).

Pulse-chase experiments which determine the time interval between the synthesis and assembly of tubulin in cultured sympathetic neurons support this hypothesis.[46] Thus, (35S)methionine-labeled tubulin partitions with the soluble fraction in cultures labeled for 5 to 10 min, whereas it is mainly associated with the cytoskeletal fraction when pulse-labeled cultures are chased for 60 to 120 min. This is a relatively short time interval when compared with the rate of slow axonal transport, suggesting that tubulin assembly occurs in relatively close proximity to its site of synthesis. Interestingly, similar results are found for neurofilament polypeptides.[46] It therefore appears that newly synthesized tubulin is polymerized in the cell body, and the assembled microtubule become tightly complexed with neurofilaments prior to their transport into the axon.

However, some evidence contradicts this view and sustains that tubulin is assembled distally at the tip of growing axons. This emerges from a study by Bamburg et al.,[47] in which microtubule-depolymerizing drugs were topically applied, using micropipettes, to different regions of cultured dorsal root ganglion cells. The sensitivity of these drugs is greatest at the tip of the growing axon by at least two orders of magnitude, suggesting that the axon tip is the major site of tubulin assembly. Consequently, it is proposed that soluble tubulin, synthesized in the neuron cell body, is transported to the end of the axon before assembling into microtubules (see also Section V.B).

Yet it remains unknown whether this mechanism is specific for growing neurites or if it also occurs in established mature processes. An alternative explanation has been proposed, emphasizing that the sensitivity of growing tips to microtubule-depolymerizing drugs might not necessarily result from being major sites of microtubule assembly, but rather from being sites at which microtubules become destabilized. Thus, tubulin could be assembled in the neuron cell body, originating colcemid-resistant microtubules which would be transported through the axon and then selectively disassembled near the growth cone to provide subunits for new assembly. This view is supported by several facts which indicate that microtubule disassembly is favored within growth cones (see Section IV.B.1).

Several studies on microtubule-associated protein (MAP)-2 elimination from growing axon-like neurites also yield contradictory results concerning the site of microtubule assembly.[48,49] (see Section IV.E).

With regard to microtubule organization within axons, the developing of the quick-freeze, freeze-fracture, deep-etch technique has allowed the observation of the three-dimensional architecture of the unfixed axonal cytoskeleton.[50,51] The axonal cytoskeleton has two regions: a thin cortical region which corresponds to the subaxolemmal cytoskeleton and the inner cytoskeleton which occupies most of the axon.

The subaxolemmal cytoskeleton has been extensively studied in the squid giant axon. It consists of a microfilament and a microtubule-associated domain.[52] Actin microfilaments are associated with the axolemma through a specialized mesh, probably composed of a spectrin-like protein such as fodrin. Microtubules run parallel to the axolemma and are embedded in a fine three-dimensional mesh of thin strands made up of a 260-kDa MAP, axolinin.[53]

This microtubule-axolinin meshwork is essential for the maintenance of membrane excitability.[54] The submembranous microtubule cytoskeleton may provide an important support to the axolemma against the mechanical stress generated by the high intracellular pressure which affects the activity of ion channels.[55] In this context, the morphological resemblance of the microtubule-axolinin meshwork of the squid giant axon to the axoplasm of the nodal part of the mammalian myelinated axons must be emphasized.[29] In these axons, ion channels are enriched in nodal regions. Thus, it appears that the microtubule-axolinin meshwork, or a very similar structure, may have a general role in maintaining membrane excitability.

The inner cytoskeleton contains a neurofilament lattice[56] and microtubule domains consisting of fascicles of microtubules which are connected to neurofilaments. Within microtubule domains, the presence of actin microfilaments and membrane organelles, including mitochondria, which are possibly conveyed by axonal transport,[51,57] has been described. There is some controversy about the fine structure of these microtubule domains. On one hand, Hirokawa[50] has observed a cross-bridging network between microtubules, as well as between microtubules and other cytoskeletal elements and membrane organelles. On the other hand, it has also been reported that axonal microtubules are embedded in a granular matrix.[58] This granular matrix may possibly correspond to a cross-linking network which appears granular because of the attachment of soluble proteins.[51]

This has been corroborated by observing the microtubule domains of axons obtained from β,β′-iminodipropionitrile (IDPN)-treated rats, in which microtubules form bundles in the center of the axons and segregate from neurofilaments which become restricted to the axonal periphery. By immunoelectron microscopy, MAP-1A and MAP-1B have been shown to be the main components of cross-bridges between microtubules in axons.[59]

It has likewise been suggested that the cross-bridges observed between membrane organelles and microtubules correspond to the microtubule-associated ATPases involved in fast axonal transport,[60] such as kinesin and dynein-like molecules (see Section V.C).

However, the chemical nature of the cross-bridges found between microtubules and neurofilaments and between microtubules and actin microfilaments is not clear. High molecular weight MAPs are the best candidates for functioning as a family of connecting links.[61,62] For example, MAP-2, which is largely restricted to dendrites, has also been detected in some axons, where it seems to be involved in cross-linking microtubules and neurofilaments.[63] MAP-3 is another candidate for cross-linking microtubules and neurofilaments, since it is enriched in axons where both microtubules and neurofilaments are present and is especially abundant in cerebellar basket cell axons, which contain many neurofilaments and only a few microtubules.[64] Fodrin, which is a brain-specific analogue of spectrin and is mainly localized to the subaxolemmal cytoskeleton, may connect microtubules with actin microfilaments and the axolemma.[65] Interestingly, fodrin and actin are prominent components of microtubule preparations obtained directly by density gradient centrifugation of brain homogenates, whereas they are only minor components in conventional microtubule preparations obtained by assembly/disassembly cycles.[65] This suggests that microtubules and actin microfilaments may interact through fodrin molecules and that this interaction is not preserved during *in vitro* cycles of assembly and disassembly.

It has been pointed out that the image of a highly stable and cross-linked cytoskeletal network corresponds to a "frozen axon" and might mask the dynamics of the axonal cytoskeleton. For instance, membrane organelles are visualized by the quick-freeze, deep-etch technique as though they were statically bound to microtubules, but are actually being transported along microtubules, as can be observed in extruded axoplasm from squid giant axons by AVEC-DIC microscopy (video enhancement applied to differential interference contrast light microscopy) (see Section V.C).

The axonal cytoskeletal polymers can be subdivided into equilibrium (or soluble) and stable polymers. Soluble polymers are in equilibrium with the monomer pool and consequently depolymerize when the monomer concentration is reduced. Stable polymers do not disassemble in response to variations in the monomer pool. Neurofilaments are stable filaments, whereas microtubules and actin microfilaments exist as both stable and equilibrium polymers. The proportion of soluble and stable polymers influences the relative plasticity and stability of the axon.[27,66,67] The proportion of stable cytoskeletal polymers increases during brain development when axonal maturation occurs. Thus, mature axons exhibit the least plasticity.

Regarding the proportion of stable and unstable microtubules, it appears that most microtubules are unstable polymers in nonneural cells. These microtubules exhibit a dynamic behavior, growing from the centrosome to the cell periphery and collapsing catastrophically, in such a way that individual microtubules are either growing or shrinking, whereas the total polymer mass remains constant.[68] In contrast, axonal microtubules are considered largely stable and are not subjected to sudden spontaneous disruptions. Most microtubules in extruded axoplasm from squid giant axons have been found to be stable in length by AVEC-DIC microscopy, whereas a minority of about 2% show dynamic length changes.[69] The particular composition of axonal microtubules, primarily their content in specific MAPs, might account for their stability.[70]

Some stable axonal microtubules are also characterized by their resistance to cold-induced depolymerization.[71,72] The balance between cold-stable and cold-labile microtubules varies in different neurons, as well as during development. For example, in adult retinal ganglion cells, about 50 to 60% of tubulin is in the form of cold/Ca^{2+}-stable polymers, whereas in immature sympathetic neurons cultured *in vitro*, only 8% of tubulin is associated with cold/Ca^{2+}-stable polymers.[71,72] It appears that cold/Ca^{2+}-stable microtubules may correspond to microtubules conveyed in the SCa component of axonal transport, while cold/Ca^{2+}-labile microtubules are usually moving faster within axons (see Section V.B).

Cold-stable microtubules may represent specialized structures that have the functional properties of microtubule-organizing centers. Ultrastructural studies performed on neurites of cultured neuronal cells indicate that cold-stable microtubule protein is in the form of short fragments of larger, generally labile microtubules.[36] There is a positive correlation between the number of stable microtubules remaining after cold treatment of cat sympathetic nerves and the extent to which the original uniform polarity orientation of axonal microtubules is recapitulated after rewarming.[35] This lends support to the hypothesis that cold-stable microtubules may act as seeds to nucleate microtubule assembly in axons.[37] Moreover, cold-stable microtubule protein purified from bovine brain seeds *in vitro* microtubule assembly under conditions that prevent the initiation of self-assembly of cold-labile microtubules.[73]

The biochemical mechanisms that confer cold stability on axonal microtubules are not entirely clear. Posttranslational modifications of tubulin and interaction of tubulin with other proteins such as neurofilaments, MAPs, and/or specific stabilizing proteins such as the so-called "STOP" proteins may contribute to the stabilization of tubulin polymers. Only the contribution of STOP proteins to microtubule stability has been thoroughly studied.

In this context, a "superstable" microtubule fraction purified from rat brain is highly enriched in a protein referred to as STOP-145.[74] When purified STOP-145 protein is added in substoichiometric amounts to cold-labile microtubules, the resulting microtubules become partly stabilized. The binding of a few STOP-145 molecules to the ends of microtubules may result in the blockage of endwise disassembly of polymers. However, this stabilization can be reversed by calcium, calmodulin, and ATP. Thus, the balance between labile and stable microtubules may be regulated by Ca^{2+}/calmodulin-dependent phosphorylation. In contrast, the binding of a large number of STOP-145 molecules generates superstable microtubules which are resistant even to freezing and sonication.[74] These superstable microtubules are probably identical or very similar to the microtubule seeds that have been described previously.[73]

Other contributions to microtubule stability include the description of posttranslationally modified forms of tubulin, e.g., acetylated tubulin, which are enriched in stable microtubules (see Section III.A). Further study is clearly required to clarify the relative contributions of tubulin and MAP modifications to microtubule stability.

B. MICROTUBULES IN THE DENDRITIC CYTOSKELETON

The dendritic cytoskeleton differs from the axonal scaffolding in some aspects, the most important of which is the fact that dendrites generally contain fewer neurofilaments than axons[75] (Figure 1).

Dendritic microtubules show two main differences when compared with axonal microtubules. First, dendritic microtubules are relatively less stable than axonal microtubules.[76] This feature, together with the paucity of neurofilaments, suggests that dendrites contain a more dynamic cytoskeleton than axons, which may be related to neuronal plasticity (see Section IV.F). Second, microtubules are quite loosely packed and highly ordered in parallel arrays within dendrites.[18] The explanation for these distinctive properties probably lies in the fact that MAPs are differently distributed in axonal and dendritic microtubules (see Section III). In particular, MAP-2 is enriched on dendritic microtubules, compared to axonal microtubules. This observation has been correlated with the fact that *in vitro* assembled microtubules containing MAP-2 and tubulin are more widely spaced than those containing tau and tubulin.[77]

On the other hand, actin microfilaments are prominent cytoskeletal elements in dendrites. It has been shown that MAP-2 can bundle actin microfilaments *in vitro* and that this interaction is modulated by MAP-2 phosphorylation.[78-80] This suggests a role for MAP-2 in connecting microtubules with actin microfilaments, in addition to its contribution to determining the organization of microtubules within dendrites. MAP-2 is abundant in dendritic spines which contain only a few microtubules but are enriched in actin microfilaments.[81] Moreover, MAP-2 phosphorylation might conceivably regulate the gel-sol transition of actin in dendritic spines and the state of microtubule assembly within dendrites, thus allowing the dendritic shape changes which underlie neuronal plasticity (see also Section IV.F).*

C. MICROTUBULES WITHIN NERVE ENDINGS

Several electron microscopy studies have shown the presence of microtubules within nerve endings.[82-87] These studies have made use of special preparative methods to prevent the microtubule disruption which occurs subsequent to conventional fixation procedures. This extreme lability of synaptic microtubules seems to depend on the presence of calcium ions as treatment with a calcium ion chelating agent such as serum albumin effectively

* A major difference between the axonal and the dendritic cytoskeleton is connected with microtubule polarity. Recently, it has been reported that microtubules in the dendrites are mixed with respect to microtubule polarity orientation, with roughly equal proportions of microtubules oriented with their fast-growing (plus) ends toward the dendrite tip or the cell body, whereas in axons all the microtubules have a uniform polarity, with their plus ends pointing to the axon tip. This difference constitutes a structural basis for organelle distribution in the neuron, as it has been suggested that certain organelles, including polyribosomes and Golgi elements, are transported toward the minus ends of the microtubules, which would facilitate their transport into dendrites. (Baas, P. W., Deitch, J. S., Black, M. M., and Banker, G. A., Polarity orientation of microtubules in hippocampal neurons: uniformity in the axon and nonuniformity in the dendrite, *Proc. Natl. Acad. Sci. U.S.A.*, 85, 8335, 1988.)

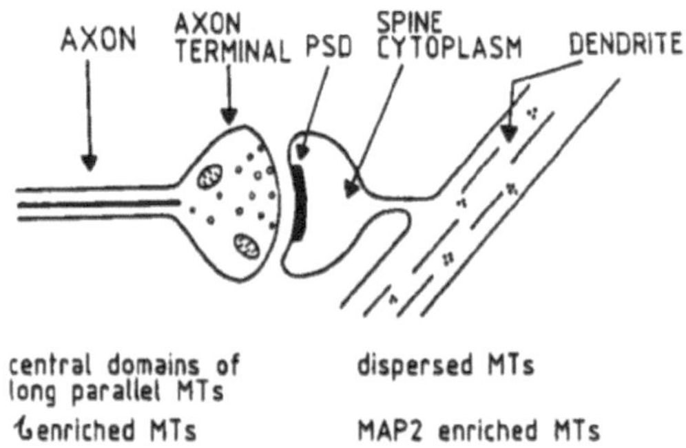

FIGURE 1. Diagrammatic representation of the specialized regions of the neuronal cytoskeleton.

preserves microtubules within nerve endings.[86,88] Synaptic microtubules appear to depolymerize so easily in the presence of calcium because of the activation of calicum-dependent proteases which digest MAPs[89] and calcium/calmodulin-dependent protein kinases (CaCAMdpk) which phosphorylate tubulin and MAPs.[90-94]

The use of albumin pretreatment in electron microscopy has revealed close structural connections between synaptic vesicles, microtubules, and presynaptic densities.[82-84,86,87] This evidence is controversial, however, as albumin may not only stabilize microtubules, but also polymerize new microtubules upon chelating calcium. Thus, the most prevalent view is that microtubules do not extend into the vesicle area of the presynaptic terminal and, therefore, vesicles must interact mainly with actin microfilaments.

There is some biochemical evidence supporting a role for microtubules and MAPs in synaptic terminals. Synapsin I, a neuronal phosphoprotein bound to small synaptic vesicles,[95] has been found associated with cytoskeletal elements including microtubules and actin microfilaments[96,97] in a phosphorylation-dependent manner.[98,99] MAP-2 might be associated with the large dense core granules found in nerve endings and other secretory cells.[100,101] Yet it remains to be established whether synapsin I and MAP-2 actually serve as phosphorylation-regulated linkers between different types of vesicles and synaptic cytoskeletal elements, actin microfilaments, and/or microtubules, and whether these interactions are involved in vesicle clustering and/or release.

III. COMPOSITION OF NEURONAL MICROTUBULES. HETEROGENEITY AND DISTRIBUTION OF MICROTUBULE PROTEINS IN THE DEVELOPING AND ADULT NERVOUS SYSTEM

A. HETEROGENEITY OF TUBULIN AND BRAIN DEVELOPMENT

The study of the polypeptide composition of microtubules prepared from brain extracts by cycles of assembly and disassembly has revealed that both the major component, tubulin, and the associated proteins, MAPs, contain multiple isoforms.

Up to 7 α-tubulin and 14 β-tubulin isoforms can be resolved in isoelectrofocusing gels.[102-110] This high degree of tubulin heterogeneity is a characteristic of the brain, as compared with other tissues.[102,108,111-113] Thus, some tubulin isoforms are common to several tissues, while others are predominantly expressed in nervous tissues. Moreover, brain-specific tubulin isoforms are largely restricted to neurons, while glial cells express mainly common isoforms.[110]

Interestingly, the same extensive tubulin heterogeneity is found in tubulin isolated from a single cultured neuron as is found in tubulin from a mass culture of peripheral nervous system neurons.[106] This suggests that tubulin heterogeneity is not primarily due to the additional contributions of different neuronal cell subpopulations, but rather may be related to the expression of different subsets of microtubules in the different domains of each neuron. It remains to be studied whether these subsets of microtubules, possibly specialized in distinctive functions (i.e., interaction with other organelles or cytoskeletal elements) might arise from the combination of specific tubulin isoforms with specific MAPs. It should be stressed that amino acid differences between tubulin isoforms are located preferentially at the carboxy-terminal region of the molecule, a domain which is involved in the binding of MAPs and regulation of microtubule assembly.[114-116] This raises the possibility that tubulin isoforms could be distributed differently in axons and dendrites, as are MAPs. Some immunofluorescence studies using monoclonal antibodies to β-tubulin isotypes support this view, as they show that two β-tubulin isotypes (Mβ2 and Mβ6) are expressed in brain in patterns noteworthily similar to those of MAP-3 and MAP-1A, respectively (Chapter 2).

Given the extraordinary variability in morphology and function among neurons, it may be expected that the requirements for specialized microtubules could be variable among different nerve cell subpopulations. Although the same neurospecific tubulin isoforms are found in different brain regions, the relative proportions of these isotubulins are variable. Furthermore, there is an apparent correlation between the sensory, associative, or motor function of defined neuronal subpopulations and the amount of each isotubulin.[117]

Tubulin heterogeneity increases during brain development.[102,104,108] This is due to the sequential appearance of neurospecific tubulin isoforms throughout fetal and postnatal development, since the common tubulin isoforms are expressed at all stages of development.

There seems to be a correlation between the degree of tubulin heterogeneity and the stage of neuronal differentiation. This lends support to the view that the particular functions depending on neurospecific tubulin isoforms are only expressed in differentiated nerve cells. It has been shown that tubulin heterogeneity is less extensive in neoplastic cells of neuronal origin than in homogenous neural populations in primary culture.[107]

The appearance of neurospecific tubulin isoforms has been studied during neuronal differentiation of teratocarcinoma and neuroblastoma cells and in long-term primary cultures of embryonic neurons.[118,119] Multipotential mouse embryonic cells contain only one α- and one β-tubulin isoform. At the time of neuronal commitment, an additional β-tubulin isoform is expressed after the induction of the corresponding gene. During the early stages of morphological differentiation and coincident with axonal extension and dendritic arborization, new α- and β-isotubulins appear by posttranslational modifications of preexisting isoforms. During the late maturation steps (neurite stabilization and synapse formation), new β-tubulin isoforms produced by gene transcription and posttranslational modifications are progressively produced.[118] Thus, it appears that brain tubulin heterogeneity is generated both at the gene level and by posttranslational modifications.

The relative amount of tubulin increases during prenatal and early postnatal brain development, thereafter decreasing in the postnatal stage[120-122] as a result of the reduction in the level of tubulin mRNAs.[123] The time period at which this decrease in tubulin content occurs corresponds to the late maturation of the brain, when neurite outgrowth and synaptogenesis are essentially complete. There is a great variability in brain development timing among different mammalian species. In rat brain, for example, neurite outgrowth occurs during a short postnatal period. Thus, the amount of rat brain tubulin increases markedly from the fetal stage until 10 to 15 d after birth and then declines gradually.[122] In other species, neuronal differentiation takes place before birth and the amount of tubulin declines after birth.

The changes in tubulin heterogeneity also seem to be complete by the time the rat is 10 to 15 d of age.[104] Thus, increased tubulin synthesis occurs at the time of neuronal growth, in parallel with the appearance of new tubulin isoforms, whereas decreased tubulin synthesis and high tubulin heterogeneity are characteristic features of a mature brain.

As mentioned above, the appearance of neurospecific tubulin isoforms is partly due to the existence of several tubulin genes.[124-126]

For example, six β-tubulin genes have been described in both mouse and human.[126] Of these, three genes, mouse Mβ3, Mβ5, and Mβ2, are expressed ubiquitously in all tissues, although at different levels. Mβ3 (which corresponds to human Hβ2) is expressed at low levels in brain and some other tissues, whereas is expressed at high levels in testis. Mβ5 (which corresponds to human Hβ2 and rat RβT.3) is expressed at high levels in brain, lung, spleen, and thymus. Mβ2 (which corresponds to human Hβ9 and rat RβT.1) is mainly expressed in brain, although is also expressed at low levels in other tissues. Two mouse β-tubulin genes, Mβ6 (which corresponds to human Hβ4) and Mβ4 (which corresponds to human Hβ5 and rat RβT.2), are brain specific.

The expression of β-tubulin genes is developmentally regulated in brains from different species.[126-130] For instance, the mRNA coded by the rat brain RβT.2 gene, which is enriched in cerebellum as compared with other brain regions, increases during postnatal development to become the predominant β-tubulin mRNA in adult cerebellum. On the other hand, the mRNA coded by the RβT.1 gene is mainly expressed during early brain development, peaking in the 1st week after birth. The expression of the ubiquitous RβT.3 gene decreases during rat brain development after birth.

Genes coding for α-tubulin are also under developmental control.[130-132] In rat brain, Tα1 mRNA is highly enriched in the embryonic nervous system as compared with adult brain, whereas T2G mRNA is expressed in many embryonic tissues, with little change in abundance during development. Moreover, Tα1 mRNA is enriched in regions with neurons actively undergoing neurite extension while T26 mRNA has a relatively homogenous distribution.[132]

It therefore appears that proteins coded by different α- and β-tubulin genes may play different roles during brain development. In particular, the expression of some tubulin mRNAs in high abundance during early development could be a simple mechanism to increase the available pool of tubulin when this is in high demand for neurite sprouting. Other neurospecific tubulin genes, i.e., RβT.2, might be expressed in certain regions of adult brain to support specialized functions performed by some microtubules. It may be easier to regulate several equivalent genes in a tissue-specific manner than a single gene when the protein serves both ubiquitous and specific functions.[132]

However, most tubulin heterogeneity is not solely the result of the existence of several genes, but rather arises from posttranslational modifications. Tubulin is modified by detyrosination, acetylation, and phosphorylation, and may also be modified through its binding to membranes. It may, in addition, be subjected to other posttranslational modifications not yet characterized.

The better-characterized tubulin modifications are detyrosination and acetylation, as the availability of monoclonal antibodies specific for acetylated, tyrosinated, and detyrosinated forms of tubulin has allowed the examination of the distribution of these isoforms during brain development.

Tubulin detyrosination (also referred to as detyrosylation or detyrosination) has been studied thoroughly. It appears that α-tubulin may be subjected to a cyclic tyrosination-detyrosination.[133-139] A tubulin-tyrosine ligase catalyzes the addition of tyrosine to the carboxy end of α tubulin.[140,141] This reaction takes place mainly on nonassembled tubulin.[135] Consequently, its function could be the replacement of the tyrosine residue, which is rapidly lost after translation, before tubulin is assembled into microtubules.[137] In fact, tyrosinated tubulin is mainly present in newly assembled microtubules.[138,139] The tyrosine residue on α tubulin is removed by a specific tubulinyl-tyrosine carboxypeptidase which acts primarily on assembled microtubules.[137,142-146] This carboxypeptidase activity can specifically modify older and stable microtubules. Thus, detyrosination is a time-dependent reaction that acts preferentially on polymerized tubulin.[138,139] Detyrosinated microtubules cannot be effectively retyrosinated, since tubulin-tyrosine ligase does not act on assembled tubulin.

During rat brain development, a decrease in the level of α-tubulin that is tyrosinated or able to act as an acceptor for tyrosination has been observed,[136] suggesting that there is an increase in the number of detyrosinated assembled microtubules during rat brain development.

Thus, it appears that tyrosinated tubulin may be enriched in newly assembled microtubules during the first stages of neuronal differentiation, whereas detyrosination may occur in parallel to axonal maturation. This has been corroborated by immunofluorescence studies[147-150] using the monoclonal antibody YL 1/2, specific for the carboxy terminus of tyrosinated α tubulin.[151] These studies indicate that the immature axons of granule cells in the developing cerebellar cortex, referred to as parallel fibers, contain tyrosinated α tubulin which disappears progressively as axonal maturation and synapse formation take place.[148] As a consequence, most tyrosinated tubulin is present in dendrites and cell bodies and is nearly absent from axonal microtubules in adult brain.[147] The time period at which detyrosination seems to occur corresponds to the transition from a growing axon to a completely mature axon. This transition does not take place in the absence of synaptogenesis. Thus, tubulin is not detyrosinated in primary cultures of cerebellar cortex in which granule cells extend axon-like processes but cannot make synapses with Purkinje cells, since these do not survive in primary culture.[149]

Tubulin acetylation occurs at a ϵ-amino group of a lysine residue in α tubulin. Acetylated α tubulin is found in most organisms in ciliary and flagellar microtubules, which are highly resistant to depolymerization,[152-154] and is also found in neurons.[155,156] Acetylation probably represents a modification of preformed microtubules, since it is inhibited by microtubule-depolymerizing drugs in cultured neuronal cells.[156] In fact, the tubulin polymer is a better substrate than the tubulin dimer for α-tubulin acetylase.[157]

Interestingly, acetylated α-tubulin appears to be largely codistributed with detyrosinated α-tubulin, being preferentially enriched in axons as shown by immunofluoresence analyses of adult rat cerebellar cortex with the specific monoclonal antibody 6-11-B-1. However, in axons of cerebellar white matter and in immature parallel fibers of 10-d-old rat cerebellum, acetylated tubulin is present together with tyrosinated tubulin, indicating that tubulin acetylation probably occurs before detyrosination during axonal outgrowth and maturation.[150] In amphibian retina, acetylated α-tubulin is also localized to the axons of horizontal and ganglion cells, as well as in the rod outer segments, which contain those microtubule arrays more resistant to depolymerizing agents.[158]

It is worth mentioning that β-tubulin might also become acetylated after long-term primary culture of mouse neurons, associated with late maturation steps.[118]

As mentioned above, tubulin is also modified by phosphorylation. β-Tubulin contains bound phosphate when purified from brain.[159-162] A β-tubulin isoform has been described which becomes phosphorylated on a serine residue at its carboxy-terminal region when neuroblastoma cells are induced to extend axon-like processes.[163,164] This phorphorylation reaction is catalyzed by a casein kinase II-related enzyme and takes place mainly on assembled tubulin.[164] Colcemid prevents and taxol enhances β-tubulin phosphorylation in cultures of neuroblastoma cells.[163] Cell fractionation studies indicate that phosphorylated tubulin is largely restricted to the cytoskeletal fraction, and is nearly absent from the soluble fraction, suggesting that phosphorylated tubulin is associated with stable axonal microtubules.

Thus, detyrosinated α-tubulin, acetylated α-tubulin, and phosphorylated β-tubulin seem to be present in stable axonal microtubule arrays. Some explanations for this association have been proposed. One involves a direct effect of tubulin modification on microtubule assembly and stability. This does not appear to be the case, at least for α-tubulin detyrosination[165,166] or acetylation.[157] Alternatively, posttranslational modifications of tubulin might enhance the interaction between modified tubulin and certain MAPs, thereby increasing microtubule stability.

Protein kinases other than casein kinase II phosphorylate tubulin *in vitro*,[93-94,167] although it is not known whether or not such reactions occur *in vivo*.

Phosphorylation of tubulin by CaCAMdpk II is of particular interest.[90-94] This kinase phosphorylates both α- and β-tubulin subunits at their carboxy-terminal regions[94] in residues different from those phosphorylated by casein kinase II.[164] Whereas tubulin phosphorylated by casein kinase II can effectively assemble into stable microtubules,[164] the phosphorylation of tubulin by CaCAMdpk II decreases its ability to polymerize and bind to MAPs[91,93,94] and enhances its association with phospholipid vesicles.[168]

This latter observation is important in view of the fact that tubulin is not only a cytosolic protein, but is also found associated with various membrane organelles such as plasma membranes, myelin sheath, postsynaptic densities, presynaptic membranes, synaptic vesicles, coated vesicles, and mitochondria.[169-178] Tubulin is synthesized on both free polyribosomes and polyribosomes bound to the rough endoplasmic reticulum;[179-181] this latter newly synthesized tubulin is incorporated into the endoplasmic reticulum membrane.[182] It appears that membrane-associated α-tubulin is an integral membrane protein since it is protected from proteolysis by the membrane and is resistant to mild detergent extraction, while β-tubulin seems to be peripherally associated with membrane-bound α-tubulin.[173,174,182] Several differences have been described between soluble and membrane tubulin, the most prominent of which is the lack of C-terminal tyrosine in membrane-bound α-tubulin.[183]

Soluble tubulin interacts with artificial membranes *in vitro*;[184-186] this binding is increased upon Ca/calmodulindependent phosphorylation (Hargreaves et al., 1986).

During rat brain development, the relative amount of membrane-bound tubulin increases in parallel with total particulate protein, differing from the decrease observed for cytosolic tubulin.[122] This indicates the possible existence of membrane-specific tubulin isoforms. The observation of some differences in the distribution of isotubulins between the particulate and cytoplasmic fractions of neuronal cells[173,175,187] lends support to this view.

As to the functional importance of membrane tubulin, it can be speculated that the translocation of membrane organelles along microtubules within axons results from the attachment of membrane-bound tubulin, or tubulin-related proteins, to microtubule-associated ATPases (see Section V.C).

Finally, it is worth emphasizing that tubulin heterogeneity, generated both at the gene level and by posttranslational modifications, seems to be conserved in all vertebrates. This might result from the selective pressure created by the interactions of tubulin isoforms with specific MAPs which would connect microtubules with other cytoskeletal elements and cellular organelles.

B. HETEROGENEITY AND DISTRIBUTION OF MICROTUBULE-ASSOCIATED PROTEINS DURING BRAIN DEVELOPMENT

1. General Properties of Microtubule-Associated Proteins

MAPs (see Table 1) were originally identified as brain proteins that copurified with tubulin through several cycles of temperature-dependent assembly and disassembly.[188-192] These proteins include two major groups: a group of high molecular weight proteins known as MAP-1 and MAP-2[61,62,189-192] and a group of 50- to 70-kDa proteins known as tau factor.[193-195] The presence has also been described of some less prominent MAPs such as the 220-kDa MAP-4[196-199] and the 180-kDa MAP-3,[64,200] which also copolymerize with brain microtubule protein preparations.

These brain MAPs are heterogeneous proteins. MAP-1 can be resolved into three polypeptides: MAP-1A, MAP-1B, and MAP-1C;[61] MAP-2[61] and MAP-3[64] appear as doublets in SDS-PAGE; MAP-4 is resolved into three components,[196] and tau factor comprises four to five closely spaced bands when separated by SDS-PAGE.[194]

It appears that the high level of heterogeneity observed in brain MAP preparations is due to variations in MAP content between different neuronal types and also to a heterogeneity of MAPs within individual neurons. With regard to the variation in MAP composition

TABLE 1
Major Classes of Microtubule-Associated Proteins

Brain MAPs	Molecular mass (SDS-PAGE) (kDa)	Distribution		
MAP-1A	350	Neurons (dendrites, axon) Glia	Increased expression kinase II)	orylated (casein
MAP-1B	325	Neurons (axon, dendrites) Glia	Increased expression (casein kinase II)	hosphorylated
MAP-2 (A,B)	270	Neurons (dendrites) Some axons and glial cells	Phosphorylated (cAM Binding to Ca/calmo	., PKC)
MAP-2C	70	Neurons (axons, dendrites)	Expression in develo	
MAP-3(A,B)	180	Glia Immature neurons Neurofilament-rich axons		
MAP-4	210—220	Glia		
Tau	55—65	Neurons (axon) Phosphorylated tau is also present in dendrites	Phosphorylated (mod ing to Ca/calmoduli	: II: PKC); bind-
Chartins	69, 72, 80		Phosphorylated	
STOP	145, 72, 56		Present in cold-stabl (CaCAMdpK)	orylated
Kinesin	120—130		Microtubule-activate protein	hograde motor
MAP-1C	300		Microtubule-activate grade motor protein	putative retro-
Other MAPs				
XMAP	215	Activated Xenopus eggs	Promotion of microtu	ends
190 Kd MAP	190	Adrenal gland		
Dynein	~400—300	Ciliary flagella	Microtubule-activate	

between neurons, the presence of different levels of tau proteins in immunoblots of different brain regions has been described.[201] As tau seems to be involved in the *in vivo* stabilization of microtubules,[202] regional variations in tau content may be indicative of corresponding variations in microtubule stability. As for MAP heterogeneity within a single neuron, tau also constitutes a case in point, as phosphorylated tau, which is not recognized by a monoclonal antibody against unphosphorylated tau, is present in cell bodies and dendrites, whereas unphosphorylated tau is restricted to axons.[203]

In addition to these well-characterized and fairly abundant proteins, minor MAPs have been identified through the selective extraction procedure developed by Duerr et al.[204] This method involves the solubilization of microtubule components in a calcium-containing buffer after first extracting soluble and membrane proteins with a nonionic detergent. With this procedure MAPs, called chartins, of 69, 72, and 80 kDa were found.[205,206]

Another class of proteins has been identified through its preferential binding to cold-stable microtubules in brain microtubule preparations; the STOP proteins[74] (see Section II.A).

A new type of MAP which preferentially promotes microtubule assembly at the microtubule plus ends has been purified from activated *Xenopus* eggs.[207] This 215-kDa protein, known as XMAP (*Xenopus* microtubule assembly protein), is accumulated during oogenesis and lost during early embryogenesis. However, there are proteins immunologically related to XMAP in adult frog brain and spinal cord.[207] It remains to be established whether similar proteins also favor assembly at the plus ends of mammalian nerve microtubules.

The use of salt-extracted taxol-stabilized microtubules[208] has allowed the detection of some new microtubule-associated proteins, such as the ATPases, possibly involved in microtubule-based motility: kinesin and MAP-1C (see Section V.C). These proteins are often termed microtubule-interacting proteins instead of MAPs, since they do not promote microtubule assembly and are only transiently bound to microtubules.

Taxol-driven microtubule assembly has also been used to detect putative MAPs in cultured cells and nonneuronal tissues. With this method, a thermostable 210-kDa MAP has been detected in HeLa[208] and CHO cells[209] and a new 190-kDa MAP has been isolated from bovine adrenal gland.[210]

There was some controversy about the presence of proteins related to brain MAPs in nonneuronal cells. It has now been clearly established that substantial amounts of brain MAPs are found only in nerve cells. Thus, antibodies raised against brain MAPs are used as tools for the differential diagnosis of neuroblastomas. As brain MAP-1 and MAP-2 are expressed in all human neuroblastoma tumors, independent of their stage of maturation, antibodies against MAP-1 and MAP-2 can distinguish grade III neuroblastomas consisting of small round undifferentiated neuroblasts with scanty cytoplasm from maligant lymphomas, Ewing's sarcomas, and rhabdomyosarcomas of similar histological appearance.[211] Yet brain MAPs, or immunologically related proteins, are present, though in significantly lower amounts, in nonneuronal cells.

These differences in MAP content may explain the singularities in the microtubule organization of nerve cells as compared with other cell types.[70] MAPs are thought to promote microtubule assembly, stabilize the resulting microtubules, and organize them with respect to one another and to other cytoskeletal components, thus playing an essential role in neuronal morphogenesis and in the organization of the neuritic cytoplasm.[70,212]

The promotion of *in vitro* assembly of tubulin has been demonstrated for MAP-1,[213] MAP-2,[190] tau,[193] and MAP-3.[214] Kinetic studies also indicate that MAP-1 and MAP-2 may be involved in the nucleation phase of microtubule assembly, whereas tau preferentially binds and stabilizes already formed polymers.[194,195,215,216] The interaction of microtubules with other cytoskeletal components (neurofilaments, actin microfilaments) and cellular organelles through MAPs has also been mentioned (see Section II).

TABLE 2
Enzymes Associated with Microtubules

Enzyme	Activity	Properties
Tubulin-tyrosine ligase	Addition of C-terminal tyrosine to α-tubulin	Binding to the amino-terminal region of β-tubulin
Tubulin-tyrosine carboxypeptidase	Cleavage of C-terminal tyrosine of α-tubulin	Preferential action on polymerized tubulin
cAMP-dependent protein kinase (type II)	Phosphorylation of MAP-2 (inhibition of microtubule assembly)	Binding of the RII subunit to MAP-2
Ca/calmodulin-dependent protein kinase	Phosphorylation of MAPs (MAP-2, tau) and tubulin (inhibition of microtubule assembly)	
Casein kinase II-like activity	Phosphorylation of MAP-1 (A,B) and β-tubulin (promotion of microtubule assembly?)	

Brain microtubule protein preparations also contain a variety of enzymes involved in posttranslational modifications of microtubule proteins (see Table 2). These enzymes are not considered MAPs, as they usually bind to MAPs rather than to tubulin. An exception is tubulin-tyrosine carboxypeptidase, which binds to polymerized tubulin and hydrolyzes the C-terminal tyrosine of tubulin.

As MAPs are phosphoproteins, a great deal of attention has been paid to the association of protein kinases with microtubules.[164,217-221,228]

The first studies on microtubule protein phosphorylation showed the presence of both cyclic nucleotide-independent and cyclic nucleotide-dependent kinase activities. The latter has been identified as the cAMP-dependent protein kinase (cAMPdpK) type II.[218,219] This kinase is associated with MAP-2 via the cAMP-binding regulatory subunits. Thus, the catalytic subunit of the enzyme can be released from MAP-2 upon cAMP incubation. cAMPdpK II holoenzyme consists of two regulatory and two catalytic subunits which become dissociated and activated after cAMP binding to the regulatory subunit. Approximately 35% of the total cAMPdpK in brain cytosol is bound to microtubules through MAP-2.[222-225]

Among the cyclic nucleotide-independent protein kinases associated with microtubules, a CaCAMdpK which phosphorylates MAP-2 and tau proteins[226,227] and a heparin-inhibited, calcium-inhibited, magnesium-activated casein kinase II-related enzyme which can use both GTP and ATP as phosphoryl donors and acts preferentially on β-tubulin and MAP-1 polypeptides have been found.[164,221,228]

2. MAP-1

MAP-1 is a highly heterogeneous entity which comprises at least three different polypeptides: MAP-1A (350 kDa) MAP-1B (320 kDa), and MAP-1C (300 kDa).[61] MAP-1C is a dynein-like ATPase not related either to MAP-1A or to MAP-1B (see Section V.C.4). MAP-1A and MAP-1B are distinct but related proteins;[61,220,228-232] both precipitate after boiling in the presence of tubulin, thus differing from tau and MAP-2, which are thermostable under these conditions. They are, however, resistant to heat treatment in tubulin-depleted MAP preparations.[233] MAP-1A and MAP-1B have different peptide maps,[235] although some similarities can be observed.[220]

Monoclonal antibodies against MAP-1A or MAP-1B do not cross-react, with the exception of 1B-3 antibody, which recognizes a phosphorylated epitope common to MAP-1A, MAP-1B, and neurofilament proteins.[234] Interestingly, a monoclonal antibody (AB18) raised

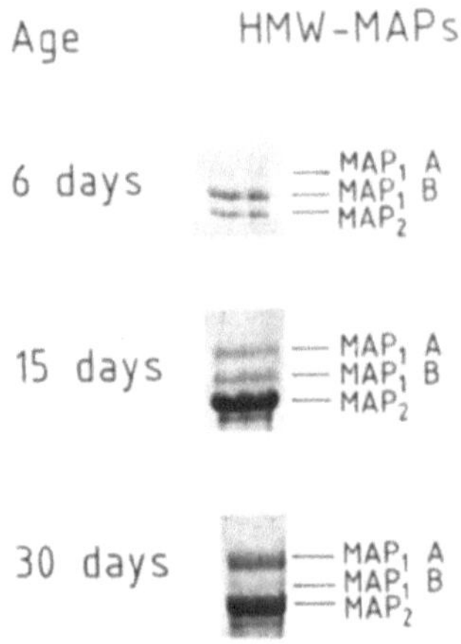

FIGURE 2. Heterogeneity of high molecular weight MAPs during the postnatal development of rat brain.

against Alzheimer's neurofibrillary tangles (see Section VI.E) recognizes a phosphorylated epitope common to MAP-1A, MAP-1B, tau, and neurofilament proteins, whereas another monoclonal antibody (AD10), also raised against neurofibrillary tangles, reacts with MAP-1A, MAP-1B, MAP-2, tau, and neurofilament proteins.[235]

Both MAP-1A and MAP-1B are phosphoproteins. A casein kinase II-related enzyme seems to be responsible for the *in vivo* phosphorylation of these proteins.[228,236] The *in vivo* phosphorylation of MAP-1B has been observed in cultured neuroblastoma cells induced to differentiate[163,228] and in juvenile rat brain upon intracraneal injection of labeled phosphate.[236] The *in vivo* phosphorylation of MAP-1A has only been detected in the adult rat brain injected with labeled phosphate.[236] It appears that phosphorylation of MAP-1 polypeptides can enhance slightly their binding to microtubules.[228,236]

Both MAP-1A and MAP-1B contain associated proteins of 28 and 30 kDa, referred to as MAP-1 light chains, which are not required for the binding of MAP-1 polypeptides to tubulin.[237]

There is some controversy about the localization of MAP-1A. Whereas Bloom et al. have described the MAP-1A as a widespread protein which is located in brain both in neurons and glia,[229,230] Huber and Matus claim that MAP-1A is present only in neurons.[238] These conflicting reports may result from the use of monoclonal antibodies recognizing different isoforms of the same molecule or particular epitopes showing a cell compartment-specific accessibility. Most MAP-1A antibodies stain neuron cell bodies, dendrites, axons, and glial cells.[229,230,239] However, these MAP-1A antibodies stain dendrites more brightly than axons.[229,239] The dendrite staining for MAP-1A is more apparent in Purkinje cells than it is in granule cells of cerebellum, a pattern opposite to that found for MAP-2.[240] By immunoelectron microscopy and a quick-freeze deep-etch study, it appears that MAP-1A constitutes mainly fuzzy filamentous cross-bridges between microtubules.[241] MAP-1A purified through a monoclonal antibody affinity column is a long, thin, filamentous and highly flexible molecule, as observed by the rotary shadowing technique.[241] When salt-extracted taxol-stabilized microtubules are decorated with affinity-purified MAP-1A, MAP-1A thin filaments protruding from microtubules and connecting them to one another are observed.[241]

The expression of MAP-1A is developmentally regulated (Figure 2). It first appears in newborn rat brain and its level increases up to postnatal day 20.[242] A similar pattern of expression is found in chicken brain.[243]

MAP-1A immunoreactivity first appears within dendrites of neurons of the neocortex[242] and hippocampus[244] around postnatal day 2. From postnatal day 3 to postnatal day 7, cerebellar MAP-1A immunoreactivity is localized mainly to white matter axons. From postnatal day 7 onwards, cerebellar MAP-1A immunoreactivity in Purkinje cell dendrites is increased.[245]

As for MAP-1B, the terminology is rather confusing. MAP-1B, as named by Bloom et al.,[231] is also referred to as MAP-1X[246] MAP 5,[247] or MAP-1C[220] by other authors.

Immunofluorescence microscopy of rat brain sections and primary cultures of rat brain cells shows that, compared to MAP-1A and MAP-2, MAP-1B is particularly prominent in axonal as well as in dendritic processes.[231] This study also indicates the presence of MAP-1B in glial cells. However, Riederer et al. have reported that MAP-5 is found only in neurons, observing MAP-5 immunoreactivity in cell bodies, axons, and dendrites.[247] Calvert and Anderton describe the almost exclusive staining of developing axons in brain sections by their monoclonal antibody against MAP-1B.[246]

Again, these conflicting reports may be attributed to the use of monoclonal antibodies recognizing different isoforms of the same molecule. With a monospecific polyclonal antibody, MAP-1B appears to be widely distributed in both neurons and glial cells, although it is more abundant in neurons, particularly within axon-like processes.[253]

The ratio of MAP-1B to tubulin in taxol-assembled microtubules from brain white matter is higher than that found in taxol-assembled microtubules from brain gray matter. In contrast, MAP-1A is found in a slightly higher ratio to tubulin in gray matter. Interestingly, MAP2 is found in a much higher ratio to tubulin in gray matter.[61] These results indicate that MAP-1B is preferentially localized in white matter structures, thus confirming the prominent immunocytochemical axonal staining.

The expression of MAP-1B is also developmentally controlled, and is more abundant in newborn than in adult rat brain.[231,246,247] This also applies to chicken brain[243] and quail retina.[248] In rat brain, there is a sharp reduction in the level of MAP-1B around postnatal day 15, coincident with the time at which axonal maturation occurs (see Figure 2). Thus, it appears that MAP-1B is more abundant in immature neuronal processes. As mentioned above, the converse is observed for MAP-1A, which suggests that MAP-1B is particularly important in neurite outgrowth while MAP-1A might be connected with neurite maturation (see Section IV.)

Both MAP-1A and MAP-1B have wide distribution outside the brain. In bovine pituitary tissue and cultured rat pituitary cells, MAP-1A and MAP-1B are major MAPs;[249] they are also prominent MAPs in bovine adrenal gland.[210] Some monoclonal and polyclonal antibodies against MAP-1 react with centrosomes in non-neuronal cells[250,251] including some antibodies against MAP-1B,[232,252,253] which suggests the presence of a protein immunologically related to brain MAP-1B in centrosomes. A 220-kDa protein which has been found in nonneural cells confined to the interphase nucleus and associated with the mitotic spindle is also immunologically related to brain MAP-1B.[253]

3. MAP-2

MAP-2 is the most abundant MAP in brain microtubule preparations and has been the most thoroughly studied. MAP-2 has an apparent molecular weight of approximately 270,000 Da in SDS-PAGE.[189] This is due to an abnormal electrophoretic behavior, since the molecular weight of MAP-2 when measured by sedimentation equilibrium and verified by gel chromatography in denaturing solvent and SDS electrophoresis at different acrylamide concentrations is 220,000.[254]

MAP-2 is thermostable and is commonly purified by heat treatment of microtubules followed by gel chromatography.[255] MAP-2 obtained by gel filtration and ion-exchange chromatography, omitting the boiling step, is indistinguishable from MAP-2 prepared by heat treatment, by the criteria of gel electrophoresis, isoelectric focusing, *S. aureus* V8 peptide mapping, gel filtration, fluorescence spectroscopy, and circular dichroism.[254] Examination of the circular dichroism of MAP-2 as a function of temperature indicates that the thermal stability of these proteins is probably due to its mostly disordered structure.[254]

MAP-2 is localized on the surface of *in vitro* assembled microtubules, forming periodic lateral projections[256,257] which are removed by mild trypsin or chymotrypsin digestion.[222,258] These projections span from 30 to 40 mm[259] or up to 80 to 90 mm[260] away from the microtubule wall.

The hydrodynamic properties, fluorescence spectroscopy, and circular dichroism of MAP-2 are compatible with MAP-2 having a largely, but not completely, disordered structure.[254] In support to this observation, nuclear magnetic resonance spectra of microtubule proteins have narrow resonances attributed to the high mobility of MAP-2 projections, and mild proteolysis of the microtubules results in the loss of both the MAP-2 projection and the narrow resonance for NMR spectra.[261]

As stated above, MAP-2 can be cleaved by limited proteolysis into a 35- to 40-kDa microtubule-binding domain and a 240-kDa projection domain.[222,258] The regulatory subunit of the cAMPdpK is associated with the projection domain of MAP-2.[223] Calcium/calmodulin binds to the microtubule-binding domain of MAP-2, generating a complex which cannot associate with tubulin.[262] It has been suggested that calcium/calmodulin has a structure similar to that of the carboxy-terminal region of tubulin which binds MAP-2.[114,264]

MAP-2 is the major substrate for endogenous phosphorylation in purified microtubule preparations.[218] Moreover, MAP-2 also contains bound phosphate when purified from brain.[225]

The phosphorylation of MAP-2 affects both the promotion of tubulin assembly[93,219,265,266] and the interaction of MAP-2 with actin microfilaments.[78,79,266] Thus, it has been suggested that it may play an important role in neuronal plasticity.[267]

The first studies of phosphorylation of MAP-2 *in vitro* were performed using the endogenous cAMPdpK or protein kinase A[218,219,222,265] which is tightly bound to the projection domain of MAP-2.[223,224] More recently, it has been suggested that CaCAMdpK, which also copurifies with brain microtubules,[226,227] also phosphorylates MAP-2 *in vitro*.[92,93,227] Moreover, insulin receptor-associated kinase and Ca^{2+}/phospholipid-dependent protein kinase C also phosphorylate MAP-2.[167,268] It appears that cAMPdpK, CaCAMdpK, Ca^{2+}/phospholipid-dependent protein kinase, and insulin receptor all phosphorylate MAP-2 in a 42-kDa peptide which contains the tubulin-binding region, but differ in the level and localization of the sites at which they phosphorylate the projection domain of MAP-2.[269]

In adult brain, MAP-2 immunoreactivity is restricted to neurons, and is mainly localized to neuronal cell bodies and dendrites.[270-274] This corresponds well with the fact that the ratio of MAP-2 to tubulin in taxol-assembled microtubules from gray matter is notably higher than that of white matter.[61]

Interestingly, MAP-2 immunoreactivity is also displayed in motor neuron axons, but not in dorsal root axons or spinal cord white matter,[63] and in astrocytes of the optic nerve periphery, but not in astrocytes of the optic tract.[275]

Mammalian adult brain MAP-2 consists of two polypeptides: MAP-2A and MAP-2B, which have very similar sizes, peptide maps, and chromatographic properties.[61] The amount of MAP-2B does not change from late embryonic stages onward, but MAP-2A first appears around postnatal day 10 and its expression rapidly increases, reaching it maximum level around postnatal day 20.[271,276]

In addition to MAP-2B, juvenile brain microtubules contain a 70-kDa protein which reacts with monoclonal antibodies raised against MAP-2.[242,276,277] This low molecular weight form of MAP-2, which has been termed MAP-2C, is thermostable, has a peptide map similar to those of high molecular weight MAP-2A and MAP-2B and polymerizes with microtubules with the same efficiency as high molecular weight MAP-2.[278] MAP-2C appears to be encoded by its own 6 Kb mRNA derived from the same gene as the 9 Kb transcripts which encode high molecular weight MAP-2.[279]

It has been suggested that MAP-2C may be a form of high molecular weight MAP-2 lacking much of its projection domain.[278,279] As the projection domains of MAP-2 appear as fine filaments which can cross-link microtubules to one another and to other cytoskeletal elements, while the microtubule-binding domain is responsible for promoting microtubule assembly, MAP-2C would stimulate tubulin polymerization without cross-linking the neuritic cytoskeleton during the early phases of neuronal differentiation.[278,279]

The fact that MAP-2C is a major component of newborn rat brain, but that its expression is greatly reduced around postnatal day 15, lends support to this view.[242] Furthermore, the widespread distribution of MAP-2 immunoreactivity within the cell bodies, dendrites, and axons of embryonic neurons may represent MAP-2C staining and correspond to a dynamic and relatively uncross-linked cytoskeleton, while the restricted dendritic distribution found in fully differentiated neurons would represent high molecular weight MAP2 staining, and a more rigid and cross-linked dendritic cytoskeleton.

According to this view, the function of juvenile low molecular weight MAP-2C would be the promotion of microtubule assembly in growing processes, and that of adult high molecular weight MAP2 would be the stabilization of mature dendritic morphology.[278,279] High molecular weight MAP-2 is believed to be involved in the interaction of microtubules with one another and with other intracellular structures such as actin microfilaments, intermediate filaments, and secretory granules.[61-63,78-80,101,280]

There is some controversy on the presence of MAP-2 or related proteins in nonneuronal tissues. Using a polyclonal antibody against MAP-2 in radioimmunoassay, Valdivia et al. found that the amount of MAP-2 in kidney, lung, spleen, and liver was 100- to 1000-fold lower than that in brain.[281] Northern blot analyses with cloned cDNA probes to MAP-2 fail to detect the expression of such gene in tissues other than brain.[282] An exception is pituitary tissue, in which the level of MAP-2 is about 10% that of brain, which is substantially higher than has been reported in other nonneuronal tissues.[249] MAP-2 has also been detected in chromaffin cells from the adrenal medulla.[283]

4. MAP-3 and MAP-4

On SDS-PAGE, MAP-3 is composed of a closely spaced pair of polypeptides, MAP-3A and MAP-3B, with apparent molecular weight of 180,000 Da which are present in both glioma and neuroblastoma cell lines. In brain, MAP-3 is present both in neurons and astrocytes. In nerve cells, its distribution is highly restricted, as it is detectable only in neurofilament-rich axons.[64]

MAP-3A predominates over MAP-3B in rat embryo brain, After birth, MAP-3B expression is increased, but the expression of both MAP-3A and MAP-3B is greatly reduced around postnatal day 15. Interestingly, MAP-3 immunoreactivity is found in neurons of the neocortex until postnatal day 10, after which is largely restricted to proliferating astroglial cells.[242] A similar pattern is observed in the cerebellum, with the disappearance of MAP-3 immunostaining from parallel fibers around postnatal day 20 and increased staining in the Bergmann astroglial cells in adult cerebellum.[245]

MAP-3 may promote microtubule assembly in neurons (in neonatal animals) and in astrocytes,[214] in addition to a putative role in cross-linking neurofilaments with microtubules.[64]

MAP-4 consists of a group of polypeptides with an apparent molecular weight of around 210,000 Da which are localized exclusively in glial cells (astrocytes and oligodendrocytes) in adult brain and in other cell types outside the nervous system.[196-199]

5. Tau Proteins

Tau proteins purified from adult brain were first described as a set of six polypeptides ranging from Mr 55 kDa to 65 kDa, which promote *in vitro* tubulin assembly.[193] The tau proteins are heat stable,[255] soluble in 2.5% perchloric acid,[284,285] and are closely related as judged from their similar peptide maps[194,195] and from their shared epitopes.[286] Similarly to MAP-2, tau proteins also bind calcium/calmodulin.[262]

Immunocytochemical studies with the monoclonal antibody tau-1 indicate that tau is largely restricted to axons.[286] Nevertheless, the determinant recognized by this antibody is also present in neuronal cell bodies and dendrites, where it is masked by phosphorylation but can be visualized after phosphatase treatment of brain sections.[203]

In addition to the 55- to 68-kDa tau proteins, sympathetic neurons (as well as differentiated rat PC12 phenochromocytoma and mouse neuroblastoma N115 cells) also contain high molecular weight tau proteins of 100 to 125 kDa which are present at only trace levels in brain.[287]

In newborn rat brain, tau consists of a single 48-kDa polypeptide.[277,288] The transition from this juvenile tau to the highly heterogeneous adult tau begins around postnatal day 15 and is completed by day 30, a time period coincident with the period of axonal maturation and synaptogenesis.[288] Juvenile tau is less efficient than adult tau in promoting microtubule assembly.[288]

Astrocytes contain two components immunologically related to tau. However, the molecular weight of these components is different from those of both juvenile and adult tau.[289] Tau-related proteins have also been bound in nonneural tissues, although at much lower levels than that found in brain.[290]

The mRNAs coding for tau proteins are heterogeneous and are developmentally regulated.[291] It has recently been reported that the cDNAs of two clones, obtained from a mouse embryonic brain gene library, code for two tau proteins which are identical in their amino terminal sequences, but differ in their carboxy terminal domains.[292] Both proteins contain common repeated sequences which correspond to the tubulin-binding peptide obtained by limited chymotryptic digestion of whole tau proteins.[293] A similar sequence is found in a cDNA coding for MAP-2 but not in a cDNA coding for MAP-1B (Cowan et al., unpublished). It seems likely that the two tau cDNAs arise from alternative splicing of a single tau gene.[292] The size of tau as predicted from the cDNA clones is 36 to 38 kDa, while mouse tau protein is 47 to 50 kDa, probably due to an anomalous electrophoretic mobility resulting from phosphorylation, protein structure, and SDS binding.[292]

The heterogeneity of tau proteins is generated at least partially at the transcriptional level. An additional degree of heterogeneity can be generated posttranscriptionally. Thus, both juvenile and adult tau proteins are phosphorylated *in vivo*.[294,295]

The *in vitro* phosphorylation of adult tau proteins has been studied extensively. Lindwall and Cole have identified two modes of phosphorylation: one (mode I) which changed tau electrophoretic mobility and one (mode II) which did not.[284,285] Tau purified from brain contains a mixture of nonphosphorylated tau and tau phosphorylated in at least mode I. Interestingly, mode I-phosphorylated tau proteins are major components of Alzheimer paired helical filaments (see Section VI.E).

Mode I phosphorylation of tau proteins seems to be due to CaCAMdpK II.[296] This phosphorylation induces a conformational change in tau since it produces a substantial decrease in electrophoretic mobility and makes a tau epitope inaccessible to the monoclonal antibody tau-1.[203] Mode I phosphorylation is modulated by the binding of phospholipids to tau proteins.[296] Lindwall and Cole have shown that mode I phosphorylation reduces the ability of tau to promote microtubule assembly[284,285] and this may be related to the fact that mode I phosphorylated tau is found associated with Alzheimer's paired helical filaments[296] (see Section VI.E).

Mode II phosphorylation of tau occurs when tau protein is incubated with (γ-^{32}P)ATP and brain microtubule protein preparations.[284,285] It appears that this mode of tau phosphorylation may be due to the action of Ca^{2+}/phospholipid-dependent protein kinase C (PKC);[297] and that mode II-phosphorylated tau is partially resistant to alkaline phosphatase.[297]

IV. ROLE OF MICROTUBULE PROTEINS IN NEURONAL MORPHOGENESIS

A. EXPERIMENTAL MODELS TO STUDY THE ROLE OF MICROTUBULE PROTEINS IN NEURONAL MORPHOGENESIS

The generation of mature neurons from undifferentiated neuroepithelial cells is a central feature in brain development. Neuroblasts proliferate within the neuroepithelium lining the ventricular system and the resulting young neuronal precursors migrate to reach their definitive loci, where they become fully differentiated. Schematically, five fundamental events are necessary for this morphological conversion: (1) axonal sprouting, (2) axonal maturation, (3) dendritic extension and arborization (4) dendritic maturation, and (5) synaptogenesis. Finally, mature neurons may be subjected to some morphological changes as a result of their synaptic activity, which has been called neuronal plasticity.

In all these phenomena, microtubules play a prominent role. As the definite shape characteristic of a mature neuron is maintained by the microtubular cytoskeleton, the generation and any change of that shape is produced through modifications in microtubule composition and/or organization.

The contribution of microtubules to neuronal morphogenesis has been thoroughly examined in nerve cells cultured *in vitro*. The studies performed on cultured neurons are easier to interpret than those performed in intact animals since the brain contain numerous neuron types and morphogenesis begins for each type at different stages of development depending on the brain region. Both primary cultures and tumour cell lines have been widely used. In fact, the foundations of tissue culture were established by Harrison, who developed a simple method to observe nerve fiber outgrowth in explants of amphibian neural tissues on plasma clots.[298,299]

Nerve cells in primary culture are immediate descendants of cells *in situ* and their characteristic features are quite similar to those of the *in situ* cells. Usually, neurons obtained from immature animals are easily adapted to *in vitro* culture. In this way, nerve cell differentiation can be observed under *in vitro* conditions. Furthermore, methods have been developed to obtain cultures of different neuron types from both the peripheral and the central nervous system which allow the study of similarities and differences in the morphogenesis of particular cell types. Finally, long-term primary cultures of neurons have been used to observe late maturation steps which are often missing in differentiated tumor cells.[118]

As compared with primary cultures, clonal cell lines derived from single tumor cells provide the opportunity of producing large homogeneous populations of cells grown under controlled, manipulable conditions. Clonal cell lines let us study the general mechanisms of nerve cell morphogenesis, which may be common to different neuron types, at a biochemical level. Clonal cell lines have been obtained mainly from three different groups of tumors: teratocarcinomas, pheochromocytomas, and neuroblastomas.

Teratocarcinomas are malignant neoplasms generally derived from early embryo or germ cells and contain pluripotent cells which are regarded as similar to the stem cells of blastocysts. From these embryonal carcinoma cells, several clonal cell lines have been obtained. Two lines, C17S1-1003 and pCC7-S-aza R1-1009, can be induced to generate fully differentiated neuronal derivatives.[119,300-302] The 1003 cell line shows an intermediate state of differentiation in which cells of neuroepithelial-like morphology form large aggregates. These aggregates contain several cavities and are similar in organization to the neuroectodermal tube.[300] In this state, 1003 cells do not extend neurites; consequently, this system has been used to study neurogenic commitment separately from later phases of differentiation.[118,119]

Pheochromocytomas are tumors of chromaffin tissue of the adrenal medulla or sympathetic ganglia. Pheochromocytoma cells, which originate from neural crest stem cells, express a typical chromaffin phenotype with characteristic catecholamine-storing granules and absence of neurites. However, both primary cultures of human pheochromocytomas and a clonal cell line (PC12) derived from a rat pheochromocytoma form axon-like processes in response to added nerve growth factor (NGF) if they are grown on adhesive substrates, such as collagen or polylysine-coated culture dishes.[303,304] PC12 cells constitute a highly useful model system to analyze the induction of neuronal differentiation by neurotrophic factors.

In particular, the use of PC12 cells has emphasized the importance of both transcription-dependent and -independent responses of neuronal precursors to neurotrophic factors.[305-307] Exposure of PC12 cells to NGF causes them to cease proliferation slowly and extend processes in 3 to 5 d, after a lag of about 18 h. Initiation of neurite outgrowth can be reversibly suppressed by inhibitors of RNA synthesis. However, when PC12 cells are "primed", i.e., treated with NGF for 1 to 2 weeks and divested of their processes by NGF deprivation of mechanical shearing, and subsequently recultured in the presence of NGF, neurite outgrowth is promoted very rapidly and is insensitive to inhibitors of RNA synthesis.[305,306] This short-term response has also been observed in primary cultures of sensory and sympathetic neurons from late embryos or neonates, since they have probably had prior *in vivo* exposure to NGF and are "primed". The use of "naive" and "primed" PC12 cells is a good system for distinguishing the contributions of gene activation and posttranslational modifications to process extension.

Neuroblastoma tumors usually arise in early childhood from neuroblast precursors of sympathetic neurons. It has been claimed that theses neoplasms may result from defective differentiation of neural crest derivatives which continue proliferating. These malignant tumors occasionally regress spontaneously, giving rise to benign ganglioneuromas, probably as a consequence of neuroblastoma cell differentiation.[308] A large number of clonal cell lines has been derived from neuroblastomas of mouse, rat, and human origin. They can be maintained *in vitro* as undifferentiated, but committed, neuroblasts or induced to differentiate by different treatments (serum deprivation, hypertonic medium, dibutyril-cAMP, prostaglandins, NGF, bromodeoxyuridine, dimethyl sulfoxide, etc.). Although the various clonal cell lines differ in their degree of maturation upon differentiation, most neuroblastoma cells become amitotic, lose their oncogenic capacity, and acquire the morphological, biochemical, and electrophysiological characteristics of mature neurons.[309-323] Neuroblastoma cell cultures have been used extensively to examine both morphologically and biochemically the early steps of neuronal morphogenesis, with particular emphasis on neurite outgrowth.[118,119,163,164,228,236,311,312] In most neuroblastoma cell lines, process extension occurs regardless of the inhibition of RNA synthesis,[311] suggesting that these cells are similar to "primed" PC112 cells.

B. ROLE OF MICROTUBULES IN NEURITE OUTGROWTH AND MATURATION

1. Microtubules and Growth Cone Motility

All neuronal processes, both axons and dendrites, grow at specialized terminal appendages called growth cones, which were first observed by Ramón y Cajal in the embryonic chick spinal cord.[12,13,324] Growth cones are involved in the primary attachment of the process to the substrate, in the assembly of surface membrane during growth, and in the guidance of the extending process to reach its target.[8]

Typical growth cones appear as fan-shaped or leaf-shaped expansions of process tips from which long thin projections called filopodia or microspikes, and broad thin expansions called lamellipodia or veils, extend. Microtubules are restricted to the central portion of the growth cone, while actin microfilaments are the major components of filopodia and lamellipodia and are found mainly beneath the plasma membrane.[325,326] In addition, growth cones contain numerous membrane organelles that are presumably derived from the Golgi apparatus. Membrane organelles are transported along microtubules within neurites until they reach the distal tip, where they fuse with the growth cone membrane in order to support the membrane extension which occurs at the growing edge.[327,328]

Process growth occurs through the formation of a veil and the subsequent maturation of the veil into a neurite.[328,329] Veils always protrude between or along microspikes, which can serve as structural supports.[328,329] The protrusive activity of veils and filopodia seems to depend on actin.[330,331] In confirmation of this, growth cone motility is inhibited by agents which interfere with actin, such as cytochalasin B, while it persists in the presence of colchicine, which abolishes neurite extension.[312]

The mechanism of growth cone motility is not known, but calcium appears to be involved, possibly through actin reorganization and/or actin-myosin interactions. In support of this theory, the growth cones of different neuronal cells have an increased calcium concentration due to the presence of a large number of calcium channels.[332-334]

The high calcium content of the growth cone may also be the reason for the delayed appearance of microtubules in the nascent process,[335] since calcium and Ca^{2+}/calmodulin-dependent phosphorylation inhibit microtubule assembly. In fact, tubulin present in growth cones is a substrate for several protein kinases, including CaCAMdpK and tyrosine-specific protein kinases.[336]

The paucity of assembled microtubules might be required for the motile activity of filopodia and lamellipodia. For example, high concentrations of taxol, a strong promoter of microtubule assembly and stabilization, disrupt growth cone motility and prevent neurite growth.[337] On the other hand, neurons cultured in sufficiently low taxol levels to allow neurite extension exhibit processes less branched than normal. Taxol may interfere with the lateral expansion of the veil that gives rise to branching.

It has been suggested that microtubules normally maintain the cross-linking of cytoskeletal elements, including actin microfilaments, thereby preventing motile activity in adjacent regions of the cell surface.[36] For instance, treatment of growing neurons with colchicine results in the progressive appearance of lateral filopodia and regions of flattened cytoplasm along the neurites.[36,338] In conclusion, the absence of microtubules within filopodia and lamellipodia seems to be essential for growth cone motility and the presence of microtubules within neurites seems to stabilize the sides of the neurites.

2. Microtubules and Process Extension

During process extension, the expanded growth cone base is being converted into a slender stable neurite while the leading growth cone continues advancing.[329] Matus has described this transition, from the functional dominance of actin filaments, which favors motility, to the dominance of microtubules, which confers rigidity, as a phase change in the neuritic cytoplasm essential for neurite outgrowth.[70] Thus, axonal outgrowth can occur in the absence of motile growth cones in cytochalasin B-treated neuronal cultures,[339,340] while it is completely abolished by microtubule-depolymerizing drugs, such as colchicine, even when growth cone motility persists.[311,312,341,342] The extension of neurites is reversibly inhibited by low concentrations of colchicine; at higher concentrations, neurites retract to the cell body.[312,341,342] This suggests that microtubules may provide the force required to extend and maintain neuronal processes. As a matter of fact, increased microtubule assembly takes place during process extension.[343-345] These newly assembled microtubules are more tightly packed and stable than microtubules in undifferentiated cells[346,347] (see Figure 3).

From a theoretical point of view, Kirschner and Mitchison have proposed a hypothesis to explain neurite outgrowth by selective stabilization of microtubules.[348] Immature neuroblasts, like most nonneuronal cells, are unpolarized cells. These cells generally contain a centrally located microtubule-organizing center (MTOC) from which unstable microtubules grow rapidly to the cell periphery with no preferred direction and are spontaneously shortened

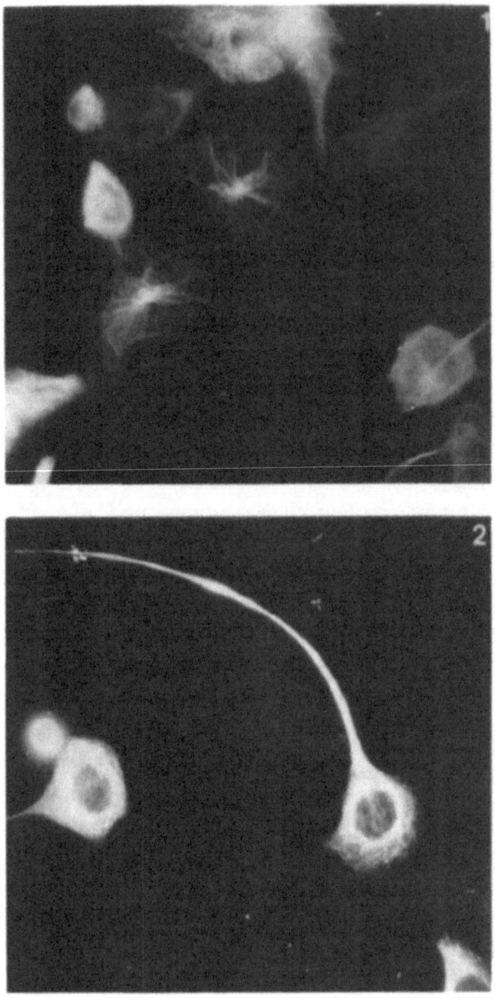

FIGURE 3. Distribution of microtubules in proliferating (1) and differentiated (2) mouse neuroblastoma cells as revealed by indirect immunofluorescence with a monoclonal antibody to tubulin.

by dynamic instability. Some cell types contain several MTOCs around the nucleus rather than a single one,[349] but the behavior of microtubules nucleated by these MTOCs is identical to that observed in cells with a single MTOC. A neurite could arise in neuroblasts as the consequence of a local extracellular signal which would activate some capping structures near the cell periphery. The region of the cell surface in which these capping structures become activated would constitute the primordium of the nascent process. Capping structures could block the ends of the microtubules directly or bind laterally near the microtubule ends, thereby preventing endwise depolymerization. The selective stabilization of capped microtubules would gradually lead to a polarization, since capped microtubules would continue growing, while all the other microtubules would remain unstable. This would let the nascent process extend. Moreover, capped microtubules could become further stabilized along their entire length by binding of MAPs and some posttranslational modifications. Therefore, the presence of MAPs and the activation of some posttranslational-processing enzymes must be required for supporting microtubule assembly and stabilization within growing neurites (see Section C).

Some experimental evidence, from immunofluorescence studies in mouse neuroblastoma cells,[350] supports the view that neurites arise by selective stabilization of a particular array of microtubules nucleated by a MTOC. In fact, the importance of selective stabilization is more clearly established than that of a MTOC.

Studies performed in N115 mouse neuroblastoma cells emphasized the role of MTOCs in neurite outgrowth,[350] which now appears as doubtful. Undifferentiated N115 neuroblastoma cells contain several MTOCs dispersed around the cell nucleus. Upon serum deprivation, these MTOCs migrate out and aggregate into a single complex near a region of the cell periphery. Finally, microtubules are orientated so that they grow primarily in a single direction from the MTOC to the cell surface. This gives rise to the extension of a process from the cell body.[350]

The orientation of growing microtubules which is observed before neurite outgrowth can be the result of the selective stabilization of capped microtubules. The nature of the capping structures involved in this stabilization remains to be established. Further stabilization of neuritic microtubules by MAPs and posttranslational modifications has also been suggested (see Section C).

The placement of the single MTOC resulting from the aggregation of previously dispersed MTOCs was thought to determine the site of neurite extension.[350] However, the results observed in neuroblastoma cell lines other than N115 do not support this interpretation, since no relationship between the location of the MTOC and the sites for neurite emergence is observed.[10,351,352] It has also been reported that mature axons do not have single MTOCs (see Section II.A). The experiments of Baas et al.[42] indicate that the initiation of neurite outgrowth might be the result of the capping of cold-stable microtubule seeds, which should be considered as the nucleation centers for microtubule assembly in extending processes (see Section II.A).

3. Microtubules and Process Maturation

Newly formed processes undergo progressive structural changes until they become converted into mature processes.[32,353,354] Computer-assisted three-dimensional reconstruction of serial electron microscopic images obtained from PC12 cells cultured in the presence of NGF has allowed the detailed study of neurite maturation.[32]

Immature axon-like processes are structurally similar to the cell body, as they contain a large number of membranous organelles and few microtubules, and have a very irregular shape and little organized internal structure as compared with mature axons. Axonal maturation in PC12 and other neuronal cells is characterized by the fact that microtubules increase in number, leading to a more cylindrical and uniform neurite shape, and to a highly ordered neuritic cytoskeleton.[32,354] Furthermore, microtubules within mature neurites become more resistant to disassembly induced by cold treatment or microtubule-depolymerizing drugs.[345-347]

Quantitative ultrastructural studies performed on serial sections of rat brains also show an increased microtubule density within Purkinge cell dendrites and within sciatic nerve fibers during the period of process maturations.[355-357] Moreover, the proportion of cold-stable microtubule protein increases fourfold during the same period.[356]

These results may explain the fact that *in vitro* microtubule assembly in rat brain extracts depends on the developmental stage.[358,359] Both the nucleation and the elongation steps of *in vitro* microtubule assembly are less efficient in supernatants prepared from immature rat brains. The addition of adult brain MAPs to such supernatants markedly stimulates microtubule assembly.[360] This clearly shows that the differences between juvenile and adult rat brain extracts are due to the increase in the amount and heterogeneity of MAPs[70,361] (see Section III.B). Thus MAPs seem to be responsible for the increased microtubule density and stability within mature axons.

As for neurofilaments, their appearance within axons parallels the increase in the proportion of stable microtubule arrays. For instance, neurofilaments appear as prominent cytoskeletal elements within mature neurites in PC12 cells after several weeks of exposure to NGF.[354] Neurites in these long-term NGF-treated cultures remain extended for up to 24 h, even in the presence of extremely high concentrations of microtubule-depolymerizing drugs which eliminate nearly all microtubules.[362] This could be due to the presence of substantial numbers of neurofilaments which might provide structural support within neuronal processes for up to 24 h, even after microtubule disassembly. Beyond this time, the progressive deterioration of neurites and cell bodies is triggered, possibly as a consequence of impaired axonal transport (see Section VI.).

These results are consistent with the view that mature axons are more stable and less plastic than developing axons. As Matus has recently emphasized, growth is the major activity of neuronal processes during brain development. Thus, growing neurites are characterized by a balance between the plasticity required for extension and the rigidity by which the appropriate form is sustained. On the other hand, the cytoskeleton becomes less flexible in mature neurons, in which the maintenance of morphology and the transport of organelles are its major activities.[70]

C. CHANGES IN MICROTUBULE COMPONENTS DURING NEURITE EXTENSION AND MATURATION

1. Expression of Microtubule Components

It has been established that microtubule mass increases in parallel with the development of neuronal processes.[343-345] An increment in the levels of microtubule components can conceivably be involved in the promotion of microtubule assembly.

The transiently enhanced expression of some tubulin genes in brain at the time of neuronal growth has been shown (see Section III.A), which has been interpreted as a mechanism to augment the available pool of tubulin required for neurite outgrowth. For example, tubulin expression is induced in response to insulin, insulin-like growth factor II and NGF, which also stimulate neurite extension in several human neuroblastoma cell lines.[363]

In PC12 cells, the tubulin level increases 2.5-fold in response to NGF. Nevertheless, tubulin accumulation occurs approximately 3 d before the boost in microtubule assembly which accompanies neurite outgrowth.[344] The fact that the total tubulin level increases faster than the polymerized tubulin level clearly shows that the amount of tubulin is not limiting for assembly. In other words, the accumulation of tubulin is necessary for neurite outgrowth, but does not itself drive microtubule assembly.[344]

This is even more apparent in certain neuroblastoma cell lines.[343,364] Both differentiated and undifferentiated neuroblastoma cells contain the same total amount of tubulin, about 4 pg per cell in N2A cells, as determined by radioimmunoassay.[343] Consequently, tubulin mass does not change during neurite extension. However, differentiated neuroblastoma cells have a microtubule pool ranging from 48 to 63% of total tubulin, whereas only 11 to 16% of total tubulin is in the microtubule pool of undifferentiated neuroblastoma cells. Thus, the amount of assembled tubulin in differentiated neuroblastoma cells is almost five times greater than in undifferentiated cells.[343]

As MAPs promote *in vitro* microtubule assembly, they are possible candidates for driving microtubule assembly during neurite extension.[344] The fact that extracts from differentiated neuroblastoma cells or mature brain tissue promote microtubule polymerization better than the extracts from the corresponding less-differentiated sources supports this view.[360,364-366]

The net increase in microtubule mass which accompanies neurite outgrowth parallels MAP expression.[344,367-369] It has thus been proposed that the induction of MAPs could be the limiting step for neurite extension. Yet, the roles performed by the different MAPs expressed during neuronal differentiation remain to be clarified.

The expression of MAPs has been studied in fetal mouse brain neuroblasts cultured in a serum-free medium.[369] These cells progressively develop a dense neuritic network. Initiation of neurite outgrowth occurs during the first 3 d of culture, while neurite elongation takes place after 3 d. High molecular weight MAPs are induced during the first culture period, while tau protein increases only after 3 d of culture.[369]

These results can be correlated with *in vitro* kinetic studies which suggest that high molecular weight MAPs may possibly be involved in the nucleation phase of microtubule assembly, while tau binds preferentially to already formed microtubules.[194,195,215,216] Thus, it appears that high molecular weight MAPs may be involved in the early promotion of microtubule assembly required for process extension, while tau might be involved in the stabilization of axonal microtubules. The fact that purified tau microinjected into nonneuronal cells results in the stabilization of microtubule arrays also supports this hypothesis.[202]

MAP expression has also been studied in clonal cell lines of neuronal origin. In PC12 cells induced to differentiate on collagen-coated dishes in the presence of NGF, the increases in microtubule mass and neurite outgrowth closely parallel 10- and 20-fold inductions of tau and MAP-1.[344] If PC12 cells are grown in the presence of NGF without access to an adhesive substrate, they fail to project processes, but both tau and MAP-1 are induced. These "primed" PC12 cells extend neurites immediately after plating on an adhesive substrate, which indicates that the lag phase observed between the addition of NGF to the culture medium and the appearance of neurites in "naive" PC12 cell cultures corresponds to the time required for induction of tau and MAP-1. However, "primed" PC12 cells fail to extend neurites when plated in the presence of cycloheximide, suggesting that other proteins in addition to tau and MAP-1 must be synthesized to produce neurite outgrowth. At present, the identity of these proteins expressed in response to substrate attachment is unknown.[344]

It is important to emphasize that MAP-1 induction has been observed in PC12 cells by several groups,[344,368] whereas other groups have failed to detect such an induction.[370] This disparity is not yet understood, although it may arise from the presence of different MAP-1 subspecies, of which only one is regulated, or from the fact that different PC12 cells clones were used in these studies.

There is some controversy with regard to the contributions of tau and MAP-1 to neurite outgrowth. Drubin et al., have reported that dibutyril cAMP, which stimulates neurite extension in PC12 cells, induces tau to a much greater extent than it induces MAP-1.[344] As neurites grown in the presence of dibutyryl cAMP are shorter and less stable than those produced by NGF,[371] it has been assumed that tau might be sufficient to promote the early phase of neurite outgrowth, while MAP-1 would be essential for neurite elongation.[344] This interpretation contradicts the observations made in primary cultures of fetal neurons suggesting that tau is involved in microtubule stabilization during neurite elongation.[369] An alternative explanation for tau induction by cAMP in PC12 cells which also supports a role for tau in neurite elongation, may be the general susceptibility of tau to many inducers unrelated to neurite extension. For example, tau is induced after crowding in PC12 cells, which does not result in MAP-1 induction nor in neurite outgrowth.[344] Thus, cAMP might promote neurite extension in PC12 cells through the rapid induction of MAP-1, and the resulting neurites may be shorter and less stable, even in the presence of high tau concentrations, due to the low extent of MAP-1 induction and the absence of other NGF-stimulated neurite extension promoters. For example, cAMP interferes with chartin phosphorylation which is essential for NGF-stimulated neurite outgrowth[372] (see Section IV.C.2).

As for other MAPs, the level of MAP-2 augments in PC12 cells treated with NGF[345] and in some neuroblastoma cells lines.[373,374] The synthesis of a 215 Kd MAP increases notably upon differentiation in some neuroblastoma cell lines.[367,375] In addition to the enhanced expression of these well-characterized MAPs, the induction of a 160-kDa microtubule assembly promoting factor in some neuroblastoma cell lines has been described.[376]

The factors which control the expression of microtubule components are not entirely known, although the involvement of some neurothrophic growth factors and hormones in the induction of MAPs has been reported.

As a case in point, NGF may be involved in MAP-1 and tau induction in NGF-responsive neurons,[344] which include neural crest-derived sensory and sympathetic neurons in the peripheral nervous system and cholinergic neurons of the basal forebrain nuclei in the central nervous system.[306,307,377]

It has also been suggested that thyroid hormones are involved in the regulation of MAP expression. Thyroid hormones are required for brain maturation, as shown by the fact that neonatal hypothyroidism leads to severe brain abnormalities which include a hypoplastic neuropile. Thyroid hormones are apparently necessary for proper neurite extension and maturation; this effect of thyroid hormones seems to occur through their action on MAP expression.[355-358,378-384] Triiodothyronine induces neurite extension and increases the synthesis of MAP-1B in neuroblastoma cells.[383] Furthermore, thyroid hormones may also induce adult tau and other adult brain MAPs at the time of axonal maturation (see Section IV.B.3), since hypothyroid animals do not show the transition from juvenile to adult brain MAPs.[358,381,382,384]

Intercellular contacts may also control MAP expression and sorting into different neuronal compartments, which is essential in the establishment of neuronal polarity (see Section IV.E). For instance, cerebellar granule cells grown in sparse monolayer culture express MAP-2 in cell bodies and neurites, while granule cells cultured at high cellular densities, conditions in which intercellular contacts are abundant, show MAP-2 entirely restricted to cell bodies.[385]

Finally, it must be emphasized that the characterization of the genes coding for MAPs and of their regulatory features is necessary to understand the contribution of the expression of MAPs to neuronal morphogenesis.

2. Posttranslational Modifications of Microtubule Components

It has been suggested that neurite extension and maturation require posttranslational modifications of microtubule components in addition to their increased synthesis. Both tubulin and MAPs are modified in different ways in parallel with the development of neuronal processes. These modifications may consequently favor microtubule assembly and stabilization within developing neurites.

Thus, posttranslational modifications of microtubule proteins may trigger microtubule assembly and neurite outgrowth in neural cells which have accumulated a critical amount of microtubule components. Some neuroblastoma cell lines which differentiate upon substrate attachment in the absence of new RNA synthesis[311] constitute a very useful system to study these modifications.

As a case in point, a MAP-1 polypeptide becomes phosphorylated when neuroblastoma cells extend axon-like processes[163,228,236] (see Figure 4). MAP-1 phosphorylation also occurs in PC12 cells induced to differentiate.[345,368,372,386] MAP-1 is heterogeneous, comprising several polypeptides; of these, only one is phosphorylated in neuroblastoma cells and has been identified as MAP-1B.[228] MAP-1B is the main phosphoprotein associated with microtubule preparations obtained from newborn rat brains labeled with $^{32}PO_4$ by a previous intracraneal injection, at a developmental stage in which most neurons are extending processes.[236] MAP-1B is phosphorylated in both serine and threonine residues[163] and its phosphopeptide pattern indicates the presence of numerous phosphorylated residues.[228] It appears that MAP-1B is phosphorylated by a casein kinase II-related enzyme present in differentiating neuronal cells.[228] The functional significance of MAP-1B phosphorylation remains unknown, although the early appearance of phosphorylated MAP-1B and its association with the cytoskeletal fraction suggest that it may favor microtubule assembly inside growing processes.

Three proteins of 69, 72, and 80 kDa, referred to as chartins,[206] also become phospho-

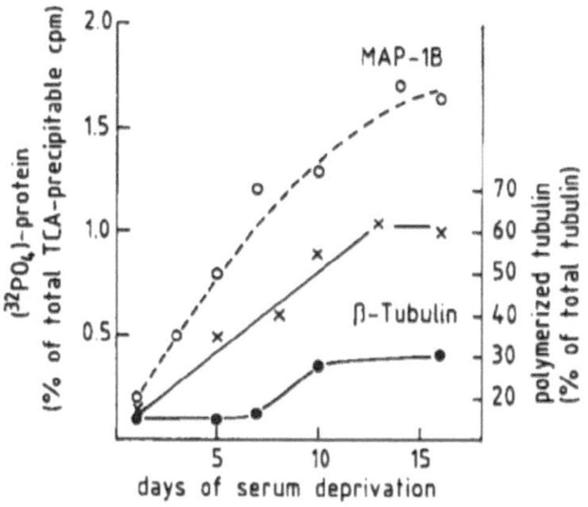

FIGURE 4. Increase in microtubule assembly and in phosphorylation of MAP-1B and β-tubulin during mouse neuroblastoma cell differentiation induced by serum withdrawal from the culture medium.

rylated during neurite outgrowth in PC12 cells.[345,362,372,386] Highly phosphorylated chartins are bound to assembled microtubules.[205,362,386] Interestingly, lithium selectively inhibits both NGF-induced neurite extension and chartin phosphorylation PC12 cells;[386] chartin phosphorylation is also inhibited by forskolin, an activator of adenylate cyclase which consequently elevates intracellular cAMP levels.[372] Forskolin significantly enhances the initial NGF-induced sprouting of short cytoplasmic extensions, while it suppresses the subsequent growth of long axon-like processes. Moreover, forskolin inhibits neurite outgrowth in primed PC12 cells. This suggests that a likely mechanism by which elevated cAMP levels inhibit neurite extension is the interference with NGF-dependent chartin phosphorylation.[372]

It can be concluded that chartin phosphorylation may be required to stabilize microtubules inside long growing neurites. Aletta and Greene have shown that microtubule-depolymerizing drugs abolish chartin phosphorylation,[362] while MAP-1B phosphorylation is practically unaltered,[163,228,362] which indicates that microtubules supply the environment necessary for maximal chartin phosphorylation.[362] However, taxol does not drive chartin phosphorylation, possibly because it blocks the interaction of chartins with microtubules.[362] In view of these data, these authors have proposed that phosphorylation of chartins bound to microtubules may favor the stabilization, elongation or bundling of microtubules, or their interaction with other cytoskeletal elements, thereby facilitating neurite growth.[362]

As for posttranslational modifications of tubulin, it appears that assembled tubulin is subjected to phosphorylation, acetylation, and detyrosination (see Section III.A). These tubulin modifications may enhance the association of modified tubulin with some MAPs, which in turn might increase polymer stability and facilitate the interaction of microtubules with other cellular elements, thus generating stable lengths of neurite behind the advancing process tip.

It has been shown that a β-tubulin isoform is phosphorylated when neuroblastoma cells are induced to extend axon-like processes[118,119,163,164,236] (see Figure 4). The β-tubulin isoform which is phosphorylated corresponds to that expressed by 1003 teratocarcinoma cells at the time of neuronal commitment.[118,119] This β-tubulin isotype, β′1, is present in committed but undifferentiated neuroblastoma cells and is converted into the phosphorylated β′2 during neuronal differentiation. Moreover, the incubation of differentiated neuroblastoma cell extracts with phosphatase results in the conversion of β′2 to β′1.[118,119] By sequencing the phosphorylated peptide, it has been found that β′1 corresponds to the product of the Mβ6 gene.[236] Tubulin phosphorylation is abolished by microtubule-depolymerizing drugs and stimulated by taxol-promoted tubulin assembly.[163] Phosphorylated tubulin is mainly associated with the cytoskeletal fraction suggesting that β-tubulin phosphorylation may modify and stabilize already-formed microtubules inside neurites. The fact that the casein kinase II-related enzyme responsible for β-tubulin phosphorylation acts preferentially on assembled tubulin supports this view.[164]

Acetylation of α-tubulin has been observed in NGF-treated PC12 cells,[156] as well as in neuroblastoma cells.[118,119] Tubulin acetylation also represents a modification of preformed microtubules since (1) it is prevented by treatment with microtubule-depolymerizing drugs, (2) acetylated tubulin is mainly found in the cytoskeletal fraction, and (3) the tubulin polymer is a better substrate than soluble tubulin for α-tubulin acetylase.[156,157]

Detyrosination of α-tubulin, which is also a modification of already formed microtubules, seems to occur in parallel with axonal matruation and synaptogenesis[149,150] (see Section III.A).

The enzymology of these posttranslational modifications of microtubule proteins has not been thoroughly studied. Thus, the mechanisms responsible for the activation of the modifying enzymes during neuronal differentiation remain to be established.

D. ROLE OF MICROTUBULE PROTEINS IN THE ESTABLISHMENT OF NEURONAL POLARITY

The establishment of neuronal polarity, which involves the separation of microstructural domains within a nerve cell, is not yet understood. Among the unsolved questions are how the branching patterns of axons and dendrites are controlled and what determines whether a given neurite becomes an axon, which grows unbranched over a long distance, or a dendrite, which branches locally near the cell body.

Axons and dendrites differ in their morphology, rate of growth, macromolecular composition, and cytoskeletal ultrastructure (see Section II.B). It has been shown that MAP-2 may partly define a somatodendritic domain, while tau is mostly present within axons (see Section III.B). The segregation of this molecule has been observed in different primary cultures.

For example, in cultured hippocampal neurons, MAP-2 immunoreactivity is present in axon-like processes during the first week of neurite formation, after which it disappears from axons and is confined to the somatodendritic compartment.[48,272] In dissociated neuronal cultures of the rat cerebrum, cells observed after plating also show equally intense MAP-2 and tau immunostaining throughout the cell body and initial neurites.[49] In fact, MAP-2 is present in neurites prior to the assembly of microtubules, thus suggesting that MAP-2 could act as a pioneer molecule which defines a process and then promotes the formation of a microtubule cytoskeleton.[387]

When neurites develop dendritic or axonal features, MAP-2 immunoreactivity disappears progressively from axon-like processes.[49] This change in subcellular localization may correspond to the transition from MAP-2C to MAP-2A expression (see Section III.B.3). The molecular mechanisms by which different MAPs are sorted into their appropriate compartments remain to be studied. One possibility is that ribosomes synthesizing MAPs might become nonrandomly distributed. The fact that certain polyribosomes are found within dendrites, while axons lack polyribosomes has been described by Bartlett and Banker.[388] Furthermore, a selective transport of some RNAs into dendrites has been recently characterized in cultured hippocampal neurons.[389]

MAP-2 immunoreactivity seems to be lost during the *in vitro* development of axons from rat hippocampal neurons in a proximodistal fashion.[48] Residual distal MAP-2 immunoreactivity within axons can be generated if MAP-2-containing microtubules are prevented from entering the axon. This supports the view that microtubules elongate at their proximal end in the growing axon. However, Kosik and Finch[49] have observed a completely different pattern in the elimination of MAP-2 from axon-like processes. MAP-2 immunoreactivity appears proximally within maturing axons, while tau immunoreactivity is observed distally in the same axons.[49] These results are consistent with the transport of unassembled microtubule components to the growing tip, where microtubule assembly would take place (see Section II.A).

Dendrite growth has also been studied in embryonic rat sympathetic neurons grown under different culture conditions.[390] After establishing the cultures in serum-free medium for 1 week, the cultures were either transferred to a serum-containing medium or maintained in serum-free medium. In serum-containing medium, sympathetic neurons become multipolar, with short tape-ring dendrites and long thin axons. These cells contain MAP-2 in their cell bodies and dendrites. Neurons maintained in serum-free medium are usually unipolar, having only a long axon-like process. In these unipolar cells, intense immunostaining for MAP-2 in axons and cell bodies is observed.[390]

It has been suggested that the axonal or dendritic nature of a neuronal process and the extend and branching of its arborization may be controlled by extracellular signals, including soluble factors and intercellular contacts.

For example, retinoic acid induces the extension of branched dendritic-like processes, while dibutyril-cAMP induces the outgrowth of unbranched axon-like processes in N2A neuroblastoma cells.[374] MAP-2 expression is enhanced upon retinoic acid treatment.[374]

Specific neuron-glia interactions have also been involved in the establishment of polarity in different neuron types.

In sensory neurons from embryonic dorsal root ganglia, a morphological transformation from an immature bipolar form to a mature pseudounipolar form is observed in the presence of satellite Schwann cells, but not in the presence of other cell types such as glioma cells, cortical astrocytes, myoblasts, or skin fibroblasts.[391] Sympathetic neurons grown *in vitro* in the absence of satellite Schwann cells become unipolar, showing an axon-like neurite, while they become multipolar in the presence of satellite Schwann cells. These multipolar sympathetic neurons show a highly branched dendrite-like MAP-2-positive neuritic compartments.[392,393]

In the central nervous system, astrocytes seem to regulate the type of processes that a nerve cell extends. Dopaminergic neurons from the mesencephalon extend axon and dendrites only when they are plated on astrocyte monolayers derived from the mesencephalon. When these neurons are plated on astrocyte monolayers obtained from the striatum, they exhibit only one long, thin, axon-like neurite.[394] Likewise, neurons from the striatum develope both axon and dendrites in the presence of striatum astrocytes, but not in the presence of mesencephalon astrocytes.[395] Moreover, AMP-2 expression is higher in neurons cultured on astrocytes derived from the same brain region than in neurons cultured on astrocytes obtained from a different brain region, and MAP-2 is localized to the dendritic compartment.[395] The molecular mechanisms leading to enhanced MAP-2 expression and dendrite extension remain to be established. It has been postulated that region-specific neurotrophic factors and local isoforms of cell and substratum adhesion molecules may affect the initiation sites for dendrites and somehow control their rate of extension.

However, some evidence contradicts this view. As a case in point, fetal hippocampal neurons cultured in serum-free medium can differentiate fully despite the absence of spatially organized environmental cues, and with no contact with neighboring neuron or glial cells.[6,48,388,396,397] Cultured hippocampal neurons extend both axons and dendrites after 1 week in culture.[48,388] After 4 weeks in culture, neurons become interconnected by an extensive network of processes which make frequent synaptic contacts with one another. Dendrites are always postsynaptic, indicating that synapses do not form indiscriminately. Thus, the establishment of neuronal polarity seems to be largely under endogenous control.[396]

The establishment of neuronal polarity in these cultures follows a highly stereotyped pattern.[6,397] After plating the cells, numerous short processes arise from the cell body, and occasionally compete with one another to become axons. First one elongates rapidly, then the other and, eventually, one becomes the definitive axon while the other retracts. After a few days, the remaining short neurites begin to grow slowly to become dendrites. In most neurons, when the axon is experimentally transected, one of the undamaged dendrite-like processes takes on axonal features, while the stump of the original axon becomes a dendrite. The new axon loses MAP-2 immunoreactivity while MAP-2 becomes apparent in the dendrite formed from the original axon. These results suggest that each of the processes initially extended by a neuron can differentiate into an axon or a dendrite, and that the assumption of axonal features by one neurite prevents the others from becoming axons.[397] The molecular mechanisms which underlie this phenomenon are completely unknown.

E. ROLE OF MICROTUBULE PROTEINS IN NEURONAL PLASTICITY

The point at which neural development ends has not been defined accurately, probably because it never ends. Thus, in both the developing and adult nervous systems, neurons undergo morphological changes in response to their synaptic input; this is referred to as neuronal plasticity. It has been proposed that these neuron shape variations give rise to modifications of neural connectivity, thereby providing the adult nervous system with a remarkable flexibility which may form the basis for learning and memory.

Repeated *in situ* imaging of adult mammalian neurons show changes in their dendritic branching; some branches retract, others elongate, and others are formed *de novo*.[398] Obviously, these events must be associated with a rearrangement of the microtubular cytoskeleton within dendrites.

It is now generally accepted that changes in neuronal cytoarchitecture may take place through posttranslational modifications of cytoskeletal proteins, in response to the different second messengers, generated by the binding of neurotransmitters to their receptors.[399] Neuronal plasticity in the visual system is a clear case in point.[267]

The developing visual system of cats has been a very useful model in the study of the biochemical basis for neuronal plasticity, since there is a critical period during which visual experience permanently molds some aspects of visual function.

The cAMP-stimulatable phosphorylation of MAP-2, a dendrite-associated protein (see Sections II.B, III.B, and IV.E), is strongly dependent on visual experience.[267] Under normal rearing conditions, *in vitro* MAP-2 phosphorylation in visual cortex homogenates increases during development. After rearing in the dark, *in vitro* MAP-2 phosphorylation in visual cortex homogenates is diminished in comparison to that of age-matched normally reared kittens, and this decrease can be reversed by a short exposure to light. Furthermore, the effect of visual stimulation on *in vitro* MAP-2 phosphorylation is specific for the visual cortex and is not observed in the auditory cortex or the lateral geniculate nucleus of the same animals. The decreased *in vitro* cAMP-stimulated MAP-2 phosphorylation after rearing in the dark probably indicates a more phosphorylated state of MAP-2 *in vivo* as compared to MAP-2 in normally reared animals. The enhanced *in vitro* cAMP-stimulated MAP-2 phosphorylation observed after exposure to light would then indicate a dephosphorylation of cAMP-stimulated sites in MAP-2 *in vivo*.[267]

In vitro phosphorylation of MAP-2 by cAMP-dependent protein kinase seems to reduce its association with microtubules, inhibiting microtubule assembly,[93,219,265,266] and abolish its interaction with actin.[78,79,266] Thus, *in vivo* cAMP-dependent MAP-2 phosphorylation may be considered responsible for the dissociation of the dendritic cytoskeletal structure which underlies the cell shape changes associated with neuronal connectivity variations. The more phosphorylated state of MAP-2 *in vivo* after dark rearing suggests the existence of a relatively uncrosslinked, malleable cytoskeleton within dendrites which would allow shape changes. Analogously, the more dephosphorylated state of MAP-2 within the light-stimulated neuronal pathways suggests that the dendrite scaffolding has been stabilized.[267]

Changes in the synaptic connectivity of the visual system are abolished after perfusing the visual cortex with 6-hydroxy-dopamine (6-OHDA), a potent neurotoxin specific for noradrenergic neurons. Visual plasticity in 6-OHDA-treated brains is restored after perfusing with norepinephrine or dibutyryl cAMP. Thus, visual plasticity seems to depend on the norepinephrine stimulation of β-adrenergic receptors in visual cortex neurons, which results in increased intracellular cAMP levels.[400] Cyclic AMP may facilitate visual cortex plasticity through MAP-2 phosphorylation. Specifically, this phosphorylation might produce the transient release of the cytoskeletal scaffolding within dendrites and dendritic spines, which could permit changes in synaptic connectivity.[267]

Interestingly, neuronal plasticity is not restricted to dendrites; for instance, the remodeling of presynaptic endings of central neurons on the cell bodies of parasympathetic ganglion cells has also been observed,[401] indicating that mechanisms similar to those described for dendrites, i.e., phosphorylation of cytoskeletal proteins, may operate within nerve terminals.

V. ROLE OF MICROTUBULES IN AXONAL TRANSPORT

A. GENERAL FEATURES OF AXONAL TRANSPORT

The neuron is a unique cell containing a long extension, the axon, which lacks ribosomes and must be supplied with proteins synthesized in the cell body. Thus, transport is the primary mechanism to endow the axon with the proteins and organelles required for axonal growth during development or regeneration, for the maintenance of axonal structure and for the synthesis of neurotransmitters at the presynaptic terminals.[1,43-45,402-407]

The necessity of an axonal transport mechanism was first appreciated by Ramón and Cajal, who established that the integrity of the axon depends on the neuron cell body.[12] When a peripheral nerve axon is severed, the segment which is disconnected from the cell body degenerates, while the remaining portion of the axon regenerates. Both axon regeneration in peripheral nerves and axon outgrowth in embryonic neuroblasts take place as a result of materials continuously supplied from the cell body to the axon.

The necessity of an axonal transport mechanism was first appreciated by Ramón y Cajal, who established that the integrity of the axon depends on the neuron cell body.[12] When a peripheral nerve axon is severed, the segment which is disconnected from the cell body degenerates, while the remaining portion of the axon regenerates. Both axon regeneration in peripheral nerves and axon outgrowth in embryonic neuroblasts take place as a result of materials continuously supplied from the cell body to the axon.

The observation of a swelling on the proximal side of the point at which a nerve has been mechanically constricted provided the first direct evidence for axonal transport.[408] This swelling resulted from the accumulation of materials synthesized in the cell body and moved in an anterograde direction upon relieving the compression. In addition to this anterograde axoplasmic transport, the fact that polio virus reached the motor neuron cell bodies from its points of inoculation in the periphery caused Weiss and Hiscoe to suggest the existence of an axonal retrograde transport mechanism.[408]

Different experimental approaches have been used in the study of axonal transport (for a review see Reference 1).

The retrograde component of axonal transport has been demonstrated through the injection of labeled materials near the terminal region of a nerve. These materials, which include tetanus toxin, NGF, horseradish peroxidase, and certain viruses, are taken up by nerve endings and transported in pinocytotic vesicles to the cell bodies. This procedure has been used widely to trace anatomic pathways, since it allows the determination of the neuron cell body from which a given axon has arisen.

The anterograde component of axonal transport has been studied mainly by radioisotopic labeling methods. The experimental systems most commonly used have been the mammalian visual system and the sciatic nerve. In the visual system, radiolabeled amino acids injected into the vitreous humor are incorporated into proteins by the retinal ganglion cells. Labeled proteins are then transported down the axons constituting the optic nerve which terminate in the lateral graniculate nucleus and the superior colliculus. As radioactive amino acids are available as precursors for only a short period, the injection is the equivalent of a pulse-labeling experiment. The movement of the pulse-labeled proteins is analyzed by removing the optic nerve, sectioning it into segments and measuring the distribution of radioactivity within these pieces at different times after injection. A similar procedure applies to the sciatic nerve.

The kinetics of axonal transport which emerges from these studies is rather complex, indicating the existence of different sets of conveyed proteins according to their rates of transport.[409-411] These groups of proteins move as coherent waves down the axons at very different rates, with no exchange of proteins between them. This has led to the proposal that transported proteins move as parts of cytological structures.

Four main components of axonal transport have been identified:

1. Slow component A (SCa). This group of proteins moves at a rate of 0.2 to 1 mm/d (0.002 to 0.01 μm/s) and consists primarily of the polypeptides forming the microtubule-neurofilament network. It ha been suggested that the principal function of SCa is to provide the stable scaffolding which supports axonal structure and gives the axon its dimensions.[43,45]
2. Slow component B (SCb). This is a quite complex group of proteins comprising more than 100 polypeptides moving at 2 to 8 mm/d (0.02 to 0.08 μm/s). Only a few of these have been identified so far, including contractile proteins (actin and myosin), cross-linking proteins (fodrin and clathrin), regulatory proteins (calmodulin), and a variety of metabolic enzymes (enolase, aldolase, and creatine phosphokinase). SCb seems to correspond to the transport of actin microfilaments and associated proteins. Some of these are soluble axonal proteins which bind specifically and reversibly to actin microfilaments.[45]
3. Intermediate component. This corresponds to mitochondria, which are conveyed along microtubules at a rate of 50 to 100 mm/d (0.6 to 1.2 μm/s).
4. Fast component. This is a rather complex group of membrane-associated proteins moving at a rate of 200 to 400 mm/d (2.4 to 4.8 μm/s) and corresponds to the bidirectional movement of membrane organelles along microtubules.[412] Thus, it actually comprises two components: an orthograde (or anterograde) and a retrograde component. It appears that each membrane organelle has a preferred direction of movement. For example, vesicles supplying presynaptic neurotransmitter-related enzymes move anterogradely, while endocytic and prelysosomal vesicles move retrogradely. The function of the fast anterograde axonal transport is the conveying of membrane organelles required for presynaptic events at the axon tip. In developing neurites, anterogradely-moving vesicles support the membrane extension which occurs at the growth cone. The retrograde component of axonal transport seems to perform two essential functions. First, it facilitates membrane recycling, since it conveys pre-lysosomal membrane organelles from the nerve ending to the cell body. Second, it provides a retrograde pathway for the transmission of information through the transport of receptors associated with membrane vesicles. These receptors can convey signal molecules from the axon tip to the neuronal soma. The retrograde transport of NGF, for example, is well documented. Laduron has proposed that the retrograde conveying of receptor-bound signal molecules may play a prominent role in neuronal development and long-term memory processing.[413]

The maintenance of the shape and functionality of neurons thus depends on axonal transport, in which microtubules are key elements, as they move within the axon as part of the slow component and also serve as tracks for the fast movement of membrane organelles.

B. SLOW AXONAL TRANSPORT OF MICROTUBULE PROTEINS

The kinetics of microtubule protein transport varies significantly in different axons.[45,407] In the optic nerve, tubulin is translocated in SCa at the same rate as neurofilament proteins, suggesting that microtubule and neurofilament proteins are tightly cross-linked. SCa-conveyed proteins are easily analysed, as they are the sole labeled proteins which remain in the optic nerve for long times after radioactive amino acid injection into the eye.

In contrast, in the axons of dorsal root ganglion and ventral motor neurons, a significant fraction of tubulin moves at a rate consistent with transport in SCb, together with actin microfilaments, although most tubulin is transported with neurofilament proteins in SCa.[414,415] The significance of these differences is not understood, but they may be connected with ontogenetic, structural and functional differences between axons.[415] Thus, the optic nerve arises from retinal ganglion cells which are derived from the rostral region of the embryonic brain, while dorsal root ganglion cells innervating sensory receptors originate from the neural crest, and ventral motor neurons innervating muscle cells arise from the caudal region of the neural tube. Regardless of its functional meaning, it must be emphasized that different subsets of microtubule proteins appear to move at different rates within certain axons.[407]

These subsets of microtubule proteins differ in some properties. For instance, most tubulin undergoing transport in SCa fails to be solubilized by cold-treatment, millimolar calcium, or high salt concentrations,[72,416,417] whereas tubulin associated with SCb is effectively solubilized by these treatments. Moreover, SCa- and SCb-transported tubulins differ in their isotubulin composition.[417]

The relative insolubility of SCa-translocated microtubule protein may result from its interaction with neurofilaments.[416,418] However, 90% of neurofilament protein is extracted from optic nerves with 2 *M* guanidine-HCl, while half of the cold-stable axonal tubulin remains insoluble, thus indicating that insolubility might be an intrinsic property of SCa-translocated tubulin.[417] Posttranslational modifications of tubulin, in addition to MAP content, may possibly account for the differences observed between SCa and SCb-conveyed microtubule proteins.

As for the axonal transport of MAPs, it has been found that tau proteins[419] and MAP-1B[420] are conveyed together with tubulin in the optic nerve.

The mechanism of slow axonal transport is unknown, though it seems to depend on the structural integrity of axonal microtubules, since it is abolished by microtubule-depolymerizing drugs.[421,422]

The "structural hypothesis" proposes that stable polymers are the moving elements in slow axonal transport.[45,402,405-407] According to this model, cytoskeletal proteins assemble in the cell body just before transport into the axon.[46] Thus, there would be a continuous neurofilament-microtubule lattice from the cell body to the axon tip which would be organized through numerous lateral cross-bridges between individual neurofilaments and microtubules (see Section II.A). Such a network would be constantly generated from newly synthesized proteins in the neuronal soma and conveyed down the axon. Eventually, the neurofilament-microtubule network would be disorganized in the synaptic terminal by calcium-dependent disassembly of microtubules and calcium-activated proteolytic degradation of neurofilaments.

As a molecular mechanism for transport, Lasek has proposed a polymer-sliding model.[407] Faster-moving SCb polymers (microfilaments and microtubules) would slide in relation to slower-moving SCa polymers (microtubules and neurofilaments). The interactions between the polymer sidearms might generate the force required for sliding. Such polymer sidearms might be ATPases similar to those mediating fast organelle transport (see Section V.C).

Some experimental evidence favors this structural hypothesis. First, microtubule and neurofilament assembly seems to take place immediately after synthesis of the corresponding proteins[46] (see Section II.A). Second, the fluorescent recovery following the photobleaching of neurite segments in PC12 cells to which fluorescent tubulin has been microinjected shows that tubulin is transported within neurites in coherent nondiffusing waves. The edges of the photobleached bands remain sharp, further indicating that most conveyed tubulin is not free to diffuse.[423] Third, the neurotoxin β,β′-iminodiproprionitrile (IDPN), which disconnects neurofilaments from microtubules, selectively blocks neurofilament movement but has little effect on the transport of tubulin, suggesting that the association of neurofilaments with microtubules is required for their conveyance down the axons.[424]

Alternatively to the structural hypothesis, some authors sustain that newly synthesized cytoskeletal proteins migrate within axons as precursors to a stationary axonal cytoskeletal network.[425,426] In this model, the neurofilament-microtubule lattice is assembled in the axon and is mainly stationary. It has been shown that only a small amount of neurofilament proteins are transported in SCa, while most neurofilaments remain stationary.[426,427] This model is also compatible with the nonuniformity and regional differentiation of the neurofilament-microtubule lattice along axons, the nodes of Ranvier consisting a clear example[29] (see Section II.A), and with the fact that tubulin assembly may occur at the axon tip[47] (see Section II.A).

It has recently been proposed that neurofilament and microtubule proteins are transported in the form of unique structural complexes which may function as supramolecular precursors for the stationary axonal cytoskeleton and could migrate within axons along stable cytoskeletal polymers.[428,429] The protein aggregates consisting of tubulin, tau, high molecular weight MAPs, fodrin, and neurofilament proteins which are found in the particulate fraction obtained after extracting conventional brain microtubule preparations with nonionic detergents and high salt concentrations are candidates for this role and have therefore been called SCAPs (slow component-associated particles).

Interestingly, isolated SCAPs contain a vanadate-inhibited microtubule-associated ATPase activity which may act as a force-generating mechanism. When isolated SCAPs are added to microtubules in the proper ratio and in the presence of ATP, an *in vitro* gelation and contraction of the microtubule preparation is observed. It appears that SCAPs bind to microtubules independently of the presence of ATP, but begin to move along microtubules only when ATP is available. The translocation of SCAPs leads to the formation of large aster-like aggregates of SCAPs and microtubules[428,429] The phenomenon of microtubule gelation and contraction was first appreciated following the addition of ATP to microtubule protein.[430] AVEC-DIC microscopy revealed that it involved the aggregation of particulate material (SCAPs) migrating along microtubules at rates close to 1 μm/min.[431]

Some features of this *in vitro* gelation of microtubules resemble slow axonal transport:

1. The rate of SCAP transport along microtubules is in the range of slow axonal transport.
2. SCAPs contain the same proteins conveyed as part of SCa.
3. Both SCAPs and SCa-transported proteins are insoluble in cold buffers containing high salt concentrations and nonionic detergents.
4. Both SCAP translocation and SCa depend on the presence of microtubules.

This would mean that axonal stationary microtubules serve as tracks for the slow movement of SCAPs in the same way as they do for membrane organelles conveyed by fast axonal transport (see Section V.C). This model calls for the existence of mechanisms such as posttranslational modifications regulating the transfer of cytoskeletal proteins from SCAPs to stationary polymers at the appropriate axonal sites. Nevertheless, it remains to be established whether SCAPs actually exist *in vivo* and are involved in slow axonal transport.

At present, it is not possible to favor any one of the proposed mechanisms for slow axonal transport: the structural hypothesis which involves the presence of sliding polymers or the stationary cytoskeleton hypothesis which involves the existence of soluble supramolecular precursors such as the so-called SCAPs.

C. FAST AXONAL TRANSPORT OF MEMBRANE ORGANELLES ALONG MICROTUBULES

1. Role of Microtubules in Fast Axonal Transport

Membrane organelles conveyed by the fast component of axonal transport can be subdivided into three groups:

1. Synaptic vesicles and membrane organelles derived from the Golgi apparatus or smooth endoplasmic reticulum and destined for insertion into the plasma membrane, all of which move anterogradely at velocities of up to 4 to 5 $\mu m\ s^{-1}$.
2. Mitochondria and other large membrane organelles, which are transported away from the cell body more slowly, at about 0.5 $\mu m\ s^{-1}$.
3. Pinocytic vesicles and prelysosomal organelles, which move retrogradely at velocities of up to 4 to 5 $\mu m\ s^{-1}$.

These three modes of fast transport are inhibited in the presence of drugs interacting with microtubules such as colchicine, vinblastine, or podophylotoxin.[432-434] Other agents which cause a decrease in the density of microtubules, such as calcium and cadmium ions, also weaken or even block fast axonal transport *in vivo*.[434] A similar effect is observed upon injection of an anti-tubulin antibody, NS-20.[435] Interestingly, other antisera against tubulin, in addition to neurofilament and actin antisera, have no effect on fast axonal transport *in vivo*. Thus, NS-20 may recognize epitopes on the microtubule binding sites for molecules involved in fast axonal transport.[435]

Electron microscopy studies have shown that the axonal cytoplasm is compartmentalized around the two major cytoskeletal elements, neurofilaments and microtubules, and that membrane organelles are concentrated within microtubule domains (see Section II.A). In fact, all membrane organelles become confined to the large central microtubule domain which is found upon exposition of rat peripheral nerves to the neurotoxin IDPN. Furthermore, autoradiograph studies performed after the application of radioactive amino acids to the cell bodies of neurons treated with IDPN show that almost all the radioactivity associated with fast-conveyed proteins is present exclusively in the microtubule domain.[436,437]

These results indicate that microtubules play a key role in the mechanism of fast axonal transport, which has been corroborated by high-resolution light microscopy.

High-resolution light microscopy involves the amplification and adjustment of video images produced by polarizing, phase-contrast, or interference microscopy to subtract background light and to enhance the contrast of the remaining image. The video enhancement of images generated by differential interference contrast (Nomarski) microscopy, called AVEC-DIC (Allen video enhancement contrast-differential interference contrast microscopy), has made the observation of subcellular structures possible, even within living cells. AVEC-DIC microscopy also allows the observation of structures which are below the limit of resolution of light microscopy, such as single microtubules and synaptic vesicles which had only been seen previously in fixed material using electron microscopy.[412,438-443]

With this technique, Allen et al.[439] first visualized the saltatory movement of large membrane organelles at velocities of up to 0.5 μm/s and the continuous movement of small vesicles at velocities close to 2 μm/s within the squid giant axon. These results agree well with the speed of fast axonal transport measured by the radioisotopic labeling method (see Section V.A), and further confirm that there are several components of membrane organelle transport. Moreover, they suggest an inverse relationship between the size of the conveyed organelles and their velocity of movement within axons.

Fast organelle transport can also be observed upon adding ATP to squid axoplasm preparations, obtained by extrusion from the axolemma, in the same way as toothpaste is squeezed.[412,441] Indeed, the rapid bidirectional translocation of organelles is better visualized when the axoplasm is gently dissociated in a buffer which also contains ATP.[440,442] Under these conditions, individual filaments detached from the bulk of axoplasm support fast organelle transport. Surprisingly, all membrane organelles, regardless of their size, move continuously and with the same velocity of about 2 μm/s along isolated filaments, suggesting that the transport machinery is common to all membrane organelles moving in a given direction. Thus, the correlation between organelle speed and size observed both *in vivo* and in intact axoplasm is probably due to the sieving of large particles by the highly cross-linked axoplasmic cytoskeleton. This sieving effect is lost after axoplasm dissociation.

By combining AVEC-DIC and electron microscopy, the individual filaments which serve as tracks for the conveyance of organelles in dissociated axoplasm have been unambiguously identified as microtubules.[443] Moreover, it is clearly seen that membrane organelles move in both directions along a single microtubule. This implies that there must be different motors conferring directionality to the translocation of organelles on an individual microtubule.[443]

Organelle translocation is strictly dependent on ATP hydrolysis, suggesting that the putative motor proteins are ATPases which reversibly connect organelles with microtubules. Thus, membrane organelles cease moving and become stably associated with microtubules when dispersed axoplasm is incubated in the presence of the nonhydrolyzable ATP analogue, AMP-PNP.[444]

It is not entirely clear at present whether fast organelle transport occurs on all the axonal microtubules or only on a subpopulation of them. When cold treatment is applied to segments of lobster giant axons, membrane organelles accumulate in rows along some, but not all, microtubules.[445] This supports the view that axons contain two distinct populations of microtubules: architectural microtubules which contribute to axonal structure by interacting with other cytoskeletal elements and transport microtubules which are the preferred substrates for organelle transport.[445]

2. *In Vitro* Assays for Microtubule-Based Motility

AVEC-DIC microscopy also allows direct visualization on a glass coverslip of *in vitro* assembled microtubules. In this system, it has been shown that isolated squid axoplasmic organelles are translocated along salt-extracted taxol-stabilized microtubules in the presence of ATP at velocities close to those reported in dispersed axoplasm. However, only a few membrane organelles move along microtubules, unless a supernatant fraction from squid axoplasm is added. In addition to greatly increasing the number of organelles moving along microtubules, axoplasmic supernatants support the translocation of carboxylated latex beads on microtubules and also cause microtubules to bind to and move along the surface of glass coverslips.[446]

The directionality of the movement of latex beads along microtubules has been examined using centrosome-nucleated microtubule arrays. Centrosomes isolated from mouse neuroblastoma cells are incubated with purified tubulin in the presence of GTP to obtain radial arrays of *in vitro* assembled microtubules with their plus ends pointing outwards.[447] These centrosome-microtubule complexes are sedimented through a glycerol cushion onto polylysine-coated glass coverslips which are then transferred to a taxol-containing buffer. Under these conditions, the gliding movement of microtubules on the glass surface is prevented. Upon addition of ATP and axoplasmic supernatants, latex bead translocation can be assayed.[442]

When an axoplasmic supernatant fraction which is obtained by a high-speed centrifugation after first removing tubulin and MAPs by taxol-induced polymerization is used, latex beads move only from the minus to the plus ends of the centrosome-nucleated microtubules. This direction corresponds to anterograde transport along microtubules in the intact axon.[448,449]

When the axoplasmic supernatant fraction is obtained by low speed centrifugation of a homogenate of the axoplasm which is obtained in the presence of 0.5% Triton-X-100 and is incubated with taxol and GTP to remove tubulin and MAPs, the latex beads are translocated along microtubules in both directions.[448] Interestingly, 1 m*M* *N*-ethylmaleimide and 25 μ*M* vanadate inhibit the retrograde bead movement, while the anterograde bead movement persists unless 100 μ*M* vanadate is added. This indicates clearly that distinct force-generating proteins are involved in anterograde and retrograde axonal transport.[448]

The movement of microtubules on a glass surface and the translocation of latex beads along radial arrays of microtubules have been used as *in vitro* assays for monitoring the purification of the motility proteins involved in fast axonal transport. In this way, a protein called kinesin has been identified as the putative anterograde motor protein present in high-speed axoplasmic supernatants, while a group of highmolecular weight dynein-like proteins are considered the retrograde motor proteins present in low-speed axoplasmic supernatants.[450]

There are, nevertheless, some differences between the bidirectional movement of latex beads and the translocation of membrane organelles along microtubules. Membrane organelles move with approximately the same velocity, about 2 μm/s, in anterograde and retrograde directions. Latex beads, however, translocate along microtubules in the retrograde direction at 1.5 μm/s, while they move in the anterograde direction more slowly, at about 0.5 μm/s. These differences in velocity probably reflect differences in the manner in which the anterograde and retrograde motor proteins interact with latex beads compared with membrane organelles.[448]

3. Kinesin: a Putative Microtubule-Associated Motor for Anterograde Organelle Transport

As the nonhydrolysable ATP analogue, AMP-PNP, causes irreversible binding of membrane organelles to microtubules in dispersed axoplasm preparations,[444] it is expected that the force-generating proteins involved in organelle transport become stably associated with microtubules in buffers containing AMP-PNP.

When high-speed axoplasmic supernatants are incubated with salt-extracted taxol-polymerized microtubules in the presence of AMP-PNP, a 110-kDa polypeptide associates with microtubules.[449] Similar 120- to 130-kDa polypeptides are found in chicken[451] and mammalian brain extracts.[449,452-454] The name of this protein, kinesin, is derived from the Greek term for movement, since it is probably involved in microtubule-based motility.

Kinesin has been purified from squid axoplasm or bovine brain extract by microtubule affinity in the presence of AMP-PNP followed by gel filtration and hydroxyapatite chromatography.[449] Both squid axoplasm and bovine brain kinesin have an approximate molecular weight of 600 to 700 kDa as determined by gel filtration, and a sedimentation coefficient of 10S. By polyacrylamide gel electrophoresis in the presence of SDS, it appears that squid kinesin is composed of 100-, 70-, and 65-kDa polypeptides in a stoichiometry of 4:1:1. An 80-kDa polypeptide is also associated in substoichiometric amounts with squid kinesin. Mammalian brain kinesin consists of 120- to 130- and 62-kDa polypeptides in a stoichiometry ratio of 2:1.[449,452,453] The presence of additional subunits has also been observed in mammalian brain kinesin.[452]

Purified kinesin induces the movement of isolated microtubules on glass surfaces and the translocation of latex beads along microtubules. Kinesin-coated latex beads always move towards the plus end of microtubules, thus suggesting that kinesin is probably the force-generating protein responsible for the anterograde component of fast axonal transport.[448-450] In support of this, the anterograde movement of latex beads along microtubules is no longer observed when a low speed axoplasmic supernatant is depleted of kinesin after passing it over an antikinesin monoclonal antibody affinity column.[448] It has furthermore been found that kinesin enhances the number of membrane organelles moving anterogradely along isolated microtubules, although this effect is not so clear, as some organelles contain bound kinesin and therefore move in the absence of added kinesin.[449,450]

Morphological examination of native kinesin has provided some clues about its possible function. Native kinesin is an elongated molecule, of about 100 nm, as observed by electron microscopy with the techniques of rotary shadowing or negative staining.[453] Kinesin is essentially a rod with a small forked domain at one end and a larger, globular, branched domain at the other. Near the middle of the rod, there is a flexible joint. The forked domain seems to contain the nucleotide-sensitive microtubule-binding domain, while the larger domain may be the site for interaction with beads and organelles.[453] This larger domain may interact with another microtubule in a nucleotide-insensitive manner, similar to its binding to negatively charged surfaces such as glass coverslips or carboxylated beads. Thus, kinesin appears to interact reversibly with a microtubule in a nucleotide-sensitive manner through its forked domain, while its larger domain binds more stably and permanently with another microtubule, a glass surface, a bead or an organelle, thereby producing the movement of microtubules along other microtubules or on glass coverslips and the translocation of beads and organelles along microtubules.

The stable attachment of kinesin to glass surfaces has been visualized directly, Thus, when a small drop of kinesin is placed on a glass coverslip, the excess is rinsed away and microtubules are then applied, the movement of these microtubules on the kinesin-coated surface, but not in other regions of the coverslip, is clearly observed, provided that ATP is present.[446,449] This further suggests that the mechanism by which kinesin powers the translocation of particles along microtubules involves its stable attachment to the particle and its nucleotide-dependent reversible binding to a microtubule.

Kinesin seems to have a reversible cycle of microtubule association-dissociation in the presence of ATP,[450] while it is bound stably to microtubules in the presence of AMP-PNP,[449] inorganic tripolyphosphate,[452] or after a complete depletion of Mg ATP.[450] The strong binding state of kinesin for microtubules which is observed after depletion of ATP (or depletion of Mg in the presence of ATP) may be similar to that which has been observed for cilliary dynein. However, AMP-PNP weakens the interaction between dynein and microtubules, while it favors the binding of kinesin to microtubules, thus indicating that kinesin belongs to a new family of microtubule-associated motility proteins.[450]

It has proposed that the ATP-dependent interaction of kinesin with microtubules can generate unidirectional movement through a cycle of attachment, motion and detachment similar to that described for dynein.[450] This mechanochemical cycle would comprise the following steps:

1. Binding of ATP to kinesin and dissociation of ATP-containing kinesin from the microtubule.
2. ATP hydrolysis in dissociated kinesin.
3. Binding of ADP-containing kinesin to the microtubule.
4. Release of ADP from kinesin bound to the microtubule and conformational change of the molecule, giving rise to the force-producing step.
5. Binding of ATP to kinesin.

Throughout this cycle, kinesin would be permanently attached to the translocated particle.

There is some variation in the ATPase activity of different kinesin preparations, depending on the exact conditions used in its purification. Thus, kinesin prepared either from bovine brain using tripolyphosphate[452] or chicken brain using AMP-PNP[451] has notably higher

ATPase activity than kinesin prepared from either squid giant axon or bovine brain using AMP-PNP.[449] These discrepancies are due possibly to the removal or retention of some regulatory cofactors of ATPase inhibitors during the purification procedures.

The ATPase activity of kinesin is greatly increased in the presence of microtubules, which suggests that it is coupled to microtubule binding and force generation. For instance, bovine brain kinesin purified using tripolyphosphate hydrolyzes Mg ATP at a rate of 60 to 80 mmol min^{-1} mg^{-1} and is stimulated about 60-fold to a rate of 4.6 μmol min^{-1} mg^{-1} in the presence of microtubules.[452]

Kinesin or related proteins are widely distributed in different tissues. By quantitative immunoblotting, it has been observed that kinesin accounts for 0.2% of the protein in neuronal tissues, while its concentration ranges from 0.05 to 0.1% of total protein in nonneural tissues including liver, epidermis, intestine, and cardiac and skeletal muscle.[454] As for its expression during development, kinesin levels undergo two important increases coinciding with the period of neurite outgrowth and with the onset of synaptic activity.[454]

The presence of kinesin in nonneural tissues raises the possibility that it is involved in microtubule-based motility in all eukaryotic cells. In these cells, kinesin may mediate the transport of Golgi-derived membrane vesicles from the perinuclear region to the cell membrane, analogous to the anterograde organelle transport observed in axons.[450] It has also been found associated with the mitotic spindle in sea urchin eggs.[455] Thus, it has been proposed that kinesin may also contribute to mitotic chromosome movements in dividing cells.[455]

4. Neuronal Dynein-Like MAPs: Force-Generating Proteins Possibly Involved in Retrograde Organelle Transport

Several authors have described the presence of high molecular weight MAPs which bind to microtubules in an ATP-dependent manner and are considered putative motility proteins in neuronal cells.[456-460]

Of these, MAP-1C has been the most thoroughly characterized.[459,460] MAP-1C has some similarities to flagellar and ciliary dyneins:

1. The electrophoretic mobility of MAP-1C in SDS-PAGE is equivalent to that of *Chlamydomonas* flagellar β-dynein.
2. In native conditions, MAP-1C has a sedimentation coefficient of 20S, similar to dynein and much greater than that of other MAPs (3-5S) or kinesin (10S).
3. Both MAP-1C and flagellar dynein bind to microtubules in the absence of ATP. ATP causes the dissociation of these molecules from microtubules.
4. The exposure of both MAP-1C and flagellar dynein to UV light in the presence of vanadate and ATP results in specific limited photocleavage of these molecules.
5. The ATPase activity of both MAP-1C and flagellar dynein is inhibited by *N*-ethylmaleimide and low concentrations of vanadate.

However, MAP-1C differs from dynein in two respects:

1. MAP-1C has lower ATPase activity than dynein. The ATPase activity of unactivated MAP-1C is about 50 mmol min^{-1} mg^{-1}, which is enhanced in the presence of microtubules up to 350 mmol min^{-1} mg^{-1}.
2. The concentration of microtubules required for maximal stimulation of MAP-1C ATPase activity is 100-fold lower than that required for the maximal activation of flagellar dynein.

On the other hand, MAP-1C differs from MAP-1A and MAP-1B not only in its ATPase activity, but also in its high resistance to proteolysis.

MAP-1C is purified from microtubules assembled in the absence of nucleotides. These microtubules are extracted with GTP to remove selectively kinesin and then incubated with ATP to extract MAP-1C. Purified MAP-1C preparations also contain, in addition to the 300-kDa major protein, some minor polypeptides of 74, 59, 57, and 53 kDa which may be associated with the major 300 Kd polypeptide.

AVEC-DIC microscopy shows the translocation of microtubules at velocities close to 1.5 $\mu m\ 5^{-1}$ on MAP-1C-coated coverslips. This gliding is inhibited by 1 m*M* *N*-ethylmaleimide or 10 μM vanadate. The directionality of this movement has been examined by observing the gliding of axonemes derived from *Chlamydomonas* flagella. Axoneme gliding occurs from the compact (or minus) to the frayed (plus) end on MAP-1C-coated slides, while it occurs in the opposite direction on kinesin-coated slides.[460] Thus, an organelle attached to MAP-1C would be driven to the proximal (minus) end of microtubules, suggesting that MAP-1C might be a motor protein involved in retrograde organelle transport.[460]

With regard to the function of a dynein-like ATPase in the propulsion of membrane vesicles along microtubules, it may be similar to the action of flagellar dynein in the sliding of microtubules. Thus, the "head" domain of a dynein-like molecule would interact reversibly with a microtubule in an ATP-dependent manner, while its nucleotide-insensitive binding domain would be stably attached to a membrane organelle. In the presence of ATP, the dynein-like protein would dissociate from the microtubule. Upon ATP hydrolysis, the resulting ADP-containing, dynein-like molecule would bind the microtubule and the release of ADP would generate the force-producing step through a conformational change of the molecule.[461,462]

At present, the relationship of MAP-1C with other putative motor proteins is not clear. The high molecular weight polypeptide described by Hollenbeck and Chapman[457] also has a sedimentation coefficient of 20S and an electrophoretic mobility similar to that of MAP-1C, but differs from MAP-1C in its ability to bundle microtubules, which has not been observed in MAP-1C preparations. The 300-kDa MAP resistant to calcium-activated proteolysis, which has been described by Koszka et al.,[458] also has dynein-like properties quite similar to those reported for MAP-1C. The high molecular weight protein extracted from squid axoplasmic vesicles by Gilbert and Sloboda[456] differs from MAP-1C in its reactivity with an anti-mammalian MAP-2 antibody. Thus, further studies are necessary to clarify the relationship between these proteins.

The presence of similar dynein-like proteins in nonneuronal cells has also been suggested. Dynein-like proteins may have a general role as retrograde motor proteins in the delivery of endocytotic vesicles to the lysosomes and in the positioning of the Golgi apparatus near the MTOC.[450]

5. Interaction of Microtubule-Associated Motor Proteins with Membrane Organelles and Mechanism of Organelle Transport

The translocation of membrane organelles along microtubules may be explained by their stable binding to motor proteins, such as kinesin and MAP-1C, which interact dynamically with microtubules in an ATP-dependent manner, thus generating the force required for organelle movement (see Sections V.C.2 and V.C.3).

The surface properties of membrane organelles, mainly the existence of specific receptors for different motor proteins, may specify the direction of movement by determining which motor protein they can bind. Thus, membrane organelles treated with trypsin do not move along microtubules, even in the presence of exogeneously added motor proteins, supporting the hypothesis that specific receptors on the organelle membrane are required to bind motor proteins.[450]

The interaction of these putative membrane receptors with motor proteins seems to be quite stable and of high affinity, since membrane vesicles separated from soluble proteins in a sucrose gradient do contain bound motor proteins and can therefore move along microtubules.[446,450,463] Indeed, around 10 to 30% of the total kinesin pool seems to be associated with membrane organelles, as determined by quantitative immunoblotting.[454]

The identity of the putative receptors for motor proteins is completely unknown. There must obviously be different receptors for kinesin and MAP-1C. The fact that these motor proteins can also cross-link two microtubules has raised the possibility that tubulin or related proteins could be receptors for motor proteins. Dynein-like proteins may possibly bind to organelles through membrane-bound tubulin or tubulin-like proteins. MAP-1C would then have two tubulin-binding sites, analogous to flagellar dynein; the ATP-dependent binding site, which would correspond to the microtubule-binding site and the ATP-independent binding site, which would correspond to the site interacting with membrane-bound tubulin. Other membrane-bound, tubulin-related proteins may be involved in the attachment of kinesin; although neither the specific binding of motor proteins to membrane organelles nor their activation have been studied in detail to date.

From an energetic point of view, few motor proteins are required to translocate an organelle. This correlates well with morphological studies showing 3-5 crossbridges linking an axoplasmic vesicle to a microtubule.[60,463]

Vale has proposed that organelles walk along microtubules following protofilament tracks.[450] Two or more cross-bridges would thus be necessary so that the binding of the vesicle to the microtubule would be maintained while a crossbridge dissociates from the microtubule. It is unknown at present whether motor proteins self-associate, forming clusters on the organelle surface, thus providing multiple cross-bridges. Clearly more research in force-generating proteins is required to propose a detailed model accounting for the fast movement of membrane organelles along microtubules.

It is also unknown whether kinesin is degraded at the axon tip or retrogradely transported in an inactivated form to the cell body. As for dynein-like proteins, they may be transported down the axon in an inactivated form and then activated to support retrograde movement without interferring with anterograde transport. It also remains to be established whether or not the activity of these motor proteins is regulated by posttranslational modifications, such as phosphorylation.

Thus, it appears that the identification of some putative motor proteins has constituted an important step in the elucidation of the mechanisms responsible for fast organelle transport which are now amenable to molecular analyses.

VI. MICROTUBULE ABNORMALITIES IN NEUROLOGICAL DISEASES

A. NEUROLOGICAL DISEASES LINKED TO MICROTUBULE ABNORMALITIES

As microtubules are critically involved in the maintenance of nerve cell shape and in axonal transport mechanisms, alterations in the organization of microtubules, their associated proteins or their interactions with other cellular elements may have profound consequences on neuronal function, giving rise to a variety of neuropathies.

For example, a decreased expression of microtubule components during the perinatal period may lead to a defective neuronal differentiation, resulting in hypoplastic neuropile and severe mental retardation, as may occur in neonatal hypothyroidism (see Section IV.C.1).

Other microtubule-related neuropathies include (1) impairment of the slow component of axonal transport, (2) impairment of fast anterograde organelle transport, (3) impairment of fast retrograde organelle transport, and (4) neurodegenerative diseases, such as senile dementia of the Alzheimer's type.

B. IMPAIRMENT OF THE SLOW COMPONENT OF AXONAL TRANSPORT

Alterations of the slow component of axonal transport can be produced by different neurotoxins and also occur in early diabetes mellitus and some inherited neuropathies.

Microtubule-disorganizing drugs such as colchicine, vinca alkaloids (vincristine and vinblastine), maytansine, and podophyllotoxin are among the agents affecting slow axonal transport, in addition to their primary effect on fast organelle transport.[421,422,464] Long-term exposition to high doses of these drugs following intracraneal injection causes the accumulation of neurofilaments within neuronal perikarya, suggesting that neurofilament transport into axons is inhibited after the disorganization of microtubules. The formation of similar perinuclear aggregates of intermediate filaments upon colchicine treatment is also observed in cultured brain cells.[280]

Neurofilaments aggregates are also found in neurons from amyotrophic lateral sclerosis and giant axonal neuropathy patients.[465,466] Not only neurons, but also other cell types, including skin fibroblasts, have intracellular masses of intermediate filaments in giant axonal neuropathy patients. In nonneuronal cells, these aggregates appear as juxtanuclear coils.[467] Such aggregates may presumably occur because intermediate filaments become disconnected from microtubules in response to disruption of cellular metabolism.[468]

Some neurotoxins, such as aluminum salts, 2,5-hexanedione, IDPN, and carbon disulfide, primarily block neurofilament transport, causing axonal swellings. These axonal swellings are found mainly in the proximal region of the IDPN-intoxicated nerves and in the distal region of the axon in hexanedione-intoxicated nerves.

The basis of IDPN neuropathy has been studied thoroughly.[57,59,63,436,437,469-472] It appears that IDPN selectively impairs neurofilament transport all along the axon, without greatly affecting tubulin or actin transport. The fast component of axonal transport also remains unaffected. Axonal swelling develop due to the retention and subsequent accumulation of neurofilaments. The failure of delivery of neurofilaments down the axon causes a reduction of the axonal caliber distal to the swellings.[469,470]

It appears that IDPN causes a segregation of neurofilaments from microtubules. Neurofilaments become restricted to the subaxolemmal space, while microtubules become confined to the central core of the axon[471,472] (see Section II.A). Following long-term exposure to IDPN, neurofilaments become distributed chaotically, eventually forming accumulations. These changes take place locally near the site of IDPN administration and are consequently independent of any alterations in the cell body. The mechanism by which IDPN produces this effect and its possible target remain unknown. Upon IDPN treatment MAP-1A and MAP-1B become confined to the central microtubule domain,[59] while MAP-2 is localized in the surrounding neurofilament domain.[63] IDPN may perhaps interfere with the association of molecules such as MAP-2 which may link microtubules to neurofilaments, thus causing the observed segregation of microtubules and neurofilaments and the partitioning of these molecules within the neurofilament domain. Alternatively, the segregation of microtubules from neurofilaments may be a consequence of structural alterations in neurofilament proteins. An example is the observation that γ-dike-tones produce the formation of pyrrole adducts and the extensive auto-oxidative crosslinking of neurofilaments.[473-475]

Axonal dwindling is also observed in experimental diabetes, which produces a reduction in the velocity and amount of protein conveyed by the SCa component of axonal transport.[476-478]

C. IMPAIRMENT OF FAST ANTEROGRADE ORGANELLE TRANSPORT

By injection into animal brains, in addition to studies with cultured neurons, it has been shown that the primary effect of microtubule-disorganizing drugs on axonal transport is the blockage of the fast component.[432-434] This effect produces an acute axonal degeneration of the Wallerian type, similar to that observed in the distal segment of a transected axon.

Wallerian degeneration may be the consequence of a block in the delivery of material by the fast anterograde component of axonal transport which is required for the renewal of the axolemma and for the maintenance of the synaptic junctions.[478]

Interestingly, the low doses of colchicine used in the treatment of patients with gout may result in a mild axonal polyneuropathy and in an accompanying myopathy.[479]

It has also been observed that colchicine has a selective neurotoxicity. For example, intrahippocampal injections destroy granule cells of the dentate gyrus, while pyramidal cells are essentially unaffected. To explain this selectivity, Steward et al., have suggested that colchicine may induce an autotoxic response which leads to neuronal death in certain neuron populations, possibly due to the accumulation of some toxic cellular product which is normally transported by a microtubule-dependent mechanism.[480]

A Wallerian-like axonal degeneration associated with a decrease in the amount of protein conveyed by the fast anterograde component of axonal transport is observed during hypoglycemia (which probably results in a deprivation of ATP, since glucose is the fuel molecule for nerve cells) and in porphyric neuropathy.[478]

D. IMPAIRMENT OF FAST RETROGRADE ORGANELLE TRANSPORT

Distal axonopathies associated with a decrease in the amount of protein conveyed by the retrograde component of fast axonal transport are induced by a variety of neurotoxins, including zinc pyridinethione, *p*-bromophenyl-acetylurea, acrylamide, and 2,5-hexanedione.[478]

A defect in retrograde organelle transport is also found in amyotrophic lateral sclerosis.[481] AVEC-DIC microscopy of motor nerves from amyotrophic lateral sclerosis (ALS) patients and control healthy subjects reveals that the number of organelles moving retrogradely per time unit is notably decreased in ALS nerves compared with those of controls. Furthermore, the mean speed of organelles moving retrogradely is significantly slower than that of controls. Anterograde organelle transport is stimulated significantly in ALS nerves, possibly as a consequence of the increased demand which arises from the progressive loss of motor neurons and the sprouting of axon branches by the residual neurons.

E. MICROTUBULE ABNORMALITIES IN NEURODEGENERATIVE DISEASES

Neurodegenerative diseases include a large and heterogeneous group of disorders, such as senile dementia of Alzheimer's type, Parkinson's disease, amyotrophic lateral sclerosis, Pick's disease, progressive supranuclear palsy, Guamanian parkinsonismdementia complex, etc. These disorders are characterized by degeneration of certain populations of neurons in various areas of the central nervous system, with relative sparing of other groups of nerve cells. Dysfunction of affected neurons is associated with several abnormal cytoskeletal structures whose presence serve as diagnostic criteria. Nerve cells exhibiting these cytoskeletal abnormalities may not be able to carry out normal functions and eventually die, leading to the progressive neuronal loss characteristic of neurodegenerative diseases.[482]

Of these diseases, senile dementia of Alzheimer's type (SDAT) has been the most thoroughly studied, since it is the most frequent neurodegenerative disorder.[483-485] SDAT is characterized by the degeneration of neurons in monoaminergic brainstem nuclei, in the basal forebrain cholinergic system and in neuronal populations within the amygdala, hippocampus, and neocortex. The etiology of SDAT is entirely unknown. Memory failure and other intellectual deficits may arise from an extensive disconnection syndrome which interrupts cortico-cortical connections and effectively isolates the hippocampus from the limbic system and the association cortex. This disconnection syndrome would be the result of the extensive neuronal degeneration observed, the cause of which is unknown, although genetic determinants, aluminum toxicity, slow virus infections, immune system dysfunction, impairment of brain metabolism, absence of neurotrophic factors, and neurotransmitter dysfunctions have been implicated.[483-485]

The major histopathological hallmark of SDAT is the presence of neurofibrillary tangles (NFTs).[484]

NFTs are argyrophilic intraneuronal inclusions scattered throughout the cerebral cortex in SDAT brains. They are particularly numerous in the hippocampus and are mainly localized in pyramidal neurons which connect the hippocampus to other regions of the cortex, basal forebrain, thalamus, and hypothalamus. The number of NFTs can be correlated with the degree of cognitive impairment in SDAT patients.

NFTs are composed of pairs of approximately 10-nm filaments helically twisted around each other, which are referred to as paired helical filaments (PHFs). PHFs are relatively insoluble in the usual protein solubilizing and denaturing agents such as SDS, urea, or guanidinium hydrochloride. Even after SDS extraction, PHFs retain their overall ultrastructural appearance, as observed in negatively stained or shadowed preparations.[486,487]

Immunochemical studies have shown that tau proteins are the main protein constituents of the PHFs.[487-496] Thus, antibodies raised against tau proteins recognize NFTs in SDAT brain sections and antibodies raised against PHFs isolated by extensive SDS extraction recognize tau proteins on Western blots.

PHF-associated tau is mainly localized to perikarya and apical dendrites of pyramidal neurons.[495,496] As normal neuronal tau is not so resistant to formalin fixation as PHF-associated tau, NFTs are prominently stained by anti-tau antibodies in fixed brain sections, while the staining of normal axons is more difficult to demonstrate.

PHF-associated tau differs from normal neuronal tau in the following aspects:

1. PHF-associated tau is partially resistant to digestion by some proteases.[491,492]
2. PHF-associated tau seems to be highly phosphorylated, since the reactivity of the monoclonal antibody tau-1 with PHF-associated tau is notably increased after phosphatase treatment.[496,497] Furthermore, antibodies specific for phosphorylated tau epitopes label NTFs *in situ*.[498] NTFs also react with monoclonal antibodies recognizing phosphorylated epitopes which are common to neurofilaments, to MAP-1 and MAP-2 and to tau.[235,499,500] The increased or abnormal phosphorylation of tau proteins may be involved in both the formation of PHFs *in vivo* and the defective *in vitro* microtubule assembly observed in SDAT brain extracts.[501] Nothing is known about the phosphorylation of tau in SDAT, although it has been proposed that it corresponds to the mode I of tau phosphorylation and may be catalyzed by Ca^{++}/calmodulin-dependent protein kinase.[296]

Ultrastructural studies on isolated PHFs demonstrate that they are unique polymers, unlike any other cytoskeletal element present in brain.[502-504] This suggests that PHFs are formed by a *de novo* polymerization of single unique structural subunit.[502-504] The identification of a modified tau polypeptide in the core of PHFs after extensive SDS treatment[505] or protease digestion[506-508] may indicate that this modified tau protein is the putative major component of PHFs. This agrees well with the fact that *in vitro* modified tau may self-assemble into polymers resembling the PHFs found in SDAT brains.[509]

Thus, it can be hypothesized that some unknown agent causes an alteration in the expression and/or posttranslational modifications of tau proteins, leading to the appearance of new tau isoforms which may be involved in the impairment of axonal transport and of microtubule assembly, and in the self assembly of tau into PHFs.

Other histopathological lesions found in SDAT brains include senile plaques, Hirano bodies and granulovacuolar degenerations.

The senile (neuritic) plaques are complex structures formed by the juxtaposition of abnormal nerve processes and glial cell processes surrounding a central amyloid core. Most nerve processes are degenerating axons filled with accumulations of straight filaments and PHFs. Neuritic plaques are found mainly in the amygdala, hippocampus and neocortex.

Amyloid deposits are also found in meningeal and intraparenchymal blood vessels. A 43 amino acid polypeptide (A4 protein) is the main component of cerebrovascular and senile plaque amyloid. This A4 protein is the cleavage product of a larger transmembrane precursor protein.[510,511]

Hirano bodies are rod-shaped eosinophilic intraneural inclusions which are quite common in the pyramidal cells of SDAT hippocampi. They are composed of paracrystalline arrays of filaments containing actin and actin-binding proteins.[512] Tau proteins also appear to be present in Hirano bodies.[513]

Granulovacuolar degenerations are intraneuronal inclusions present in pyramidal neurons of the hippocampus. They appear as clear vacuoles with central dark inclusions and seem to be derived from cellular autophagic mechanisms.[514] These vacuoles are immunostained by antibodies against tubulin, tau and neurofilaments.[515,516]

Intraneuronal fibrillary inclusions also appear in brains from patients affected by other neurodegenerative disorders. These include Pick bodies, characteristic of Pick disease, and Lewy bodies, the hallmark of Parkinson's disease. They are not so well characterized as NFTs of SDAT, although immunochemical data suggest the presence of tau epitopes in Pick bodies[517] and of neurofilament epitopes in Lewy bodies.[518]

The research into the molecular mechanisms which are responsible for the cytoskeletal abnormalities associated with neurodegenerative diseases now constitutes a major focus of attention in neurobiology, and significant progress will probably take place in the next few years.

REFERENCES

1. **Brady, S. T. and Lasek, R. J.,** Axonal transport: a cell-biological method for studying proteins that associate with the cytoskeleton, *Methods Cell Biol.*, 25, 365, 1982.
2. **Mascagno, E. R.,** Genetic approaches to invertebrate neurogenesis, *Curr. Top. Dev. Biol.*, 15, 319, 1980.
3. **Scott, B. E., Engelbert, V. E., and Fisher, K. C.,** Morphological and electrophysiological characteristics of dissociated chick embryonic spinal ganglion cells in culture, *Exp. Neurol.*, 23, 230, 1969.
4. **Banker, G. A. and Cowan, W. M.,** Rat hippocampal neurons in dispersed cell culture, *Brain Res.* 126, 397, 1977.
5. **Kriegstein, A. R. and Dichter, M. A.,** Morphological classification of rat cortical neurons in cell culture, *J. Neurosci.*, 3, 1634, 1983.
6. **Dotti, C. G., Sullivan, C. A., and Banker, G. A.,** The establishment of polarity by hippocampal neurons in culture, *J. Neurosci.*, 8, 1454, 1988.
7. **Solomon, F.,** Specification of cell morphology by endogenous determinants, *Cell,* 90, 547, 1981.
8. **Letourneau, P. C.,** Nerve fiber growth and its regulation by extrinsic factors, *Curr. Top. Neurobiol.*, 1, 213, 1982.
9. **Prochiantz, A.,** Neuronal growth and shape, *Dev. Neurosci.*, 7, 189, 1985.
10. **Krystosek, A.,** Cytoplasmic and extracellular control of neurite formation, in *The Cytoskeleton in Cell Differentiation and Development*, Maccioni, R. B. and Arechaga, J., Eds., IRL Press, Oxford, 1987, 209.
11. **Ramón y Cajal, S.,** Variaciones morfológicas normales y patológicas del retículo neurofibrilar, *Trab. Lab. Invest. Biol.*, 3, 1, 1904.
12. **Ramón y Cajal, S.,** *Degeneration and Regeneration of the Nervous System* (translated by R. M. May), Oxford University Press, London, 1928.
13. **Ramón y Cajal, S.,** *Studies on Vertebrate Neurogenesis* (translated by L. Guth), Charles C Thomas, Springfield, IL, 1929.
14. **De Renyi, G. S.,** Structure of cells in tissues as revealed by microdissection: observations of neurofibrils in living nervous tissue of lobster *(Homarus americanus)*, *J. Comp. Neurol.*, 48, 441, 1929.
15. **Palay, S. L.,** Synapses in the central nervous system, *J. Biophys. Biochem. Cytol.*, 2, 193, 1956.
16. **Boycott, B. B., Gray, E. G., and Guillery, R. W.,** Synaptic structure and alteration with environmental temperature: a study by light and electron microscopy of the central nervous system of lizards, *Proc. Roy. Soc. Lond. B.*, 152, 151, 1961.

17. **Potter, H. D.,** The distribution of neurofibrils coextensive with microtubules and neurofilaments in dendrites and axons of the tectum, cerebellum and palium of the frog, *J. Comp. Neurol.*, 143, 385, 1971.
18. **Peters, A., Palay, S. L., and De Webster, H.,** *The Fine Structure of the Nervous System: the Neurons and Supporting Cells*, W. B. Saunders, Philadelphia, 1976.
19. **Traub, P.,** *Intermediate Filaments. A Review*, Springer-Verlag, Vienna, 1985.
20. **Shaw, G.,** Neurofilaments: abundant but mysterious neuronal structures, *Bioessays*, 4, 161, 1986.
21. **Lasek, R. J., Oblinger, M. M., and Drake, P. F.,** Molecular biology of neuronal geometry: expression of neurofilament genes influences axonal diameter, *Cold Spring Harbor Symp. Quant. Biol.*, 48, 731, 1984.
22. **Hoffman, P. N., Griffin, J. W., and Price, D. L.,** Control of axonal caliber by neurofilament transport, *J. Cell Biol.*, 99, 705, 1984.
23. **Hoffman, P. N., Cleveland, D. W., Griffin, J. W., Landes, P. W., Cowan, N. J., and Price, D. L.,** Neurofilament gene expression: a major determinant of axonal caliber, *Proc. Natl. Acad. Sci. U.S.A.*, 84, 3472, 1987.
24. **Bertolini, B., Monaco, G., and Rossi, A.,** Ultrastructure of a regular arrangement of microtubules and neurofilaments, *J. Ultrastruct. Res.*, 33, 173, 1970.
25. **Shaw, G., Osborn, M., and Weber, K.,** Arrangement of neurofilaments, microtubules and microfilament-associated proteins in cultured dorsal root ganglion cells, *Eur. J. Cell biol.*, 24, 20, 1981.
26. **Jacobs, M. and Thomas, C.,** The organization of 10-nm filaments and microtubules in embryonic neurons from spinal ganglia, *J. Neurocytol.*, 11, 657, 1982.
27. **Lasek, R. J. and Morris, J. R.,** The microtubule-neurofilament network: the balance between plasticity and stability in the nervous system, in *Biological Functions of Microtubules and Related Structures*, Sakai, H., Mohri, H., and Borisy, G. G., Eds., Academic Press, Tokyo, 1982, 329.
28. **Gilbert, D. S.,** Axoplasm architecture and physical properties as seen in the Myxicola giant axon, *J. Physiol.*, 253, 257, 1975.
29. **Tsukita, S. and Ishikawa, H.,** The cytoskeleton in myelinated axons: serial section study, *Biomed. Res.*, 2, 424, 1981.
30. **Bray, D. and Bunge, M. B.,** Serial analysis of microtubules in cultured rat sensory neuron axons, *J. Neurocytol.*, 10, 589, 1981.
31. **Chalfie, M. and Thomson, J. N.,** Organization of neuronal microtubules in the nematode *C. elegans*, *J. Cell Biol.*, 82, 278, 1979.
32. **Jacobs, J. R. and Stevens, J. K.,** Changes in the organization of the neuritic cytoskeleton during NGF-activated differentiation of PC12 cells: A serial electron-microscopic study of the development and control of neurite shape, *J. Cell. Biol.*, 103, 895, 1986.
33. **Rodriguez Echandia, E. L. and Piezzi, R. S.,** Microtubules in the nerve fibers of the toad Bufo arenarum. Effect of low temperature on the sciatic nerve, *J. Cell Biol.*, 39, 491, 1968.
34. **Banks, P., Mayor, D., and Oven, T.,** Effects of low temperatures on microtubules in the non-myelinated axons of post-ganglionic sympathetic nerves, *Brain Res.*, 83, 277, 1975.
35. **Heidemann, S., Hamborg, M., Thomas, S., Song, B., Dindley, S., and Chu, D.,** Spatial organization of axonal microtubules, *J. Cell. Biol.*, 99, 1289, 1984.
36. **Joshi, H. C., Baas, P., Chu, D. T., and Heidemann, S. R.,** The cytoskeleton of neurites after microtubule depolymerization, *Exp. Cell Res.*, 163, 233, 1986.
37. **Baas, P. W. and Heidemann, S. R.,** Microtubule reassembly from nucleating fragments during the regrowth of amputated neurites, *J. Cell Biol.*, 103, 917, 1987.
38. **Burton, P. R. and Paige, J. L.,** Polarity of axoplasmic microtubules in the olfactory nerve of the frog, *Proc. Natl. Acad. Sci. U.S.A.*, 78, 3269, 1981.
39. **Filliatreu, G. and DiGiamberardino, L.,** Microtubule polarity in myelinated axons as studied after decoration with tubulin, *Biol. Cell.*, 42, 69, 1981.
40. **Heidemann, S. R., Landers, J. M., and Hamborg, M. A.,** Polarity orientation of axonal microtubules, *J. Cell Biol.*, 91, 661, 1981.
41. **Viancour, T. A. and Forman, D. S.,** Polarity orientation of microtubules in squid and lobster axons, *J. Neurocytol.*, 16, 69, 1987.
42. **Baas, P. W., White, L. A., and Heidemann, S. R.,** Microtubule polarity reversal accompanies regrowth of amputated neurites, *Proc. Natl. Acad. Sci. U.S.A.*, 84, 5272, 1987.
43. **Black, M. M. and Lasek, R. J.,** Slow components of axonal transport: two cytoskeletal networks, *J. Cell Biol.*, 86, 616, 1980.
44. **Lasek, R. J. and Brady, S. T.,** The axon: a prototype for studying expressional cytoplasms, *Cold Spring Harbor Symp. Quant. Biol.*, 46, 113, 1982.
45. **Lasek, R. J., Gainer, J. A., and Brady, S. T.,** Axonal transport of the cytoplasmic matrix, *J. Cell Biol.*, 99, 212, 1984.
46. **Black, M. M., Keyser, P., and Sobel, E.,** Interval between the synthesis and assembly of cytoskeletal proteins in cultured neurons, *J. Neurosci.*, 6, 1004, 1986.
47. **Bamburg, J. R., Bray, D., and Chapman, K.,** Assembly of microtubules at the tip of growing axons, *Nature*, 321, 788, 1986.

48. **Caceres, A., Banker, G., and Binder, L.,** Immunocytochemical localization of tubulin and MAP-2 during development of hippocampal neurons in culture, *J. Neurosci.*, 6, 714, 1986.
49. **Kosik, K. S. and Finch, E. A.,** MAP-2 and tau segregate into dendritic and axonal domains after the elaboration of morphologically distinct neurites: an immunocytochemical study of cultured rat cerebrum, *J. Neurosci.*, 7, 3142, 1987.
50. **Hirokawa, N.,** Cross-linker system between neurofilaments, microtubules and membranous organelles in frog axons revealed by the quick-freeze, deep-etching methods, *J. Cell Biol.*, 94, 129, 1982.
51. **Hirokawa, N.,** Quick-freeze, deep-etch visualization of the axonal cytoskeleton, *Trends Neurosci.*, 9, 67, 1986.
52. **Tsukita, S., Tsukita, S., Kobayashi, T., and Matsumoto, G.,** Subaxolemmal cytoskeleton in squid giant axon: morphological identification of microtubule- and microfilament-associated domains of axolemma, *J. Cell Biol.*, 102, 1710, 1986.
53. **Kobayashi, T., Tsukita, S., Tsukita, S., Yamamoto, Y., and Matsumoto, G.,** Subaxolemmal cytoskeleton in squid giant axon: biochemical analysis of microtubules, microfilaments and their associated high molecular weight proteins, *J. Cell Biol.*, 102, 1699, 1986.
54. **Matsumoto, G., Ichikawa, M., Tasaki, A., Murofushi, H., and Sakai, H.,** Axonal microtubules necessary for generation of sodium current in squid giant axons. Pharmacological study on sodium current and restoration of sodium current by microtubule proteins and 260 Kd protein, *J. Membrane Biol.*, 77, 77, 1983.
55. **Terakawa, S. and Nakayama, T.,** Are axoplasmic microtubules necessary for membrane excitation?, *J. Membrane Biol.*, 85, 65, 1985.
56. **Hirokawa, N., Glicksman, M. A., and Willard, M. B.,** Organization of mammalian neurofilament polypeptides within the neuronal cytoskeleton, *J. Cell Biol.*, 98, 1523, 1984.
57. **Papasozomenos, S. and Payne, M. R.,** Actin immunoreactivity localizes with segregated microtubules and membranous organelles and in the subaxolemmal region in the ββ′ iminodipropionitrile axon, *J. Neurosci.*, 6, 3483, 1986.
58. **Schnapp, B. J. and Reese, T. S.,** Cytoplasmic structure in rapid-frozen axons, *J. Cell Biol.*, 94, 667, 1982.
59. **Hirokawa, N., Bloom, G. S., and Vallee, R. B.,** Cytoskeletal architecture and immunocytochemical localization of microtubule-associated proteins in regions of axons associated with rapid axonal transport. The ββ′-iminodipropionitrile-intoxicated axon as a model system, *J. Cell Biol.*, 101, 227, 1985.
60. **Miller, R. H. and Lasek, R. J.,** Cross bridges mediate anterogranule and retrograde vesicle transport along microtubules in squid axoplasm, *J. Cell Biol.*, 101, 2181, 1985.
61. **Vallee, R. B. and Bloom, G. S.,** High-molecular-weight microtubule-associated proteins, *Mod. Cell Biol.*, 3, 21, 1984.
62. **Wiche, G.,** High-molecular-weight microtubule-associated proteins: an ubiquitous family of cytoskeletal connecting links, *Trends Biochem. Sci.*, 10, 67, 1985.
63. **Papasozomenos, S. C., Binder, L. I., Bender, P., and Payne, M. R.,** Microtubule-associated protein MAP2 within axons of spinal motor neurons: association with microtubules and neurofilaments in normal and ββ′-iminodipropionitrile (IDPN)-treated axons, *J. Cell Biol.*, 100, 74, 1985.
64. **Huber, G., Alaimo-Beuret, D., and Matus, A.,** MAP-3: characterization of a novel microtubule-associated protein, *J. Cell Biol.*, 100, 496, 1985.
65. **Fach, B. L., Graham, S. F., and Keates, R. A. B.,** Association of fodrin with brain microtubules, *Can. J. Biochem. Cell Biol.*, 63, 372, 1985.
66. **Morris, J. R. and Lasek, R. J.,** Stable polymers of the axonal cytoskeleton. The axoplasmic ghost, *J. Cell. Biol.*, 92, 192, 1982.
67. **Morris, J. R. and Lasek, R. J.,** Monomer-polymer equilibria in the axon: direct measurement of tubulin and actin as polymer and monomer in axoplasm, *J. Cell Biol.*, 98, 2064, 1984.
68. **Schulze, E. and Kirschner, M.,** Microtubule dynamics in interphase cells, *J. Cell Biol.*, 102, 1020, 1986.
69. **Seitz-Tutter, D., Langford, G. M., and Weiss, D. G.,** Dynamic instability of native microtubules from squid axons is rare and independent of gliding and vesicle transport, *Exp. Cell. Res.*, 178, 504, 1988.
70. **Matus, A.,** Microtubule-associated proteins: their potential role in determining neuronal morphology, *Ann. Rev. Neurosci.*, 11, 29, 1988.
71. **Black, M. H., Cochram, J. M., and Kurdyla, J. T.,** Solubility properties of neuronal tubulin: evidence for labile and stable microtubules, *Brain Res.*, 295, 255, 1984.
72. **Brady, S., Tytell, M., and Lasek, R. J.,** Axonal transport and axonal tubulin: biochemical evidence for cold stability, *J. Cell Biol.*, 99, 1716, 1984.
73. **Job, D. and Margolis, R. L.,** Isolation from bovine brain of a super-stable microtubule subpopulation with microtubule seeding activity, *Biochemistry*, 23, 3025, 1984.
74. **Job, D., Rauch, C. T., and Margolis, R. L.,** High concentrations of STOP protein induce a microtubule super-stable state, *Biochem. Biophys. Res. Commun.*, 148, 429, 1987.
75. **Sasaki, S., Stevens, S. K., and Bodick, N.,** Serial reconstruction of microtubule arrays within dendrites of the cat retinal ganglion cell: the cytoskeleton of a vertebrate dendrite, *Brain Res.*, 259, 193, 1983.

76. **Burgoyne, R. D., Gray, E. G., Sullivan, K., and Baron, J.,** Depolymerization of dendritic microtubules following incubation of rat brain cortical slices, *Neurosci. Lett.*, 31, 81, 1982.
77. **Black, M. M.,** Comparison of the effects of MAP-2 and tau on the packing density of *in vitro* assembled microtubules, *Proc. Natl. Acad. Sci. U.S.A.*, 84, 7783, 1987.
78. **Selden, S. C. and Pollard, T. D.,** Phosphorylation of microtubule-associated proteins regulates their interaction with actin filaments, *J. Biol. Chem.*, 258, 7064, 1983.
79. **Pollard, T. D., Selden, S. C., and Maupin, P.,** Interaction of actin filaments with microtubules, *J. Cell Biol.*, 99, 33s, 1984.
80. **Sattilaro, W.,** Interaction of microtubule-associated protein MAP-2 with actin filaments, *Biochemistry*, 25, 2003, 1986.
81. **Caceres, A., Payne, M. R., Binder, L. I., and Steward, O.,** Immunohistochemical localization of actin and microtubule-associated protein MAP-2 in dendritic spines, *Proc. Natl. Acad. Sci. U.S.A.*, 80, 1738, 1983.
82. **Gray, E. G.,** Presynaptic microtubules and their association with synaptic vesicles, *Proc. R. Soc. Lond. Biol. Sci.*, 190, 369, 1975.
83. **Gray, E. G.,** Microtubules in synapses of the retina, *J. Neurocytol.*, 5, 361, 1976.
84. **Gray, E. G.,** Problems of understanding the substructure of synapses, *Progr. Brain Res.*, 45, 207, 1976.
85. **Chan, K. Y. and Bunt, A. H.,** Association between mitochondria and microtubules in synaptosomes and axon terminals of cerebral cortex, *J. Neurocytol.*, 7, 137, 1978.
86. **Gray, E. G., Westrum, L. E., Burgoyne, R. D., and Barron, J.,** Synaptic organization and neuron microtubule distribution, *Cell Tissue Res.*, 226, 579, 1982.
87. **Westrum, L. E., Gray, E. G., Burgoyne, R. D., and Barron, J.,** Synaptic development and microtubule organization, *Cell Tissue Res.*, 231, 93, 1983.
88. **Gordon-Weeks, P. R., Burgoyne, R. D., and Gray, E. G.,** Presynaptic microtubules: organization and assembly-disassembly, *Neuroscience*, 7, 739, 1982.
89. **Burgoyne, R. D. and Cumming, R.,** Evidence for the presence of high-Mr MAPs and their Ca^{2+}-dependent proteolysis in synaptosomal cytosol, *FEBS Lett.*, 146, 274, 1982.
90. **Burke, B. E. and De Lorenzo, R. J.,** Ca^{2+} and calmodulin-stimulated endogenous phosphorylation of neurotubulin, *Proc. Natl. Acad. Sci. U.S.A.*, 78, 991, 1981.
91. **Yamauchi, T. and Fujisawa, H.,** Disassembly of microtubules by the action of calmodulin-dependent protein kinase II which occurs only in the brain tissues, *Biochem. Biophys. Res. Commun.*, 110, 287, 1983.
92. **Goldenring, J. R., Gonzalez, B., McGuire, J. S., Jr., and De Lorenzo, R. J.,** Purification and characterization of a calmodulin-dependent kinase from rat brain cytosol able to phosphorylate tubulin and microtubule-associated proteins, *J. Biol. Chem.*, 258, 12632, 1983.
93. **Yamamoto, H., Fukunaga, K., Goto, S., Tanaka, E., and Miyamoto, E.,** Ca^{2+}/calmodulin-dependent regulation of microtubule formation via phosphorylation of MAP-2, tau factor and tubulin, and comparison with the cyclic AMP-dependent phosphorylation, *J. Neurochem.*, 44, 759, 1985.
94. **Wandosell, F., Serrano, L., Hernández, M. A., and Avila, J.,** Phosphorylation of tubulin by a calmodulin-dependent protein kinase, *J. Biol. Chem.*, 261, 10332, 1986.
95. **De Camilli, P. and Greengard, P.,** Synapsin I: a synaptic vesicle associated neuronal phosphoprotein, *Biochem. Pharmacol.*, 35, 4349, 1986.
96. **Goldenring, J. R., Lashir, R. S., Vallano, M. L., Veda, T., Saito, S., Sternberg, N. H., Steinberg, L. A., and De Lorenzo, R. J.,** Association of synapsin I with neuronal cytoskeleton, *J. Biol. Chem.*, 261, 8495, 1986.
97. **Baines, A. J. and Bennett, V.,** Synapsin I is a microtubule-bundling protein, *Nature*, 319, 145, 1986.
98. **Petrucci, T. and Morrow, J. S.,** Synapsin I: an actin-bundling protein under phosphorylation control, *J. Cell Biol.*, 105, 1355, 1987.
99. **Bahler, M. and Greengard, P.,** Synapsin I bundles F-actin in a phosphorylation-dependent manner, *Nature*, 326, 704, 1987.
100. **Sherline, P., Lee, Y. C., and Jacobs, L. S.,** Binding of microtubules to pituitary secretory granules and secretory granule membranes, *J. Cell Biol.*, 72, 380, 1977.
101. **Suprenant, K. A. and Dentler, W. L.,** Association between endocrine pancreatic secretory granules and *in vitro* assembled microtubules is dependent upon microtubule-associated proteins, *J. Cell Biol.*, 93, 164, 1982.
102. **Gozes, I. and Littauer, U. Z.,** Tubulin microheterogeneity increases with rat brain development, *Nature*, 276, 411, 1978.
103. **Marotta, C. A., Harris, J. L., and Gilbert, J. M.,** Characterization of multiple forms of brain tubulin subunits, *J. Neurochem.*, 30, 1431, 1978.
104. **Dahl, J. L. and Weibel, V. J.,** Changes in tubulin heterogeneity during postnatal development of rat brain, *Biochem. Biophys. Res. Commun.*, 86, 822, 1979.
105. **George, H. J., Misra, L., Field, D. J., and Lee, J. C.,** Polymorphism of brain tubulin, *Biochemistry*, 20, 2402, 1981.

106. **Gozes, I. and Sweadner, K. J.,** Multiple tubulin forms are expressed by a single neurone, *Nature*, 294, 477, 1981.
107. **Gozes, I.,** Tubulin in the nervous system, *Neurochem. Int.*, 4, 101, 1982.
108. **Denoulet, P., Eddé, B., Jeantet, C., and Gros, F.,** Evolution of tubulin heterogeneity during mouse brain development, *Biochemie*, 64, 165, 1982.
109. **Wolff, A., Denoulet, P., and Jeantet, C.,** High level of tubulin microheterogeneity in the mouse brain, *Neurosci. Lett.*, 31, 323, 1982.
110. **Moura-Neto, V., Mallat, M., Jeantet, C., and Prochiantz, A.,** Microheterogeneity of tubulin proteins in neuronal and glial cells from the mouse brain in culture, *EMBO J.*, 2, 1243, 1983.
111. **Denoulet, P., Jeantet, C., and Gros, F.,** Tubulin microheterogeneity during mouse liver development, *Biochem. Biophys. Res. Commun.*, 105, 806, 1982.
112. **Díez, J. C., Little, M., and Avila, J.,** Quantitative determination, isolation and characterization of pig lung tubulin, *Biochem. J.*, 219, 277, 1984.
113. **Díez, J. C. and Avila, J.,** Homogeneity of lung tubulin isoforms during lung maturation, *Biochemie*, 69, 1059, 1983.
114. **Serrano, L., Avila, J., and Maccioni, R. B.,** Controlled proteolysis of tubulin by subtilisin: localization of the site for MAP-2 interaction, *Biochemistry*, 23, 4675, 1984.
115. **Serrano, L., de la Torre, J., Maccioni, R. B., and Avila, J.,** Involvement of the carboxy-terminal domain of tubulin in microtubule assembly and regulation, *Proc. Natl. Acad. Sci. U.S.A.*, 81, 5989, 1984.
116. **Serrano, L., Montejo, E., Hernández, M. A., and Avila, J.,** Localization of the tubulin binding site for tau protein, *Eur. J. Biochem.*, 153, 595, 1988.
117. **Tran, L. H., Binet, S., and Meininger, V.,** Regional specificity of tubulin heterogeneity in adult mouse brain, *Neurochem. Int.*, 10, 121, 1987.
118. **Denoulet, P., Eddé, B., Koulakoff, A., Netter, Y., and Gros, F.,** Isotubulin expression during mouse neuronal differentiation, in *The Cytoskeleton in Cell Differentiation and Development*, Maccioni, R. B. and Arechaga, J., Eds. IRL Press, Oxford, 1987, 221.
119. **Eddé, B., De Nechaud, B., Denoulet, P., and Gros, F.,** Control of isotubulin expression during neuronal differentiation of mouse neuroblastoma and teratocarcinoma cell lines, *Dev. Biol.*, 123, 549, 1987.
120. **Dutton, G. R. and Barondes, S.,** Microtubular protein: synthesis and metabolism in developing brain, *Science*, 166, 1637, 1969.
121. **Bamburg, J. R., Shooter, E. M., and Wilson, L.,** Developmental changes in microtubule protein of chick brain, *Biochemistry*, 12, 1476, 1973.
122. **Schmitt, H., Gozes, I., and Littauer, U. Z.,** Decrease in levels and rates of synthesis of tubulin and actin in developing rat brain, *Brain Res.*, 121, 327, 1977.
123. **Gozes, I., Walker, M. D., Kaye, A. M., and Littauer, U. Z.,** Synthesis of tubulin and actin by neural and glial nuclear preparations from developing rat brain, *J. Biol. Chem.*, 252, 1819, 1977.
124. **Cleveland, D. W. and Sullivan, K. F.,** Molecular biology and genetics of tubulin, *Ann. Rev. Biochem.*, 54, 331, 1985.
125. **Cleveland, D. W.,** The multitubulin hypothesis revisited: what have we learned?, *J. Cell Biol.*, 104, 381, 1987.
126. **Cowan, N. J., Lewis, S. A., Sarkar, S., and Gu, W.,** Functional versatility of mammalian β-tubulin isotypes, *The Cytoskeleton in Cell Differentiation and Development*, Maccioni, R. B. and Arechaga, J., Eds., IRL Press, Oxford, 1987, 157.
127. **Bond, J. F. and Farmer, S. R.,** Regulation of tubulin and actin mRNA production in rat brain: expression of new tubulin mRNAs with development, *Mol. Cell Biol.*, 3, 1333, 1983.
128. **Bond, J. F., Robinson, G. S., and Farmer, S. R.,** Differential expression of two neural cell-specific β-tubulin mRNAs during rat brain development, *Mol. Cell Biol.*, 4, 1313, 1984.
129. **Havercroft, J. C. and Cleveland, D. W.,** Programmed expression of β-tubulin genes during development and differentiation of the chicken, *J. Cell Biol.*, 99, 1927, 1984.
130. **Lewis, S. A., Gwo-ShuLee, M., and Cowan, N. J.,** Five mouse tubulin isotypes and their regulated expression during development, *J. Cell Biol.*, 101, 852, 1985.
131. **Ginzburg, J., Teichmann, A., Griffin, W. S. T., and Littauer, U. Z.,** Differential expression of α-tubulin in rat cerebellum as revealed by in situ hybridization, *FEBS Lett.*, 194, 161, 1981.
132. **Miller, F. D., Naus, C. G., Durand, M., Bloom, F. E. and Milner, R. J.,** Isotypes of α-tubulin are differentially regulated during neuronal maturation, *J. Cell Biol.*, 105, 3065, 1987.
133. **Barra, H. S., Arce, C. A., Rodriguez, J. A., and Caputto, R.,** Some common properties of the protein that incorporates tyrosine as a single unit into microtubule proteins, *Biochem. Biophys. Res. Commun.*, 60, 1384, 1974.
134. **Rodriguez, J. A. and Borisy, G. G.,** Modification of the C-terminus of brain tubulin during development, *Biochem. Biophys. Res. Commun.*, 83, 579, 1978.
135. **Arce, C. A., Hallak, M. E., Rodriguez, J. A., Barra, H. S. and Caputto, R.,** Capability of tubulin and microtubules to incorporate and to release tyrosine and phenylalanine and the effects of the incorporation of these amino acids on tubulin assembly, *J. Neurochem.*, 31, 205, 1978.

136. **Barra, H. S., Arce, C. A., and Caputto, R.,** Total tubulin and its aminoacylated and non-aminoacylated forms during the development of the brain, *Eur. J. Biochem.*, 169, 439, 1980.
137. **Thompson, W. C.,** The cyclic tyrosination/detyrosination of α-tubulin, *Methods Cell Biol.*, 24, 235, 1982.
138. **Gundersen, G. G., Khawaja, S., and Bulinski, J. C.,** Post-polymerization detyrosination of α-tubulin: a mechanism for subcellular differentiation of microtubules, *J. Cell Biol.*, 105, 251, 1987.
139. **Webster, D. R., Gundersen, G. G., Bulinski, J. C., and Borisy, G. G.,** Assembly and turnover of detyrosinated tubulin *in vivo*, *J. Cell Biol.*, 105, 265, 1987.
140. **Raybin, D. and Flavin, M.,** Enzyme which specifically adds tyrosine to the α-chain of tubulin, *Biochemistry*, 16, 2189, 1977.
141. **Schroder, H. C., Wehland, J., and Weber, K.,** Purification of brain tubulin-tyrosine ligase by biochemical and immunological methods, *J. Cell Biol.*, 100, 276, 1985.
142. **Argaraña, C. E., Barra, H. S., and Caputto, R.,** Release of (^{14}C)tyrosine from tubulinyl-(^{14}C)tyrosine by brain extract. Separation of a carboxypeptidase from tubulin-tyrosine ligase, *Mol. Cell Biochem.*, 19, 17, 1978.
143. **Argaraña, C. E., Barra, H. S., and Caputto, R.,** Tubulinyl-tyrosine carboxypeptidase from chicken brain: properties and partial purification, *J. Neurochem.*, 34, 114, 1980.
144. **Kumar, N. and Flavin, M.,** Preferential action of a brain detyrosinolating carboxypeptidase on polymerized tubulin, *J. Biol. Chem.*, 256, 7678, 1981.
145. **Barra, H. S., Argaraña, C. E., and Caputto, R.,** Enzymatic detyrosination of tubulin tyrosinated in rat brain slices and extracts, *J. Neurochem.*, 38, 112, 1982.
146. **Arce, C. A. and Barra, H. S.,** Association of tubulinyl-tyrosine carboxypeptidase with microtubules, *FEBS Lett.*, 157, 75, 1983.
147. **Cumming, R., Burgoyne, R. D., and Lytton, N. A.,** Axonal subpopulations in the central nervous system demonstrated using monoclonal antibodies against α-tubulin, *Eur. J. Cell Biol.*, 31, 241, 1983.
148. **Cumming, R., Burgoyne, R. D., and Lytton, N. A.,** Immunocytochemical demonstration of α-tubulin modification during axonal maturation in the cerebellar cortex, *J. Cell Biol.*, 98, 347, 1984.
149. **Cumming, R., Burgoyne, R. D., and Lytton, N. A.,** Immunofluorescence distribution of α-tubulin, β-tubulin and MAP-2 during *in vitro* maturation of cerebellar granule cells neurons, *Neuroscience*, 12, 775, 1984.
150. **Cambray-Deakin, M. A. and Burgoyne, R. D.,** Posttranslational modifications of α tubulin: acetylated and detyrosinated forms in axons of rat cerebellum, *J. Cell Biol.*, 104, 1569, 1987.
151. **Wehland, J., Schroder, H. C., and Weber, K.,** Amino acid sequence requirements in the epitope recognized by the α-tubulin specific rat monoclonal antibody YL1/2, *EMBO J.*, 3, 1295, 1984.
152. **L'Hernault, S. W. and Rosenbaum, J. L.,** Chlamydomonas α-tubulin is posttranslationally modified in the flagella during flagellar assembly, *J. Cell Biol.*, 96, 258, 1983.
153. **L'Hernault, S. W. and Rosenbaum, J. L.,** Chlamydomonas α-tubulin is posttranslationally modified by acetylation on the ϵ-amino group of a lysine, *Biochemistry*, 24, 473, 1985.
154. **Piperno, G. and Fuller, M. T.,** Monoclonal antibodies specific for an acetylated form of α-tubulin recognize antigens in cilia and flagella from a variety of organisms, *J. Cell Biol.*, 101, 2085, 1985.
155. **Black, M. M. and Kurdyla, J. T.,** Posttranslational modification of tubulin in cultured neurons, *Brain Res.*, 289, 355, 1983.
156. **Black, M. M. and Keyser, P.,** Acetylation of α-tubulin in cultured neurons and the induction of α-tubulin acetylation in PC12 cells by treatment with NGF, *J. Neurosci.*, 7, 1837, 1987.
157. **Maruta, H., Greer, K., and Rosenbaum, J. L.,** The acetylation of α-tubulin and its relationship to the assembly and disassembly of microtubules, *J. Cell Biol.*, 103, 571, 1986.
158. **Sale, W. S., Besharse, J. C., and Piperno, G.,** Distribution of acetylated α-tubulin in retina and *in vitro*-assembled microtubules, *Cell Motil. Cytosk.*, 9, 243, 1988.
159. **Eipper, B. A.,** Rat brain microtubule protein: purification and determination of covalently bound phosphate and carbohydrate, *Proc. Natl. Acad. Sci. U.S.A.*, 69, 2283, 1972.
160. **Eipper, B. A.,** Rat brain tubulin and protein kinase activity, *J. Biol. Chem.*, 249, 1398, 1974.
161. **Eipper, B. A.,** Properties of rat brain tubulin, *J. Biol. Chem.*, 249, 1407, 1974.
162. **Sandoval, I. V. and Cuatrecasas, P.,** Protein kinase associated with tubulin: affinity chromatography and properties, *Biochemistry*, 13, 3424, 1975.
163. **Gard, D. L. and Kirschner, M. W.,** A polymer-dependent increase in phosphorylation of β-tubulin accompanies differentiation of a mouse neuroblastoma cell line, *J. Cell Biol.*, 100, 764, 1985.
164. **Serrano, L., Díaz-Nido, J., Wandosell, F., and Avila, J.,** Tubulin phosphorylation by casein kinase II is similar to that found *in vivo*, *J. Cell Biol.*, 105, 1731, 1987.
165. **Raybin, D. and Flavin, M.,** Modification of tubulin by tyrosination in cells and extracts and its effect on assembly *in vitro*, *J. Cell Biol.*, 73, 492, 1977.
166. **Khawaja, S., Gundersen, G. G., and Bulinski, J. C.,** Enhanced stability of microtubules enriched in detyrosinated tubulin is not a direct function of detyrosination level, *J. Cell Biol.*, 106, 141, 1988.

167. **Kadowaki, T., Fujita-Yamaguchi, Y., Nishida, E., Takaku, F., Akiyama, T., Kathuria, S., Akamura, Y., and Kasuga, M.,** Phosphorylation of tubulin and microtubule-associated proteins by the purified insulin receptor kinase, *J. Biol. Chem.*, 260, 4016, 1985.
168. **Hargreaves, A., Wandosell, F., and Avila, J.,** Phosphorylation of tubulin enhances its interaction with membranes, *Nature*, 323, 827, 1986.
169. **Battacharyya, B. and Wolff, J.,** Membrane-bound tubulin in brain and thyroid tissue, *J. Biol. Chem.*, 250, 7639, 1975.
170. **Estridge, M.,** Polypeptides similar to α and β subunits of tubulin are exposed on the neuronal surface, *Nature*, 268, 60, 1977.
171. **Feit, H., Kelly, P., and Cotman, C. W.,** Identification of a protein related to tubulin in the postsynaptic density, *Proc. Natl. Acad. Sci. U.S.A.*, 74, 1047, 1977.
172. **Gozes, I., and Richter-Landsberg, C.,** Identification of tubulin associated with rat brain myelin, *FEBS Lett.*, 95, 169, 1978.
173. **Gozes, I., and Littauer, U. Z.,** The α-subunit of tubulin is preferentially associated with brain presynaptic membrane, *FEBS Lett.*, 99, 86, 1979.
174. **Zisapel, N., Levi, M., and Gozes, I.,** Tubulin: an integral protein of mammalian synaptic vesicle membranes, *J. Neurochem.*, 34, 26, 1980.
175. **De Nechaud, B., Wolff, A., Jeantet, C., and Bourre, J. M.,** Characterization of tubulin in mouse brain myelin, *J. Neurochem.*, 41, 1538, 1983.
176. **Pfeffer, S. R., Drubin, D. G., and Kelly, R. B.,** Identification of three coated vesicle components as α- and β-tubulin linked to a phosphorylated 50,000-dalton polypeptide, *J. Cell Biol.*, 97, 40, 1983.
177. **Wiedemmann, B. and Himms, L. T.,** Tubulin is a major protein constituent of bovine brain coated vesicles, *Biochem. Biophys. Res. Commun.*, 115, 303, 1983.
178. **Hargreaves, A. and Avila, J.,** Localization and characterization of tubulin-like proteins associated with brain mitochondria: the presence of a membrane-specific isoform, *J. Neurochem.*, 45, 490, 1985.
179. **Floor, E. R., Gilbert, J. M., Nowak, T. S., Jr.,** Evidence for the synthesis of tubulin on membrane-bound and free ribosomes from rat forebrain, *Biochim. Biophys., Acta*, 442, 285, 1976.
180. **Sato, M., Yoshida, Y., and Takahashi, Y.,** *In vitro* biosynthesis of tubulin on total free and membrane bound polysomes from the developing rat brain, *J. Neurochem.*, 31, 1361, 1978.
181. **Cleveland, D. W., Kirschner, M. W., and Cowan, N. J.,** Isolation of separate mRNAs for α- and β-tubulin and characterization of the corresponding *in vitro* translation products, *Cell*, 15, 1021,
182. **Soifer, D., and Czosnek, H.,** Association of newly synthesized tubulin with brain microsomal membranes, *J. Neurochem.*, 35, 1128, 1980.
183. **Nath, J., and Flavin, M.,** A structural difference between cytoplasmic and membrane-bound tubulin of brain, *FEBS Lett.*, 95, 335, 1978.
184. **Caron, J. and Berlin, R. D.,** Interaction of microtubule proteins with phospholipid vesicles, *J. Cell Biol.*, 81, 665, 1979.
185. **Caron, J. and Berlin, R. D.,** Dynamic interactions between microtubules and artificial membranes, *Biochemistry*, 26, 3681, 1987.
186. **Klausner, R. D., Kumar, N., Weinstein, J. N., Blumenthal, R., and Flavin, M.,** Interaction of tubulin with membrane vesicles. I. Association with vesicles at the phase transition, *J. Biol. Chem.*, 256, 5879, 1981.
187. **Chan, K. Y. and Baxter, C. F.,** Compartments of tubulin and tubulin-like proteins in differentiating neuroblastoma cells, *Brain Res.*, 174, 135, 1979.
188. **Murphy, D. B. and Borisy, G. G.,** Association of high-molecular weight proteins with microtubules and their role in microtubule assembly *in vitro*, *Proc. Natl. Acad. Sci. U.S.A.*, 72, 2696, 1975.
189. **Borisy, G. G., Marcum, J. M., Olmsted, J. B., Murphy, D. B., and Johnson, K. A.,** Purification of tubulin and associated high molecular weight proteins from porcine brain and characterization of microtubule assembly *in vitro*, *Ann. N.Y. Acad. Sci.*, 253, 107, 1975.
190. **Sloboda, R. D., Dentler, W. L., and Rosenbaum, J. L.,** Microtubule-associated proteins and the stimulation of tubulin assembly *in vitro*, *Biochemistry*, 15, 4497, 1976.
191. **Murphy, D. B., Johnson, K. A., and Borisy, G. G.,** Role of tubulin-associated proteins in microtubule nucleation and elongation, *J. Mol. Biol.*, 117, 33, 1977.
192. **Murphy, D. B., Vallee, R. B., and Borisy, G. G.,** Identity and polymerization-stimulatory activity of the non-tubulin proteins associated with microtubules, *Biochemistry*, 16, 2598, 1977.
193. **Weingarten, M. D., Lockwood, A. H., Hwo, S. Y., and Kirschner, M. W.,** A protein factor essential for microtubule assembly, *Proc. Natl. Acad. Sci. U.S.A.*, 72, 1858, 1975.
194. **Cleveland, D. W., Hwo, S. Y., and Kirschner, M. W.,** Purification of tau, a microtubule associated protein that induces assembly of microtubules from purified tubulin, *J. Mol. Biol.*, 116, 207, 1977.
195. **Cleveland, D. W., Hwo, S. Y., and Kirschner, M. W.,** Physical and chemical properties of purified tau factor and the role of tau in microtubule assembly, *J. Mol. Biol.*, 116, 227, 1977.
196. **Parysek, L. M., Asnes, C. F., and Olmsted, J. B.,** MAP-4: occurrence in mouse tissues, *J. Cell Biol.*, 99, 1309, 1984.

197. **Parysek, L. M., Asnes, C. F., and Olmsted, J. B.,** MAP-4: a microtubule-associated protein specific for a subset of tissue microtubules, *J. Cell Biol.*, 99, 2287, 1984.
198. **Parysek, L. M., DelCerro, M., and Olmsted, J. B.,** Microtubule-associated protein 4 antibody: a new marker for astroglia and oligodendroglia, *Neuroscience*, 15, 869, 1985.
199. **Olmsted, J. B., Asnes, C. F., Parysek, L. M., Lyon, H. D., and Kidder, G. M.,** Distribution of MAP-4 in cells are in adult and developing tissues, *Ann. N.Y. Acad. Sci.*, 466, 292, 1986.
200. **Matus, A., Huber, G., and Bernhardt, R.,** Neuronal microdifferentiation, *Cold Spring Harbor Symp. Quant. Biol.*, 48, 775, 1983.
201. **Drubin, D. G., Caput, D., and Kirschner, M.,** Studies in the expression of the microtubule-associated protein, tau, during mouse brain development, with newly isolated complimentary DNA probes, *J. Cell Biol.*, 98, 1090, 1984.
202. **Drubin, D. G. and Kirschner, M. W.,** Tau protein function in living cells, *J. Cell Biol.*, 103, 2739, 1986.
203. **Papasozomenos, S. C. and Binder, L. I.,** Phosphorylation determines two distinct species of tau in the central nervous system, *Cell Motil. Cytoskel.*, 8, 210, 1987.
204. **Duerr, A., Pallas, D., and Solomon, F.,** Molecular analysis of cytoplasmic microtubules in situ: identification of both widespread and specific proteins, *Cell*, 24, 203, 1981.
205. **Pallas, D., and Solomon, F.,** Cytoplasmic microtubule-associated proteins: phosphorylation at novel sites is correlated with their incorporation into assembled microtubules, *Cell*, 30, 407, 1982.
206. **Magendantz, M. and Solomon, F.,** Analyzing the components of microtubules: antibodies against chartins, associated proteins from cultured cells, *Proc. Natl. Acad. Sci. U.S.A.*, 82, 6581, 1985.
207. **Gard, D. L. and Kirschner, M. W.,** A microtubule-associated protein from Xenopus eggs that specifically promotes assembly at the plus-end, *J. Cell Biol.*, 105, 2203, 1987.
208. **Vallee, R. B.,** A taxol-dependent procedure for the isolation of microtubules and microtubule-associated proteins, *J. Cell Biol.*, 92, 435, 1982.
209. **Brady, R. C. and Cabral, F. R.,** Identification of a 210-kDa microtubule-associated in chinese hamster ovary cells, *J. Cell Biol.*, 101, 30a (abstr.), 1985.
210. **Kotani, S., Murofushi, H., Maekawa, S., Sato, C., and Sakai, H.,** Characterization of microtubule-associated proteins isolated from bovine adrenal gland, *Eur. J. Biochem.*, 156, 23, 1986.
211. **Artlieb, U., Krepler, R., and Wiche, G.,** Expression of microtubule-associated proteins, MAP-1 and MAP-2, in human neuroblastomas and differential diagnosis of immature neuroblasts, *Lab. Invest.*, 53, 684, 1985.
212. **Olmsted, J. B.,** Microtubule-associated proteins, *Ann. Rev. Cell Biol.*, 2, 421, 1986.
213. **Kuznetsov, S. A., Rodinov, V. I., Gelfand, V. I., and Rosenblat, V. A.,** Microtubule-associated protein MAP-1 promotes microtubule assembly *in vitro*, *FEBS Lett.*, 135, 241, 1981.
214. **Huber, G., Pehling, G., and Matus, A.,** The novel microtubule-associated protein MAP-3 contributes to the *in vitro* assembly of brain microtubules, *J. Biol. Chem.*, 261, 2270, 1986.
215. **Carlier, M. F., Simon, C., and Pantaloni, D.,** Polymorphism of tubulin oligomers in the presence of microtubule-associated proteins. Implications in microtubule assembly, *Biochemistry*, 23, 1582, 1984.
216. **Díez, J. C., de la Torre, J., and Avila, J.,** Differential association of the different brain microtubule proteins in different *in vitro* assembly condition, *Biochim. Biophys. Acta*, 838, 32, 1985.
217. **Soifer, D., Laszlo, A., Mack, K., Scotto, J., and Siconolfi, L.,** The association of a cyclic AMP-dependent protein kinase activity with microtubule protein, *Ann. N.Y. Acad. Sci.*, 253, 598, 1975.
218. **Sloboda, R. D., Rudolph, S. A., Rosenbaum, J. L., and Greengard, P.,** Cyclic AMP-dependent endogenous phosphorylation of a microtubule-associated protein, *Proc. Natl. Acad. Sci. U.S.A.*, 72, 177, 1975.
219. **Jameson, L., Frey, T., Zeeberg, B., Dalldorf, F., and Caplow, M.,** Inhibition of microtubule assembly by phosphorylation of microtubule-associated proteins, *Biochemistry*, 19, 2472, 1980.
220. **Herrmann, H., Dulton, M. S., and Wiche, G.,** Microheterogeneity of microtubule-associated proteins, MAP-1 and MAP-2, and differential phosphorylation of individual subcomponents, *J. Biol. Chem.*, 260, 5795, 1985.
221. **Díaz-Nido, J., Serrano, L., and Avila, J.,** Differential phosphorylation of microtubule proteins by ATP and GTP, *Mol. Cell Biochem.*, 79, 73, 1988.
222. **Vallee, R. B.,** Structure and phosphorylation of MAP-2, *Proc. Natl. Acad. Sci. U.S.A.*, 77, 3206, 1980.
223. **Vallee, R. B., DiBartolomeis, M. J., and Therkauf, W. E.,** A protein kinase bound to the projection portion of MAP-2, *J. Cell Biol.*, 90, 568, 1981.
224. **Therkauf, W. E. and Vallee, R. B.,** Molecular characterization of the cAMP-dependent protein kinase bound to microtubule-associated protein 2, *J. Biol. Chem.*, 257, 3284, 1982.
225. **Therkauf, W. E., and Vallee, R. B.,** Extensive cAMP-dependent and cAMP-independent phosphorylation of microtubule-associated protein 2, *J. Biol. Chem.*, 258, 7883, 1983.
226. **Larson, R. E., Goldenring, J. R., Vallano, M. L., and DeLorenzo, R. J.,** Identification of endogenous calmodulin-dependent kinase and calmodulin binding proteins in cold-stable microtubule preparations from rat brain, *J. Neurochem.*, 44, 1566, 1985.

227. **Vallano, M. L., Goldenring, J. R., Buckholz, T. M., and Larson, R. E.,** Separation of endogenous calmodulin and cAMP-dependent kinases from microtubule preparations, *Proc. Natl. Acad. Sci. U.S.A.*, 82, 3202, 1985.
228. **Díaz-Nido, J., Serrano, L., Méndez, E., and Avila, J.,** A casein kinase II-related activity is involved in phosphorylation of microtubule-associated protein MAP-1B during neuroblastoma cell differentiation, *J. Cell Biol.*, 106, 2057, 1988.
229. **Bloom, G. S., Schoenfeld, T. A., and Vallee, R. B.,** Widespread distribution of the major polypeptide of MAP1 in the nervous system, *J. Cell Biol.*, 98, 320, 1984.
230. **Bloom, G. S., Luca, F. C., and Vallee, R. B.,** Widespread cellular distribution of MAP-1A in the mitotic spindle and on interphase microtubules, *J. Cell Biol.*, 98, 331, 1984.
231. **Bloom, G. S., Luca, F. C., and Vallee, R. B.,** Microtubule-associated protein 1B: identification of a major component of the neuronal cytoskeleton, *Proc. Natl. Acad. Sci. U.S.A.*, 82, 5404, 1985.
232. **Vallee, R. B. and Luca, F. C.,** Light chain content and other criteria for the identification of MAP-1A and MAP-1B, in *Microtubules and Microtubule Inhibitors 1985*, De Brabander, M. and De Mey, J., Eds., Elsevier, Amsterdam, 1985.
233. **Vallee, R. B.,** On the use of heat stability as a criterion for the identification of microtubule-associated proteins, *Biochem. Biophys. Res. Commun.*, 133, 128, 1985.
234. **Luca, F. C., Bloom, G. S., and Vallee, R. B.,** A monoclonal antibody that cross-reacts with phosphorylated epitopes on two microtubule-associated proteins and two neurofilament polypeptides, *Proc. Natl. Acad. Sci. U.S.A.*, 83, 1006, 1986.
235. **Ksiezak-Reding, H. and Yen, S.-H.,** Two monoclonal antibodies recognize Alzheimer's neurofibrillary tangles, neurofilament, and microtubule-associated proteins, *J. Neurochem.*, 48, 455, 1987.
236. **Avila, J., Serrano, L., Hernández, M. A., and Díaz-Nido, J.,** Phosphorylation of neuronal microtubule proteins, *Protoplasma*, 145, 82, 1988.
237. **Vallee, R. B. and Davis, S. E.,** Low molecular weight microtubule-associated proteins are light chains of microtubule-associated protein 1, *Proc. Natl. Acad. Sci. U.S.A.*, 80, 1342, 1983.
238. **Huber, G. and Matus, A.,** Immunohistochemical localization of microtubule-associated protein 1 in rat cerebellum using monoclonal antibodies, *J. Cell Biol.*, 98, 777, 1984.
239. **Shiomura, Y. and Hirokawa, N.,** Colocalization of MAP-1A and MAP-2 on neuronal microtubules *in situ* as revealed with double-label immunofluorescence microscopy, *J. Cell Biol.*, 104, 1575, 1987.
240. **Huber, G. and Matus, A.,** Differences in the cellular distributions of two microtubule-associated proteins, MAP-1 and MAP-2, in rat brain, *J. Neurosci.*, 4, 151, 1984.
241. **Shiomura, Y. and Hirokawa, N.,** The molecular structure of MAP-1A *in vivo* and *in vitro*. An immunoelectron microscopy and quick-freeze, deep-etch study, *J. Neurosci.*, 7, 1461, 1987.
242. **Riederer, B. and Matus, A.,** Differential expression of distinct microtubule-associated proteins during brain development, *Proc. Natl. Acad. Sci. U.S.A.*, 82, 6006, 1985.
243. **Kumagai, H., Imazawa, M., and Miyamoto, K.,** Developmental changes in components of chick brain microtubule-associated protein MAP-1 and tau proteins, *J. Biochem.*, 97, 529, 1985.
244. **Matus, A. and Riederer, B.,** Microtubule-associated proteins in the developing brain, *Ann. N.Y. Acad. Sci.*, 466, 167, 1986.
245. **Bernhardt, R., Huber, G., and Matus, A.,** Differences in the developmental patterns of three microtubule-associated proteins in brain, *J. Neurosci.*, 5, 977, 1985.
246. **Calvert, R. and Anderton, B. H.,** A microtubule associated protein MAP-1 (X) which is expressed at elevated levels during development of rat cerebellum, *EMBO J.*, 4, 1171, 1985.
247. **Riederer, B., Cohen, R., and Matus, A.,** MAP-5: a novel brain microtubule-associated protein under strong developmental regulation, *J. Neurocytol.*, 15, 763, 1986.
248. **Tucker, R. P. and Matus, A.,** Developmental regulation of two microtubule-associated proteins (MAP-2 and MAP-5) in the embryonic avian retina, *Development*, 101, 535, 1987.
249. **Bloom, G. S., Luca, F. C., and Vallee, R. B.,** Identification of high molecular weight microtubule-associated proteins in anterior pituitary tissue and cells using taxol-dependent purification combined with microtubule-associated protein specific antibodies, *Biochemistry*, 24, 4185, 1985.
250. **Mascardo, R. M., Sherline, P., and Weatherbee, J.,** Localization of high molecular weight microtubule-associated proteins MAP-1 and MAP-2 in HeLa microtubule organizing center, *Cytobios*, 35, 113, 1982.
251. **De Mey, J., Aerts, F., Moeremans, M., Gevens, G., Dancels, G., and De Brabander, M.,** Anti-MAP 1 reacts with the centrosomes, kinetochores, midbody and spindle of mitotic PTK 2 cells, *J. Cell Biol.*, 99, 447, 1984.
252. **Bonifacino, J. S., Klausner, R. D., and Sandoval, I. V.,** A widely distributed nuclear protein immunologically related to the microtubule-associated protein MAP-1 is associated with the mitotic spindle, *Proc. Natl. Acad. Sci. U.S.A.*, 82, 1146, 1985.
253. **Díaz-Nido, J. and Avila, J.,** Characterization of proteins immunologically related to brain microtubule-associated protein MAP-1B in nonneural cells, *J. Cell. Sci.*, 92, 607, 1989.

254. **Hernández, M. A., Avila, J., and Andreu, J. M.,** Physicochemical characterization of the heat-stable microtubule-associated protein MAP-2, *Eur. J. Biochem.*, 154, 41, 1986.
255. **Herzog, W. and Weber, K.,** Fractionation of brain microtubule-associated proteins. Isolation of two different proteins which stimulate tubulin polymerization *in vitro*, *Eur. J. Biochem.*, 92, 1, 1978.
256. **Amos, L. A.,** Arrangement of high molecular weight associated proteins on purified mammalian brain microtubules, *J. Cell Biol.*, 72, 642, 1977.
257. **Kim, H., Binder, L. I., and Rosenbaum, J. L.,** The periodic association of MAP-2 with brain microtubules *in vitro*, *J. Cell Biol.*, 80, 266, 1979.
258. **Vallee, R. B. and Borisy, G. G.,** Removal of the projections from cytoplasmic microtubules *in vitro* by digestion with trypsin, *J. Biol. Chem.*, 252, 377, 1977.
259. **Zingsheim, H. P., Herzog, W., and Weber, K.,** Differences in surface morphology of microtubules reconstituted from pure brain tubulin using two different microtubule-associated proteins: the high molecular weight MAP-2 proteins and tau proteins, *Eur. J. Cell Biol.*, 19, 175, 1979.
260. **Voter, W. A. and Erickson, H. P.,** Electron microscopy of MAP-2, *J. Ultrastruct. Res.*, 80, 374, 1982.
261. **Woody, R. W., Clark, D. C., Roberts, G. C. K., Martin, S. R., and Bayley, P. M.,** Molecular flexibility in microtubule proteins: proton nuclear magnetic resonance characterization, *Biochemistry*, 22, 2186.
262. **Lee, Y. C. and Wolff, J.,** Calmodulin binds to both MAP-2 and tau proteins, *J. Biol. Chem.*, 259, 1226, 1984.
263. **Lee, Y. C. and Wolff, J.,** The calmodulin binding domain on MAP-2, *J. Biol. Chem.*, 259, 8041, 1984.
264. **Serrano, L., Valencia, A., Caballero, R., and Avila, J.,** Localization of the high affinity calcium-binding site on tubulin molecule, *J. Biol. Chem.*, 261, 7076, 1986.
265. **Murthy, A. and Flavin, M.,** Microtubule assembly using the microtubule-associated protein MAP-2 prepared in defined stated of phosphorylation with protein kinase and phosphatase, *Eur. J. Biochem.*, 137, 37, 1983.
266. **Nishida, E., Kotani, S., Kuwaki, T., and Sakai, H.,** Phosphorylation of microtubule-associated proteins controls both microtubule assembly and MAP-actin interaction, in *Biological Functions of Microtubules and Related Structures*, Sakai, H., Mohri, H., and Borisy, G. G., Eds., Academic Press, Tokyo, 1982, 285.
267. **Aoki, C. and Siekevitz, P.,** Ontogenetic changes in the cAMP-stimulatable phosphorylation of cat visual cortex proteins, particularly of MAP-2: effect of normal and dark rearing and the exposure to light, *J. Neurosci.*, 5, 2465, 1985.
268. **Tsuyama, S., Bramblett, G. T., Huang, K.-P., and Flavin, M.,** Calcium/phospholipid-dependent protein kinase recognizes sites in MAP-2 which are phosphorylated in living brain and are not accessible to other kinases, *J. Biol. Chem.*, 261, 4110, 1986.
269. **Hernández, M. A., Wandosell, F., and Avila, J.,** Localization of the phosphorylation sites for different kinases in the microtubule-associated protein MAP-2, *J. Neurochem.*, 48, 85, 1987.
270. **Bernhardt, R. and Matus, A.,** Light and electron microscopic studies of the distribution of MAP-2 in rat brain: a difference between the dendritic and axonal cytoskeletons, *J. Comp. Neurol.*, 226, 203, 1984.
271. **Burgoyne, R. D. and Cumming, R.,** Ontogeny of MAP-2 in rat cerebellum: differential expression of the doublet polypeptides, *Neuroscience*, 11, 156, 1984.
272. **Caceres, A., Banker, G. A., Steward, O., Binder, L., and Payne, M.,** MAP-2 is localized to the dendrites of hippocampal neurons which develop in culture, *Dev. Brain Res.*, 13, 314, 1984.
273. **Caceres, A., Binder, L. I., Payne, M. R., Bender, P., Rebhun, L., and Steward, O.,** Differential subcellular localization of tubulin and the microtubule-associated protein MAP-2 in brain tissue as revealed by immunohistochemistry with monoclonal hybridoma antibodies, *J. Neurosci.*, 4, 394, 1984.
274. **DeCamilli, P., Miller, P. E., Navone, F., Theurkauf, W. E., and Vallee, R. B.,** Distribution of microtubule-associated protein 2 in the nervous system of the rat studied by immunofluorescence, *Neuroscience*, 11, 817, 1984.
275. **Papasozomenos, S. C. and Binder, L. I.,** Microtubule-associated protein 2 (MAP-2) is present in astrocytes of the optic nerve but absent from astrocytes of the optic tract, *J. Neurosci.*, 6, 1748, 1986.
276. **Binder, L. I., Frankfurter, A., Him, H., Caceres, A., Payne, M. R., and Rebhun, L. I.,** Heterogeneity of MAP-2 during rat brain development, *Proc. Natl. Acad. Sci. U.S.A.*, 81, 5613, 1984.
277. **Couchie, D. and Nunez, J.,** Immunological characterizations of microtubule-associated proteins specific for the immature brain, *FEBS Lett.*, 188, 331, 1985.
278. **Garner, C. C., Brugg, B., and Matus, A.,** A 70-kilodalton microtubule-associated protein (MAP-2c), related to MAP-2, *J. Neurochem.*, 70, 609, 1988.
279. **Garner, C. C. and Matus, A.,** Different forms of microtubule-associated protein 2 are encoded by separate mRNA transcripts, *J. Cell Biol.*, 106, 779, 1988.
280. **Bloom, G. S. and Vallee, R. B.,** Association of MAP-2 with microtubules and intermediate filaments in cultured brain cells, *J. Cell Biol.*, 96, 1523, 1983.

281. **Valdivia, M. M., Avila, J., Coll, J., Colaço, C., and Sandoval, I. V.,** Quantitation and characterization of the microtubule-associated protein MAP-2 in porcine tissues and its isolation from porcine and human cell lines, *Biochem. Biophys. Res. Commun.*, 105, 1241, 1982.
282. **Lewis, S. A., Villasante, A., Sherline, P., and Cowan, N.,** Brain specific expression of MAP-2 detected using a cloned cDNA probe, *J. Cell Biol.*, 102, 2098, 1986.
283. **Burgoyne, R. D. and Norman, K. M.,** Presence of microtubule-associated protein 2 in chromaffin cells, *Neuroscience*, 14, 955, 1985.
284. **Lindwall, G. and Cole, R. D.,** Phosphorylation affects the ability of tau proteins to promote microtubule assembly, *J. Biochem.*, 259, 5301, 1984.
285. **Lindwall, G. and Cole, R. D.,** The purification of tau protein and the occurrence of two phosphorylation states of tau in brain, *J. Biol. Chem.*, 259, 12241, 1984.
286. **Binder, L. I., Frankfurter, A., and Rebhum, L. I.,** The distribution of tau polypeptides in the mammalian central nervous system, *J. Cell Biol.*, 101, 1371, 1985.
287. **Peng, I., Binder, L. I., and Black, M. M.,** Cultured neurons contain a variety of microtubule-associated proteins, *Brain Res.*, 361, 200, 1985.
288. **Nunez, J., Couchie, D., Brion, J. P., and Tardy, M.,** Expression of juvenile and adult tau proteins during neuronal differentiation, in *The Cytoskeleton in Cell Differentiation and Development*, Maccioni, R. B. and Arechaga, J., Eds., IRL Press, Oxford, 1987, 193.
289. **Couchie, D., Fages, C., Bridoux, A. M., Rolland, B., Tardy, M., and Nunez, J.,** Microtubule-associated proteins and *in vitro* astrocyte differentiation, *J. Cell Biol.*, 101, 2095, 1985.
290. **Montejo de Garcini, E., Díez, J. C., and Avila, J.,** Quantitation and characterization of tau factor in porcine tissues, *Biochim. Biophys. Acta*, 881, 456, 1986.
291. **Ginzburg, I., Scherson, T., Giveon, D., Behar, L., and Littauer, U. Z.,** Modulation of mRNA for microtubule associated proteins during brain development, *Proc. Natl. Acad. Sci. U.S.A.*, 79, 4892, 1982.
292. **Lee, G., Cowan, N., and Kirschner, M.,** The primary structure and heterogeneity of tau protein from mouse brain, *Science*, 239, 285, 1988.
293. **Aizawa, H., Kawasaki, H., Murofushi, H., Kotani, S., Suzuki, K., and Sakai, H.,** Microtubule-binding domain of tau proteins, *J. Biol. Chem.*, 263, 7703, 1988.
294. **Pierre, M. and Nunez, J.,** Multisite phosphorylation of tau proteins from rat brain, *Biochem. Biophys. Res. Commun.*, 115, 212, 1983.
295. **Butler, M. and Shelanski, M. L.,** Microheterogeneity of microtubule-associated tau proteins is due to differences in phosphorylation, *J. Neurochem.*, 47, 1517, 1986.
296. **Baudier, J. and Cole, R. D.,** Phosphorylation of tau proteins to a state like that in Alzheimer's brain is catalyzed by a calcium/calmodulin-dependent kinase and modulated by phospholipids, *J. Biol. Chem.*, 262, 17577, 1987.
297. **Baudier, J., Lee, S.-H., and Cole, R. D.,** Separation of the different microtubule-associated tau protein species from bovine brain and their mode II phosphorylation by Ca^{2+}/phospholipid-dependent protein kinase C, *J. Biol. Chem.*, 262, 17584, 1987.
298. **Harrison, R. G.,** Observations on the living developing nerve fiber, *Proc. Soc. Exp. Biol. Med.*, 4, 140, 1907.
299. **Harrison, R. G.,** The outgrowth of the nerve fiber as a mode of protoplasmic movement, *J. Exp. Zool.*, 9, 787, 1910.
300. **Darmon, M., Bottenstein, J., and Sato, G.,** Neural differentiation following culture of embryonal carcinoma cells in a serum-free defined medium, *Dev. Biol.*, 85, 463, 1981.
301. **Pfeiffer, S., Jakob, H., Mikoshiba, K., Dubois, P., Guenet, J. L., Nicolas, J. F., Guillard, J., Chevance, L. G., and Jacob, F.,** Differentiation of a teratocarcinoma cell line: preferential development of cholinergic neurons, *J. Cell Biol.*, 88, 57, 1981.
302. **Eddé, B., Jakob, M., and Darmon, M.,** Two specific markers for neural differentiation of embryonal carcinoma cells, *EMBO J.*, 2, 1473.
303. **Greene, L. A. and Tischler, A. S.,** Establishment of a moradrenergic clonal line of rat adrenal pheochromocytoma cells which respond to nerve growth factor, *Proc. Natl. Acad. Sci. U.S.A.*, 73, 2424, 1976.
304. **Greene, L. A. and Tischler, A. S.,** PC12 pheochromocytoma cultures in neurobiological research, *Adv. Cell Neurobiol.*, 3, 374, 1982.
305. **Burstein, D. E. and Greene, L. A.,** Evidence for both RNA synthesis-dependent and independent pathways in stimulation of neurite outgrowth by nerve growth factor, *Proc. Natl. Acad. Sci. U.S.A.*, 75, 6059, 1978.
306. **Greene, L. A.,** The importance of both early and delayed responses in the biological action of nerve growth factor, *Trends Neurosci.*, 7, 91, 1984.
307. **Levi-Montalcini, R. and Calissano, P.,** NGF as a paradigm for other polypeptide growth factors, *Trends Neurosci.*, 9, 473, 1986.
308. **Pochedly, C.,** The biological vagaries of neuroblastoma, in *Neuroblastoma*, Pochedly, C., Ed., Edward Arnold, London, 1977, 273.

309. **Augusti-Tocco, G. and Sato, G.**, Establishment of functional clonal lines of neurons from mouse neuroblastoma, *Proc. Natl. Acad. Sci. U.S.A.*, 64, 311, 1969.
310. **Schubert, D. and Jacob, F.**, 5-bromodeoxyuridine induced differentiation of a neuroblastoma cell line, *Proc. Natl. Acad. Sci. U.S.A.*, 67, 247, 1970.
311. **Seeds, N. W., Gilman, A. G., Amano, T., and Nirenberg, M.**, Regulation of axon formation by clonal lines of a neural tumor, *Proc. Natl. Acad. Sci. U.S.A.*, 66, 160, 1970.
312. **Yamada, K. M., Spooner, B. S., and Wessels, M. K.**, Axon growth: role of microfilaments and microtubules, *Proc. Natl. Acad. Sci. U.S.A.*, 66, 1206, 1970.
313. **Furmanski, P., Silverman, D. J., and Lubin, M.**, Expression of differentiated functions in mouse neuroblastoma mediated by dibutyryl cyclic AMP, *Nature*, 233, 413, 1971.
314. **Prasad, N. K.**, Morphological differentiation induced by prostaglandins of mouse neuroblastoma cells in culture, *Nature New Biol.*, 236, 49, 1972.
315. **Prasad, N. K.**, Differentiation of neuroblastoma cells in culture, *Biol. Rev. Cambridge Phil. Soc.*, 50, 129, 1975.
316. **Shubert, D., Harris, A. J., Heinemann, S., Kadakoro, Y., Patrick, J., and Steinbeck, J. H.**, Differentiation of clonal cell lines of nerve and muscle, *Curr. Top. Neurobiol.*, 1, 55, 1973.
317. **McMorris, F. A., Nelson, P. G., and Ruddle, F. H.**, Contribution of clonal systems to neurobiology, *Neurosci. Res. Prog. Bull.*, 11, 411, 1973.
318. **Ross, J., Olmsted, J. B., and Rosenbaum, J. L.**, The ultrastructure of mouse neuroblastoma cells in tissue culture, *Tissue Cell*, 7, 107, 1975.
319. **Kimhi, Y., Palfrey, C., Spector, I., Barak, Y., and Littauer, U. Z.**, Maturation of neuroblastoma cells in the presence of dimethylsulfoxide, *Proc. Natl. Acad. Sci. U.S.A.*, 73, 462, 1976.
320. **Littauer, U. Z., Palfrey, C., Kimhi, Y., and Spector, I.**, Induction of differentiation in mouse neuroblastoma cells, *Natl. Cancer Inst. Monogr.*, 48, 333, 1978.
321. **Littauer, U. Z., Giovanni, M. Y., and Glick, M. C.**, Differentiation of human neuroblastoma cells in culture, *Biochem. Biophys. Res. Commun.*, 88, 933, 1979.
322. **Portier, M. M., Croizat, B., Berthelot, F., Eddé, B., Paulin, D., and Gros, F.**, Molecular approach to the study of neural function and differentiation, in *Gene Expression in Normal and Transformed Cells*, Celis, J. E. and Bravo, R., Eds., Plenum Press, London, 1983.
323. **Jensen, L. M.**, Phenotypic differentiation of aphidicolin-selected human neuroblastoma cultures after long-term exposure to nerve growth factor, *Dev. Biol.*, 120, 56, 1987.
324. **Ramón y Cajal, S.**, Sur l'origine et les ramifications des fibres nerveuses de la moelle embryonaire, *Anat. Anzerger*, 5, 609, 1890.
325. **Yamada, K. M., Spooner, B. S., and Wessells, N. K.**, Ultrastructure and function of growth cones and axons of cultured nerve cells, *J. Cell Biol.*, 49, 614, 1971.
326. **Bunge, M. B.**, Fine structure of nerve fibers and growth cones of isolated sympathetic neurons in culture, *J. Cell Biol.*, 56, 713, 1973.
327. **Landis, S. C.**, Neuronal growth cones, *Ann. Rev. Physiol.*, 45, 567, 1983.
328. **Goldberg, D. J. and Burmeister, D. W.**, Stages in axon formation: observations of growth of Aplysia axons in culture using video-enhanced contrast-differential interference contrast microscopy, *J. Cell Biol.*, 103, 1921, 1986.
329. **Aletta, J. M. and Greene, L. A.**, Growth cone configuration and advance: a time-lapse study using video-enhanced differential interference contrast microscopy, *N. Neurosci.*, 8, 1425, 1988.
330. **Bray, D.**, The mechanism of growth cone movements, *Neurosci. Res. Prog. Bull.*, 20, 821, 1982.
331. **Bray, D. and Chapman, K.**, Analysis of microspike movement on the neuronal growth cone, *J. Neurosci.*, 5, 3204, 1985.
332. **Bolsover, S. R. and Spector, I.**, Measurement of calcium transients in the soma, neurites and growth cones of single cultured neurons, *J. Neurosci.*, 6, 1934, 1986.
333. **Connor, J. A.**, Digital imaging of free calcium changes and of spatial gradients in growing processes in single mammalian central nervous system neurons, *Proc. Natl. Acad. Sci. U.S.A.*, 83, 6179, 1986.
334. **Mattson, M. P. and Kater, S. B.**, Calcium regulation of neurite elongation and growth cone motility, *J. Neurosci.*, 7, 4034, 1987.
335. **Cambray-Deakin, M. A., Morgan, A., and Burgoyne, R. D.**, Sequential appearance of cytoskeletal components during the early stages of neurite outgrowth from cerebellar granule cells *in vitro*, *Dev. Brain Res.*, 37, 197, 1987.
336. **Cheng, N. and Sahyoun, N.**, The growth cone cytoskeleton: glycoprotein association, calmodulin binding, and tyrosine/serine phosphorylation of tubulin, *J. Biol. Chem.*, 263, 3935, 1988.
337. **Letourneau, P. C. and Ressler, A. H.**, Inhibition of neurite initiation and growth by taxol, *J. Cell Biol.*, 98, 1355, 1984.
338. **Bray, D., Thomas, C., and Shaw, G.**, Growth cone formation in cultures of sensory neurons, *Proc. Natl. Acad. Sci. U.S.A.*, 75, 5226, 1978.

339. **Marsh, L. and Letourneau, P. C.,** Growth of neurites without filopodial or lamellipodial activity in the presence of cytochalasin B, *J. Cell Biol.*, 99, 2041, 1984.
340. **Bentley, D. and Toroian-Raymond, A.,** Disoriented pathfinding by pioneer neurone growth cones deprived of filopodia by cytochalasin treatment, *Nature*, 323, 712, 1986.
341. **Daniels, M. P.,** Colchicine inhibition of nerve fiber formation *in vitro*, *J. Cell Biol.*, 53, 164, 1972.
342. **Daniels, M. P.,** The role of microtubules in the growth and stabilization of nerve fibers, *Ann. N.Y. Acad. Sci.*, 253, 535, 1975.
343. **Olmsted, J. B.,** Tubulin pools in differentiating neuroblastoma cells, *J. Cell Biol.*, 89, 418, 1981.
344. **Drubin, D. G., Feinstein, S. C., Shooter, E. M., and Kirschner, M. W.,** Nerve growth factor-induced neurite outgrowth in PC12 cells involves the coordinate induction of microtubule assembly and assembly promoting factors, *J. Cell Biol.*, 101, 1799, 1985.
345. **Black, M. M., Aletta, J. M., and Greene, L. A.,** Regulation of microtubule composition and stability during NGF-promoted neurite outgrowth, *J. Cell Biol.*, 103, 545, 1986.
346. **Black, M. M. and Greene, L. A.,** Changes in the colchicine susceptibility of microtubules associated with neurite outgrowth: studies with NGF-responsive PC12 pheochromocytoma cells, *J. Cell Biol.*, 95, 379, 1982.
347. **Heidemann, S. R., Joshi, H. C., Schechter, A., Fletcher, J. R., and Bothwell, M.,** Synergistic effects of cAMP and NGF of neurite outgrowth and microtubule stability of PC12 cells, *J. Cell Biol.*, 100, 916, 1985.
348. **Kirschner, M. W. and Mitchison, T.,** Beyond self-assembly, from microtubules to morphogenesis, *Cell*, 45, 329, 1986.
349. **Spiegelman, B. M., Lopata, M. A., and Kirschner, M. W.,** Multiple sites for the initiation of microtubule assembly in mammalian cells, *Cell*, 16, 239, 1979.
350. **Spiegelman, B. M., Lopata, M. A., and Kirschner, M. W.,** Aggregation of microtubule initiation sites preceding neurite outgrowth in mouse neuroblastoma cells, *Cell*, 16, 253, 1979.
351. **Brinkley, B. R., Cox, S. M., and Fistel, S. H.,** Organizing centers for cell processes, *Neurosci. Res. Prog. Bull.*, 19, 108, 1981.
352. **Sharp, G. A., Weber, K., and Osborn, M.,** Centriole number and process formation in established neuroblastoma cells and primary dorsal root ganglion neurons, *Eur. J. Cell Biol.*, 29, 97, 1982.
353. **Tennyson, V. M.,** The fine structure of the axon and growth cone of the dorsal root neuroblast of the rabbit embryo, *J. Cell Biol.*, 44, 62, 1970.
354. **Tischler, A. S. and Greene, L. A.,** Morphologic and cytochemical properties of a clonal line of rat adrenal pheochromocytoma cells which respond to nerve growth factor, *Lab. Invest.*, 39, 77, 1978.
355. **Faivre, C., Legrand, C., and Rabié, A.,** Effects of thyroid deficiency and corrective effects of thyroxine on microtubules and mitochondria in cerebellar purkinje cell dendrites of developing rats, *Dev. Brain Res.*, 8, 21, 1983.
356. **Faivre, C., Legrand, C., and Rabié, A.,** The microtubular apparatus of cerebellar purkinje cells during postnatal development of the rat: the density and cold-stability of microtubules increase with age and are sensitive to thyroid hormone, *Int. J. Dev. Neurosci.*, 3, 559, 1985.
357. **Marc, C. and Rabié, A.,** Microtubules and neurofilaments in the sciatic nerve fibers of the developing rat: effects of thyroid deficiency, *Int. J. Dev. Neurosci.*, 3, 353, 1985.
358. **Fellous, A., Lennon, A. M., Francon, J., and Nunez, J.,** Thyroid hormones and neurotubule assembly *in vitro* during brain development, *Eur. J. Biochem.*, 101, 365, 1979.
359. **Lennon, A. M., Francon, J., Fellous, A., and Nunez, J.,** Rat, mouse and guinea-pig brain development and microtubule assembly, *J. Neurochem.*, 35, 804, 1980.
360. **Francon, J., Lennon, A. M., Fellous, A., Mareck, A., Pierre, M., and Nunez, J.,** Heterogeneity of MAPs and brain development, *Eur. J. Biochem.*, 129, 465, 1982.
361. **Nunez, J.,** Differential expression of microtubule components during brain development, *Dev. Neurosci.*, 8, 125, 1986.
362. **Aletta, J. M. and Greene, L. A.,** Sequential phosphorylation of chartin microtubule-associated proteins is regulated by the presence of microtubules, *J. Cell Biol.*, 105, 277, 1987.
363. **Mill, J. F., Chao, M. V., and Ishii, D. N.,** Insulin, insulin-like growth factor II and nerve growth factor effects on tubulin mRNA levels and neurite formation, *Proc. Natl. Acad. Sci. U.S.A.*, 82, 7126, 1985.
364. **Morgan, J. L. and Seeds, N. W.,** Tubulin constancy during morphological differentiation of mouse neuroblastoma cells, *J. Cell Biol.*, 67, 136, 1975.
365. **Seeds, N. W. and Maccioni, R. B.,** Proteins from morphologically differentiated neuroblastoma cells promote tubulin polymerization, *J. Cell Biol.*, 76, 547, 1978.
366. **Biocca, S., Cattaneo, A., and Calissano, P.,** A macromolecular structures favouring microtubule assembly in NGF differentiated pheochromocytoma cells (PC12), *EMBO J.*, 2, 643,1983.
367. **Olmsted, J. B. and Lyon, H. D.,** A microtubule-associated protein specific to differentiated neuroblastoma cells, *J. Biol. Chem.*, 256, 3507, 1981.

368. **Greene, L. A., Liem, R. K. H., and Shelanski, M. L.,** Regulation of a high molecular weight microtubule-associated protein in PC12 cells by nerve growth factor, *J. Cell Biol.*, 96, 76, 1983.
369. **Couchie, D., Faivre-Bauman, A., Puymirat, J., Guilleminot, J., Tixier-Vidal, A., and Nunez, J.,** Expression of microtubule-associated proteins during the early stages of neurite extension by brain neurons cultured in a defined medium, *J. Neurochem.*, 47, 1255, 1986.
370. **Lewis, S. A., Sherline, P., and Cowan, N.,** A cloned DNA encoding MAP-1 detects a single copy gene in mouse, and a brain abundant mRNA whose level decreases with development, *J. Cell Biol.*, 102, 2106, 1986.
371. **Gunning, P. W., Landreth, G. E., Bothwell, M. A., Shooter, E. M.,** Differential and synergistic actions of NGF and cAMP in PC12 cells, *J. Cell Biol.*, 89, 240, 1981.
372. **Greene, L. A., Drexier, S. A., Connolly, J. L., Rukenstein, A., and Green, S. H.,** Selective inhibition of responses to nerve growth factor and of microtubule-associated protein phosphorylation by activators of adenylate cyclase, *J. Cell Biol.*, 103, 1967, 1986.
373. **Izant, J. G. and McIntosh, J. R.,** Microtubule-associated proteins: a monoclonal antibody to brain MAP-2 binds to differentiated neurons, *Proc. Natl. Acad. Sci. U.S.A.*, 77, 4741, 1980.
374. **Fischer, I., Shea, T. B., Sapirstein, U. S., and Kosik, K. S.,** Expression and distribution of MAP-2 in neuroblastoma and primary neuronal cells, *Dev. Brain Res.*, 25, 94, 1986.
375. **Olmsted, J. B., Cox, J. V., Asnes, C. F., Parysek, L. M., and Lyon, H. D.,** Cellular regulation of microtubule organization, *J. Cell Biol.*, 99, 28s, 1984.
376. **Seeds, N. W., Hinds, K. R., and Maccioni, R. B.,** Characterization of a microtubule assembly promoting factor in differentiated neuroblastoma cells, in *The Cytoskeleton in Cell Differentiation and Development*, Maccioni, R. B. and Arechaga, J., Eds., IRL Press, Oxford, 1987, 187.
377. **Thoenen, H., Bandtlow, C., and Heumann, R.,** The physiological functions of nerve growth factor in the central nervous system: comparison with the periphery, *Rev. Physiol. Biochem. Pharmacol.*, 109, 146, 1987.
378. **Eayrs, J. T.,** Influence of the thyroid on the central nervous system, *Br. Med. Bull.*, 16, 122, 1960.
379. **Puymirat, J., Loudes, C., Faivre-Bauman, A., Tixier-Vidal, A., and Bourre, J. M.,** Expression of neural functions by mouse fetal hypothalamic cells cultured in hormonally defined medium, *Cold Spring Harbor Conf. Cell Prolif.*, 9, 1033, 1982.
380. **Puymirat, J., Barret, A., Vigny, A., Loudes, C., Faivre-Bauman, A., and Tixier-Vidal, A.,** Triiodothyronine enhances the morphological maturation of dopaminergic neurons from fetal mouse hypothalamus cultured in serum-free medium, *Neuroscience*, 10, 801, 1983.
381. **Nunez, J.,** Effects of thyroid hormones during brain differentiation, *Mol. Cell Endocrinol.*, 37, 125, 1984.
382. **Nunez, J.,** Microtubules and brain development: the effects of thyroid hormones, *Neurochem. Int.*, 7, 959, 1985.
383. **Hargreaves, A., Yusta, B., Aranda, A., Avila, J., and Pascual, A.,** Triiodothyronine (T3) induces neurite formation and increases synthesis of a protein related to MAP-1B in cultured cells of neuronal origin, *Dev. Brain Res.*, 38, 141, 1988.
384. **Francon, J., Fellous, A., Lennon, A. M., and Nunez, J.,** Is thyrosine a regulatory signal for neurotubule assembly during brain development?, *Nature*, 266, 188, 1977.
385. **Alaimo-Beuret, D. and Matus, A.,** Changes in the cytoplasmic distribution of MAP-2 during the differentiation of cultured cerebellar granule cells, *Neuroscience*, 14, 1103, 1988.
386. **Burstein, D. E., Seely, P. J., and Greene, L. A.,** Lithium ion inhibits NGF-induced neurite outgrowth and phosphorylation of NGF-modulated microtubule-associated proteins, *J. Cell Biol.*, 101, 862, 1985.
387. **Bernhardt, R. and Matus, A.,** Initial phase of dendrite growth: evidence of the involvement of high molecular weight microtubule-associated proteins before the appearance of tubulin, *J. Cell Biol.*, 92, 589, 1982.
388. **Bartlett, W. P. and Banker, G. A.,** An electron microscopic study of the development of axons and dendrites by hippocampal neurons in culture. I. Cells which develop without intercellular contacts, *J. Neurosci.*, 4, 1944, 1984.
389. **Davis, L., Banker, G. A., and Steward, O.,** Selective dendritic transport of RNA in hippocampal neurons in culture, *Nature*, 330, 477, 1987.
390. **Higgins, D., Waxman, A., and Banker, G.,** The distribution of MAP-2 changes when dendritic growth is induced in rat sympathetic neurons *in vitro*, *Neuroscience*, 24, 583, 1988.
391. **Mudge, A. W.,** Schwann cells induce morphological transformation of sensory neurons *in vitro*, *Nature*, 309, 367, 1984.
392. **Allen, M. M. and Higgins, D.,** Ganglionic non-neural cells induce dendritic growth in cultured sympathetic neurons, *Soc. Neurosci. Abstr.*, 12, 159, (abstr.), 1986.
393. **Bruckenstein, D. and Higgins, D.,** Environmental determinations of dendritic growth in cultured sympathetic neurons, *Trans. Ann. Soc. Neurochem. Abstr.*, 17, 197, 1986.
394. **Dennis-Donini, S., Glowinski, J., and Prochiantz, A.,** Glial heterogeneity may define the three-dimensional shape of mouse mesencephalic dopaminergic neurons, *Nature*, 307, 641, 1984.

395. **Chamak, B., Fellous, A., Glowinski, J., and Prochiantz, A.,** MAP-2 expression and neurite outgrowth and branching are coregulated through region-specific neuroastroglial interactions, *J. Neurosci.*, 7, 3163, 1987.
396. **Bartlett, W. P. and Banker, G. A.,** An electron microscopic study of the development of axons and dendrites by hippocampal neurons in culture. II. Synaptic relationships, *J. Neurosci.*, 4, 1954, 1984.
397. **Dotti, C. G. and Banker, G. A.,** Experimentally induced alteration in the polarity of developing neurons, *Nature*, 330, 254, 1987.
398. **Purves, D. and Hadley, R. D.,** Changes in the dendritic branching of adult mammalian neurons revealed by repeated imaging *in situ*, *Nature*, 315, 404, 1985.
399. **Mattson, M. P.,** Neurotransmitters in the regulation of neuronal cytoarchitecture, *Brain Res. Rev.*, 13, 179, 1988.
400. **Kasamatsu, T.,** A possible role for cyclic nucleotides in plasticity of visual cortex, *Soc. Neurosci. Abstr.*, 6, 494 (abstr.), 1980.
401. **Purves, D., Voyvodic, J. T., Magrassi, L., and Yawo, H.,** Nerve terminal remodelling visualized in living mice by repeated examination of the same neuron, *Science*, 238, 1122, 1987.
402. **Lasek, R. J. and Hoffman, P. N.,** The neuronal cytoskeleton, axonal transport and axonal growth, *Cold Spring Harbor Conf. Cell Prolif.*, 3, 1021, 1976.
403. **Wilson, D. L. and Stone, G. C.,** Axoplasmic transport of proteins, *Ann. Rev. Biophys. Bioeng.*, 8, 27, 1979.
404. **Grafstein, B. and Forman, D. S.,** Intracellular transport in neurons, *Physiol. Rev.*, 60, 1167, 1980.
405. **Lasek, R. J.,** Axonal transport: a dynamic view of neuronal structure, *Trends Neurosci.*, 3, 87, 1980.
406. **Lasek, R. J.,** Translocation of the cytoskeleton in neurons and axonal growth, *Phil. Trans. R. Soc. Lond. B*, 299, 313, 1982.
407. **Lasek, R. J.,** Polymer sliding in axons, *J. Cell Sci.*, (Suppl. 5), 161, 1986.
408. **Weiss, P. and Hiscoe, H. B.,** Experiments on the mechanism of nerve growth, *J. Exp. Zool.*, 107, 315, 1948.
409. **Willard, M., Cowan, W. M., and Vagelos, P. R.,** The polypeptide compositions of intraxonally transported proteins: evidence for four transport velocities, *Proc. Natl. Acad. Sci. U.S.A.*, 71, 2183, 1979.
410. **Tytell, M., Black, M. M., Gainer, J., and Lasek, R. J.,** Axonal transport: each rate component reflects the movement of distinct macromolecular complexes, *Science*, 214, 179, 1981.
411. **Baitinger, C., Levine, J., Lorenz, T., Simon, C., Skene, P., and Willard, M.,** Characteristics of axonally-transported proteins, in *Axoplasmic Transport*, Weiss, D. G., Ed., Springer-Verlag, Berlin, 1982, 110.
412. **Brady, S. T., Lasek, R. J., and Allen, R. D.,** Videomicroscopy of fast axonal transport in extruded axoplasm: a new model for study of molecular mechanisms, *Cell Motil.*, 5, 81, 1985.
413. **Laduron, P. M.,** Axonal transport of neuroreceptors: possible involvement in long-term memory, *Neuroscience*, 22, 767, 1987.
414. **McQuarrie, I. G., Brady, S. T., and Lasek, R. J.,** Diversity in the axonal transport of structural proteins: major differences between optic and spinal axons in the rat, *J. Neurosci.*, 6, 1593, 1986.
415. **Oblinger, M. M., Brady, S. T., McQuarrie, I. G., and Lasek, R. J.,** Cytotypic differences in the protein composition of the axonally transported cytoskeleton in mammalian nerves, *J. Neurosci.*, 7, 453, 1987.
416. **Tashiro, T., Kunokawa, M., and Komiya, Y.,** Two populations of axonally transported tubulin differentiated by their interactions with neurofilaments, *J. Neurochem.*, 43, 1220, 1984.
417. **Brady, S. T. and Black, M. M.,** Axonal transport of microtubule proteins: cyto-typic variation of tubulin and MAPs in neurons, *Ann. N.Y. Acad. Sci.*, 466, 199, 1986.
418. **Tashiro, T., and Komiya, Y.,** Two distinct components of tubulin transport in sensory axons of the rat recognized by dimethyl sulfoxide treatment, *Biomed. Res.*, 4, 443, 1983.
419. **Tytell, M., Brady, S. T., and Lasek, R. J.,** Axonal transport of a subclass of tau proteins: evidence for the regional differentiation of microtubules in neurons, *Proc. Natl. Acad. Sci. U.S.A.*, 81, 1570, 1984.
420. **Fischer, I., Lewis, S. E., Binder, L. I., and Nixon, R. A.,** Axonal transport of a high molecular weight microtubule-associated protein, *J. Cell Biol.*, 105, 203a (abstr.), 1987.
421. **Hanson, M. and Edstrom, A.,** Mitosis inhibitors and axonal transport, *Int. Rev. Cytol. Suppl.*, 7, 373, 1978.
422. **Komiya, Y.,** Recovery of slow axonal transport after colchicine treatment on rat sciatic nerve, *J. Neurobiol.*, 14, 87, 1983.
423. **Keith, C. H.,** Slow transport of tubulin in the neurites of differentiated PC12 cells, *Science*, 235, 337, 1987.
424. **Griffin, J. W., Hoffman, P. N., and Price, D. L.,** Axonal transport in iminodipropionitrile neuropathy, in *Axoplasmic Transport in Physiology and Pathology*, Weiss, D. G. and Gorio, A., Eds., Springer-Verlag, Berlin, 1982, 109.

425. **Weiss, D. G.,** The mechanism of axoplasmic transport, in *Axoplasmic Transport,* Iqbal, Z., Ed., CRC Press, Boca Raton, FL, 1986.
426. **Nixon, R. A.,** The axonal transport of cytoskeletal proteins: a reappraisal, in *Axonal Transport,* Alan R. Liss, New York, 1987, 175.
427. **Nixon, R. A. and Loguinenko, K. B.,** Multiple fates of newly synthesized neurofilament proteins: evidence for a stationary neurofilament network distributed nonuniformly along axons of retinal ganglion cell neurons, *J. Cell Biol.,* 162, 647, 1986.
428. **Weisenberg, R. C., Gao, B., Awodi, S., and Glynn, J.,** Microtubule gelation-contraction *in vitro* and slow axonal transport, in *Axonal Transport,* Alan R. Liss, New York, 1987, 165.
429. **Weisenberg, R. C., Flynn, J., Gao, B., Awodi, S., Skel, F., Goodman, S. R., and Riederer, B. M.,** Microtubule gelation-contraction: essential components and relation to slow axonal transport, *Science,* 238, 1119, 1987.
430. **Weisenberg, R. C. and Cianci, C.,** ATP-induced gelation-contraction of microtubules assembled *in vitro, J. Cell Biol.,* 99, 1527, 1984.
431. **Weisenberg, R. C., Day Allen, R., and Inoué, S.,** ATP-dependent formation and motility of aster-like structures with isolated calf brain microtubule proteins, *Proc. Natl. Acad. Sci. U.S.A.,* 83, 1728, 1986.
432. **McClure, W. O.,** Effects of drugs on axoplasmic transport, *Adv. Pharmacol. Chemother.,* 10, 185, 1972.
433. **Paulson, J. C. and McClure, W. O.,** Inhibition of axoplasmic transport by colchicine, podophyllotoxin and vinblastine: an effect on microtubules, *Ann. N.Y. Acad. Sci.,* 253, 517, 1975.
434. **Gan, S.-D., Fan, M.-M., and He, G.-P.,** The role of microtubules in axoplasmic transport *in vivo, Brain Res.,* 369, 75, 1986.
435. **Johnston, K. M., Connolly, J. A., and VanderKooy, D.,** Inhibition of axonal transport *in vivo* by a tubulin-specific antibody, *Brain Res.,* 385, 38, 1986.
436. **Papasozomenos, S. C., Yoon, M., Crane, R., Autilio-Gambetti, L., and Gambetti, P.,** Redistribution of proteins of fast axonal transport following administration of ββ'-iminodipropionitrile: a quantitative autoradiographic study, *J. Cell Biol.,* 95, 672, 1982.
437. **Griffin, J. W., Fahnestock, K. E., Price, D. L., and Hoffman, P. N.,** Microtubule-neurofilament seggregation produced by iminodipropionitrile: evidence for the association of fast axonal transport with microtubules, *J. Neurosci.,* 3, 557, 1983.
438. **Allen, R. D., Allen, N. S., and Travis, J. L.,** Video-enhanced differential interference contrast (AVEC-DIC) microscopy: a new method capable of analyzing microtubule-related movement in the reticulopodial network of *Allogromia laticollaris, Cell Motil.,* 1, 291, 1981.
439. **Allen, R. D., Metuzals, J., Tasaki, I., Brady, S. T., and Gilbert, S. P.,** Fast axonal transport in squid giant axon, *Science,* 218, 1127, 1982.
440. **Allen, R. D., Weiss, D. G., Hayden, J. H., Brown, D. T., Fujinaka, H., and Simpson, M.,** Gliding movement of and bidirectional organelle transport along single native microtubules from squid axoplasm: evidence for an active role of microtubules in cytoplasmic transport, *J. Cell Biol.,* 100, 1736, 1985.
441. **Brady, S. T., Lasek, R. J., and Allen, R. D.,** Fast axonal transport in extruded axoplasm form squid giant axon, *Science,* 218, 1129,1982.
442. **Vale, R. D., Schnapp, B. J., Reese, T. S., and Sheetz, M. P.,** Movement of organelles along filaments dissociated form axoplasm of the squid giant axon, *Cell,* 40, 449, 1985.
443. **Schnapp, B. J., Vale, R. D., Sheetz, M. P., and Reese, T. S.,** Single microtubules from squid axoplasm support bidirectional movement of organelles, *Cell,* 40, 455, 1985.
444. **Lasek, R. J. and Brady, S. T.,** Attachment of transported vesicles to axonal microtubules is facilitated by AMP-PNP, *Nature,* 316, 645, 1985.
445. **Miller, R. H., Lasek, R. J., and Katz, M. J.,** Preferred microtubules for vesicle transport in lobster axons, *Science,* 235, 220, 1987.
446. **Vale, R. D., Schnapp, B. J., Reese, T. S., and Sheetz, M. P.,** Organelle, bead and microtubule translocation promoted by soluble factors from squid giant axon, *Cell,* 40, 559, 1985.
447. **Mitchison, T. and Kirschner, M.,** Microtubule assembly nucleated by isolated centrosomes, *Nature,* 312, 232, 1984.
448. **Vale, R. D., Schnapp, B. J., Mitchison, T., Steurer, E., Reese, T. S., and Sheetz, M. P.,** Different axoplasmic proteins generate movement in opposite directions along microtubules *in vitro, Cell,* 43, 623, 1985.
449. **Vale, R. D., Reese, T. S., and Sheetz, M. P.,** Identification of a novel force-generating protein (kinesin) involved in microtubule-based motility, *Cell,* 41, 39, 1985.
450. **Vale, R. D.,** Intracellular transport using microtubule-based motors, *Ann. Rev. Cell Biol.,* 3, 347, 1987.
451. **Brady, S. T.,** A novel brain ATPase with properties expected for the fast axonal transport motor, *Nature,* 317, 73, 1985.
452. **Kuznetsov, S. A. and Gelfand, V. I.,** Bovine brain kinesin is a microtubule-associated ATPase, *Proc. Natl. Acad. Sci. U.S.A.,* 83, 8530, 1986.
453. **Amos, L. A.,** Kinesin from pig brain studied by electron microscopy, *J. Cell Sci.,* 87, 105, 1987.

454. **Hollebeck, P. J.,** The cellular and subcellular distribution of kinesin, *Protoplasma*, 145, 145, 1988.
455. **Scholey, J. M., Porter, M. E., Grissom, P. M., and McIntosh, J. R.,** Identification of kinesin in sea urchins eggs, and evidence for its localization in the mitotic spindle, *Nature*, 318, 483, 1985.
456. **Gilbert, S. P. and Sloboda, R. D.,** Identification of a MAP-2-like ATP-binding protein associated with axoplasmic vesicles that translocate on isolated microtubules, *J. Cell Biol.*, 103, 947, 1986.
457. **Hollenbeck, P. J. and Chapman, K.,** A novel microtubule-associated protein from mammalian nerve show ATP-dependent binding to microtubules, *J. Cell Biol.*, 103, 1539, 1986.
458. **Koszka, C., Foisner, R., Seyfert, H. M., and Wiche, G.,** Isolation of a Ca^{2+}-protease resistant high Mr microtubule binding protein from mammalian brain: characterization of properties partially expected for a dynein-like molecule, *Protoplasma*, 138, 54, 1987.
459. **Paschal, B. M., Shpetner, H. S., and Vallee, R. B.,** MAP-1C is a microtubule-activated ATPase which translocates microtubules *in vitro* and has dynein-like properties, *J. Cell Biol.*, 105, 1273, 1987.
460. **Paschal, B. M. and Vallee, R. B.,** Retrograde transport by the microtubule-associated protein MAP-1C, *Nature*, 330, 181, 1987.
461. **Johnson, K. A.,** The pathways of ATP hydrolysis by dynein, *J. Biol. Chem.*, 258, 13825, 1983.
462. **Johnson, K. A., Porter, M. E., and Shimizu, T.,** Mechanism of force production for microtubule-dependent movements, *J. Cell Biol.*, 99, 132s, 1984.
463. **Gilbert, S. P., Allen, R. D., and Sloboda, R. D.,** Translocation of vesicles from squid axoplasm on flagellar microtubules, *Nature*, 325, 245, 1985.
464. **Ghetti, B.,** Induction of neurofibrillary degeneration following treatment with maytansine *in vivo*, *Brain Res.*, 163, 9, 1979.
465. **Hirano, A., Donnenfleld, H., Sasaki, S., and Nakano, I.,** Fine structural observation of neurofilamentous changes in amyotrophic lateral sclerosis, *J. Neuropathol. Exp. Neurol.*, 43, 461, 1984.
466. **Asbury, A. K., Gade, M. K., Baringer, J. R., and Berg, B. O.,** Giant axonal neuropathy — a unique case with segmental neurofilamentous masses, *Acta Neuropathol.*, 20, 237, 1978.
467. **Pena, S. D. J.,** Giant axonal neuropathy: intermediate filament aggregates in cultured skin fibroblasts, *Neurology*, 31, 1470, 1981.
468. **Klymkowsky, M. W., Christian, E., Ham, R. G., Plummer, D. J., and Sprenger, F.,** Giant axonal neuropathy mutations, intermediate filaments, and cellular metabolism, in *Intrinsic Determinants of Neuronal Form and Function*, Alan R. Liss, New York, 1988, 441.
469. **Griffin, J. W., Hoffman, P. N., Clark, A. W., Carroll, P. T., and Price, D. L.,** Slow axonal transport of neurofilament proteins: impairment by β,β′ iminodipropionitrile administration, *Science*, 202, 633, 1978.
470. **Clark, A. W., Griffin, J. W., and Price, D. L.,** The axonal pathology in chronic IDPN intoxication, *J. Neuropathol. Exp. Neurol.*, 39, 42, 1980.
471. **Papasozomenos, S. C., Autilio-Gambetti, L., and Gambetti, P.,** Reorganization of axoplasmic organelles following β β′ iminodipropionitrile administration, *J. Cell Biol.*, 91, 866, 1981.
472. **Griffin, J. W., Fahnestock, K. E., Price, D. L., and Cork, L. C.,** Cytoskeletal disorganization induced by local application of ββ′ iminodipropionitrile and 2,5-hexanedione, *Ann. Neurol.*, 14, 55, 1983.
473. **Graham, D. G.,** Hexane neuropathy: a proposal for the pathogenesis of a hazard of occupational exposure and inhalant abuse, *Chem. Biol. Interact.*, 32, 339, 1980.
474. **Graham, D. G., Szakal-Quin, G., Priest, J. W., and Anthony, D. C.,** *In vitro* evidence that covalent crosslinking of neurofilaments occurs in γ-diketone neuropathy, *Proc. Natl. Acad. Sci. U.S.A.*, 81, 4979, 1984.
475. **DeCaprio, A. P.,** *N*-Hexane neurotoxicity: a mechanism involving pyrrole adduct formation in axonal cytoskeletal protein, *Neurotoxicology*, 8, 199, 1987.
476. **Jakobsen, J. and Sidenius, P.,** Decreased axonal transport of structural proteins in streptozotocin diabetic rats, *J. Clin. Invest.*, 66, 292, 1980.
477. **Vitadello, M., Filliatreau, G., Gupont, J. L., Hassig, R., Gorio, A., and DiGiamberardino, L.,** Altered axonal transport of cytoskeletal proteins in the mutant diabetic mouse, *J. Neurochem.*, 45, 860, 1985.
478. **Jakobsen, J., Sidenius, P., and Braendgaard, H.,** A proposal for the classification of neuropathies according to their axonal transport abnormalities, *J. Neurol. Neurosurg. Psych.*, 49, 986, 1986.
479. **Kuncl, R. W., Duncan, G., Watson, D., Alderson, K., Rogawski, M. A., and Peper, M.,** Colchicine myopathy and neuropathy, *N. Engl. J. Med.*, 316, 1562, 1987.
480. **Steward, O., Goldschmidt, R. B., and Sutula, T.,** Neurotoxicity of colchicine and other tubulin-binding agents: a selective vulnerability of certain neurons to the disruption of microtubules, *Life Sci.*, 35, 43, 1984.
481. **Breuer, A. C., Lynn, M. P., Atkinson, M. B., Chou, S. M., Wilbourn, A. J., Marks, K. E., Culvet, J. E., and Fleegler, E. J.,** Fast axonal transport in amyotrophic lateral sclerosis: an intra-axonal organelle traffic analysis, *Neurology*, 37, 738, 1987.
482. **Goldman, J. E. and Yen, S.-H.,** Cytoskeletal protein abnormalities in neurodegenerative diseases, *Ann. Neurol.*, 19, 209, 1986.
483. **Wurtman, R. J.,** Alzheimer's disease, *Sci. Am.*, 252, 48, 1985.

484. **Perry, R. H.,** Recent advances in neuropathology of Alzheimer's disease, *Br. Med. Bull.*, 42, 34, 1986.
485. **Katzman, R.,** Alzheimer's disease, *N. Engl. J. Med.*, 314, 964, 1986.
486. **Selkoe, D. J., Ihara, Y., and Salazu, F. J.,** Alzheimer's disease: insolubility of partially purified paired helical filaments in sodium dodecyl sulfate and urea, *Science,* 213, 1243, 1982.
487. **Iqbal, K., Zaidi, T., Thompson, C. H., Merz, P. A., and Wisniewski, H. M.,** Alzheimer paired helical filaments: bulk isolation, solubility and protein composition, *Acta Neuropathol.*, 62, 167, 1984.
488. **Brion, J. P., Van den Bosch de Aguilar, P., and Flament-Durand, J.,** Senile dementia of Alzheimer type: morphological and immunocytochemical studies, *Adv. Appl. Neurol. Sci.*, 2, 164, 1985.
489. **Brion, J. P., Couck, A. M., Passareiro, H., and Flament-Durand, J.,** Neurofibrillary tangles of Alzheimer's disease: an immunohistochemical study, *J. Submicroscop. Cytol.*, 17, 89, 1985.
490. **Brion, J. P., Passareiro, H., Nunez, J., and Flament-Durand, J.,** Mise en évidence de la protéine tau an niveau des lésions de dégénérescence neurofibrillaire de la maladie d'Alzheimer, *Arch. Biol.*, 95, 229, 1985.
491. **Nukina, N. and Ihara, Y.,** Proteolytic fragments of Alzheimer's paired helical filaments, *J. Biochem.*, 98, 1715, 1985.
492. **Nukina, N. and Ihara, Y.,** One of the antigenic determination of paired helical filaments is related to tau proteins, *J. Biochem.*, 99, 1541, 1986.
493. **Delacourte, A. and Defossez, A.,** Alzheimer's disease: tau proteins, the promoting factors of microtubule assembly, are major components of paired helical filaments, *J. Neurol. Sci.*, 76, 173, 1986.
494. **Grundke-Iqbal, I., Iqbal, K., Quinlan, M., Tung, Y.-C., Zaidi, M. S., and Wisniewski, H. M.,** Microtubule-associated protein tau. A component of Alzheimer paired helical filaments, *J. Biol. Chem.*, 261, 6084, 1986.
495. **Kosik, K. S., Joachim, C. L., and Selkoe, D. J.,** Microtubule-associated protein tau is a major antigenic component of paired helical filaments in Alzheimer disease, *Proc. Natl. Acad. Sci. U.S.A.*, 83, 4044, 1986.
496. **Wood, J. G., Mirra, S. S., Pollock, N. J., and Binder, L. I.,** Neurofibrillary tangles of Alzheimer disease share antigenic determinants with the axonal microtubule-associated protein tau, *Proc. Natl. Acad. Sci. U.S.A.*, 83, 4040, 1986.
497. **Grunde-Iqbal, I., Iqbal, K., Tung, Y.-C., Quinlam, M., Wisniewski, H. M., and Binder, L. I.,** Abnormal phosphorylation of the microtubule-associated protein tau in Alzheimer cytoskeletal pathology, *Proc. Natl. Acad. Sci. U.S.A.*, 83, 4913, 1986.
498. **Ihara, Y., Nukina, N., Miura, R., and Ogawaira, M.,** Phosphorylated tau protein is integrated into paired helical filaments in Alzheimer's disease, *J. Biochem.*, 99, 1807, 1986.
499. **Ksiezak-Reding, H., Dickson, D. W., Davies, P., and Yen, S.-H.,** Recognition of tau epitopes by anti-neurofilament antibodies that bind to Alzheimer neurofibrillary tangles, *Proc. Natl. Acad. Sci. U.S.A.*, 84, 3410, 1987.
500. **Nukina, N., Kosik, K. S., and Selkoe, D. J.,** Recognition of Alzheimer paried helical fialements by monoclonal neurofilament antibodies is due to crossreaction with tau protein, *Proc. Natl. Acad. Sci. U.S.A.*, 84, 3415, 1987.
501. **Iqbal, K., Grundke-Iqbal, I., Zaidi, T., Merz, P. A., Wen, G. Y., Shaikh, S. S., Wisniewski, H. M., Alafuzoff, I., and Winbald, B.,** Defective brain microtubule assembly in Alzheimer's disease, *Lancet*, ii, 421, 1986.
502. **Wischik, C. M., Crowther, R. A., Stewart, M., and Roth, M.,** Subunit structures of paired helical filaments on Alzheimer's disease, *J. Cell Biol.*, 100, 1905, 1985.
503. **Crowther, R. A., Wischik, C. M., and Stewart, M.,** Analysis of the structure of paired helical filaments, *Proc. EMSA*, 43, 734, 1985.
504. **Wischik, C. M. and Crowther, R. A.,** Subunit structure of the Alzheimer tangle, *Br. Med. Bull.*, 42, 57, 1986.
505. **Nieto, A., Correas, I., Montejo de Garcini, E., and Avila, J.,** A modified form of microtubule-associated tau protein is the main component of paired helical filaments of Alzheimer's disease, *Biochem. Biophys. Res. Commun.*, 154, 660, 1988.
506. **Wischik, C. M., Novak, M., Edwards, P. C., Klug, A., Tichelaar, W., and Crowther, R. A.,** Structural characterization of the core of the paired helical filament of Alzheimer disease, *Proc. Natl. Acad. Sci. U.S.A.*, 85, 4884, 1988.
507. **Wischik, C. M., Novak, N., Thgersen, H. C., Edwards, P. C., Renswick, M. J., Jakes, R., Walker, J. E., Milstein, C., Roth, M., and Klug, A.,** Isolation of a fragment of tau derived from the core of the paired helical filament of Alzheimer disease, *Proc. Natl. Acad. Sci. U.S.A.*, 85, 4506, 1988.
508. **Goedert, M., Wischik, C. M., Crowther, R. A., Walker, J. E., and Klug, A.,** Cloning and sequencing of the cDNA encoding a core protein of the paired helical filament of Alzheimer diseases: Identification as the microtubule-associated protein tau, *Proc. Natl. Acad. Sci. U.S.A.*, 85, 4051, 1988.
509. **Montejo de Garcini, E. and Avila, J.,** *In vitro* conditions for the self-polymerization of the microtubule-associated protein tau factor, *J. Biochem.*, 102, 1415, 1987.

510. **Masters, C. L., Simms, G., Weinman, N. A., Multhaup, G., McDonald, B. L., and Beyreuther, K.,** Amyloid plaque core protein in Alzheimer disease and Down syndrome, *Proc. Natl. Acad. Sci. U.S.A.*, 82, 4245, 1985.
511. **Kang, J., Lemaire, H.-G., Unterbeck, A., Salbaum, J. M., Master, C. L., Grzeschik, K.-H., Multhamp, G., Beyreuther, K., and Muller-Hill, B.,** The precursor of Alzheimer's disease amyloid A4 protein resembles a cell-surface receptor, *Nature*, 325, 733, 1987.
512. **Galloway, P. G., Perry, G., and Gambetti, P.,** Hirano body filaments contain actin and actin-associated proteins, *J. Neuropathol. Exp. Neurol.*, 46, 185, 1987.
513. **Galloway, P. G., Perry, G., Kosik, K. S., and Gamgetti, P.,** Hirano bodies contain tau protein, *Brain Res.*, 403, 337, 1987.
514. **Wisniewski, H. M. and Terry, R. D.,** Morphology of the aging brain, human and animal, *Prog. Brain Res.*, 40, 167, 1973.
515. **Price, D. J., Struble, R. G., Altschuler, R. J., Cassanova, M. F., Cork, L. C., and Murphy, D. B.,** Aggregation of tubulin in neurons in Alzheimer's disease, *J. Neuropathol. Exp. Neurol.*, 44, 366 (abstr.), 1985.
516. **Dickson, D. W., Ksiezak-Reding, H., Davies, P., and Yen, S.-H.,** A monoclonal antibody that recognizes a phosphorylated epitope in Alzheimer neurofibrillary tangles, neurofilaments and tau proteins immunostains granulovacuolar degeneration, *Acta Neuropathol.*, 73, 254, 1987.
517. **Perry, G., Stewart, D., Friedman, R., Manetto, V., Autilio-Gambetti, L., and Gambetti, P.,** Filaments of Pick's bodies contain altered cytoskeletal elements, *Ann. J. Pathol.*, 127, 559, 1987.
518. **Galloway, P. G., Perry, G., and Gambetti, P.,** Lewy bodies share epitopes with neurofilaments, tubulin and neurofibrillary tangles, *Int. Congr. Neuropathol.*, 10, 298, 1986.

INDEX

A

D

H

I

O

P

T

U